Lecture Notes in Computer Science 13248

More information about this series at https://link.springer.com/bookseries/558

Uday Kiran Rage · Vikram Goyal ·
P. Krishna Reddy (Eds.)

Database Systems for Advanced Applications

DASFAA 2022 International Workshops

BDMS, BDQM, GDMA, IWBT, MAQTDS, and PMBD
Virtual Event, April 11–14, 2022
Proceedings

Editors
Uday Kiran Rage
University of Aizu
Aizu, Japan

Vikram Goyal
Indraprastha Institute of Information Technology, New Delhi
Delhi, India

P. Krishna Reddy
Data Sciences and Analytics Center
International Institute of Information Technology
Hyderabad, Telangana, India

ISSN 0302-9743 ISSN 1611-3349 (electronic)
Lecture Notes in Computer Science
ISBN 978-3-031-11216-4 ISBN 978-3-031-11217-1 (eBook)
https://doi.org/10.1007/978-3-031-11217-1

This Springer imprint is published by the registered company Springer Nature Switzerland AG
The registered company address is: Gewerbestrasse 11, 6330 Cham, Switzerland

Preface

It is our great pleasure to introduce the workshop proceedings of the 27th International Conference on Database Systems for Advanced Applications (DASFAA 2022), held during April 11–14, 2022, in Hyderabad, India. The entire conference was organized online due to the outbreak of COVID-19 across the world. DASFAA provides a leading international forum for discussing the latest research on database systems and advanced applications. The conference's long history has established the event as a premier research conference in the field of databases.

As a part of DASFAA 2022 program, the following six workshops were selected by the workshop co-chairs: the 1st Workshop on Pattern mining and Machine learning in Big complex Databases (PMBD 2022), the 6th International Workshop on Graph Data Management and Analysis (GDMA 2022), the 2nd International Workshop on Blockchain Technologies (IWBT 2022), the 8th International Workshop on Big Data Management and Service (BDMS 2022), the first workshop on Managing Air Quality Through Data Science (MAQTDS 2022), and the 7th International Workshop on Big Data Quality Management (BDQM 2022).

The review process of workshop papers followed a double-blind two-tiered review system following the tradition of DASFAA. Almost all the valid submissions were reviewed by at least three Program Committee (PC) members.

Out of a total of 65 valid research track submissions, 30 submissions were accepted as full papers (acceptance rate of 46.1%). These papers were presented on the first day of the conference. All six workshops were held on April 11, 2022, in conjunction with DASFAA 2022.

We would like to thank all workshop organizers and reviewers for their hard work in providing us with thoughtful and comprehensive reviews and recommendations. Many thanks to the authors who submitted their papers to the workshops. We would like to express our sincere thanks to Maya Ramanath, Wookey Lee, and Sanjay Kumar Madria for helping in the selection of workshops. We also thank Springer for their financial support and publishing the workshop proceedings. We thank Google for the sponsorship. We feel indebted to the DASFAA Steering Committee for its continuing guidance.

We appreciate the hosting organization IIIT Hyderabad, which is celebrating its silver jubilee in 2022. We thank the researchers at the Data Sciences and Analytics Center (DSAC) and the Kohli Center on Intelligent Systems (KCIS) at IIIT Hyderabad for their support. We also thank the administration and staff of IIIT Hyderabad for their help.

We hope that the readers of the proceedings find the content interesting, rewarding, and beneficial to their research.

April 2022

R. Uday Kiran
Vikram Goyal
P. Krishna Reddy

Organization

DASFAA 2022 was organized by IIIT Hyderabad, Hyderabad, Telangana, India.

Steering Committee Chair

Lei Chen	Hong Kong University of Science and Technology, Hong Kong

Honorary Chairs

P. J. Narayanan	IIIT Hyderabad, India
S. Sudarshan	IIT Bombay, India
Masaru Kitsuregawa	University of Tokyo, Japan

Steering Committee Vice Chair

Stephane Bressan	National University of Singapore, Singapore

Steering Committee Treasurer

Yasushi Sakurai	Osaka University, Japan

Steering Committee Secretary

Kyuseok Shim	Seoul National University, South Korea

General Chairs

P. Krishna Reddy	IIIT Hyderabad, India
Mukesh Mohania	IIIT Delhi, India
Anirban Mondal	Ashoka University, India

Program Committee Chairs

Arnab Bhattacharya	IIT Kanpur, India
Lee Mong Li Janice	National University of Singapore, Singapore
Divyakant Agrawal	University of California, Santa Barbara, USA

Steering Committee

Zhiyong Peng	Wuhan University, China
Zhanhuai Li	Northwestern Polytechnical University, China
Krishna Reddy	IIIT Hyderabad, India
Yunmook Nah	DKU, South Korea
Wenjia Zhang	University of New South Wales, Australia
Zi Huang	University of Queensland, Australia
Guoliang Li	Tsinghua University, China
Sourav Bhowmick	Nanyang Technological University, Singapore
Atsuyuki Morishima	University of Tsukaba, Japan
Sang-Won Lee	SKKU, South Korea
Yang-Sae Moon	Kangwon National University, South Korea

Industry Track Chairs

Prasad M. Deshpande	Google, India
Daxin Jiang	Microsoft, China
Rajasekar Krishnamurthy	Adobe, USA

Demo Chairs

Rajeev Gupta	Microsoft, India
Koichi Takeda	Nagoya University, Japan
Ladjel Bellatreche	ENSMA, France

PhD Consortium Chairs

Vikram Pudi	IIIT Hyderabad, India
Srinath Srinivasa	IIIT Bangalore, India
Philippe Fournier-Viger	Shenzen University, China

Panel Chairs

Jayant Haritsa	Indian Institute of Science, India
Reynold Cheng	Hong Kong University, China
Georgia Koutrika	Athena Research Center, Greece

Sponsorship Chair

P. Krishna Reddy	IIIT Hyderabad, India

Publication Chairs

Vikram Goyal	IIIT Delhi, India
R. Uday Kiran	University of Aizu, Japan

Workshop Chairs

Maya Ramanath	IIT Delhi, India
Wookey Lee	Inha University, South Korea
Sanjay Kumar Madria	Missouri Institute of Technology, USA

Tutorial Chairs

P. Sreenivasa Kumar	IIT Madras, India
Jixue Liu	University of South Australia, Australia
Takahiro Hara	Osaka University, Japan

Publicity Chairs

Raj Sharma	Goldman Sachs, India
Jamshid Bagherzadeh Mohasefi	Urmia University, Iran
Nazha Selmaoui-Folcher	University of New Caledonia, New Caledonia

Organizing Committee

Lini Thomas	IIIT Hyderabad, India
Satish Narayana Srirama	University of Hyderabad, India
Manish Singh	IIT Hyderabad, India
P. Radha Krishna	NIT Warangal, India
Sonali Agrawal	IIIT Allahabad, India
V. Ravi	IDRBT, India

Organizing Chairs for PMBD

Philippe Fournier-Viger	Shenzhen University, China
Mourad Nouioua	Harbin Institute of Technology, China
Hamido Fujita	Iwate Prefectural University, Japan
Lin Zhang	Tencent, China
Vincent S. Tseng	National Chiao Tung University, Taiwan

Organizing Chairs for IWBT

Sanjay Chaudhary	Ahmedabad University, India
Krishnasuri Narayanam	IBM Research, India

Organizing Chairs for GDMA

Lei Zou	Peking University, China
Xiaowang Zhang	Tianjin University, China
Weiguo Zheng	Fudan University, China

Organizing Chairs for BDMS

Kai Zheng	University of Electronic Science and Technology of China, China
Xiaoling Wang	East China Normal University, China
An Liu	Soochow University, China

Organizing Chairs for BDQM

Xiaoou Ding	Harbin Institute of Technology, China
Xueli Liu	Tianjin University, China

Organizing Chairs for MAQTDS

Girish Agrawal	O.P. Jindal Global University, India
Jai Ganesh	Mphasis, India

Program Committee for PMBD

Jaroslav Frnda	University of Zilina, Slovakia
Pinar Karagoz	Middle East Technical University, Turkey
Amirat Hanane	University of Laghoaut, Algeria
M. Saqib Nawaz	Peking University, China
Yun Sing Koh	University of Auckland, New Zealand
Tin Truong	Dalat University, Vietnam
Wei Song	North China University of Technology, China
Farid Nouioua	LSIS, CNRS, France
Moulay Akhloufi	Université de Moncton, Canada
Srikumar Krishnamoorthy	Indian Institute of Management, India
Tzung-Pei Hong	National University of Kaohsiung, Taiwan
Siddharth Dawar	Indraprastha Institute of Information Technology, India

Program Committee for IWBT

Devesh Jinwala	SVNIT Surat, India
Minoru Kuribayashi	Okayama University, Japan
Mehul Raval	Ahmedabad University, India
Iyyanki V. Murali Krishna	RCI, DRDO, India
Ratnik Gandhi	Sentrana, Canada
Sriram Birudavolu	NASSCOM-DSCI Cybersecurity Centre of Excellence, India
Sridhar Vedhanabatla	DSCI, India
Gaurav Somani	Central University of Rajasthan, India
Mansukh Savaliya	VGEC Ahmedabad, India
Amit Ganatra	CHARUSAT, India
Mandar Chaudhary	eBay, USA
Vikas Jaiman	Maastricht University, The Netherlands

Program Committee for GDMA

Guohui Xiao	Free University of Bozen-Bolzano, Italy
Chengzhi Piao	The Chinese University of Hong Kong, China
Peng Peng	Hunan University, China
Yu Liu	Beijing Jiaotong University, China
Jing Wang	Fudan University, China
Youhuan Li	Hunan University, China
Meng Wang	Southeast University, China
Liang Hong	Wuhan University, China
Tieyun Qian	Wuhan University, China
Gong Cheng	Nanjing University, China

Program Committee for BDMS

Muhammad Aamir Cheema	Monash University, Australia
Xuanjing Huang	Fudan University, China
Yan Wang	Macquarie University, Australia
Xiaochun Yang	Northeastern University, China
Kun Yue	Yunnan University, China
Dell Zhang	Birbeck, University of London, UK
Xiao Zhang	Renmin University of China, China
Bolong Zheng	Huazhong University of Science and Technology, China
Wendi Ji	East China Normal University, China
Qizhi Liu	Nanjing University, China

Bin Mu	Tongji University, China
Yaqian Zhou	Fudan University, China

Program Committee for BDQM

Chengliang Chai	Tsinghua University, China
Shaoxu Song	Tsinghua University, China
Jiannan Wang	Simon Fraser University, Canada
Yajun Yang	Tianjin University, China
Chen Ye	Hangzhou Dianzi University, China
Feng Zhang	Renmin University of China, China
Kaiqi Zhang	Harbin Institute of Technology, China
Wenjie Zhang	University of New South Wales, Australia
Zhaonian Zou	Harbin Institute of Technology, China

Program Committee for MAQTDS

Girish Agrawal	O.P. Jindal Global University, India
Geetam Tiwari	IIT Delhi, India
Bakul Budhiraja	Queen's University Belfast, UK
Raja Sengupta	McGill University, Canada
Prasad Pathak	FLAME University, India

Sponsoring Institutions

IIIT Hyderabad, India
Google, India

Contents

PMDB

An Algorithm for Mining Fixed-Length High Utility Itemsets 3
Le Wang

A Novel Method to Create Synthetic Samples with Autoencoder Multi-layer Extreme Learning Machine 21
Yulin He, Qihang Huang, Shengsheng Xu, and Joshua Zhexue Huang

Pattern Mining: Current Challenges and Opportunities 34
Philippe Fournier-Viger, Wensheng Gan, Youxi Wu, Mourad Nouioua, Wei Song, Tin Truong, and Hai Duong

Why Not to Trust Big Data: Discussing Statistical Paradoxes 50
Rahul Sharma, Minakshi Kaushik, Sijo Arakkal Peious, Mahtab Shahin, Ankit Vidyarthi, Prayag Tiwari, and Dirk Draheim

Localized Metric Learning for Large Multi-class Extremely Imbalanced Face Database 64
Seba Susan and Ashu Kaushik

Top-*k* Dominating Queries on Incremental Datasets 79
Jimmy Ming-Tai Wu, Ke Wang, and Jerry Chun-Wei Lin

IWBT

Collaborative Blockchain Based Distributed Denial of Service Attack Mitigation Approach with IP Reputation System 91
Darshi Patel and Dhiren Patel

Model-Driven Development of Distributed Ledger Applications 104
Piero Fraternali, Sergio Luis Herrera Gonzalez, Matteo Frigerio, and Mattia Righetti

Towards a Blockchain Solution for Customs Duty-Related Fraud 120
Christopher G. Harris

Securing Cookies/Sessions Through Non-fungible Tokens 135
Kaushal Shah, Uday Khokhariya, Nidhay Pancholi, Shambhavi Kumar, and Keyur Parmar

GDMA

Chinese Spelling Error Detection and Correction Based on Knowledge Graph ... 149
Ximin Sun, Jing Zhou, Shuai Wang, Huichao Li, Jiangkai Jia, and Jiazheng Zhu

Construction and Application of Event Logic Graph: A Survey ... 160
Bin Zhang, Ximin Sun, Xiaoming Li, Dan Liu, Shuai Wang, and Jiangkai Jia

Enhancing Low-Resource Languages Question Answering with Syntactic Graph ... 175
Linjuan Wu, Jiazheng Zhu, Xiaowang Zhang, Zhiqiang Zhuang, and ZhiYong Feng

Profile Consistency Discrimination ... 189
Jing Zhou, Ximin Sun, Shuai Wang, Jiangkai Jia, Huichao Li, Mingda Wang, and Shuyi Li

BDMS

H-V: An Improved Coding Layout Based on Erasure Coded Storage System ... 203
Tiantong Mu, Ying Song, Mingjie Yang, Bo Wang, and Jiacheng Zhao

Astral: An Autoencoder-Based Model for Pedestrian Trajectory Prediction of Variable-Length ... 214
Yupeng Diao, Yiteng Su, Ximu Zeng, Xu Chen, Shuncheng Liu, and Han Su

A Survey on Spatiotemporal Data Processing Techniques in Smart Urban Rail ... 229
Li Jian, Huanran Zheng, Bofeng Chen, Tingliang Zhou, Hui Chen, and Yanjun Li

Fast Vehicle Track Counting in Traffic Video ... 244
Ruoyan Qi, Ying Liu, Zhongshuai Zhang, Xiaochun Yang, Guoren Wang, and Yingshuo Jiang

TSummary: A Traffic Summarization System Using Semantic Words ... 257
Xu Chen, Ximu Zeng, Shuncheng Liu, Zhi Xu, Yuyang Xia, Ruyi Lai, and Han Su

Attention-Cooperated Reinforcement Learning for Multi-agent Path Planning ... 272
Jinchao Ma and Defu Lian

Big Data-Driven Stable Task Allocation in Ride-Hailing Services 291
Jingwei Lv, Nan Zhou, and Shuzhen Yao

Weighted Mean-Field Multi-Agent Reinforcement Learning via Reward Attribution Decomposition 301
Tingyu Wu, Wenhao Li, Bo Jin, Wei Zhang, and Xiangfeng Wang

BDQM

Evaluating Presto and SparkSQL with TPC-DS 319
Yinhao Hong, Sheng Du, and Jianquan Leng

Optimizing the Age of Sensed Information in Cyber-Physical Systems 330
Yinlong Li, Siyao Cheng, Feng Li, Jie Liu, and Hanling Wu

Aggregate Query Result Correctness Using Pattern Tables 347
Nitish Yadav, Ayushi Malhotra, Sakshee Patel, and Minal Bhise

Time Series Data Quality Enhancing Based on Pattern Alignment 363
Jianping Huang, Hao Chen, Hongkai Wang, Jun Feng, Liangying Peng, Zheng Liang, Hongzhi Wang, Tianlan Fan, and Tianren Yu

Research on Feature Extraction Method of Data Quality Intelligent Detection 376
Weiwei Liu, Shuya Lei, Xiaokun Zheng, and Xiao Liang

MAQTDS

Big Data Resources to Support Research Opportunities on Air Pollution Analysis in India 389
Sarath K. Guttikunda

Air Quality Data Collection in Hyderabad Using Low-Cost Sensors: Initial Experiences 402
N. Chandra Shekar, A. Srinivas Reddy, P. Krishna Reddy, Anirban Mondal, and Girish Agrawal

Visualizing Spatio-temporal Variation of Ambient Air Pollution in Four Small Towns in India 417
Girish Agrawal, Hifzur Rahman, Anirban Mondal, and P. Krishna Reddy

Author Index 437

PMDB

An Algorithm for Mining Fixed-Length High Utility Itemsets

Le Wang(✉)

Ningbo University of Finance and Economics, Ningbo 315175, Zhejiang, China
wangleboro@gmail.com

Abstract. High utility pattern/itemset mining is a hotspot of data mining. Different from the traditional frequent pattern, high utility pattern takes into consideration not only the number of items in the transaction, but also the weight of these items, such as profit and price. Hence the computational complexity of this mining algorithm is higher than the traditional frequent pattern mining. Thus, one essential topic of this field is to reduce the search space and improve the mining efficiency. Constraint on pattern length can effectively reduce algorithm search space while fulfill a certain kind of actual requirement. Addressing fixed length high utility pattern mining, we propose a novel algorithm, called HUIK (High Utility Itemsets with K-length Miner), that first compresses transaction data into a tree, then recursively searches high utility patterns with designated length using a pattern growth approach. An effective pruning strategy is also proposed to reduce the number of candidate items on the compressed tree, to further reduce the search space and improve algorithm efficiency. The performance of the algorithm HUIK is verified on six classical datasets. Experimental results verify that the proposed algorithm has a significant improvement in time efficiency, especially for long datasets and dense datasets.

Keywords: Data mining · High utility itemsets · Pattern growth · Frequent pattern

1 Introduction

High Utility Pattern/Itemset (HUP/HUI) mining is a hot topic in data mining [1–5]. It is derived from frequent pattern mining, but it is different from frequent pattern mining. Frequent pattern mining takes each item of a transaction itemset as binary, i.e., it does not consider the internal utility value (quantity) and external utility value (such as importance, profit, price, etc.) of each item in a transaction itemset [6]. HUP mining introduces the internal utility value and external utility value of items into pattern mining. HUP has been applied to many fields, and its commercial value has been reflected in many applications, including website click stream analysis [7, 8], mobile commerce environment [9], retail store cross-marketing commercial value [10], genetic recombination and other applications [11].

Yao et al. [12] proposed the related definitions and mathematical model of HUP mining. The task of mining HUPs is to find all patterns whose utility value is not less than

U. K. Rage et al. (Eds.): DASFAA 2022 Workshops, LNCS 13248, pp. 3–20, 2022.
https://doi.org/10.1007/978-3-031-11217-1_1

a user-specified minimum utility value (threshold). The pruning strategy of traditional frequent pattern mining algorithms is not applicable in HUP mining because a superset of a low utility pattern may be a HUP, which makes search space of mining algorithms larger than frequent pattern mining, which makes HUP mining is much more difficult than frequent pattern mining.

Although there has been a lot of research on HUP mining [8, 13–17], there is still a relatively large search space. In order to provide the useful HUPs for users, Fournier-Viger et al. proposed a length-constrained HUPs mining algorithm FHM+ based on FHM [2, 18], which can mine specified length HUPs, and it can effectively remove out some patterns that users are not interested in. When this algorithm calculates the estimated utility value of the patterns, the utility value of the transaction with an unsuitable length can be excluded, and the number of candidates can be reduced. Thus, this algorithm effectively improves its efficiency of mining HUPs.

Aiming at this kind of HUPs mining with length constraints, we propose a new mining algorithm HUIKM (High Utility Itemsets with K-length Miner) for improving the performance of this kind algorithms in this paper. This algorithm firstly maps a data to a tree, and then mines HUPs from the tree based on the method of pattern-growth; at the same time, an effective pruning strategy is given to reduce the search space. In the experiment, classical sparse and dense datasets are used to evaluate the performance of HUIKM. The experimental results show that the time efficiency of the algorithm has been greatly improved.

The contributions of this paper include:

- We designed a tree structure to maintain data information, from which the utility value of any specified-length HUP can be retrieved.
- We designed a new algorithm based on tree and pattern-growth for mining the specified-length HUPs.
- We performed an extensive experiment on classical datasets under different situations, and compared HUIKM with FHM+, EFIM and ULBMiner.

The structure of this paper is as follows: The second section gives the problem description & related definitions, and related works. The third section describes the proposed algorithm HUIKM. The fourth section conducts experimental tests. The fifth section gives the conclusions.

2 Background

In this section, we give the related definitions and work of the HUP mining.

2.1 Problem Description and Definitions

In this paper, we adopt definitions similar to those presented in the previous works [2, 7, 10, 13, 14, 18]. Given a *utility-valued transaction dataset* $D = \{T_1, T_2, T_3, \ldots, T_n\}$, which contains n transactions and m unique items $I = \{i_1, i_2, \ldots, i_m\}$. A transaction T_d ($d = 1, 2, 3, \ldots, n$) contains one or more unique items in I, and is also called a transaction

Table 1. An example of a transaction dataset

TID	Transaction	*TU*
T_1	(A,4) (C,3) (F,1)	47
T_2	(C,1) (D,4) (E,10)	58
T_3	(A,4) (B,4) (D,2) (E,6)	54
T_4	(A,1) (E,1)	6
T_5	(A,6) (B,2) (D,2) (E,1)	46
T_6	(A,3) (B,3) (D,1) (G,1)	30
T_7	(B,2) (G,2)	10
T_8	(A,3) (B,7) (C,1) (E,3)	49

Table 2. Profits

Item	Profit
A	4
B	3
C	10
D	7
E	2
F	1
G	2

itemset; e.g., $T_1 = \{(A,4)\ (C,3)\ (F,1)\}$. Each item i_j in a transaction T_d is attached with a quantity which is called *internal utility* (denoted as $q(i_j,T_d)$); e.g., $q(A,T_1) = 4$, $q(C,T_1) = 3$ and $q(F,T_1) = 1$ in Table 1. An item i_j has a unit profit $p(i_j)$, which is called *external utility*, e.g. $p(A) = 4$ in Table 2. $|D|$ represents the size of the dataset D, i.e., the number of transactions in a dataset D; and $|T_d|$ represents the number of items in a transaction T_d, i.e., the length of transaction.

Definition 1. The utility value of the item i_j in a transaction T_d is denoted as $U(i_j, T_d)$, and is defined as:

$$U(i_j, T_d) = \mathrm{p}(i_j) \times \mathrm{q}(i_j, T_d) \tag{1}$$

For example, in Table 1 and 2, $U(A,T_1) = 4 \times 4 = 16$, $U(C,T_1) = 10 \times 3 = 30$, and $U(F,T_1) = 1 \times 1 = 1$.

Definition 2. The utility value of itemset X in a transaction T_d is denoted as $U(X, T_d)$, and is defined as:

$$U(X, T_d) = \begin{cases} \sum_{i_j \in X} U(i_j, T_d), & \text{if } X \subseteq T_d \\ 0, & \text{else} \end{cases} \tag{2}$$

For example, in Table 1 and 2, $U(\{AC\}, T_1) = 46$, $U(\{AF\}, T_1) = 17$.

Definition 3. The utility value of itemset X in a dataset D is denoted as $U(X)$, and is defined as:

$$U(X) = \sum_{T_d \in D \wedge X \subseteq T_d} U(X, T_d) \tag{3}$$

For example, in Table 1 and 2, $U(\{AC\}) = U(\{AC\}, T_1) + U(\{AC\}, T_6) = 46 + 22 = 68$.

Definition 4. The utility value of transaction T_d is denoted as $TU(T_d)$, and is defined as:

$$TU(T_d) = \sum_{i_j \in T_d} U(i_j, T_d) \tag{4}$$

For example, in Table 1 and 2, $TU(T_1) = U(\mathrm{A}, T_1) + U(\mathrm{C}, T_1) + U(\mathrm{F}, T_1) = 16 + 30 + 1 = 47$.

Definition 5. The utility of the dataset D is denoted as TU, and is defined as:

$$TU = \sum_{T_d \in D} TU(T_d) \tag{5}$$

For example, in Table 1 and 2, $TU = 47 + 58 + 54 + 6 + 46 + 30 + 10 + 49 = 300$.

Definition 6. The *transaction-weighted utility value* of itemset X is denoted as $TWU(X)$ (also called TWU value), and is defined as:

$$TWU(X) = \sum_{T_d \in D \wedge T_d \supseteq X} TU(T_d) \tag{6}$$

Definition 7. The *minimum utility threshold* δ is a user-specified percentile of total transaction utility values of the given dataset D; so the *minimum utility value, MinU* (also called a user-specified minimum utility value), is defined as:

$$MinU = TU \times \delta \tag{7}$$

Definition 8. An itemset is called a high utility pattern/itemset (HUP/HUI) if its utility value is not less than the minimum utility value. A HUP is called fixed-length pattern if its length meets a user-specified length k, is also denoted as HUPK/HUIK (High Utility Pattern/Itemset with K-length).

Given a transaction database D, the task of mining HUPK aims at finding all HUPKs from the dataset D. Mining HUPKs from a database also refers to finding all itemsets whose utility values are not less than a user-specified minimum utility value and whose lengths meet a user-specified value.

2.2 Related Work

There has been a lot of research on HUP mining [8, 13–17]. The most typical method is to mine HUPs by two phases. The first phase generates candidate itemsets of HUPs by over-estimating the utility value of patterns; in the second phase, the dataset is scanned to calculate the utility value of each candidate. The two-phase method often produces a large number of candidates in the first phase (even including the non-existing itemsets of dataset), which not only requires a lot of space to store candidates, but also causes a huge amount of calculation in the second stage to get the utility value of each candidate. The typical two-phase algorithms mainly include Two-Phase [13], IHUP [8], UP-Growth and UP-Growth+ [19], and MU-Growth [16], etc.

In order to avoid generating candidates, more mining algorithms have been proposed, such as HUI-Miner [14], FHM [2], HUP-Miner [17], ULB-Miner [20], and d2HUP [21]. These algorithms do not maintain a large number of candidates, and are also called one-phase algorithms. It is possible to directly calculate whether each pattern is a HUP. HUI-Miner firstly proposed the utility-list structure for mining HUPs. Then FHM, HUP-Miner, and ULB-Miner have been proposed based on the utility-list structure. The algorithm FHM applied a depth-first search to find HUPs, and was shown to be up to seven times faster than HUI-Miner. D2HUP [21] directly found HUPs using the method of pattern-growth, maintained a database using a hyper structure, and was shown to be up to one order of magnitude faster than the algorithm UP-Growth. Zida et al. [3] proposed a new algorithm EFIM, which applied two new upper bounds of utility values to reduce the search space of the algorithm, and U-List structure to maintain a dataset; the experimental results showed that the performance of EFIM has been greatly improved. ULB-Miner [20] extended FHM and HUI-Mine, reduced the memory and runtime usage of the algorithm by utilizing a utility list buffer structure. The time efficiency of mining HUPs has been continuously improved.

Based on HUPs mining, several variant algorithms have been proposed, e.g., high average-utility patterns mining [22, 23], Top-K high utility patterns mining [4, 24], and HUPs mining from data stream [25]. Most of these studies mainly apply methods of one-phase or two-phase.

Fournier-Viger et al. [18] proposed a length-constrained HUPs mining algorithm FHM+ based on FIIM, which can mine specified length HUPs, and it can effectively remove out some patterns that users are not interested in. When this algorithm calculates the estimated utility value of the patterns, the utility value of the transaction with an unsuitable length can be excluded, and the number of candidates can be reduced. Thus, this algorithm effectively improves its efficiency of mining HUPs.

FP-Growth is a good method for frequent pattern mining, which can compress data into a tree effectively and find all frequent patterns efficiently. Since the utility data contains more information than the data that FP-Growth processes, FP-Growth cannot be used to mine HUPs. However, it is a good method to compress the data into a tree, it can reduce search space and improve search efficiency. In this paper, we propose an algorithm HUIKM based on FP-Growth for mining fix-length HUPs.

3 Algorithm HUIKM

The algorithm HUIKM mainly consists of two steps, as shown in Algorithm 1: (1) the transaction dataset is mapped to a tree, (2) HUPKs are mined from the tree by a pattern-growth method.

Algorithm 1: The algorithm HUIKM

```
Input   : D: transactions data;
          minutil: the user-specified minimum utility value;
          k: the user-specified length of high utility itemsets.
Output  : HUIs
// create a Tree T and a header Table H
CreateGTree(D,minutil,k) ;
// find all HUIs,which lenth is k, from the Tree T
MHUIK(T,H,base-itemset,k) ;
```

3.1 Create a Tree and a Header Table

In the process of mining pattern, the algorithm needs to repeatedly scan the dataset to calculate the utility value of patterns. If the identical transactions in the dataset can be compressed together as much as possible, this can greatly reduce the search space of the algorithm. In order to be able to effectively compress transaction itemsets (as much as possible to compress the identical transactions or items to an identical node or branch on the tree), we propose an algorithm, called HUIKM. HUIKM streamlines each transaction itemset, deletes the non-candidate items of each transaction. The same transactions are compressed into a branch, and itemsets with the same preorders are also compressed together, which can effectively reduce the search space.

In order to further describe the algorithm HUIK and ensure the algorithm accuracy, we begin by giving the following definitions:

Definition 9. The maximum utility sum of k items in a transaction itemset T_d is called maximum transaction utility (denoted as $TUK(T_d, k)$), and is defined by

$$TUK(T_d, k) = \begin{cases} \max(\sum_{j=1}^{k} U(x_{i_j}, T_d) \mid (1 \leq i_1 < i_2 < \cdots < i_k \leq |T_d|) \wedge (x_{i_j} \in T_d)) \ if |T_d| \geq k \\ 0, \quad else \end{cases} \tag{8}$$

For example, in Table 1 and 2, $\boldsymbol{TUK(T_1,1)}$ = max(U(A, T_1), U(C, T_1), U(F, T_1)) = max(16, 30, 1) = 30, $\boldsymbol{TUK(T_1,2)}$ = max(U(AC, T_1), U(AF, T_1), U(CF, T_1)) = max(46, 17, 31) = 46, $\boldsymbol{TUK(T_1,3)}$ = max(U(ACF, T_1,)) = max(47) = 47.

Definition 10. The maximum transaction utility of an itemset X in a transaction T_d is denoted as $TUK(X,T_d, k)$, and is defined by

$$TUK(X, T_d, k) = \begin{cases} U(X, T_d) + TUK(T_d - X, k - |X|) \ if \ X \subseteq T_d \wedge |T_d| \geq k \\ 0, \quad else \end{cases} \tag{9}$$

For example, in Table 1 and 2, $\boldsymbol{TUK(\{C\},T_1,2)}$ = U(C, T_1) + TUK({T_1 - {C},1) = 30 + max(U(A, T_1), U(F, T_1)) = max(16,1) = 46, $\boldsymbol{TUK(\{F\},T_1,2)}$ = U(F, T_1) + TUK({T_1 - {F},1) = 1 + max(16,30) = 31, $\boldsymbol{TUK(\{C\},T_2,2)}$ = U(C, T_2) + TUK({T_2 - {C},1) = 10 + max(28,20) = 38, $\boldsymbol{TUK(\{C\},T_8,2)}$ = U(C, T_8) + TUK({T_8 - {C},1) = 10 + max(U(A, T_8), U(B, T_8), U(E, T_8)) = 10 + max(12, 21, 6) = 31.

Definition 11. The maximum transaction utility of an itemset X in a dataset D is called maximum transaction weight utility (denoted as *TWUK(X,k)*), and is defined by

$$TWUK(X,k) = \sum_{T_d \in D \wedge T_d \supseteq X} TUK(X, T_d, k) \tag{10}$$

For example, in Table 1 and 2, ***TWUK*(\{C\},2)** $= TUK(\{C\},T_1,2) + TUK(\{C\},T_2,2) + TUK(\{C\},T_8,2) = 46 + 38 + 31 = 115$.

Definition 12. An itemset/item X is called a candidate (or promising itemset/item) for HUPK if its TWUK value is not less than a user-specified minimum utility value, otherwise it is an non-candidate (or unpromising itemset/item).

Definition 13. Given a transaction itemset $T_d = \{x_1, x_2, \cdots, x_i, \cdots\}$, and an ordered subset $X = \{x_i, x_{i_1}, x_{i_2}, \cdots, x_{i_j}\}$ of itemset T_d, then itemset $\{x_1, x_2, \cdots, x_{i-1}\}$ is named remain transaction-itemset of X in T_d; the maximum remain-transaction utility of an itemset X in a remain transaction-itemset T_d is denoted as ***RTWUK***(X,T_d,k), and is defined by

$$RTWUK(X, T_d, k) = \begin{cases} U(X, T_d) + TUK(\{x_1, x_2, ..., x_{i-1}\}, k - |X|) \ if\ X \subseteq T_d \wedge |T_d| \geq k \\ 0, \qquad else \end{cases} \tag{11}$$

The maximum remain-transaction utility of an itemset X in a dataset D is denoted as ***RTWUK***(X,T_d,k), and is defined by

$$RTWUK(X,k) = \sum_{T_d \in D \wedge X \in T_d} RTWUK(x_i, T_d, k) \tag{12}$$

For example, in Table 1 and 2, ***RTWUK*(\{C\},T_1,2)** $= U(\{C\}, T_1) + TUK(\{T_1 - \{AC\},1) = 30 + \max(U(F, T_1)) = 31$, ***RTWUK*(\{C\},$T_2$,2)** $= U(\{C\}, T_2) + TUK(\{T_2 - \{C\},1) = 10 + \max(U(D, T_2), U(E, T_2)) = 10 + 28 = 38$, ***RTWUK*(\{C\},$T_8$,2)** $= U(\{C\}, T_8) + TUK(\{T_8 - \{ABC\},1) = 10 + \max(U(E, T_8)) = 10 + 6 = 16$. ***RTWUK*(\{C\}, 2)** $= RTWUK(\{C\},T_1,2) + RTWUK(\{C\},T_2,2) + RTWUK(\{C\},T_8,2) = 31 + 38 + 16 = 85$.

Definition 14. An itemset/item X is called a candidate (or promising itemset/item) for HUPK if its RTWUK value is not less than a user-specified minimum utility value, otherwise it is an non-candidate (or unpromising itemset/item).

Property 1. Let *TWUK* or *RTWUK* value of an itemset X be not less than a user-specified minimum utility value, then any non-empty subset of X is also a promising itemset; let *TWUK* or *RTWUK* value of an itemset X be less than a user-specified minimum utility value, then any superset of X is also an unpromising itemset.

Proof: An itemset Y_1 is any non-empty subset of X, and an itemset Y_2, whose length is less than k, is any superset of X. According to definition 11 and 13, $TWUK(Y_1, k) > TWUK(X, k) > TWUK(Y_2, k)$ and $RTWUK(Y_1, k) >$

$RTWUK(X, k) > RTWUK(Y_2, k)$; thus, any non-empty subset of X is a promising itemset if TWUK or RTWUK value of itemset X is not less than a user-specified minimum utility value; any superset of X is an unpromising itemset if *TWUK* or *RTWUK* value of itemset X is less than a user-specified minimum utility value.

When the HUIKM algorithm creates the header table, it uses property 1 to find all candidates and save them in the header table. The *TWUK* value of property 1 is less than the *TWU* value, so property 1 can effectively reduce the number of candidates, i.e., reduce the subsequent search space. At the same time, the HUIKM algorithm uses property 1 to reduce the number of subsequent processing items when adding the transaction itemset to the tree, and saves the *RTWUK* value to the original *TWUK* value of the header table. The steps of HUIKM are as follows:

Step 1: Create a header table: calculate the TWUK value of each item by one scan of a dataset, maintain the candidates into a header table H, and sort them by the ascending order of TWUK value;
Step 2: The TWUK value of each item in the header table H is re-assigned to 0;
Step 3: Remove the non-candidates from each transaction itemset, sort remain items of each transaction in the order of the header table;
Step 4: If the length of the processed itemset is less than the user-specified length *k*, go back the above step to process next transaction itemset; otherwise, add this processed transaction itemset to a tree, and save the utility information to last node of the itemset. The *RTWUK* value and utility value of each item of this itemset is accumulated into the header table H.

The structure of a header table: "Item" records item name; "twuk/rtwuk" records the TWUK value of each item when creating the header table, and records the *RTWUK* value of each item when creating a tree; "utililty" records the utility value sum of each item and a base-itemset; "link" records the nodes of each item in a tree.

The tree node structure of the algorithm HUIKM is as follows: each node records the item name and child nodes, but the last node of each itemset on the tree also includes: "piu" is a list that records the utility of each item of the itemset that is from the node to the root node; "bu" records the utility value of the base itemset (see Example 2 in Sect. 3.2. The base item set is empty when the first tree is created. Thus, this value is not recorded when the first tree is created).

Example 1: We take the data in Table 1 and Table 2 as an example to illustrate the tree creation process (set the minimum utility threshold to 60 and k to 3):

(1) Create the header table as shown in Fig. 1(a).
(2) Process transaction T_1: After deleting the non-candidate "F" from the transaction itemset, the length of remaining itemset of the transaction (2) is less than the value of $k(3)$. Thus, there is no need to add the processed itemset to the tree.
(3) Process transaction T_2: delete non-candidate items, and sort them to get the itemset {E, D, C} and their utility values, then add the itemset to the tree T, and save the utility value of each item into the last node of this itemset (*piu*), the result is shown in Fig. 1(b). When adding each item to the tree, the utility value of each item and

the utility value of the base-itemset (*bu*) are accumulated to the utility of the header table (the initial value of the base-itemset is empty). When adding item "C" to the tree, its *RTWUK* value (58) is added to the *twuk* of the header table, and its node is added to the link of the header table; the result is shown in Fig. 1(b).

(4) The transaction itemset T_3 is handled in the same way: the tree is shown in Fig. 1(c) after adding T_3 to the tree. When adding items "D" and "B", their *RTWUK* value (42) is added to the header table, and their nodes are also added to the link of the header table. The result is shown in the header table in Fig. 1(c).

(5) The remaining transactions are added into the tree by the same way. When adding T_5 to the tree, only the utility values of the items are added to node "B", and the *RTWUK* values of "D" and "B" are accumulated to the header table.

(6) The result is shown in Fig. 1(d) after adding all transactions to the tree.

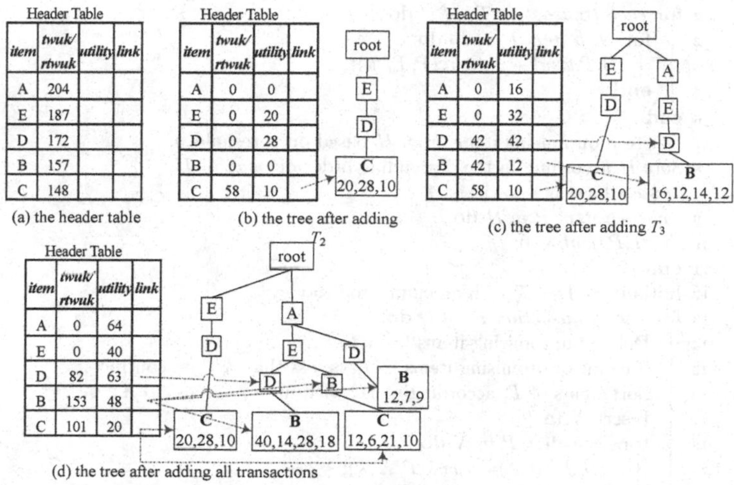

Fig. 1. A case of creating a tree and a header table

The algorithm of creating the header table and tree is shown in Algorithm 2. First, create a header table via one scan of dataset. The processing steps are as follows:

(1) Calculate the TWUK value of each item (lines 2–6) and store it into a header table H;

(2) Remove these items, whose TWUK values are less than the minimum utility value, from the header table (line 7);

(3) Sort the items of the header table H by the descending order of TWUK values(line 8);

(4) Re-assign the TWUK value of each item in H to 0 (lines 9–11).

Second, add each transaction T_d to a tree. The processing steps are as follows:

(1) Remove non-candidates from a transaction itemset T_d (line 14);
(2) Process next transaction if the length of itemset T_d is less than k (line 15);
(3) Sort items of T_d by the order of H (line 16);
(4) Insert T_d to the tree, accumulate the *RTWUK* value of item in T_d into the header table H (line 19), accumulate the sum of utility values of item and base-itemset into H (line 20), and add the link of new node to H (line 21).

Algorithm 2: The CreateGTree Procedure

```
   Input  : D: transactions data, k: the length of high utility
            itemsets.
   Output : a tree T
   // First scan of the database D
1  Initiate a header table H containing the fields of item, twuk, and
   links;
2  for each transaction T_d of D do
3  |  for each item P in T_d do
4  |  |  H.P.twuk+ = twuk(P, T_d, k);
5  |  end
6  end
7  Delete unpromising items from H based on threshold η;
8  Sort H according in the descending order of twuk of H;
   // Second scan of the database D
9  for each item P in H do
10 |  H.P.twuk = 0;
11 end
12 Initialize a Tree T with an empty root node;
13 for each transaction T_d of D do
14 |  Delete unpromising items from T_d;
15 |  if count of promising items in T_d is less than k then continue;;
16 |  Sort items of T_d according to H, with utility values, to X;
17 |  Insert X to T;
18 |  for each item P in X do
19 |  |  H.P.twuk+ = rtwuk(P, X, k);
20 |  |  H.P.utility+ = (U(P, X) + bu);
   |  |  // bu: N.bu
21 |  |  New node is added to H.P.link;
22 |  end
23 end
```

3.2 Ming HUIK from a Tree

Algorithm HUIKM will process each item of the header table after creating a header table and a tree in Sect. 3.1. It is necessary to determine whether the base-itemset of each item in the header table is a HUIK, and it is necessary to determine whether there is a HUI in its superset. If the superset may include a HUIK, it is necessary to create a sub tree and a sub header table, and then recursively process the new header table and tree. The steps of processing each item P in the header table are as follows:

Step 1: If the *RTWUK* value of item *P* is less than the minimum utility value, skip to step 4 otherwise, add item *P* to the base-itemset. The base-itemset is a HUIK if its length and utility values are not less than the user-specified values.
Step 2: If the length of the base-itemset is less than the user-specified value, create a new sub tree and a sub header table by scanning the tree via the link of item *P* in the header table, and recursively process the new header table and tree if the new header table is not empty.
Step 3: Remove the item *P* from the base-itemset;
Step 4: Process the utility information of item *P* as follows: accumulate or copy the utility of the base-itemset and list *piu* to parent node, and delete the nodes of item *P* from the tree.

Algorithm 3: MHUIK

```
Input   : T: a tree, H: a header table, base-itemset, k: the length
          of high utility itemsets.
Output  : HUIs
1  for each item P in H (with a bottom-up sequence) do
       // Step 1: Generate HUIs and create sub TN-tree
2      if H.P.rtwuk ≥ min_util then
3          base-itemset = base-itemset ∪{P};
4          if |base − itemset| == k and H.P.utility ≥ min_util then
               // Calculate BU and NU
5              Copy base-itemset to HUIs ; // generate one HUI
6          end
7          if |base − itemset| < k then
8              Create a sub tree subT and a header table subH for
                 base-itemset;
9              MHUIK(subT, subH, base-itemset,k) ; // recursive call
10         end
11         Remove item P from itemset base-itemset;
12     end
       // Step 2: modify the utility information of the last item of itemset
13     foreach node N for item H.P in T do
14         remove utility value of item P from list N.piu;
15         if N.parent.bu==NULL then
16             N.parent.bu = N.bu;
17             N.parent.piu = N.piu;
18         else
19             N.parent.bu = N.parent.bu + N.bu;
20             N.parent.piu = N.parent.piu + N.piu;
21         end
22         Remove node N from T;
23     end
24 end
25 return HUIs;
```

Algorithm 3 is to mine HUIKs from a tree. The algorithm handles each item of a header table from the last item. The processing steps of each item consist mostly of two steps. (1) Determine whether each item will be a candidate for HUIKs (lines 2–12); (2) Move the utility information of nodes to the parent node (lines 13–23).

The first step consists mostly of the following sub-steps: (1) If the *RTWUK* value of an item is less than the minimum utility value *MinU*, move the utility information of nodes named this item and process the next item; otherwise, proceed to the next step. (2) Add the current item to the *base-itemset* (line 3; known as base itemset or prefix itemset). (3) If the length of the *base-itemset* equals the user specified minimum length value *k* and the utility of this *base-itemset* is not less than *MinU*, this *base-itemset* is a HUIK (lines 4–6). (4) If he length of the *base-itemset* is less than *k*, create a sub header table and a sub tree for *base-itemset* (or the current item), and recursively process the sub header table and the sub tree (lines 7–10). (5) Remove the current item from the *base-itemset*.

The second step consists the following sub-steps: (1) Delete the utility value of the current item from the list *piu* of nodes whose name is the current item. (2) If the parent of the current nodes does not include the utility information, move the utility information of the current nodes to the parent node (line 16–17; refer to Example 2 for details); otherwise, accumulate the utility information of the current nodes to the parent node (lines 19–20). (3) Remove the current nodes from the tree (line 22).

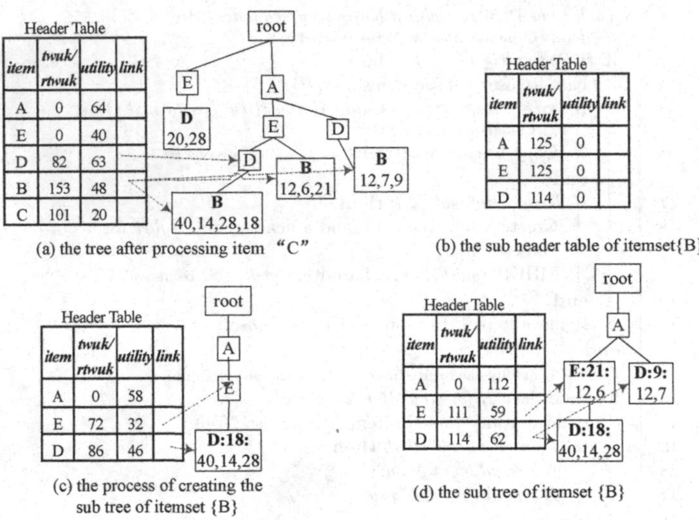

(a) the tree after processing item "C"

(b) the sub header table of itemset{B}

(c) the process of creating the sub tree of itemset {B}

(d) the sub tree of itemset {B}

Fig. 2. A case of creating a sub-tree and a header table

Example 2: We take the tree and header table in Fig. 1(d) as an example to illustrate the steps of mining HUIKs from a tree. We will process each item of the header table from the last item. (1) The *RTWUK* value (101) of the item "C" is not less than 60, so add item "C" into the base-itemset (initial value is empty), i.e., base-itemset = {C}. (2) The base-itemset is not a HUIK because its length (1) is not equal to 3. (3) Since the length of the base-itemset is less than 3, create a sub header table and a tree for this base-itemset by the following: get two itemsets {E,D,C} and {A,E,B,C}, and their utility information by scanning the tree in Fig. 1(d), and use these two itemsets and their utility information to create a sub header table. The sub header does not contain any items whose *TWUK* value is not less than 60. Thus, we do not need to create a sub tree. (4) Remove item "C" from the base-itemset. (5) Remove the utility the base-itemset and list *piu* to parent node (nodes "D"and "B"), the result is shown in Fig. 2(a). (6) Remove the nodes "C" from the tree.

We process the item "B" of the header table by the same way: (1) Create a sub header table by scanning three branches on the tree, the result is shown in Fig. 2(b). (2) Get the first itemset {A,E,D,B} from the branch in Fig. 2(a), sort the itemset by the order of the sub header table, and add this ordered itemset to a sub tree, the result is shown in Fig. 2(c), the utility list of the itemset {40,14,28} and the utility of the base-itiemset (bu) are stored into the node "D" in Fig. 2(c); we do not modify the *RTWU* value of in the sub header table and modify the utility value ($58 = 40 + 18$) of item "A" in the header table when the item "A" of this itemset is added to the tee; modify the RTWUK value ($72 = 40 + 14 + 18$) and the utility value ($32 = 14 + 18$) in the header table when the item "E" of this itemset is added to the tee; modify the RTWUK value ($86 = 40 + 28 + 18$) and the utility value ($46 = 28 + 18$) in the header table when the item "D" of this itemset is added to the tee. (3) According to Algorithm 3, we recursively process the header table in Fig. 2(d), and mine 3 HUIKs ({B,D,E:62}, {B,D,A:114}, {B,E,A:111}). (4)After processing the item "B" in Fig. 2(a), process the next item "D" of the header table in Fig. 2(a) and get a HUIK {D,E,A:82}. (5) We do not need to process the other items "A" and "E" in the header table in Fig. 2(a) since the *RTWUK* values of these two items are less than 60. Finally, we find four HUIKs from the dataset: {B,D,E:62},{B,D,A:114}, {B,E,A:111} and {D,E,A:82}.

4 Experimental Results

Table 3. Dataset characteristics

Dataset	Distinct items (#)	Avg. trans. length (AS)	Transactions (#)	Type
Chainstore	46,086	7.2	1,112,949	Very sparse, short transactions
Pumsb	2,111	74	49,046	Sparse, very long transactions
Retail	16,470	10.3	88,162	Very sparse, short transactions
Connect	129	43	67,557	Dense, long transactions
Mushroom	119	23	8,124	Dense, moderately long transactions
Chess	76	37	3,196	Dense, long transactions

In order to assess the performance of HUIKM, we compare the performance of HUIKM with FHM+ [18], EFIM [3] and ULBMiner [20]. But EFIM and ULBMiner mine all HUIs; so we firstly mine HUIs using EFIM and ULBMiner, and then find fixed-length HUIs. We call the revised algorithms EFIM_k and ULBMiner_K respectively. All algorithms are implemented in JAVA programming language. FHM+, EFIM and ULBMiner are downloaded from the SPMF website (http://www.philippe-fournier-viger.com/spmf/) [26]. The experimental data are six classical datasets downloaded from the SPMF website. The characteristics of these six datasets are shown in Table 3. Experimental platform: Windows 7 operating system, 16G memory, Intel(R) Core(TM) i7-6500 CPU @ 2.50 GHz.

In each experiment, the four algorithms find the same HUIKs, and the results are shown in Fig. 3. According to Theorem 1, deleting non-candidates from a header table or a sub header table does not result in the loss of HUIKs. Algorithm HUIKM uses pattern-growth for mining HUIKs; when processing header table or sub header table, the items that have been already processed do not appear again in subsequent processing, so the processed items do not need to be considered, that is, the *RTWUK* value is the *TWUK* value that does not include the utility value of processed items. Thus, algorithm HUIKM can find all HUIKs.

The experiment firstly compared the running time of the two algorithms on six datasets. On each dataset, the larger the fixed-length *k* value, the larger the *TWUK* value of each item, the more candidates will be retained in the header table, the greater the amount of computation the algorithm will be and the more time-consuming the algorithm will be. There are generally fewer long patterns on sparse datasets, while there are relatively many long patterns on dense datasets. In order to compare the running time of the algorithm under different *k* values, *k* takes 2 and 3 on two sparse datasets, and 4 and 6 on dense datasets, respectively. The higher the *k* value, the more items the header table will have, i.e. the greater the search space of the algorithm, the less efficient the algorithm will run, as shown in Fig. 4. But the larger the value of k, the more patterns will not be, as shown in Fig. 3(a). For example, on the dataset Chainstore, the higher *k* value, the fewer HUIKs produced. It is obvious from Fig. 4 that the proposed algorithm is more efficient than FHM under various *k* values.

Figure 4 shows the running time of two algorithms under minimum utility thresholds and various *k*-values. It is obvious from Fig. 4 that HUIKM is more efficient than FHM under various *k* values and various minimum utility thresholds. The efficiency of the algorithm is increased to 2–3 orders of magnitude on dense datasets. HUIKM can reduce the search space and improve the performance by the following methods: (1) it uses the *TWUK* values to create header tables, and the *TWUK* value is not more than the *TWU* value, so this can obviously reduce the number of items of the header table; (2) Items closing to the root node is not considered for mining HUIKs; (3) HUIKM uses the *RTWUK* value instead of *TWUK* value to determine if there will be one HUIK or more HUIKs. HUIKM effectively reduces the number of items of the header table using three methods above, and effectively reduces the amount of computation. At the same time, it can also be seen from Fig. 4 that the running time of HUIKM increases with the decrease of threshold, the increase is relatively flat, especially on the dataset containing long transactions, HUIKM reduces the number of items in the header table

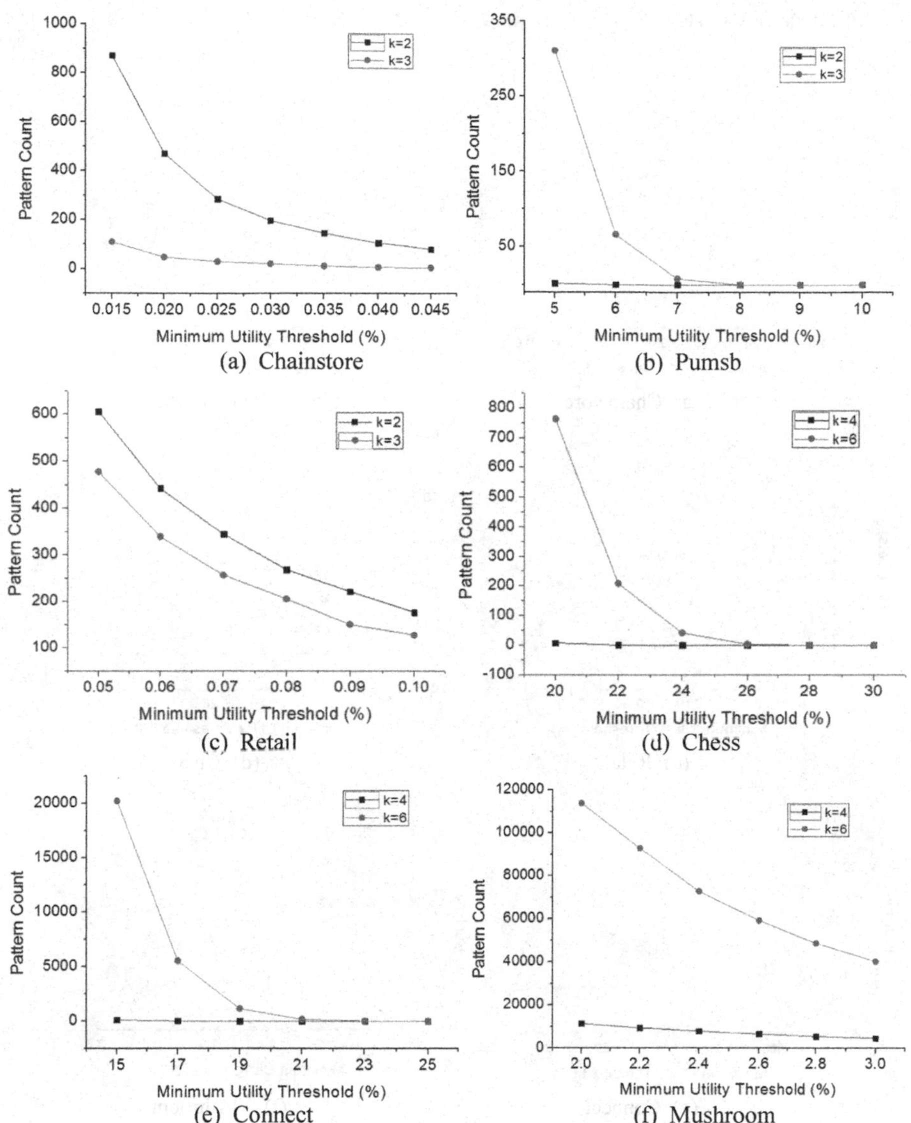

(a) Chainstore (b) Pumsb

(c) Retail (d) Chess

(e) Connect (f) Mushroom

Fig. 3. Number of patterns

more obviously in the long dataset, and HUIKM can effectively compress the data into the tree on the dense dataset, thereby reducing the search space of the algorithm, and the running time can be increased to 2–3 orders of magnitude on the dense dataset, as shown in Fig. 4(c).

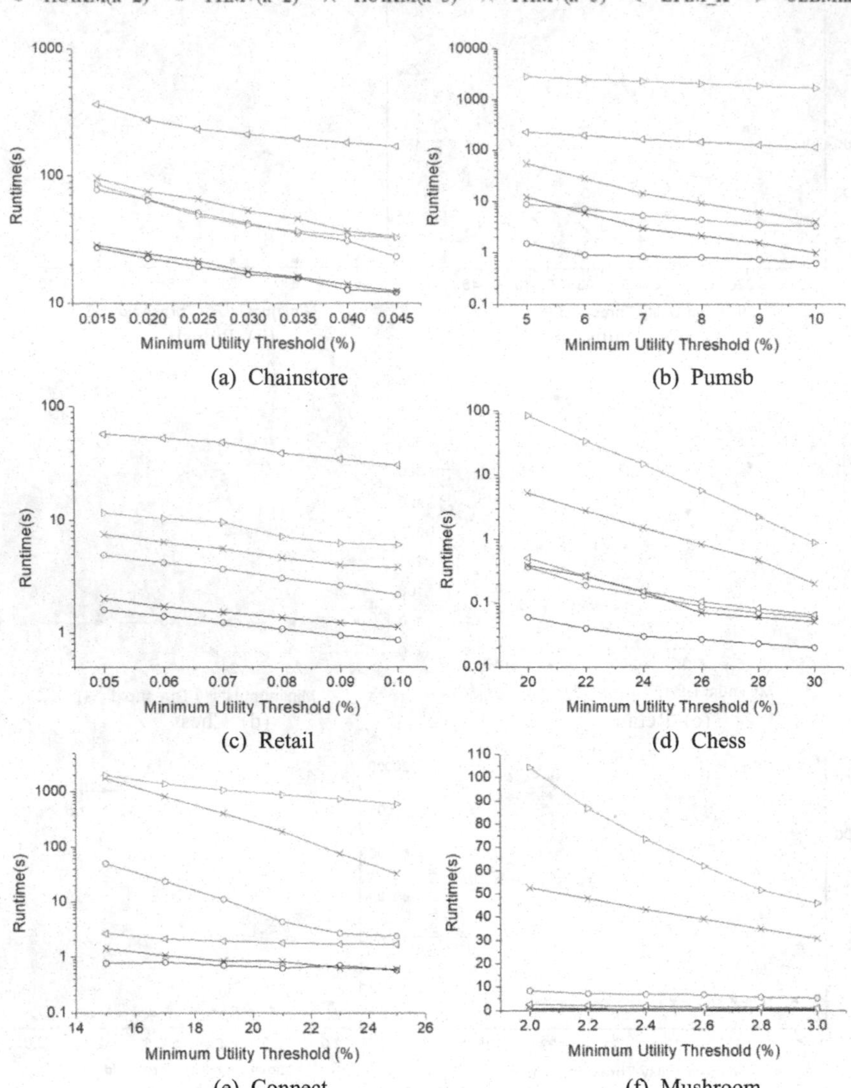

Fig. 4. Running time on different datasets

5 Conclusion

In this paper, we propose an algorithm for mining HUIKs through the study of the HUP mining algorithm. This algorithm firstly compresses a data effectively into a tree, and then uses the method of pattern-growth to recursively find all HUIKs; HUIKM effectively reduces the search space and improves the efficiency of the algorithm since this algorithm uses the tree structure for maintaining the data and uses the *RTWUK* value to reduce the number of header tables. In our experiments, we compare the performance of HUIKM with FHM+, EFIM and ULBMiner on six classical datasets. The experimental results show that the running time of HUIKM is greatly improved on different datasets, especially on the long and dense datasets, which can be increased to 2–3 orders of magnitude.

But HUIKM costs much time in creating sub trees and header table, we hope to further optimize it in the next step. In the future, we will also extend the approach in sequential data to reveal high utility sequential pattern.

Acknowledgement. This work is partially supported by the Zhejiang Philosophy and Social Science Project (19GXSZ49YB).

References

1. Lin, C., Lan, G., Hong, T.: Mining high utility itemsets for transaction deletion in a dynamic database. Intell. Data Anal. **19**(1), 43–55 (2015)
2. Fournier-Viger, P., Wu, C.-W., Zida, S., Tseng, V.S.: FHM: faster high-utility itemset mining using estimated utility co-occurrence pruning. In: Andreasen, T., Christiansen, H., Cubero, J.-C., Raś, Z.W. (eds.) ISMIS 2014. LNCS (LNAI), vol. 8502, pp. 83–92. Springer, Cham (2014). https://doi.org/10.1007/978-3-319-08326-1_9
3. Zida, S., et al.: EFIM: a fast and memory efficient algorithm for high-utility itemset mining. Knowl. Inf. Syst. **51**(2), 595–625 (2017)
4. Han, X., Liu, X., Li, J., Gao, H.: Efficient top-k high utility itemset mining on massive data. Inf. Sci. **557**, 382–406 (2021)
5. Verma, A., Dawar, S., Kumar, R., Navathe, S., Goyal, V.: High-utility and diverse itemset mining. Appl. Intell. **51**(7), 4649–4663 (2021). https://doi.org/10.1007/s10489-020-02063-x
6. Agrawal, R., Imielinski, T., Swami, A.. Mining association rules between sets of items in large databases. In: ACM SIGMOD International Conference on Management of Data, Washington, DC, United States. ACM (1993)
7. Li, H., et al.: Fast and memory efficient mining of high utility itemsets in data streams. In: 2008 Eighth IEEE International Conference on Data Mining. IEEE (2008)
8. Ahmed, C.F., et al.: Efficient tree structures for high utility pattern mining in incremental databases. IEEE Trans. Knowl. Data Eng. **21**(12), 1708–1721 (2009)
9. Shie, B.-E., Hsiao, H.-F., Tseng, V.S.: Efficient algorithms for discovering high utility user behavior patterns in mobile commerce environments. Knowl. Inf. Syst. **37**(2), 363–387 (2012). https://doi.org/10.1007/s10115-012-0483-z
10. Li, Y., Yeh, J., Chang, C.: Isolated items discarding strategy for discovering high utility itemsets. Data Knowl. Eng. **64**(1), 198–217 (2008)
11. Zihayat, M., Davoudi, H., An, A.: Mining significant high utility gene regulation sequential patterns. BMC Syst. Biol. **11**(6), 109 (2017)
12. Yao, H., Hamilton, H.J., Butz, G.J.: A foundational approach to mining itemset utilities from databases. In: 4th SIAM International Conference on Data Mining (ICDM 2004), Lake Buena Vista, FL, United States (2004)
13. Liu, Y., Liao, W.K., Choudhary, A.: A two-phase algorithm for fast discovery of high utility itemsets. In: Ho, T.B., Cheung, D., Liu, H. (eds.) Advances in Knowledge Discovery and Data Mining, pp. 689–695. Springer, Heidelberg (2005). https://doi.org/10.1007/11430919_79
14. Liu, M., Qu, J.: Mining high utility itemsets without candidate generation. In: 21st ACM International Conference on Information and Knowledge Management (CIKM 2012), Maui, HI, United States. Association for Computing Machinery (2012)

15. Lan, G., et al.: Applying the maximum utility measure in high utility sequential pattern mining. Expert Syst. Appl. **41**(11), 5071–5081 (2014)
16. Yun, U., Ryang, H., Ryu, K.H.: High utility itemset mining with techniques for reducing overestimated utilities and pruning candidates. Expert Syst. Appl. **41**(8), 3861–3878 (2014)
17. Krishnamoorthy, S.: Pruning strategies for mining high utility itemsets. Expert Syst. Appl. **2015**(42), 2371–2381 (2015)
18. Fournier-Viger, P., Lin, J.-W., Duong, Q.-H., Dam, T.-L.: FHM$+$: faster high-utility itemset mining using length upper-bound reduction. In: Fujita, H., Ali, M., Selamat, A., Sasaki, J., Kurematsu, M. (eds.) IEA/AIE 2016. LNCS (LNAI), vol. 9799, pp. 115–127. Springer, Cham (2016). https://doi.org/10.1007/978-3-319-42007-3_11
19. Tseng, V.S., et al.: Efficient algorithms for mining high utility itemsets from transactional databases. IEEE Trans. Knowl. Data Eng. **25**(8), 1772–1786 (2013)
20. Duong, Q.-H., Fournier-Viger, P., Ramampiaro, H., Nørvåg, K., Dam, T.-L.: Efficient high utility itemset mining using buffered utility-lists. Appl. Intell. **48**(7), 1859–1877 (2017). https://doi.org/10.1007/s10489-017-1057-2
21. Liu, J., Wang, K., Fung, B.C.: Direct discovery of high utility itemsets without candidate generation. In: 2012 IEEE 12th International Conference on Data Mining. IEEE (2012)
22. Kim, J., et al.: One scan based high average-utility pattern mining in static and dynamic databases. Futur. Gener. Comput. Syst. **111**, 143–158 (2020)
23. Truong, T., et al.: Efficient high average-utility itemset mining using novel vertical weak upper-bounds. Knowl.-Based Syst. **183**, 104847 (2019)
24. Krishnamoorthy, S.: Mining top-k high utility itemsets with effective threshold raising strategies. Expert Syst. Appl. **117**, 148–165 (2019)
25. Nam, H., et al.: Efficient approach of recent high utility stream pattern mining with indexed list structure and pruning strategy considering arrival times of transactions. Inf. Sci. **529**, 1–27 (2020)
26. Fournier-Viger, P., et al.: The SPMF open-source data mining library version 2. In: Berendt, B., Bringmann, B., Fromont, É., Garriga, G., Miettinen, P., Tatti, N., Tresp, V. (eds.) ECML PKDD 2016. LNCS (LNAI), vol. 9853, pp. 36–40. Springer, Cham (2016). https://doi.org/10.1007/978-3-319-46131-1_8

A Novel Method to Create Synthetic Samples with Autoencoder Multi-layer Extreme Learning Machine

Yulin He[1,2(✉)], Qihang Huang[2], Shengsheng Xu[2], and Joshua Zhexue Huang[1,2]

[1] Guangdong Laboratory of Artificial Intelligence and Digital Economy (SZ), Shenzhen 518107, China

[2] College of Computer Science and Software Engineering, Shenzhen University, Shenzhen 518060, China

{yulinhe,zx.huang}@szu.edu.cn, {1900271056,2070276067}@email.szu.edu.cn

Abstract. The imbalanced classification is an important branch of supervised learning and plays the important roles in many application fields. Compared with the sophisticated improvements on classification algorithms, it is easier to obtain the good performance by synthesizing the minority class samples so that the classification algorithms can be trained based on the balanced data sets. In consideration of the strong representation ability of multi-layer extreme learning machine (MLELM), this paper proposes a new method to create the synthetic minority class samples based on auto-encoder ML-ELM (simplified as AE-MLELM-SynMin). Firstly, an AE-MLELM is trained to obtain the deep feature encodings of original minority class samples. Secondly, the crossover and mutation operations are preformed on the original deep feature encodings and a number of new deep feature encodings are generated. Thirdly, the synthetic minority class samples are created by transforming the new deep feature encodings with AE-MLELM. Finally, the persuasive experiments are conducted to demonstrate the effectiveness of AE-MLELM-SynMin method. The experimental results show that our method can obtain the better imbalanced classification performance than SMOTE, Borderline-SMOTE, Random-SMOTE, and SMOTE-IPF methods.

Keywords: Imbalanced classification · Minority class · Synthetic samples · SMOTE · Autoencoder MLELM

1 Introduction

The imbalanced classification or imbalanced learning is an important branch of data mining and machine learning, which has the broad applications in the actual fields [1], e.g., defect detection, fault detection, medical diagnosis, fraud detection. The imbalanced classification problem means that a classifier is constructed based on a labeled data set, where there are the obvious differences among the numbers of samples belonging to different classes. For example, the

U. K. Rage et al. (Eds.): DASFAA 2022 Workshops, LNCS 13248, pp. 21–33, 2022.
https://doi.org/10.1007/978-3-031-11217-1_2

defective samples in defect detection application always account for a small part of the total sample, while the qualified samples account for the majority. The traditional classification algorithms are designed for the balanced data sets. For the imbalanced data sets, the traditional classification methods can not effectively learn from the minority class samples. Although their classification accuracies are still relatively high for imbalanced classification problems, the higher classification accuracies usually cannot effectively reflect the effectiveness of classification algorithms. Because for an imbalanced classification data set, if the classification method judges all samples as normal ones, that is to say, all the abnormal samples are also judged as normal ones, then the final classification accuracy will be high, but such results are meaningless for many practical applications.

At present, the studies on how to solve the imbalanced classification problems mainly focus on the algorithm-driven methods [2–6] and data-driven methods [7–12]. The algorithm-driven method is to construct an algorithm which places more emphasis on the minority class samples. The data-driven method is mainly to make the number of different labeled training samples more balanced. The two simplest methods are random over-sampling and random under-sampling. The former supplements the minority class samples by randomly selecting the minority class samples repeatedly, while the latter randomly selects a small number of majority class samples. In most cases, the over-sampling method is better than the under-sampling method. This is because the under-sampling method eliminates the valuable data. In the actual scenarios, the data are scarce and valuable, so the over-sampling method is often a better choice. Compared with the method based on the algorithm level, the method based on the data level has the characteristics of simple implementation and low computational complexity, so the focus of this research is mainly on the data level method.

The classic data-level method is the synthetic minority over-sampling technology (SMOTE) [7]. SMOTE does not simply copy the samples. Its principle is as follows. For each minority sample, a sample is randomly selected from its k nearest neighbors and then a sample point is randomly selected on the line of two samples. Because of SMOTE's well imbalanced learning performances, it attracts the widespread attention from scholars. Han et al. [8] proposed the Borderline-SMOTE method. The main idea is to oversample the minority class samples on the boundary between the majority class and minority class samples. Dong et al. [9] proposed a more general Random-SMOTE method to generate new sample points among three sample points. Compared with the classic SMOTE method, the distribution of sample points generated by this method is more uniform and this method can greatly improve the sparseness of sample space. Sáezet al. [10] introduced an Iterative-Partitioning Filter (IPF) to extend the SMOTE algorithm, the SMOTE-IPF algorithm, which overcomes the problems caused by noisy and boundary samples in imbalanced data sets. Lee et al. [13] proposed a Guassian-SMOTE algorithm that combined Gaussian probability distribution with SMOTE algorithm. This method obeies Gaussian probability distribution when SMOTE method is used to create minority class samples. Douzas et al. [14] proposed the k-means-SMOTE algorithm combining k-means and SMOTE algorithm. Based on the k-means algorithm, the method generates more samples in the sparse area of minority class samples than in the dense area of minority class samples.

The experimental results show that these SMOTE variants create minority class samples to achieve the better results on general classifiers. However, the comprehensive analysis shows that there are two major defects in this kind of method. The first defect is the inconsistency of probability distribution between the original samples and the synthetic samples and the second defect is the loss of information amount. This is because the new sample points generated by SMOTE method are randomly selected on the line of two minority class sample points, which causes the new sample points generated to be too fixed and results in insufficient information amount. Although some variants of SMOTE method improves performance to some extent, they does not improve the two shortcomings mentioned earlier.

In this paper, we propose a synthetic minority class technology based on auto-encoder multi-layer extreme learning machine (AE-MLELM), abbreviated as AE-MLELM-SynMin. AE-MLELM has a good representation capability. And the intermediate hidden layer is another representation of input layer [15]. The crossover and mutation operations are conducted on the hidden-layer output matrix and then the hidden layer is restored to get new minority class samples. Experimental results show that our method has the better performance than SMOTE, Borderline-SMOTE, Random-SMOTE, and SMOTE-IPF. The remainder of this paper is arranged as follows. Section 2 introduces the classic SMOTE method and analyzes its defects. Section 3 presents the proposed AE-MLELM-SynMin method in details. Section 4 shows the experimental results. Section 5 concludes this paper.

2 SMOTE Method

SMOTE is a data-driven imbalanced data classification method. It uses the method of synthesizing minority class samples to expand the number of minority class samples until the number of samples in each class is roughly equal. Its basic principle is described below. For each sample $\mathrm{v}_n^{(k)}, n = 1, 2, \cdots, \mathcal{N}_{\min k}$ in the k-th minority class data set $\mathbb{D}_{\min k}$, its $\mathcal{P}$ nearest neighbors are $\mathrm{v}_{n1}^{(k)}, \mathrm{v}_{n2}^{(k)}, \cdots, \mathrm{v}_{n\mathcal{P}}^{(k)}$, a point $\mathrm{v}_{np}^{(k)}$ is randomly selected from these sample points, and the synthesized new minority class sample is denoted as $\bar{\mathrm{v}}^{(k)}$. The new sample point $\bar{v}_d^{(k)}$ is calculated as

$$\bar{v}_d^{(k)} = v_{nd}^{(k)} + \lambda \left[v_{npd}^{(k)} - v_{nd}^{(k)} \right], d = 1, 2, \cdots, \mathcal{D}, \tag{1}$$

where $\lambda \in [0, 1]$ is a random number that obeys a uniform distribution. We repeat the above-mentioned process many times until the enough samples are synthesized for the k-th minority class.

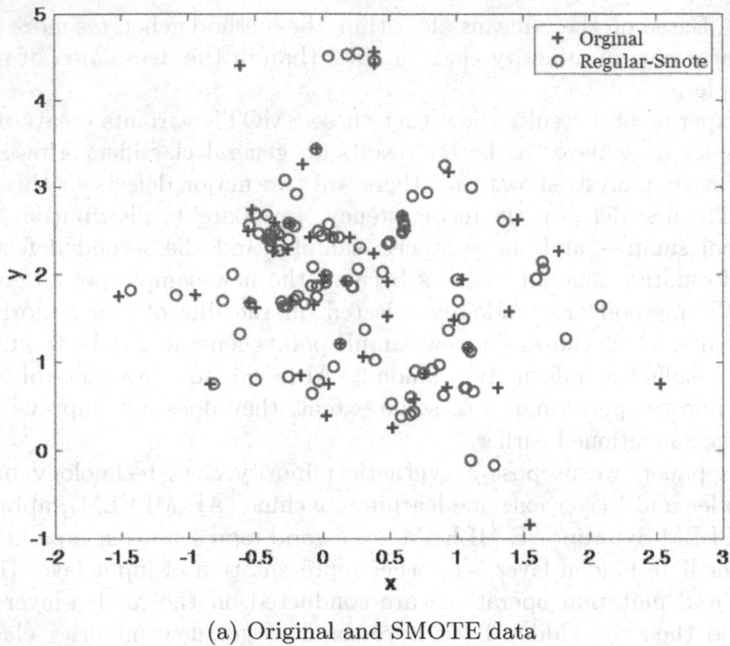

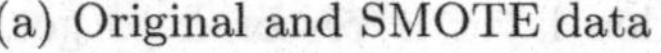
(a) Original and SMOTE data

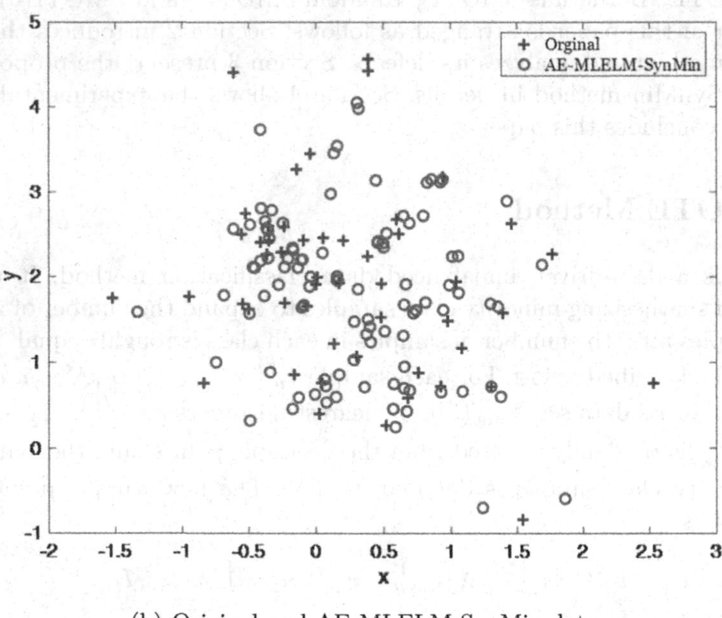

(b) Original and AE-MLELM-SynMin data

Fig. 1. Comparison among original, SMOTE, and AE-MLELM-SynMin data

The related experiments show that the SMOTE method and its variants have the good performance, but the more detailed analysis indicates that these SMOTE-based methods have some inherent defects. SMOTE has the problem of insufficient information amount for the synthesized new samples. In order to explain the shortcoming of SMOTE method, we provide a legend to show the distributions of original and synthesized samples. Figure 1(a) shows 50 original samples and 100 samples synthesized by SMOTE method. Figure 1(b) shows 50 original samples and 100 samples synthesized by the AE-MLELM-SynMin method. It can be seen from the figure that the new samples synthesized by SMOTE are mostly distributed near the original sample points. The new sample points synthesized by AE-MLELM-SynMin method are more scattered and have more information amount. The experiment in Sect. 4 also confirms this observation.

3 Proposed AE-MLELM-SynMin Method

Due to the problems of fixed samples and insufficient information amount generated by classical SMOTE algorithm and its variants, we introduce the synthesis minority class technology based on AE-MLELM, namely AE-MLELM-SynMin.

3.1 Training AE-MLELM

For the k-th minority class, an AE-MLELM, denoted as AE-MLELM$_k$ is firstly trained. AE-MLELM is a special MLELM with the same input and output. Its input and output expressions are shown as

$$\mathrm{X}_{\min k} = \begin{bmatrix} v_{11}^{(k)} & v_{12}^{(k)} & \cdots & v_{1\mathcal{D}}^{(k)} \\ v_{21}^{(k)} & v_{22}^{(k)} & \cdots & v_{2\mathcal{D}}^{(k)} \\ \vdots & \vdots & \ddots & \vdots \\ v_{\mathcal{N}_{\min k},1}^{(k)} & v_{\mathcal{N}_{\min k},2}^{(k)} & \cdots & v_{\mathcal{N}_{\min k},\mathcal{D}}^{(k)} \end{bmatrix}. \tag{2}$$

For MLELM, each of its hidden layers is an AE-MLELM with the same input and output. We can then calculate the hidden-layer output matrix for each layer. As for the calculation of the first hidden layer output matrix $\mathbf{H}^1$, since its input and output matrices are both $\mathrm{X}_{\min k}$, the first hidden layer output matrix $\mathbf{H}^1$ is obtained by the single hidden layer extreme learning machine with both input and output nodes, then the calculation formula of $\mathbf{H}^1$ is

$$\mathbf{H}^1 = g\left(\mathrm{X}\left(\beta^1\right)^T\right), \tag{3}$$

where β^1 is the weight of the output layer of the first layer and $g(*)$ is the activation function.

Similarly, for the i-th$(i = 1, 2, \cdots, n)$ hidden layer, its input and output matrices are the output matrix of the hidden layer of the upper layer $\mathbf{H}^{i-1}$, and the output matrix of the hidden layer is $\mathbf{H}^i$. The calculation formula of the output matrix of the hidden layer $\mathbf{H}^i$ is

$$\mathbf{H}^i = g\left(\mathbf{H}^{i-1}\left(\boldsymbol{\beta}^i\right)^T\right), \tag{4}$$

where $\boldsymbol{\beta}^i$ is the output-layer weight. And we derive the calculation formula of the output matrix of the last hidden layer as

$$\mathbf{H}^l = g\left(\mathbf{H}^{l-1}\left(\boldsymbol{\beta}^l\right)^T\right). \tag{5}$$

Then we calculate the output layer weight of the last hidden layer with the following formula

$$\boldsymbol{\beta}^{l+1} = \left(\mathbf{H}^l\right)^\dagger \mathbf{X}, \tag{6}$$

where $\left(\mathbf{H}^l\right)^\dagger$ is the Moore-Penrose inverse [16] of $\mathbf{H}^l$. AE-MLELM$_k$ uses $\boldsymbol{\beta}^{l+1}$ to decode $\mathrm{X}_{\min k}$.

3.2 Conducting Crossover and Mutation Operations

After the training of AE − MLELM$_k$ is finished, the crossover and mutation operations for the last hidden-layer output matrix $\mathrm{H}^l_{\min k}$ are conducted and the changed hidden-layer output matrix $\overline{\mathrm{H}}^l_{\min k}$ is expressed as

$$\overline{\mathrm{H}}^l_{\min k} = \begin{bmatrix} \bar{h}^{(k)}_{11} & \bar{h}^{(k)}_{12} & \cdots & \bar{h}^{(k)}_{1\mathcal{L}} \\ \bar{h}^{(k)}_{21} & \bar{h}^{(k)}_{22} & \cdots & \bar{h}^{(k)}_{2\mathcal{L}} \\ \vdots & \vdots & \ddots & \vdots \\ \bar{h}^{(k)}_{\mathcal{N}_{\min k},1} & \bar{h}^{(k)}_{\mathcal{N}_{\min k},2} & \cdots & \bar{h}^{(k)}_{\mathcal{N}_{\min k},\mathcal{L}} \end{bmatrix}. \tag{7}$$

- We randomly select two lines $h^{(k)}_m$ and $h^{(k)}_n$ from $\mathrm{H}^l_{\min k}$ to perform the crossover operation. The crossover rule is shown in the following formula

$$\begin{cases} \mathrm{h}^{(k)}_m = \left(h^{(k)}_{m1}, \cdots, h^{(k)}_{ml'}, h^{(k)}_{m,l'+1}, \cdots, h^{(k)}_{m\mathcal{L}}\right) \\ \mathrm{h}^{(k)}_n = \left(h^{(k)}_{n1}, \cdots, h^{(k)}_{nl'}, h^{(k)}_{n,l'+1}, \cdots, h^{(k)}_{n\mathcal{L}}\right) \end{cases} \xrightarrow{\text{Crossover}}$$

$$\left\{ \begin{aligned} \bar{\mathrm{h}}^{(k)}_i &= (\underbrace{\bar{h}^{(k)}_{i1}, \cdots, \bar{h}^{(k)}_{il'}}_{h^{(k)}_{m1}, \cdots, h^{(k)}_{ml'}}, \underbrace{\bar{h}^{(k)}_{i,l'+1}, \cdots, \bar{h}^{(k)}_{i\mathcal{L}}}_{h^{(k)}_{n,l'+1}, \cdots, h^{(k)}_{n\mathcal{L}}}) \\ \bar{\mathrm{h}}^{(k)}_j &= (\underbrace{\bar{h}^{(k)}_{j1}, \cdots, \bar{h}^{(k)}_{jl'}}_{h^{(k)}_{n1}, \cdots, h^{(k)}_{nl'}}, \underbrace{\bar{h}^{(k)}_{j,l'+1}, \cdots, \bar{h}^{(k)}_{j\mathcal{L}}}_{h^{(k)}_{m,l'+1}, \cdots, h^{(k)}_{m\mathcal{L}}}) \end{aligned} \right\}, \tag{8}$$

where $\mathrm{h}^{(k)}_m, \mathrm{h}^{(k)}_n \in \mathrm{H}_{\min k}$, and $\bar{\mathrm{h}}^{(k)}_i, \bar{\mathrm{h}}^{(k)}_j \in \overline{\mathrm{H}}_{\min k}$, and l' is the crossover position. In this paper, l' can be set in the intervals $[0.1\mathcal{L}, 0.4\mathcal{L}]$, $[0.4\mathcal{L}, 0.6\mathcal{L}]$ and $[0.6\mathcal{L}, 0.9\mathcal{L}]$.

- For the mutation operation, the mutation location l' is randomly selected for $\forall h_m^{(k)} \in \mathrm{H}^l_{\min k}$. The mutation operation is conducted as the following formula:

$$
\mathrm{h}_m^{(k)} = \left(h_{m1}^{(k)}, \cdots, h_{ml'}^{(k)}, h_{m,l'+1}^{(k)}, \cdots, h_{m\mathcal{L}}^{(k)}\right) \xrightarrow{\text{Mutation}}
$$

$$
\bar{h}_i^{(k)} = \begin{cases}
(\underbrace{\bar{h}_{il'}^{(k)}}_{1-h_{ml'}^{(k)}}, \underbrace{\bar{h}_{i,l'+1}^{(k)}, \cdots, \bar{h}_{i\mathcal{L}}^{(k)}}_{h_{m,l'+1}^{(k)}, \cdots, h_{m\mathcal{L}}^{(k)}}), \\
\qquad l' = 1 \\
(\underbrace{\bar{h}_{i1}^{(k)}, \cdots, \bar{h}_{il'-1}^{(k)}}_{h_{m1}^{(k)}, \cdots, h_{m,l'-1}^{(k)}}, \underbrace{\bar{h}_{il'}^{(k)}}_{1-h_{ml'}^{(k)}}, \underbrace{\bar{h}_{i,l'+1}^{(k)}, \cdots, \bar{h}_{i\mathcal{L}}^{(k)}}_{h_{m,l'+1}^{(k)}, \cdots, h_{m\mathcal{L}}^{(k)}}), \\
\qquad l' \in (1, \mathcal{L}) \\
(\underbrace{\bar{h}_{i1}^{(k)}, \cdots, \bar{h}_{il'-1}^{(k)}}_{h_{m1}^{(k)}, \cdots, h_{m,l'-1}^{(k)}}, \underbrace{\bar{h}_{il'}^{(k)}}_{1-h_{ml'}^{(k)}}), \\
\qquad l' = \mathcal{L}
\end{cases} \tag{9}
$$

where $\overline{\mathrm{h}}_i^{(k)} \in \overline{\mathrm{H}}_{\min k}$, and $\bar{h}_{il'}^{(k)} = 1 - h_{ml'}^{(k)} \in (0, 1)$.

Algorithm 1. AE-MLELM-SynMin algorithm

Input The original minority class data sets $\mathrm{X}_{\min 1}, \mathrm{X}_{\min 2}, \cdots, \mathrm{X}_{\min \mathcal{K}}$;
Output The synthetic minority class data sets $\overline{\mathrm{X}}_{\min 1}, \overline{\mathrm{X}}_{\min 2}, \cdots, \overline{\mathrm{X}}_{\min \mathcal{K}}$;

for $k = 1; k <= \mathcal{K}; k++$ **do**
    Training AE-MLELM$_k$:
    Calculating H^i for each layer according to Eq. (5);
    Calculating the output layer weights matrix $\beta^{l+1}_{\min k}$;
    Conducting crossover and mutation operations based on the last hidden-layer matrix $\mathrm{H}^l_{\min k}$;
    while The number of synthetic hidden-layer output vectors does not reach $\overline{\mathcal{N}}_{\min k}$ **do**
        Crossover operation as shown in Eq. (8);
        Mutation operation as shown in Eq. (9);
    end while
    Creating synthetic data set $\overline{\mathrm{X}}_{\min k}$ based on synthetic hidden-layer output matrix $\overline{\mathrm{H}}^l_{\min k}$ according to Eq. (10);
end for

3.3 Creating Synthetic Samples

Finally, the minority class samples are synthesized. After the synthetic hidden layer output matrix is obtained, the generation formula of synthesized minority class samples is

$$
\overline{\mathrm{X}}_{\min k} = \overline{\mathrm{H}}^l_{\min k} \beta^{l+1}_{\min k}, \tag{10}
$$

where $\beta^{l+1}_{\min k}$ and $\overline{\mathrm{H}}^{l}_{\min k}$ are obtained from the first and second steps of the algorithm. Algorithm 1 gives the pseudo-code of AE-MLELM-SynMin. For each of minority classes, the above-mentioned steps are repeated until the sufficient samples are synthesized.

4 Experiments

In this section, we mainly use two experiments to verify the feasibility and effectiveness of the AE-MLELM-SynMin algorithm. These two experiments involve the comparative analysis of information amounts between SMOTE and AE-MLELM-SynMin data and the comparison among SMOTE [7], Borderline-SMOTE [8], Random-SMOTE [9] and SMOTE-IPF [10] algorithms based on decision tree classifier.

4.1 Experiment Setting

AE-MLELM-SynMin is implemented with the Python programming language and SMOTE, Borderline-SMOTE, Random-SMOTE and SMOTE-IPF algorithms are downloaded from the Python Package Index (PyPI)[1], which is a

Table 1. Details of 10 binary classification data sets

Data set	Attribute	Sample	Majority class	Minority class	Imbalance ratio
ecoli1	7	336	259	77	3.36
ecoli2	7	336	284	52	5.46
yeast1	8	1484	1055	429	2.46
yeast5	8	1484	1440	44	32.73
glass0	9	214	144	70	2.06
glass4	9	214	201	13	15.47
vowel0	13	988	898	90	9.98
segment0	19	2308	1979	329	6.02
vehicle0	18	846	647	199	3.25
new-thyroid1	5	215	180	35	5.14

Table 2. Information amount comparison among original data, SMOTE data, and AE-MLELM-SynMin data

Data set size	1000	2000	3000	4000	5000	6000	7000	8000	9000	10000
Original data	1.765	1.772	1.756	1.762	1.760	1.762	1.764	1.764	1.767	1.760
SMOTE data	1.763	1.771	1.754	1.761	1.759	1.761	1.763	1.764	1.767	1.760
AE-MLELM-SynMin data	2.783	2.485	2.381	2.285	2.427	2.293	2.270	2.303	2.228	2.270

[1] https://pypi.org/

software library for the Python programming language. In the experiments, the number of nearest neighbors are all set to 5 for these SMOTE algorithm variants. The number of iterations and percentage of samples removed for SMOTE-IPF are set as 3 and 1%, respectively.

For AE-MLELM-SynMin, we set (1) the crossover-mutation factor $\xi = 2$, that is, the ratio of crossover and mutation in the synthesized samples is 2:1; (2) The position intervals of crossover operation are $[0.4\mathcal{L}, 0.6\mathcal{L}]$, $[0.1\mathcal{L}, 0.4\mathcal{L}]$ and $[0.6\mathcal{L}, 0.9\mathcal{L}]$; and (3) the number of positions to be changed in a mutation operation is generated randomly on the interval $[0.001\mathcal{L}, 0.1\mathcal{L}]$ and the interval of mutation point is $[1, \mathcal{L}]$, where $\mathcal{L}$ is the number of hidden-layer nodes of AE-MLELM. We get the real data sets from KEEL-dataset repository [17] including 10 binary classification data sets in Table 1.

For the settings of hidden layers and node number of AE-MLELM, we set the depth as 5 and node number of each layer as 1000. In this experiment, we choose the decision tree algorithm as a classifier. The reason is that decision tree algorithm generally works well on class-imbalanced data.

4.2 Information Amount Analysis of SMOTE and AE-MLELM-SynMin

In order to explain the effectiveness of AE-MLELM-SynMin method, we use the resubstitution entropy to measure the information amount of dataset [18]. In the experiment, We randomly create 10 different original data sets with sizes ranging from 1000 to 10000 in step of 1000 and synthesize the corresponding 10 SMOTE data sets and 10 AE-MLELM-SynMin data sets, respectively. Then, we calculate the information amounts of original data, SMOTE data and AE-MLELM-SynMin data. We repeat the synthetic sample 10 times for each original data set and calculate the average value of the information amounts. For a given data set

$$\begin{aligned} \mathbb{A} = \{\mathrm{a}_n \mid \mathrm{a}_n = (a_{n1}, a_{n2}, \cdots, a_{n\mathcal{D}}), a_{md} \in \Re, \\ n = 1, 2, \cdots, \mathcal{N}, d = 1, 2, \cdots, \mathcal{D}\}' \end{aligned} \tag{11}$$

its entropy is calculated as

$$\mathrm{Info}(\mathbb{A}) = \frac{1}{\mathcal{D}} \sum_{d=1}^{\mathcal{D}} \mathrm{Ent}\left(\mathrm{A}_d\right), \tag{12}$$

where A_d is the d-th attribute of data set $\mathbb{A}$. The re-substitution entropy of data set $\{a_{1d}, a_{2d}, \cdots, a_{\mathcal{N}d}\}$ is calculated as

$$\mathrm{Ent}\left(\mathrm{A}_d\right) = -\sum_{n=1}^{\mathcal{N}} \mathrm{In}\left[\hat{p}_{-n}\left(a_{nd}\right)\right], \tag{13}$$

where

$$\hat{p}_{-n}\left(a_{nd}\right) = \frac{1}{\mathcal{N}-1} \sum_{\substack{m=1 \\ m \neq n}}^{\mathcal{N}} \frac{1}{\sqrt{2\pi} h} \exp\left[-\frac{1}{2}\left(\frac{a_{nd} - a_{md}}{h}\right)^2\right] \tag{14}$$

Table 3. Comparative results for binary classification problems based on decision tree classifier

Data set	Algorithm							
	AE-MLELM-SynMin(2021)				SMOTE(2002)			
	Auc	G-mean	F1	Accuracy	Auc	G-mean	F1	Accuracy
ecoli1	0.881 ± 0.037	0.879 ± 0.039	0.768 ± 0.050	0.877 ± 0.031	0.884 ± 0.042	0.883 ± 0.044	0.773 ± 0.058	0.880 ± 0.033
ecoli2	**0.880 ± 0.047**	**0.875 ± 0.052**	**0.780 ± 0.067**	**0.928 ± 0.023**	:0.870 ± 0.051	0.865 ± 0.058	0.753 ± 0.075	0.916 ± 0.029
yeast1	0.687 ± 0.027	0.676 ± 0.039	0.557 ± 0.036	0.708 ± 0.052	**0.698 ± 0.020**	**0.694 ± 0.021**	**0.571** ± 0.026	**0.728 ± 0.022**
yeast5	**0.959 ± 0.029**	**0.958 ± 0.030**	0.570 ± 0.071	0.956 ± 0.013	0.948 ± 0.046	0.946 ± 0.050	0.695 ± 0.068	0.976 ± 0.008
glass0	**0.796 ± 0.050**	**0.786 ± 0.058**	**0.707 ± 0.063**	0.762 ± 0.049	0.783 ± 0.053	0.779 ± 0.053	0.696 ± 0.061	0.766 ± 0.050
glass4	**0.861 ± 0.103**	**0.843 ± 0.143**	0.618 ± 0.159	0.939 ± 0.032	0.831 ± 0.112	0.804 ± 0.158	0.620 ± 0.172	0.948 ± 0.025
vowel0	**0.966 ± 0.017**	**0.966 ± 0.017**	**0.837 ± 0.048**	**0.965 ± 0.013**	0.947 ± 0.025	0.946 ± 0.026	0.813 ± 0.051	0.960 ± 0.013
segment0	**0.987 ± 0.009**	**0.987 ± 0.009**	0.970 ± 0.013	0.991±0.004	0.986 ± 0.008	0.986 ± 0.008	0.969 ± 0.012	**0.991 ± 0.003**
vehicle0	0.915 ± 0.030	0.913 ± 0.031	0.806 ± 0.047	0.889 ± 0.036	0.921 ± 0.023	0.921 ± 0.023	0.850 ± 0.032	0.923 ± 0.017
new-thyroid1	**0.966 ± 0.033**	**0.966 ± 0.035**	0.914 ± 0.060	0.969 ± 0.023	0.946 ± 0.047	0.944 ± 0.050	0.908 ± 0.068	0.969±0.024
Data set	Algorithm							
	Borderline-SMOTE(2005)				Random-SMOTE(2011)			
	Auc	G-mean	F1	Accuracy	Auc	G-mean	F1	Accuracy
ecoli1	**0.893 ± 0.035**	**0.891 ± 0.035**	**0.779 ± 0.053**	0.88 ± 0.034	0.883 ± 0.041	0.881 ± 0.042	0.771 ± 0.053	0.879 ± 0.030
ecoli2	0.852 ± 0.053	0.844 ± 0.061	0.740 ± 0.080	0.915 ± 0.030	0.875 ± 0.05	0.871 ± 0.054	0.753 ± 0.073	0.914 ± 0.031
yeast1	0.697±0.022	0.692 ± 0.025	0.571 ± 0.027	0.723 ± 0.036	**0.698 ± 0.020**	0.693 ± 0.021	**0.571 ± 0.026**	0.726 ± 0.021
yeast5	0.949 ± 0.049	0.947 ± 0.055	0.683 ± 0.076	0.975±0.008	0.945 ± 0.044	0.943 ± 0.047	**0.704 ± 0.069**	**0.977 ± 0.007**
glass0	0.761 ± 0.051	0.756 ± 0.052	0.670 ± 0.060	0.746±0.048	0.786 ± 0.049	0.782 ± 0.050	0.700 ± 0.057	**0.768 ± 0.049**
glass4	0.851 ± 0.134	0.815 ± 0.211	**0.667±0.227**	**0.957 ± 0.027**	0.831 ± 0.114	0.801 ± 0.173	0.618 ± 0.186	0.947 ± 0.026
vowel0	0.948 ± 0.029	0.948 ± 0.030	0.813 ± 0.055	0.96 ± 0.014	0.941 ± 0.030	0.940 ± 0.031	0.801 ± 0.052	0.958 ± 0.013
segment0	0.956 ± 0.019	0.955 ± 0.020	0.929 ± 0.027	0.980 ± 0.008	**0.986 ± 0.007**	**0.986 ± 0.007**	0.969 ± 0.012	0.991 ± 0.004
vehicle0	**0.924±0.022**	**0.924±0.023**	0.857 ± 0.031	0.927 ± 0.017	0.921 ± 0.022	0.92 ± 0.023	**0.859 ± 0.03**	**0.93 ± 0.015**
new-thyroid1	0.952 ± 0.052	0.950 ± 0.057	0.921 ± 0.072	0.974 ± 0.023	0.950 ± 0.045	0.948 ± 0.049	**0.921 ± 0.061**	**0.974 ± 0.020**
Data set	Algorithm							
	SMOTE-IPF(2015)				/			
	Auc	G-mean	F1	Accuracy	/	/	/	/
ecoli1	0.886 ± 0.040	0.884 ± 0.041	0.778 ± 0.058	**0.884 ± 0.033**	/	/	/	/
ecoli2	0.865 ± 0.050	0.860 ± 0.055	0.746 ± 0.077	0.913 ± 0.031	/	/	/	/
yeast1	0.698 ± 0.021	0.693 ± 0.023	0.571 ± 0.027	0.727 ± 0.023	/	/	/	/
yeast5	0.946 ± 0.041	0.944 ± 0.044	0.700 ± 0.068	**0.977 ± 0.007**	/	/	/	/
glass0	0.782±0.047	0.778 ± 0.048	0.695 ± 0.055	0.765 ± 0.046	/	/	/	/
glass4	0.834±0.120	0.802 ± 0.187	0.614 ± 0.182	0.947 ± 0.025	/	/	/	/
vowel0	0.946 ± 0.027	0.946 ± 0.029	0.813 ± 0.050	0.960 ± 0.013	/	/	/	/
segment0	0.986 ± 0.008	0.986 ± 0.008	**0.970 ± 0.012**	0.991 ± 0.004	/	/	/	/
vehicle0	0.919 ± 0.023	0.918 ± 0.024	0.856 ± 0.029	0.928 ± 0.015	/	/	/	/
new-thyroid1	0.952 ± 0.046	0.950 ± 0.049	0.921 ± 0.063	**0.974 ± 0.020**	/	/	/	/

is the leave-one-out cross-validation kernel density estimator, $h > 0$ is the bandwidth parameter. And we set h as 0.2 in this experiment.

Table 2 shows the information amount comparison results of original data, SMOTE data and AE-MLELM-SynMin data. In Table 2, we can draw the following conclusions. First, AE-MLELM-SynMin data has the highest information amount. Second, original data and SMOTE data have the same information amount, and AE-MLELM-SynMin data has 1.2 to 1.6 times more information amount than original data or SMOTE data. So the data synthesized by AE-MLELM-SynMin has a large increase in the amount of information, and finally can achieve a better effect on the general classifier.

4.3 Comparison Among AE-MLELM-SynMin, SMOTE, Borderline-SMOTE, Random-SMOTE, and SMOTE-IPF

In this experiment, we use the decision tree classifier downloaded from the scikit-learn Python machine learning library[2] to classify the datasets created by AE-MLELM-SynMin, SMOTE, Borderline-SMOTE, Random-SMOTE and SMOTE-IPF, and compare their imbalanced classification performances. Table 1 shows the 10 binary data sets used in this experiment. The specific process of this experiment is that we first randomly divide each original data set into 70% training set and 30% testing set and then fill the minority samples created by AE-MLELM-SynMin and SMOTE variants for the training set. Finally we use the trained classifier to predict the testing set and evaluate the experimental results with the four indicators of AUC [19], G-mean [20], F1 [21] and accuracy. The above process is repeated 400 times and take the average of 400 AUCs, G-means, F1s and accuracies as the final experimental result.

The classification results of decision tree are shown in Table 3, which lists the AUC, G-mean, F1 and accuracy of decision tree classifiers corresponding to AE-MLELM-SynMin, SMOTE, Borderline-SMOTE, Random-SMOTE and SMOTE-IPF methods. We can draw several conclusions from the comparison of bold markers in Table 3. Firstly, AE-MLELM-SynMin method has the better AUCs and G-means compared with other SMOTE-based methods on 7 data sets. Secondly, the AE-MLELM-SynMin method achieves the better F1s on 3 data sets. In general, the performance of AE-MLELM-SynMin method is better than other SMOTE-based methods when dealing with the imbalanced classification problems.

5 Conclusion

The proposed AE-MLELM-SynMin method was constructed based on AE-MLELM, which introduced the ideas of crossover and mutation in evolutionary algorithm. This method had the advantages of easy understanding and simple implementation. The experimental results showed that, compared with other SMOTE-based methods, the AE-MLELM-SynMin method can improve the information amount of synthesized samples and thus obtained the better imbalanced classification performance than SMOTE-based methods.

Acknowledgement. The authors would like to thank the chairs and anonymous reviewers whose meticulous readings and valuable suggestions help them to improve this paper significantly. This paper was supported by National Natural Science Foundation of China (61972261) and Basic Research Foundation of Shenzhen (JCYJ 20210324093609026, JCYJ 20200813091134001), and Scientific Research Foundation of Shenzhen University for Newly-introduced Teachers (860/000002110628).

[2] https://scikit-learn.org/stable/modules/tree.html.

References

1. Japkowicz, N., Stephen, S.: The class imbalance problem: a systematic study. Intell. Data Anal. **6**(5), 429–449 (2002)
2. Díez-Pastor, J.F., Rodríguez, J.J., García-Osorio, C., Kuncheva, L.I.: Random balance: ensembles of variable priors classifiers for imbalanced data. Knowl.-Based Syst. **85**, 96–111 (2015)
3. Liu, X., Wu, J., Zhou, Z.: Exploratory undersampling for class-imbalance learning. IEEE Trans. Syst. Man Cybernet Part B (Cybernetics) **39**(2), 539–550 (2009)
4. Sun, Y., Kamel, M.S., Wong, A.K.C., Wang, Y.: Cost-sensitive boosting for classification of imbalanced data. Pattern Recogn. **40**(12), 3358–3378 (2007)
5. Tan, S.: Neighbor-weighted K-nearest neighbor for unbalanced text corpus. Expert Syst. Appl. **28**(4), 667–671 (2005)
6. Zong, W.W., Huang, G.B., Chen, Y.Q.: Weighted extreme learning machine for imbalance learning. Neurocomputing **101**, 229–242 (2013)
7. Chawla, N.V., Bowyer, K.W., Hall, L.O., Kegelmeyer, W.P.: SMOTE: synthetic minority over-sampling technique. J. Artif. Intell. Res. **16**(1), 321–357 (2002)
8. Han, H., Wang, W.Y., Mao, B.H.: Borderline-SMOTE: a new over-sampling method in imbalanced data sets learning. Lect. Notes Comput. Sci. **3644**, 878–887 (2005)
9. Dong, Y.J., Wang, X.H.: A new over-sampling approach: Random-SMOTE for learning from imbalanced data sets. In: Proceedings of the 5th International Conference on Knowledge Science, Engineering and Management, vol. 10, pp. 343–352 (2011)
10. Sáez, J.A., Luengo, J., Stefanowski, J., Herrera, F.: SMOTE-IPF: addressing the noisy and borderline examples problem in imbalanced classification by a re-sampling method with filtering. Inf. Sci. **291**, 184–203 (2015)
11. Calleja, J.L., Fuentes, O.: A Distance-based over-sampling method for learning from imbalanced data sets. In: Proceedings of the Twentieth International Florida Artificial Intelligence Research Society Conference (2007)
12. Puntumapon, K., Waiyamai, K.: A pruning-based approach for searching precise and generalized region for synthetic minority over-sampling. In: Tan, P.-N., Chawla, S., Ho, C.K., Bailey, J. (eds.) PAKDD 2012. LNCS (LNAI), vol. 7302, pp. 371–382. Springer, Heidelberg (2012). https://doi.org/10.1007/978-3-642-30220-6_31
13. Lee, H., Kim, J., Kim, S.: Gaussian-based SMOTE algorithm for solving skewed class distributions. Int. J. Fuzzy Logic Intell. Syst. **17**, 229–234 (2017)
14. Douzas, G., Bacao, F., Last, F.: Improving imbalanced learning through a heuristic oversampling method based on k-means and SMOTE. Inf. Sci. **465**, 1–20 (2018)
15. Kasun, L., Zhou, H.M., Huang, G.B., Vong, C.M.: Representational Learning with ELMs for Big Data. IEEE Intell. Syst. **28**, 31–34 (2013)
16. Lu, S.X., Wang, X., Zhang, G.Q., Zhou, X.: Effective algorithms of the Moore-Penrose inverse matrices for extreme learning machine. Intell. Data Anal. **19**, 743–760 (2015)
17. Alcala-Fdez, I., et al.: KEEL data-mining software tool: data set repository, integration of algorithms and experimental analysis framework. J. Multiple-Valu. Logic Soft Comput. **17**, 255–287 (2010)
18. He, Y.L., Liu, J.N.K., Wang, X.Z., Hu, Y.X.: Optimal bandwidth selection for re-substitution entropy estimation. Appl. Math. Comput. **219**(8), 3425–3460 (2012)

19. Hand, D.J., Till, R.J.: A simple Generalisation of the area under the ROC curve for multiple class classification problems. Mach. Learn. **45**(2), 171–186 (2001)
20. Sun, Y., Kamel, M.S., Wang, Y.: Boosting for learning multiple classes with imbalanced class distribution. In: Proceedings of the Sixth International Conference on Data Mining, pp. 592–602 (2006)
21. Lipton, Z.C., Elkan, C., Naryanaswamy, B.: Optimal thresholding of classifiers to maximize F1 measure. In: Proceedings of Machine Learning and Knowledge Discovery in Databases, pp. 225–239 (2014)

Pattern Mining: Current Challenges and Opportunities

Philippe Fournier-Viger[1(✉)], Wensheng Gan[2], Youxi Wu[3], Mourad Nouioua[4], Wei Song[5], Tin Truong[6], and Hai Duong[6]

[1] Shenzhen University, Shenzhen, China
philfv@szu.edu.cn
[2] Jinan University, Guangzhou, China
[3] Hebei University of Technology, Tianjin, China
[4] University of Bordj Bou Arreridj, Bordj Bou Arreridj, Algeria
[5] North China University of Technology, Beijing, China
songwei@ncut.edu.cn
[6] Dalat University, Dalat, Vietnam
{tintc,haidv}@dlu.edu.vn

Abstract. Pattern mining is a key subfield of data mining that aims at developing algorithms to discover interesting patterns in databases. The discovered patterns can be used to help understanding the data and also to perform other tasks such as classification and prediction. After more than two decades of research in this field, great advances have been achieved in terms of theory, algorithms, and applications. However, there still remains many important challenges to be solved and also many unexplored areas. Based on this observations, this paper provides an overview of six key challenges that are promising topics for research and describe some interesting opportunities. Those challenges were identified by researchers from the field, and are: (1) mining patterns in complex graph data, (2) targeted pattern mining, (3) repetitive sequential pattern mining, (4) incremental, stream, and interactive pattern mining, (5) heuristic pattern mining, and (6) mining interesting patterns.

Keywords: Data mining · Pattern mining · Challenges · Opportunities

1 Introduction

Nowadays, large amounts of data of various types are stored in databases of various organizations. Hence, it has become important for many organizations to develop automatic or semi-automatic tools to analyze data. Pattern mining is a subfield of data mining that aims at identifying interesting and useful patterns in data. The aim is to find patterns that are easily interpretable by users, and thus can help in understanding the data. Patterns can be used to support decision-making but also to perform other tasks such as classification, clustering and prediction.

U. K. Rage et al. (Eds.): DASFAA 2022 Workshops, LNCS 13248, pp. 34–49, 2022.
https://doi.org/10.1007/978-3-031-11217-1_3

Pattern mining research started more than two decades ago. While initial studies have focused on discovering frequent patterns on data such as shopping data, the field has rapidly changed to consider other data types and pattern types. Also, major improvements have been made to algorithms and data structures to improve efficiency, scalability, and provide more features.

This paper provides an overview of key challenges and opportunities in pattern mining, that deserve more attention. To write this paper, seven researchers from the field of pattern mining were invited to write about a key challenge of their choice. **Six challenges have been identified:**

1. C1: Mining patterns in complex graph data (*by P. Fournier-Viger*)
2. C2: Targeted pattern mining (*by W. Gan*)
3. C3: Repetitive sequential pattern mining (*by Y. Wu*)
4. C4: Interactive pattern mining (*by M. Nouioua*)
5. C5: Heuristic pattern mining (*by W. Song*)
6. C6: Mining interesting patterns (*by T. Truong and H. Duong*)

The rest of this paper is organized as follows. The Sects. 2 to 6 describe the six challenges. Then, Sect. 7 draws a conclusion.

2 C1: Mining Patterns in Complex Graph Data

The first studies on frequent pattern mining have focused on analyzing transaction databases, which are tables of records described using binary attributes. Although this data representation has many applications, it remains very simple and thus it is unsuitable for many applications where complex data must be analyzed. Hence, a current trend in pattern mining is to develop algorithms to analyze complex data such as temporal data, spatial data, time series and graphs. Graph data has attracted the attention of many researchers in recent years [13,18] as it can encode various types of information such as social links in friendship-based social networks, chemical molecules, roads between cities, flights between airports, and co-authorship of papers in academia.

Initial studies on graph pattern mining have focused on *frequent subgraph mining*, which aims at finding connected subgraphs that are common to many graphs of a graph database, or that appear frequently in a single graph [18]. In the original problem, the input graphs have a rather simple form. They are static, have vertices and edges which can each have at most one label, the graphs must be connected, self-loops are forbidden (an edge from a vertex to itself), and there can be at most a single edge between two vertices. This simple representation restricts the applications of frequent subgraph mining. To broaden the applications of pattern mining in graphs, the following challenges must be solved.

Handling More Complex Types of Graphs. A key challenge is to consider richer types of graphs such as those shown in Fig. 1 [13]. Those are *directed graphs* (where edges may have directions), *weighted graphs* (where numbers are assigned to edges to indicate the strength of relationships), *attributed graphs* (where each node can be described using many categorical or nominal attributes), *multi-labeled graphs* (where edges and vertices may have multiple labels), and multi-relational graphs (graphs where multiple edges of different types may connect nodes) [13]. For some domains such as social network analysis, handling rich graph data is crucial. For example, a user profile (vertex) on a social network may be described using many attributes and persons have various types of relationships (edges).

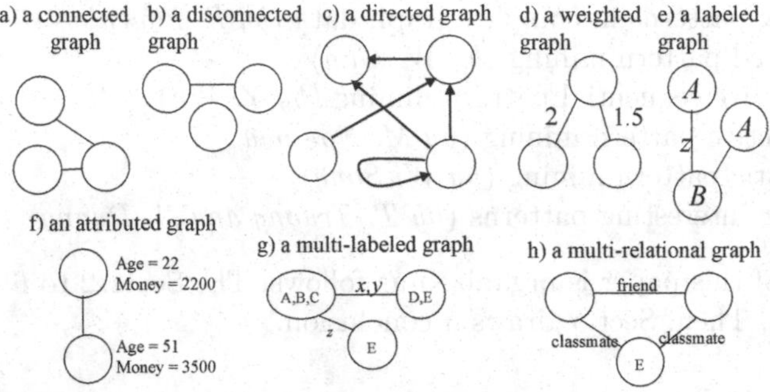

Fig. 1. Different types of graphs

Handling Dynamic Graphs. Another key challenge is to mine patterns in graphs that change over time. There are three main types of changes: *topological changes* (edges may be added or removed), *label evolution* (node labels may change over time), and a mix of both two cases [13]. Many algorithms for analyzing dynamic graphs adopt a snapshot model where graphs are observed at different timestamps. Other models of time should also be studied such as events having a duration (each event has a start and end time and may overlap with other events). A related research direction is to design algorithms to update patterns incrementally when new data arrives, or to process evolving graphs in real-time as a stream. It is also possible to search for different types of temporal patterns in a dynamic graph such as sequences of changes, periodic patterns (a pattern that is repeating itself over time) and trending patterns (a pattern that has an increasing trend over time) and attribute rules [12].

Discovering Specialized Types of Graphs. Another important challenge in graph pattern mining is to design algorithms that are specialized for mining specific patterns rather than more general patterns. For example, algorithms have been designed to mine sub-trees [29] or paths instead of more complex graphs. The benefit of solving some more specialized graph pattern mining problem is that more efficient algorithm can be developed due to specialized optimizations.

Discovering Novel Pattern Types. Many recent studies have focused on finding new pattern types or to use new criteria to select patterns. For instance, a trend is to design algorithms to find statistically significant patterns based on statistical tests, or to use correlation measures to filter out spurious patterns.

Handling Multi-modal Data. Another interesting research direction in graph pattern mining is to combine graph data with other data types to perform a joint analysis of this data. Multi-modal data refers to data of different modes such as graph data, combined with video data and audio data.

Developing Solutions to Applied Graph Pattern Mining Problems. Another important research topic in graph mining is to design specialized solutions to address the needs of some given applications. For instance, it was shown to be advantageous to develop custom algorithms to analyze alarm data from a computer network rather than using generic graph mining algorithms [12]. By designing a tailored solution, better performance may be obtained and more interesting patterns (e.g. by using custom measures to select patterns).

3 C2: Targeted Pattern Mining

The current pattern mining literature provides various methods to find all interesting patterns using several parameters. In other words, most of the pattern mining algorithms aims at discovering the complete set of patterns (i.e., itemsets, rules, sequences, graphs, etc.) that satisfy predefined thresholds. However, in general, a huge number of discovered patterns may not be interesting, which are usually based on variations special interest. To filter out redundant information and obtain concise results, *targeted pattern mining* (TPM, or called *targeted pattern search*) provides a different solution to the classic pattern mining problem. To be specific, instead of discovering a large number of patterns that may not be the target ones, users in TPM could input a single or several targets at a time and then discover/query the desired patterns containing the input target [26,44]. Therefore, the interactive TPM method can return the concise queries with the user-defined targets.

In summary, the goal of TPM is to discover a particular subset or group that contain one or several special patterns. TPM is often more interesting and reliable for finite samples resulting in potential finite subset. Different from traditional pattern mining algorithms, TPM is computationally much more difficult to compute the subset from the potential search space. How to estimate special subset but not the all patterns satisfy the given parameters is quite challenging.

Several definitions about targeted pattern have been provided before, including targeted frequent itemset mining [20], targeted sequence mining [5], targeted high-utility itemset mining [26], and targeted high-utility sequence mining [44]. Up to now, several frequency-based or utility-driven TPM models have been developed, as briefly reviewed below.

Targeted FIM and ARM Algorithms. Kubat et al. [20] studied specialized queries of frequent itemsets in a transaction database. All the rules (i.e., targeted queries) can be extracted from the designed Itemset-Tree. Here the user-specified itemset is antecedent in rules. An improved Itemset-Tree [14] was designed for quickly querying frequent itemsets during the operation process. Both algorithms adopt the minimum confidence and support measurement, and the improved Itemset-Tree can be updated incrementally with new transactions. For multitude-targeted mining, the guided FP-growth [30] was designed to determine the frequency of each given itemset based on target Itemset-Tree. After that, a constraint-based ARM query model [1] was also introduced for exploratory analysis of diverse clinical databases.

Targeted SPM Algorithms. The sequential ordering of items is commonly seen in real-life applications. To handle the sequence data that is more complex than transaction data, Chueh et al. [8] reversed the original sequences to discover targeted sequential patterns with time intervals. Based on the definition of targeted SPM, Chand et al. [5] proposed a novel SPM algorithm to discover patterns with checking whether they satisfied the recency and monetary constraint and also were target-oriented. However, the target pattern in this approach is defined in the end of each sequence. A goal-oriented algorithm [7] can extract the transaction activities before losing the customer. By utilizing TPM, this algorithm can handle the problem of determining whether a customer is leaving and toward a specific goal.

Utility-Driven TPM Algorithms. Previous TPM algorithms mainly adopt the measurement of frequency and confidence, but them do not involve the concept of utility [15], which is helpful for discovering more informative patterns and knowledge. Recently, Miao et al. [26] are the first to introduce a targeted high-utility itemset querying model (abbreviated as TargetUM). TargetUM introduced several key definitions and formulated the problem of mining the desired set of high-utility itemsets containing given target items. A utility-based trie tree was designed to index and query target itemsets on-the-fly. Consider the sequence data, Zhang et al. [44] introduced targeted high-utility sequence querying problem and proposed the TUSQ algorithm. Targeted utility-chain and two novel upper bounds on utility measurement (namely suffix remain utility and terminated descendants utility) are proposed in the TUSQ model.

Several open problems of targeted pattern mining/search and interesting directions (including but not limited to) in the future are highlighted in detail below. It is important to note that these open problems are also widespread in other pattern mining tasks.

- **What type of data to be mined**. As we know, there are many types of data in real world, such as transaction data, sequence, streaming data, spatiotemporal data, complex event, time-series, text and web, multi-media, graphs, social network, and uncertain data. How to design effective TPM algorithms to deal with these data is very urgent and more challenging.
- **What kind of pattern or knowledge to be mined**. For example, there are two categories, descriptive vs. predictive data mining, which is based on different kind of knowledge. As reviewed before, itemset, sequence, rule, graph, and event are the different kinds of patterns that are extracted from various types of data. However, few TPM algorithms can discover these kinds of patterns.

- **More effective data structure.** According to the current studies, the indexing and searching in TPM are more challenging than that of traditional pattern mining. In particular, when dealing with big data, we need more effective data structures to store rich information from data.
- **More powerful strategies.** Due to the difficulty, the search space of TPM has an explosion. Thus, how to reduce the search space using powerful pruning strategies (w.r.t. upper bounds) plays a key role in improving the performance of the TPM algorithm.
- **Different applications.** In general, there are many applications of data mining methods, including discrimination, association analysis, classification, clustering, trend/deviation, outlier detection, etc. It is clear that different application requires a special solution of TPM.
- **Visualization.** It is interesting that the data and mining results will be displayed automatically in search process. In the future, there are many opportunities to increase the interpretability of the results, the ease of use of the model, and the interactivity of the mining process.

To summary, targeted pattern mining/search is difficult and quite different from previous mining methods. In the future, there are many opportunities and interesting work in this research field.

4 C3: Repetitive Sequential Pattern Mining

Sequential pattern mining (SPM) has been used in keyphrase extraction [42] and feature selection [41]. The goal of SPM is to discover interesting subsequences (also called patterns). The most common problem is to mine frequent patterns whose supports are no less than a user-defined parameter called *minsup*. The definitions are as follows.

Definition 1. (sequence and sequence database) Suppose we have a set of items σ. A sequence S is an ordered list of itemsets $S = \{s_1, s_2, \cdots, s_n\}$, where s_i is an itemset, which is a subset of σ. A sequence database is composed of k sequences, i.e. $SDB = \{S_1, S_2, \cdots, S_k\}$.

Example 1. For a sale dataset, suppose there are five products: a, b, c, d, and e, i.e. $\sigma = \{a, b, c, d, e\}$. Suppose customer 1 first purchased items a, b, and c, then bought a, b, and e, then purchased c, then bought (a, b, d), and e, then purchased a and c, and finally bought (a, c) and e. The shopping sequence of customer 1 is $S_1 = \{s_1, s_2, s_3, s_4, s_5, s_6\} = \{(a, b, c), (a, b, e), (c), (a, b, d, e), (a, c), (a, c, e)\}$. Similarly, we assume that for customer 2, $S_2 = \{s_1, s_2\} = \{(a, b, d), (c)\}$. Thus, the sequence database is $SDB = \{S_1, S_2\}$.

This kind of sequence format is quite general since the sequence is an ordered list of itemsets, which means that each itemset contains one or more items. Thus, such sequence is called a sequence with itemsets. But for many applications, the data is represented as an ordered list of items called a sequence with items, which means that each itemset contains only one item, e.g. DNA sequences, protein sequences, virus sequences, and time series. For example, "attaaagg" is a segment of the SARS-CoV-2 virus.

Definition 2. (pattern and occurrence) A pattern $P = p_1, p_2, \cdots, p_m$ is also a sequence. A pattern P is a subsequence of a sequence $S = \{s_1, s_2, \cdots, s_n\}$ if and only if $p_1 \subseteq s_{i_1}, p_2 \subseteq s_{i_2}, \cdots, p_m \subseteq s_{i_m}$, and $1 \leq i_1 < i_2 < \cdots < i_m \leq n$. $I = <i_1, i_2, \cdots, i_m>$ is an occurrence of pattern P in sequence S.

Example 2. Pattern $P = \{(a, b), (c)\}$ occurs in sequences S_1 and S_2. For example, <1,3> is an occurrence of pattern P in sequence S_1, since $p_1 = (a, b) \subseteq s_1 = (a, b, c)$ and $p_2 = (c) \subseteq s_3 = (c)$.

Definition 3. (support and frequent pattern) The support is the number of occurrences of a pattern P in a sequence database *SDB*, represented as $sup(P, SDB)$. If the support is no less than the predefined threshold *minsup*, then the pattern is called a frequent pattern.

Classical SPM cares if a pattern occurs in a sequence or not, but it ignores a pattern's repetitions in a sequence. For example, P occurs in both S_1 and S_2. Thus, the support of P in *SDB* is 2. However, pattern P occurs many times in sequence S_1. If we neglect the repetition, many important interesting patterns will be lost. However, researchers mainly focused on mining the repetitive patterns in a sequence database with items, rather than in a sequence database with itemsets. Various methods have been investigated to mine various kinds of patterns such as patterns without gap [6], patterns with self-adaptive gap [41], and patterns with gap constraint [21,27,28,40]. An illustrative example is given.

Example 3. Suppose we have a sequence $S = s_1s_2s_3s_4s_5s_6s_7s_8 = aabababa$.

(1) **Pattern without gap**: Pattern without gap is also called consecutive subsequences [6], i.e. for occurrence $I = <i_1, i_2, \cdots, i_m>$, it requires that $i_2 = i_1 + 1$, $i_3 = i_2 + 1$, $\cdots$, $i_m = i_{m-1} + 1$. For example, there are two occurrences of pattern $P = p_1p_2p_3$ = aba in sequence S: <2,3,4> and <6,7,8>. The advantage of this method is that it is easy to calculate the support. However, the restriction is too strict, which will lead to the loss of a lot of important information.
(2) **Pattern with self-adaptive gap** [41]: It means that there is no constraint on the occurrence. For example, <1,7,8> is an occurrence of $P = aba$ in S. The advantage of this method is that users do not need any prior knowledge and it is easy to find the characteristics of the sequence database. However, there are too many occurrences, which will lead to difficulties in analyzing the results.
(3) **Pattern with gap constraint**: In this case, users should predefine a gap = [M, N], and for each occurrence, it needs to satisfy that $M \leq i_k - i_{k-1} - 1 \leq N$ $(1 < k \leq m)$, where M and N are the minimum and maximum wildcards. This method can prune some meaningless occurrences. For example,

if gap = [0,2], <1,3,4> is an occurrence of pattern $P = aba$ in sequence S, since $p_1 = s_1 = a$, $p_2 = s_3 = b$, $p_3 = s_4 = a$, and both 1,3 and 3,4 satisfy the gap constraint [0,2], while <1,5,6> is not an occurrence, since $5 - 1 - 1 = 3 > 2$. This approach not only is more challenging, but also has many types. As far as we know there are four types: no condition [27], the one-off condition [21], the nonoverlapping condition [40], and the disjoint condition [28].

- No condition means that each item can be reused [27]. Therefore, all ten occurrences of $P = aba$ with gap [0,2] in sequence S are acceptable under no condition: <1,3,4>, <1,3,6>, <2,3,4>, <2,3,6>, <2,5,6>, <2,5,8>, <4,5,6>, <4,5,8>, <4,7,8>, and <6,7,8>.
- The one-off condition means that each item can be used at most once [21]. Therefore, <1,3,4> and <2,5,6> are two occurrences of P in S which satisfy the one-off condition, while <1,3,4> and <4,5,6> do not.
- The nonoverlapping condition means that each item cannot be reused by the same p_j, but can be reused by different p_j [40]. <1,3,4> and <4,5,6> satisfy the nonoverlapping condition, since in <1,3,4>, p_3 matches $s4$ and in <4,5,6> p_1 matches s_4. However, <1,3,6> and <2,3,4> do not satisfy the nonoverlapping condition, since in both occurrences, p_2 matches s_3. Hence, there are three occurrences of P in S under the nonoverlapping condition: <1,3,4>, <4,5,6>, and <6,7,8>.
- The disjoint condition means that the maximum position of an occurrence should be less than the minimum position of the next occurrence [28]. For example, there are two occurrences of P in S: <1,3,4> and <6,7,8>.

Although the four conditions are very similar, their characteristics are different. The advantages of no condition are that the support can be calculated in polynomial time and it is a complete mining approach. However, this mining approach does not satisfy the Apriori property and has to apply the Apriori-like strategy to generate candidate patterns. For the one-off condition, although it satisfies the Apriori property, it cannot exactly calculate its support, since it is an NP-Hard problem. Therefore, the mining approach is an approximate mining approach. Although both the nonoverlapping condition and the disjoint condition satisfy the Apriori property, and the support can be calculated in polynomial time [39], the disjoint condition is easier to calculate than the nonoverlapping condition and may lose some feasible occurrences.

If we apply the four conditions in a sequence database with itemsets, it is a more challenging task, since the support calculation and candidate generation are significantly different from those for a sequence database with items. For the support calculation, in a sequence with items, we require that $p_j = s_{i_j}$, while in a sequence with itemsets, we require that $p_j \subseteq s_{i_j}$. For candidate generation, in a sequence database with items, we only apply S-Concatenation to generate candidates, while in a sequence database with itemsets, we adopt S-Concatenation and I-Concatenation to generate candidates.

Hence, the following tasks should be further investigated in sequence databases with itemsets. 1). What are the computational complexities of calculating the supports under different conditions? 2). Given a database with itemsets, how to design effective mining algorithms for these conditions? 3). If the dataset is dynamic or a stream database, how to design effective mining algorithms? 4). A variety of SPM methods were proposed to meet different requirements, such as closed SPM, maximal SPM, top-k SPM, compressing SPM, co-occurrence SPM, rare SPM, negative SPM, tri-partition SPM, and high utility SPM. However, most of them neglect the repetitions and consider sequence databases with itemsets. If the repetitions cannot be neglected, how to design effective mining algorithms? 5). For a specific problem, there are many approaches to solve it. However, what is the best approach? For example, for a sequence classification problem, there are many methods to extract the features, such as frequent patterns and contrast patterns under the four conditions. However, which one is the best approach?

5 C4: Incremental, Stream and Interactive Pattern Mining

A key limitation of traditional pattern mining algorithms such as Apriori and FP-Growth is that they are batch algorithms. This means that if the input database is updated, the user needs to run again the algorithm to get new results even if the database is slightly changed. Consequently, classical algorithms are inefficient for various real applications where databases are dynamics. To address this challenge, various approaches have been adopted which can be roughly classified into three categories: (1) Incremental pattern mining algorithms, (2) Stream pattern mining algorithms and (3) Interactive pattern mining algorithms.

Incremental pattern mining algorithms are designed to update the set of discovered patterns once the database is updated by inserting or deleting some transactions. To avoid repetitively scanning the database, a strategy is to use a buffer that contains the set of almost frequent itemsets in memory [19,23]. **Stream pattern mining algorithms** are designed to deal with databases that change in real-time and where new data may arrive at a very high speed. These algorithms aim to process transactions quickly to return an approximate set of patterns rather that the complete set. Two representative algorithms for incremental pattern mining are estDec and estDec+ [32]. estDec employs a lexicographic tree structure called a prefix tree to identify and maintain significant itemsets from an online data stream. Significant itemsets are itemsets that may be frequent itemsets in the near future. It has been observed that the size of the prefix tree, which is located in the main memory, becomes very large as the number of significant itemsets increases. Thus, if the size of the prefix tree becomes larger than the available memory space, estDec fails to identify new significant itemsets. As a result, the accuracy of estDec results is degraded [32]. estDec+ and other algorithms have been designed to solve this problem.

Interactive pattern mining tries to handle dynamic databases differently by injecting users preferences, users feedback or user targeted queries, into the mining process [3,4,14,20,22]. In contrast with incremental and stream pattern mining where algorithms aim to maintain and update a large set of patterns that may be uninteresting to users, interactive pattern mining algorithms focus only on some specific sets of patterns that are needed by the user. Besides, several approaches have been designed which can generally be classified in three categories: (1) Targeted querying based approaches, (2) Users feed-backs based approaches and (3) Visualization based approaches.

Targeted Querying Based Approaches. These approaches let the user search for patterns containing specific items by sending some targeted queries to the system to search for interesting patterns. Then, the system interacts and tries to give quick answers to the user queries [14,20,22]. See Sect. 3 for more details.

Users Feedback Based Approaches. Users feedback based approaches are more interactive comparing with targeted querying based approaches. The key idea is to progressively address feedback sent by users during the mining process. Bhuiyan et al. [4] proposed an interactive pattern mining system that is based on the sampling of frequent patterns from hidden datasets. Hidden datasets exist in various real applications where the data owner and the data analyst is no necessarily the same entity. Thus, the data analyst may not have the full access to the data and the data owner has to maintain the confidentiality of the data by providing to analysts only some samples from data that would be beneficial to him but without giving him the possibility to reconstruct the entire dataset from the given samples [4]. The proposed interactive systems aims to continuously update effective sampling distributions by binary feedback from the users. The proposed system works as follows: Using a Markov Chain Monte Carlo (MCMC) sampling method, the system return a small set of frequent patterns (samples) to each analysts (user). Then, each analyst sends a feedback about its associated samples. The feedback used in this method is a simple feedback where the response of a user on a pattern is to indicate if this pattern is interesting or uninteresting. The system defines a scoring function based on users' feedback and updates each sampling distribution taking into consideration its corresponding user's interests. Following these steps, the proposed system can progressively address the user preferences so that the data remains confidential. Experiments on itemset and graph mining datasets demonstrate the usefulness of the proposed system. Based on the same approach, an improved version of this system was proposed [3]. Besides, authors have adopted a better scoring function for graph data by using graph topology and new improved feedback mechanisms, namely, periodic feedback and conditional periodic feedback.

Another common problem in pattern mining that motivates researchers to design interactive pattern discovery tools is the problem of pattern explosion [11]. More precisely, traditional pattern mining algorithms discover a large number of patterns, of which many are redundant or similar. As a result, the analyst or the data expert should invest substantial efforts to look for the desired patterns which is not an easy task. To overcome this limitation, an interactive pattern discovery framework was proposed [11] for two mining tasks, frequent itemset mining and subgroup discovery. The proposed framework consists of three steps: (1) Mining patterns, (2) Interacting with the user and (3) learning user-specific pattern interestingness. Besides, The user is only asked to rank small sets of patterns, while a ranking function is inferred from users feedback using preference learning techniques. In the experimental results, it has been demonstrated that the system was able to learn accurate pattern rankings for both mining tasks.

Visualization Based Approaches. Another important aspect to design a good interactive pattern mining system is the visualisation aspect. More precisely, data visualisation techniques play an important role in making the discovered knowledge understandable and interpretable by humans [17]. In fact, the output of the implemented algorithms is presented to the user only in a textual form, which may impose many limitations such as the difficulty to identify similar patterns and the difficulty to understand the relation between patterns. There are various visualization techniques for different forms of patterns. For instance, researchers [2], have used a lattice based representation based on the Hasse diagram to visualise the output of frequent itemset mining. All possible itemsets can be represented in the diagram and the frequent itemsets are highlighted in bold. Other visualisation techniques have been used to efficiently present itemsets to the user such as pixel based visualization and tree based visualisation [17]. As for itemsets mining, various visualization tools were proposed for the other pattern mining problems such as mining association rules, mining sequential patterns and mining episodes. The reader can refer to [17] where a detailed survey that present the visualisation techniques designed for each mining task.

6 C5: Heuristic Pattern Mining

Since the emergence of the subfield, the excessive number of results caused by combinatorial explosion has been the most fundamental problem encountered in pattern mining. Although many efficient pattern mining algorithms have been proposed, the excessive results still lead to a high computational cost, and the high computational cost required to mine all exact patterns is not proportional to the actionability of the mining results. Occasionally, the result of a large number of discovered patterns can even lead to decision makers not knowing how to use them. Furthermore, for application fields such as recommender systems, it is not necessary to use all exact patterns.

To solve this problem, heuristic pattern mining (HPM) algorithms have been developed to identify an approximate subset of all patterns within a reasonable time. Inspired by biological [9] and physical [35] phenomena, heuristic methods are effective for solving combinatorial problems such as pattern mining. Compared with exact pattern mining algorithms, HPM algorithms are efficient and do not need any domain knowledge in advance. Several heuristic algorithms are used in the subfield of pattern mining, such as the genetic algorithm (GA) [9], particle swarm optimization (PSO) [24], artificial bee colony (ABC) [33], cross-entropy (CE) [35], and bat algorithm (BA) [34]. In addition to achieving high efficiency, these HPM algorithms can also discover a sufficient number of exact patterns. From the perspective of the key components of the entire mining process, the following challenges are summarized.

Identifying the Appropriate Objective. For heuristic methods, a fitness function is the objective function used to measure the performance of each individual to determine which will survive and reproduce into the next generation. In existing HPM algorithms, typical measures, for example, support [9] and utility [34], are used as fitness functions directly. Standard fitness functions are easy to implement; however, for more specific patterns (closed pattern or top-k pattern), flexible fitness functions must be applied, which are more difficult to implement. Furthermore, in complex scenarios, for example, choosing sets of products that are both economical and fast to deliver, using multi-objective fitness functions [45] is also a challenging issue.

Speed up the Mining Process. Determining how to narrow the search space is a general key issue in algorithm design. HPM algorithms also have this problem. The downward closure property is the most widely used principle in HPM algorithms. Specifically, support is used for frequent itemsets [9], and transaction-weighted utilization is used for high utility itemsets [34]. Other strategies include using the lengths of the discovered patterns to generate promising patterns [33] and developing a tree structure to avoid invalid combinations [24]. Considering the characteristics of heuristic methods and the resulting patterns, it would be interesting to develop a new data structure or pruning strategies to improve mining efficiency.

Diversifying the Discovered Results. Heuristic methods are likely to fall into local optima. This is also true for HPM algorithms. When HPM algorithms fall into local optimal values, it is difficult to produce new patterns in subsequent iterations. Inspired by the mutation operator of the GA, randomly generating some individuals in the next generation is the most important approach to avoid falling into local optima [9]. Although determining how to increase the diversity of results is difficult, a first step is to measure the diversity of results, such as the bit edit distance [35].

Designing a General Framework. Pattern mining is different from problems in which there are relatively few best values, all patterns have support/utilities or other measures no lower than the minimum threshold are the mining targets. In addition to the GA, PSO, ABC, CE, and BA, other heuristic methods such as the ant colony system and dolphin echolocation optimization have been used for pattern mining recently. However, attempting to use each heuristic method individually to mine patterns is not only expensive but also infeasible. Therefore, integrating all the objectives, processes, and results into a general HPM framework is a promising approach [10,34].

At the present time, HPM mainly focuses on itemsets. Thus, discovering a complex pattern, for example, sequential pattern or graph pattern [31], using heuristic methods is challenging.

7 C6: Mining Interesting Patterns

The problem of Interesting Pattern Mining (IPM) plays an important role in Data Mining. The goal is to discover patterns of interest to users in databases

(DBs) where interest is measured using functions. One of the first interestingness functions used in IPM is the support function to mine frequent patterns (FPM) in binary DBs. A pattern is said to be frequent in a binary DB if the number of its appearances in transactions of the database (or its support) is no less than a user-predefined minimum support threshold. For the special measure, in order to efficiently solve the combinatorial explosion in FPM, a nice property of the support, the Downward Closure or Anti-Monotonicity - AM, has been applied. This property states that if a pattern is not frequent (infrequent), all its super-patterns are also infrequent, or the whole branch rooted at the infrequent pattern (on the prefix search tree) can be pruned immediately.

However, the support measure is not suitable for all applications. Thus, other interestingness functions have been designed to find important patterns that may be rare but useful or interesting for many real-life applications. Some of the most popular kinds are utility functions of patterns in quantitative DBs (QDBs). Utility functions can be used for example to find the most profitable purchase patterns in customer transactions. Note that the support can be seen as a special utility function. A simple QDB is called a quantitative transaction DB (QTDB), where each (input) transaction is a quantitative itemset (a set of quantitative items). A more general QDB is quantitative sequence DB (QSDB) of which each input quantitative sequence consists of a sequence of quantitative itemsets.

Moreover, a key challenge in the problem of high utility pattern mining (HUPM) is that such utility functions usually do not satisfy the AM property. To overcome this challenge, we need to devise upper bounds or weak upper bounds (on the utilities) that satisfy the AM property or weaker (such as Anti-Monotone like - AML) ones. In this context, given a utility function u of patterns, a function ub is said to be an upper bound (UB) on u if $ub(x) \geq u(x)$ for any pattern x. And a function wub is said to be a weak upper bound (WUB) on u if $wub(x) \geq u(y)$ for any extension pattern y of x. Usually, given a (W)UB, the tighter (W)UB is, the stronger its pruning ability is. The effort and time for devising good and tight (W)UBs is often very long.

For example, in the first problem of high utility itemset mining (HUIM) on a QTDB D, the utility u of an itemset A is defined as the summation of its utilities $u(A,T)$ in all transactions T of D containing A, where the utility $u(A,T)$ of A in T is the summation of utilities of items of A appearing in T. Similarly, for the second problem of high average utility itemset mining (HAUIM) in a QTDB D, the average utility au of an itemset A is defined as the utility $u(A)$ divided by its length $length(A)$. From the first time 2004 [43] (2009) where HUIM (HAUIM, respectively) was proposed, it took more than 8 (10) years to obtain good tighter UBs based on the remaining utility [25] (WUBs based on vertical representation of QTDB [37], respectively). It is worthy to note that for the average utility au, besides UBs (on it), there are many WUBs, which are much tighter than the UBs. The number of WUBs found so far is about five times more than that of UBs, and devising such good WUBs requires much effort and time.

For the more general problems of high utility sequence mining (HUSM) on a QSDB D, because each sequence α may appear multiple times in an input quantitative sequence (IQS) Ψ of D, there are many ways to define the utility of α in Ψ. There are two popular kinds of such utilities, denoted as $u_{max}(\alpha, \Psi)$ and $u_{min}(\alpha, \Psi)$, that are respectively defined as the maximum and minimum values among utilities of occurrences of α in Ψ. Then, $u_{max}(\alpha)$ and $u_{min}(\alpha)$ are respectively the summation of $u_{max}(\alpha, \Psi)$ and $u_{min}(\alpha, \Psi)$ of α in all IQSs Ψ containing α. Similarly, there are two other kinds of utilities named $au_{max}(\alpha)$ and $au_{min}(\alpha)$ that are respectively defined as $u_{max}(\alpha)$ and $u_{min}(\alpha)$ divided by length of α. For the first (third) utility u_{max} (au_{max}), to find good UBs [16] (WUBs [36], respectively), it took about 10 years (8 years, respectively). Furthermore, devising such UBs (for example on u_{max}) without mathematically proving it strictly may lead to inexactness in corresponding algorithms [16].

For the new second (fourth) utility u_{min} (au_{min}), the time for devising good UBs (WUBs) on it has decreased significantly only in one paper (e.g. [38] for au_{min}). Thus, from the theoretical results presented in the paper, a natural and useful question that has been raised is how to propose a generic framework for the IPM problem according to any new interestingness function, and a general and simple method to quickly design (W)UBs on functions using weeks instead of years? In more details, given a QSDB D and a new interestingness function itr that may not satisfy AM and a user-specified minimum interestingness threshold mi, the corresponding IPM problem is to mine the set $\{\alpha | itr(\alpha) \geq mi\}$ of all highly interesting patterns. The *first* question is how to quickly devise (W)UBs on itr so that they are as tight as possible and have anti-monotone-like properties? The goal of these requirements is to allow significantly reducing the search space. The *second* question is how to transform checking the anti-monotone-like properties of itr in the whole D into simpler one in each input quantitative sequence? Moreover, these theoretical results must be proven strictly in mathematical language. Then, the main challenge that aims at significantly reducing time for devising good (W)UBs on itr will be solved.

8 Conclusion

The field of pattern mining has been rapidly changing. This paper has provided an overview of six key challenges, each identified by a researcher from the field.

References

1. Abeysinghe, R., Cui, L.: Query-constraint-based mining of association rules for exploratory analysis of clinical datasets in the national sleep research resource. BMC Med. Inform. Decis. Making **18**(2), 58 (2018)
2. Alsallakh, B., Micallef, L., Aigner, W., Hauser, H., Miksch, S., Rodgers, P.: The state-of-the-art of set visualization. In: Computer Graphics Forum, vol. 35, pp. 234–260. Wiley Online Library (2016)
3. Bhuiyan, M., Hasan, M.A.: Interactive knowledge discovery from hidden data through sampling of frequent patterns. Statist. Anal. Data Mining ASA Data Sci. J. **9**(4), 205–229 (2016)
4. Bhuiyan, M., Mukhopadhyay, S., Hasan, M.A.: Interactive pattern mining on hidden data: a sampling-based solution. In: Proceedings of the 21st ACM International Conference on Information and Knowledge Management, pp. 95–104 (2012)

5. Chand, C., Thakkar, A., Ganatra, A.: Target oriented sequential pattern mining using recency and monetary constraints. Int. J. Comput. App. **45**(10), 12–18 (2012)
6. Chen, M.S., Park, J.S., Yu, P.S.: Efficient data mining for path traversal patterns. IEEE Trans. Knowl. Data Eng. **10**(2), 209–221 (1998)
7. Chiang, D.A., Wang, Y.F., Lee, S.L., Lin, C.J.: Goal-oriented sequential pattern for network banking churn analysis. Expert Syst. App. **25**(3), 293–302 (2003)
8. Chueh, H.E., et al.: Mining target-oriented sequential patterns with time-intervals. Int. J. Comput. Sci. Inf. Technol. **2**(4), 113–123 (2010)
9. Djenouri, Y., Comuzzi, M.: Combining apriori heuristic and bio-inspired algorithms for solving the frequent itemsets mining problem. Inf. Sci **420**, 1–15 (2017)
10. Djenouri, Y., Djenouri, D., Belhadi, A., Fournier-Viger, P., Lin, J.C.-W.: A new framework for metaheuristic-based frequent itemset mining. Appl. Intell. **48**(12), 4775–4791 (2018). https://doi.org/10.1007/s10489-018-1245-8
11. Dzyuba, V., Leeuwen, M.v., Nijssen, S., De Raedt, L.: Interactive learning of pattern rankings. Int. J. Artif. Intell. Tools **23**(06), 1460026 (2014)
12. Fournier-Viger, P., Cheng, C., Cheng, Z., Lin, J.C., Selmaoui-Folcher, N.: Mining significant trend sequences in dynamic attributed graphs. Knowl. Based Syst. **182**, 104797 (2019)
13. Fournier-Viger, P., et al.: A survey of pattern mining in dynamic graphs. Wiley Interdiscip. Rev. Data Min. Knowl. Discov. **10**(6), e1372 (2020)
14. Fournier-Viger, P., Mwamikazi, E., Gueniche, T., Faghihi, U.: MEIT: memory efficient itemset tree for targeted association rule mining. In: Motoda, H., Wu, Z., Cao, L., Zaiane, O., Yao, M., Wang, W. (eds.) ADMA 2013. LNCS (LNAI), vol. 8347, pp. 95–106. Springer, Heidelberg (2013). https://doi.org/10.1007/978-3-642-53917-6_9
15. Gan, W., et al.: A survey of utility-oriented pattern mining. IEEE Trans. Knowl. Data Eng. **33**(4), 1306–1327 (2021)
16. Gan, W., et al.: ProUM: projection-based utility mining on sequence data. Inf. Sci. **513**, 222–240 (2020)
17. Jentner, W., Keim, D.A.: Visualization and visual analytic techniques for patterns. In: High-Utility Pattern Mining, pp. 303–337 (2019)
18. Jiang, C., Coenen, F., Zito, M.: A survey of frequent subgraph mining algorithms. Knowl. Eng. Rev. **28**, 75–105 (2013)
19. Koh, J.-L., Shieh, S.-F.: An efficient approach for maintaining association rules based on adjusting FP-tree structures. In: Lee, Y.J., Li, J., Whang, K.-Y., Lee, D. (eds.) DASFAA 2004. LNCS, vol. 2973, pp. 417–424. Springer, Heidelberg (2004). https://doi.org/10.1007/978-3-540-24571-1_38
20. Kubat, M., Hafez, A., Raghavan, V.V., Lekkala, J.R., Chen, W.K.: Itemset trees for targeted association querying. IEEE Trans. Knowl. Data Eng. **15**(6), 1522–1534 (2003)
21. Lam, H.T., Morchen, F., Fradkin, D., Calders, T.: Mining compressing sequential patterns. Statist. Anal. Data Mining ASA Data Sci. J. **7**(1), 34–52 (2014)
22. Li, X., Li, J., Fournier-Viger, P., Nawaz, M.S., Yao, J., Lin, J.C.W.: Mining productive itemsets in dynamic databases. IEEE Access **8**, 140122–140144 (2020)
23. Lin, C.W., Hong, T.P., Lu, W.H.: The pre-FUFP algorithm for incremental mining. Expert Syst. App. **36**(5), 9498–9505 (2009)
24. Lin, J.C.W., Yang, L., Fournier-Viger, P., Hong, T.P., Voznak, M.: A binary PSO approach to mine high-utility itemsets. Soft Comput. **21**(17), 5103–5121 (2017)
25. Liu, M., Qu, J.: Mining high utility itemsets without candidate generation. In: Proceedings of the 21st ACM International Conference on Information and Knowledge Management, pp. 55–64 (2012)

26. Miao, J., Wan, S., Gan, W., Sun, J., Chen, J.: TargetUM: targeted high-utility itemset querying. arXiv preprint arXiv:2111.00309 (2021)
27. Min, F., Zhang, Z.H., Zhai, W.J., Shen, R.P.: Frequent pattern discovery with tri-partition alphabets. Inf. Sci. **507**, 715–732 (2020)
28. Ouarem, O., Nouioua, F., Fournier-Viger, P.: Mining episode rules from event sequences under non-overlapping frequency. In: Fujita, H., Selamat, A., Lin, J.C.-W., Ali, M. (eds.) IEA/AIE 2021. LNCS (LNAI), vol. 12798, pp. 73–85. Springer, Cham (2021). https://doi.org/10.1007/978-3-030-79457-6_7
29. Qu, W., Yan, D., Guo, G., Wang, X., Zou, L., Zhou, Y.: Parallel mining of frequent subtree patterns. In: Qin, L., et al. (eds.) SFDI/LSGDA -2020. CCIS, vol. 1281, pp. 18–32. Springer, Cham (2020). https://doi.org/10.1007/978-3-030-61133-0_2
30. Shabtay, L., Yaari, R., Dattner, I.: A guided FP-growth algorithm for multitude-targeted mining of big data. arXiv preprint arXiv:1803.06632 (2018)
31. Shelokar, P., Quirin, A., Cordón, O.: Three-objective subgraph mining using multiobjective evolutionary programming. Comput. Syst. Sci **80**(1), 16–26 (2014)
32. Shin, S.J., Lee, D.S., Lee, W.S.: CP-tree: an adaptive synopsis structure for compressing frequent itemsets over online data streams. Inf. Sci. **278**, 559–576 (2014)
33. Song, W., Huang, C.: Discovering high utility itemsets based on the artificial bee colony algorithm. In: Phung, D., Tseng, V.S., Webb, G.I., Ho, B., Ganji, M., Rashidi, L. (eds.) PAKDD 2018. LNCS (LNAI), vol. 10939, pp. 3–14. Springer, Cham (2018). https://doi.org/10.1007/978-3-319-93040-4_1
34. Song, W., Huang, C.: Mining high utility itemsets using bio-inspired algorithms: a diverse optimal value framework. IEEE Access **6**, 19568–19582 (2018)
35. Song, W., Zheng, C., Huang, C., Liu, L.: Heuristically mining the top-k high-utility itemsets with cross-entropy optimization. Appl. Intell. 1–16 (2021). https://doi.org/10.1007/s10489-021-02576-z
36. Truong, T., Duong, H., Le, B., Fournier-Viger, P.: EHAUSM: an efficient algorithm for high average utility sequence mining. Inf. Sci. **515**, 302–323 (2020)
37. Truong, T., Duong, H., Le, B., Fournier-Viger, P., Yun, U.: Efficient high average-utility itemset mining using novel vertical weak upper-bounds. Knowl. Based Syst. **183**, 104847 (2019)
38. Truong, T., Duong, H., Le, B., Fournier-Viger, P., Yun, U.: Frequent high minimum average utility sequence mining with constraints in dynamic databases using efficient pruning strategies. Appl. Intell. **52**, 1–23 (2021)
39. Wu, Y., Shen, C., Jiang, H., Wu, X.: Strict pattern matching under non-overlapping condition. Sci. China Inf. Sci. **50**(1), 012101 (2017)
40. Wu, Y., Tong, Y., Zhu, X., Wu, X.: NOSEP: nonoverlapping sequence pattern mining with gap constraints. IEEE Trans. Cybern. **48**(10), 2809–2822 (2018)
41. Wu, Y., Wang, Y., Li, Y., Zhu, X., Wu, X.: Self-adaptive nonoverlapping contrast sequential pattern mining. IEEE Trans. Cybern. (2021)
42. Xie, F., Wu, X., Zhu, X.: Efficient sequential pattern mining with wildcards for keyphrase extraction. Knowl. Based Syst. **115**, 27–39 (2017)
43. Yao, H., Hamilton, H.J., Butz, C.J.: A foundational approach to mining itemset utilities from databases. In: Proceedings of the 2004 SIAM International Conference on Data Mining, pp. 482–486. SIAM (2004)
44. Zhang, C., Du, Z., Dai, Q., Gan, W., Weng, J., Yu, P.S.: TUSQ: targeted high-utility sequence querying. arXiv preprint arXiv:2103.16615 (2021)
45. Zhang, L., Fu, G., Cheng, F., Qiu, J., Su, Y.: A multi-objective evolutionary approach for mining frequent and high utility itemsets. Appl. Soft Comput. **62**, 974–986 (2018)

Why Not to Trust Big Data: Discussing Statistical Paradoxes

Rahul Sharma[1(✉)], Minakshi Kaushik[1], Sijo Arakkal Peious[1], Mahtab Shahin[1], Ankit Vidyarthi[2], Prayag Tiwari[3], and Dirk Draheim[1]

[1] Information Systems Group, Tallinn University of Technology, Akademia tee 15a, 12618 Tallinn, Estonia
{rahul.sharma,minakshi.kaushik,sijo.arakkal,mahtab.shahin, dirk.draheim}@taltech.ee

[2] Jaypee Institute of Information Technology, Noida, India

[3] Department of Computer Science, Aalto University, Espoo, Finland
prayag.tiwari@aalto.fi

Abstract. Big data is driving the growth of businesses, data is the money, big data is the fuel of the twenty-first century, and there are many other claims over Big Data. Can we, however, rely on big data blindly? What happens if the training data set of a machine learning module is incorrect and contains a statistical paradox? Data, like fossil fuels, is valuable, but it must be refined carefully for the best results. Statistical paradoxes are difficult to observe in datasets, but they are significant to analyse in every small or big dataset. In this paper, we discuss the role of statistical paradoxes on Big data. Mainly we discuss the impact of Berkson's paradox and Simpson's paradox on different types of data and demonstrate how they affect big data. We provide that statistical paradoxes are more common in a variety of data and they lead to wrong conclusions potentially with harmful consequences. Experiments on two real-world datasets and a case study indicate that statistical paradoxes are severely harmful to big data and automatic data analysis techniques.

Keywords: Big data · Artificial intelligence · Machine learning · Data science · Simpson's paradox · Explainable AI

1 Introduction

Data has always been critical in making decisions. Earlier, statistics and mathematics have been used to draw insights from data. However, in the last two decades, with the emergence of social media and big data technologies, data science, artificial intelligence (AI) and machine learning (ML) techniques have gained massive ground in practice and theory. These decision support techniques

This work has been partially conducted in the project "ICT programme" which was supported by the European Union through the European Social Fund.

U. K. Rage et al. (Eds.): DASFAA 2022 Workshops, LNCS 13248, pp. 50–63, 2022.
https://doi.org/10.1007/978-3-031-11217-1_4

are being used widely to develop intelligent applications and acquire deeper insights from structured and unstructured data. In most AI use cases, ML based trained artificial systems provide fast and accurate outcomes; however, it does not guarantee accurate results for every use case in real life. Moreover, like statistics, understanding causal relationships and evaluating the existence of statistical paradoxes in the training dataset is not in the mainstream data science, AI and ML application scenarios. AI, machine learning, and big data are now widely used in medical sciences, social sciences, and politics, and they have a direct or indirect impact on human life and decisions. Therefore, understanding causal relationships and evaluating the existence of statistical paradoxes is essential for fair decision making [13,14].

Statistical reasoning and probability theory are the foundation of many AI, big data and data science techniques, e.g., random forest [7], support vector machines [11], etc. Therefore, it is usual to have causal relationships and statistical paradoxes in these decision support techniques. A paradox can be a statement that leads to an apparent self-contradictory conclusion. Even the most well-known and documented paradoxes frequently confound domain specialists because they fundamentally violate common sense.

There are many statistical paradoxes (e.g., Simpson's Paradox, Berkson's Paradox, Latent Variables, Law of Unintended Consequences, Tea Leaf Paradox, etc.). Statistical paradoxes are not new to be discussed in statistics and mathematics; expert mathematicians and statisticians adequately addressed the severe impact of paradoxes. However, in modern decision support techniques, specifically AI and data science, causal relationships, data fallacies and statistical paradoxes are not appropriately addressed. In this article, we discuss the impact of Berkson's Paradox, Yule-Simpson paradox and causal inference on big data. We highlight several hidden problems in data that are not yet discussed in big data mining. We use two benchmark datasets for machine learning and a case study to demonstrate the existence of Simpson's paradox in different types of data.

The paper is organised as follows. In Sect. 2, we discuss why not trust data science, AI, ML and big data. In Sec. 3, we discuss two statistical paradoxes and discuss their impacts on big data mining. In Sect. 4, we use two benchmark datasets for machine learning to demonstrate the effects of Simpson's paradox. In Sect. 5 we provide a case study to analyse the impact of Simpson's paradox in real life. Finally, a discussion and conclusion is provided in Sect. 6 and Sect. 7.

2 Why Not to Trust on Data Science, AI, ML and Big Data

In AI, ML and Data Science, observing trends, mean and correlation between two variables for making decisions is not always correct. E.g., suppose in a city, the Covid-19 infection rate of smokers is less than the infection rate of the non-smokers. Can we claim that smoking prevent Covid-19? It is a perfect case of poor data science where all the variables and features in the dataset are not appropriately observed. In today's world, data literacy may not seem exciting when compared to machine learning algorithms or big data mining, but it should be the foundation for all data mining processes.

Datasets, irrespective of their size and type, are not self-explanatory. It's all numbers and statistics responsible for creating stories out of datasets. Therefore, it's essential to validate a dataset statistically and evaluate the existence of any statistical paradoxes. AI, ML, big data and data science based techniques generate knowledge from data. Therefore, decision support techniques are easily prone to statistical paradoxes and can not be trusted.

3 Statistical Paradoxes

Statistical paradoxes aren't something that hasn't been discussed before. These terms are widely used in statistics and have been around for over a century. Statistical paradoxes are fundamentally related with various statistical challenges and mathematical logic including causal inference [27,28], the ecological fallacy [24,31], Lord's paradox [36], propensity score matching [32], suppressor variables [10], conditional independence [12], partial correlations [16], p-technique [8] and mediator variables [26]. The instances of statistical paradoxes specifically Simpson's paradox have been discussed in various data mining techniques [17], e.g., association rule mining [2] and numerical association rule mining [20,21,34]

More recently, Kügelgen et al. [37] pointed out the importance of statistical analysis of real data and demonstrated instances of Simpson's paradox in Covid-19 data analysis. They provide that the overall case fatality rate (CFR) was higher in Italy than in China. However, in every age group, the fatality rate was higher in China than in Italy. These observations raise many questions on the accuracy of data and its analysis. Heather et al. [25] have addressed the existence of Simpson's paradox. In psychological science, Kievit et al. [22] examined the instances of Simpson's paradox. Alipourfard et al. [3] have discovered the existence of Simpson's paradox in social data and behavioural data [4]. Therefore, understanding data, especially big data, is more critical than its processing. In the following two sections, we discuss Berkson's paradox and Yule-Simpson's Paradox to demonstrate their vast impact on big datasets.

3.1 Berkson Paradox

Berkson's paradox can make it appear as if there is a relationship between two independent variables when there is no relationship between the variables. In 1946, despite diabetes being a risk factor for cholecystitis, Berkson [5] observed a negative correlation between cholecystitis and diabetes in hospital patients. Berkson state that If at least one of two independent events occurs, they become conditionally dependent. In other words, two independent events become conditionally dependent, given that at least one of them occurs. Statistically, Berkson's paradox and Simpson's paradox are very close to each other. Berkson's paradox is a type of selection bias caused by systematically observing some events more than others.

$$if\ 0 < P(A) < 1,\ 0 < P(B) < 1\ and \tag{1}$$

$$P(A|B) = P(A)\ then \tag{2}$$

$$P(A|B, A \cup B) = P(A)\ Hence \tag{3}$$

$$P(A|B, A \cup B) > P(A) \tag{4}$$

As given in Eq. 1 to Eq. 4, $P(A|B)$, a conditional probability, is the probability of observing event A given that B is true. The probability of A given both B and $(A\ or\ B)$ is smaller than the probability of A given $(A\ or\ B)$.

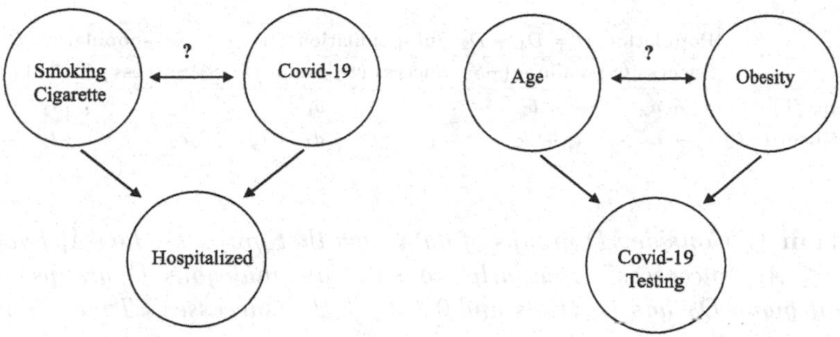

Fig. 1. Berkson's paradox: two noticeable example of Covid-19 which introduce a collider.

As we all know, smoking cigarettes is a well-known risk factor for respiratory diseases. However, recently Wenzel T. [9] observed a negative co-relation between Covid-19 severity and smoking cigarettes. In another observation, Griffith et al. [18] describe it as a Collider Bias or Berkson's paradox. In Fig. 1, we demonstrate an example of collider. Here Smoking cigarettes, Covid-19 are two independent variables, but they collide with another random variable, hospitalised. Here, the variable hospitalised is collider for both smoking cigarettes and Covid-19.

3.2 Yule-Simpson's Paradox

In the year 1899, Karl Pearson et al. [29] demonstrated a statistical paradox in marginal and partial associations between continuous variables. Later in 1903, Udny Yule [38] explained "the theory of association of attributes in statistics" and revealed the existence of an association paradox with categorical variables.

In a technical paper published in 1951 [33], Edward H. Simpson described the phenomenon of reversing results. However, in 1972, Colin R. Blyth coined the term "Simpsons Paradox" [6]. Therefore, this paradox is known with different names and it is popular as the Yule-Simpson effect, amalgamation paradox, or reversal paradox [30].

We start the discussion on the paradox by using the real-world dataset from Simpson's article [33]. In this example, analysis for medical treatment is demonstrated. Table 1 summarises the effect of the medical treatment for the entire population ($N = 52$) as well as for men and women separately in subgroups. The treatment appears effective for both male and female subgroups; however, the treatment seems ineffective at the whole population level.

Table 1. 2×2 contingency table with sub population groups D1 and D2.

	Population $D = D_1 + D_2$		Sub-population D_1		Sub-population D_2	
	Success (S)	Failure ($\neg S$)	Success (S)	Failure ($\neg S$)	Success (S)	Failure ($\neg S$)
Treatment (T)	$a_1 + a_2$	$b_1 + b_2$	a_1	b_1	a_2	b_2
No Treatment ($\neg T$)	$c_1 + c_2$	$d_1 + d_2$	c_1	d_1	c_2	d_2

Definition 1. *Consider D groups of data such that group D_1 has A_i trials and $0 \le a_i \le A_i$ "successes". Similarly, consider an analogous D groups of data such that group D_2 has B_i trials and $0 \le b_i \le B_i$ "successes" Then, Simpson's paradox occurs if*

$$\frac{a_1}{A_1} \ge \frac{b_1}{B_1} and \frac{a_2}{A_i} \ge \frac{b_2}{B_2} \text{ for all } i = 1, 2, \ldots, n \text{ but } \frac{\sum_{i=1}^{n} a_i}{\sum_{i=1}^{n} A_i} \le \frac{\sum_{i=1}^{n} b_i}{\sum_{i=1}^{n} B_i} \tag{5}$$

we use the following example to show how this equation works.

$$\frac{10}{20} > \frac{30}{70} and \frac{10}{50} > \frac{10}{60} \text{ but } \frac{10+10}{20+50} < \frac{30+10}{70+60}, \tag{6}$$

We could also flip the inequalities and still have the paradox since A and B are chosen arbitrarily.

Classically the paradox is expressed via contingency tables. Let a 2×2 contingency table for treatment (T) and success (S) in the ith sub-population is represented by a four-dimensional vector of real numbers $D = (a_1, b_1, a_2, b_2)$. Then

$$D = \sum_{i=1}^{N} D_i = \left(\sum a_i, \sum b_i, \sum c_i, \sum d_i\right) \tag{7}$$

is the aggregate dataset over N sub populations. This can be read as given in Table 1.

We can also demonstrate the Simpson's paradox scenario via probability theory and conditional probabilities. Let $T = treatment$, $S = successful$, $M = Male$, and $F = Female$ then,

$$\mathsf{P}(S \mid T) = \mathsf{P}(S \mid \neg T) \tag{8}$$

$$\mathsf{P}(S \mid T, M) > \mathsf{P}(S \mid \neg T, M) \tag{9}$$

$$\mathsf{P}(S \mid T, \neg M) > \mathsf{P}(S \mid \neg T, \neg M) \tag{10}$$

Based on Eq. 8, 9 and 10, one should use the treatment or not? As per the success rate for the male and female population, the treatment is a success, but overall, the treatment is a failure. This reversal of results between groups population and the total population has been referred to as Simpson's Paradox. In statistics, this concept has been discussed widely and named differently by several authors [29,38].

4 Existence of Simpson's Paradox in Big Data

Simpson's paradox can exist in any dataset irrespective of its size and type [23]. The paradox demonstrates the importance of having human experts in the loop to examine and query Big Data results. In this section, we present datasets to analyse the presence and implications of Simpson's paradox on big data.

To identify an instance of the Simpson paradox in a continuous dataset with n continuous variable and m discrete variables, we can compute a correlation matrix $(n \times n)$ for all the data. Then for m discrete variable with k_m levels, an additional $(n \times n)$ matrix needs to be calculated for each level of variables as follows. Therefore, we need to calculate the $1 + \sum_i^m = k_i$ correlation matrices of size $(n \times n)$ and compare it with the lower half of $\sum_i^m = k_i$ for subgroup levels. We have also discussed the measures to find the impact of one numerical variable to another numerical variable [19].

4.1 Datasets

We use the iris dataset and miles per gallon (mpg) dataset, the two benchmark datasets for machine learning to demonstrate the presence of Simpson's paradox in data.

Iris Dataset: Ronald Fisher introduced the iris dataset in a research paper [15]. It consists three types of iris species (Setosa, Versicolor, Virginicare), each with 50 data samples. The species names are categorical attributes, length and width are continuous attributes.

In order to identify the existence of Simpson's paradox in the iris datasets, we first visualise the relationship between the length and width of each pair of candidate attributes. As shown in Fig. 2, in the iris dataset, we identify the

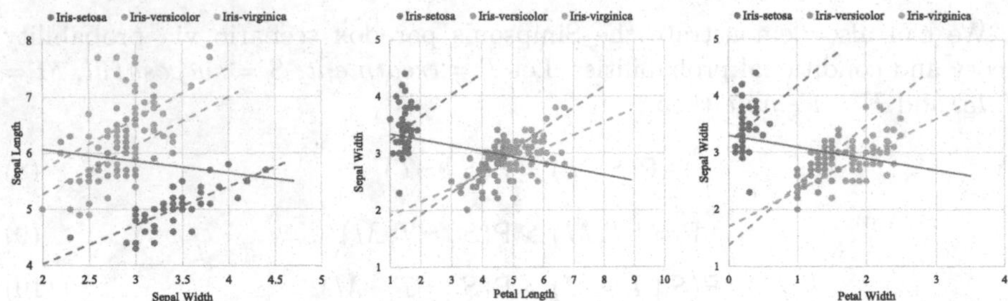

Fig. 2. Simpson's paradox in Iris dataset: there is a positive correlation between the three pairs of sepal length and petal width for the Iris-setosa, Iris-versicolor and Iris-virginicare (dashed lines). However, the overall trend for the length and width for the entire population is negative (solid red line) in all three combinations. (Color figure online)

existence of Simpson's paradox for three pairs of measurements. 1. sepal length and width, 2. sepal width and petal length, and 3. sepal width and petal width.

In Fig. 2, the correlation between sepal width and sepal length is positive (dashed line) for each species. However, the correlation between sepal width and sepal length for the entire population is negative (solid red trend line). Similarly, the pair of petal length, width, and the pair of petal width and sepal width have positive trends for each species; however, the overall trend for the length and width for the entire population is negative in both cases. Therefore, this is a clear case of Simpson's paradox in the iris dataset.

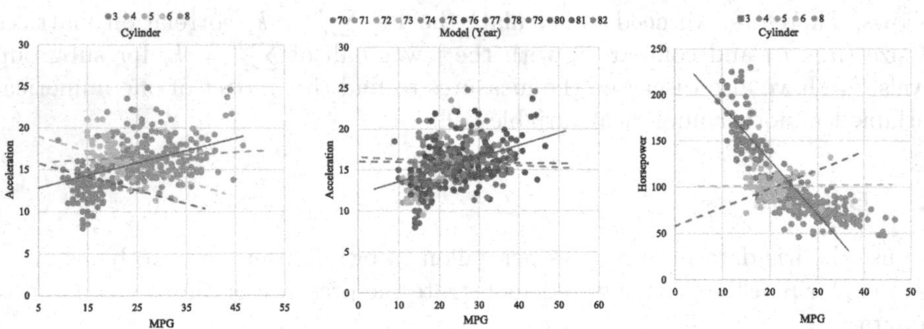

Fig. 3. Simpson's paradox in auto MPG dataset: there is a negative correlation between MPG and acceleration for three cylinders engines and six cylinders engines; however, the overall trend between MPG and acceleration is positive (solid red line). Similarly, the overall trend is positive for MPG and acceleration with respect to the model year. However, the overall trend between MPG and horsepower according to the engine cylinders is negative. (Color figure online)

The MPG Dataset: Ross Quinlan used the Auto MPG dataset in 1993 [30]. The dataset contains 398 automobile records from 1970 to 1982, including the vehicle's name, MPG, number of cylinders, horsepower, and weight. The dataset includes three multi-valued discrete attributes and five continuous attributes.

In order to identify the existence of Simpson's paradox in the MPG datasets, we visualise the relationship between MPG, acceleration and horsepower for two categorical attributes (number of cylinders and model year). The goal of analysing the dataset is to know the factors that influence each car's overall fuel consumption. The dataset consists of fuel consumption in mpg, horsepower, number of cylinders, displacement, weight, and acceleration.

In the MPG dataset, we identify the existence of Simpson's paradox in three pairs of measurements. 1. MPG with acceleration according to the engine cylinders, 2. MPG with acceleration with respect to their model year, and 3. MPG with horsepower according to the engine cylinders. In the Fig. 3, it is visualised that there is a negative correlation between MPG and acceleration for three cylinders engines and six cylinders engines; however, the overall trend between MPG and acceleration is positive (solid red line). Similarly, the overall trend is opposite for MPG with acceleration with respect to the model year and MPG with horsepower according to the engine cylinders.

5 Analysis Simpson's Paradox in Real Life: A Case Study

The case study is from the California Department of developmental services (CDDS), United States of America [35]. As per the annual reports published by the department, the average annual expenditures on Hispanic residents were approximately one-third (1/3) of the average expenditures on White non-Hispanic residents. According to the marginal analysis, it was a solid gender discrimination case. However, a conditional analysis of ethnicity and age found no evidence of ethnic discrimination. Furthermore, except for one age group, the trends were completely opposite. The average annual expenditures on White non-Hispanic residents were less than the expenditures on Hispanic residents. Therefore, it is a perfect case of Simpson's paradox in real life.

Table 2. Number of residents by ethnicity and percentage of expenditures.

Ethnicity	Sum of Expend. ($)	% of Expend.	# of Residents	% of Residents
American Indian	145753	0.81	4	0.4
Asian	2372616	13.13	129	12.9
Black	1232191	6.82	59	5.9
Hispanic	4160654	23.03	376	37.6
Multi Race	115875	0.64	26	2.6
Native Hawaiian	128347	0.71	3	0.3
Other	6633	0.04	2	0.2
White not Hispanic	9903717	54.82	401	40.1
Total	18065786	100%	1000	100%

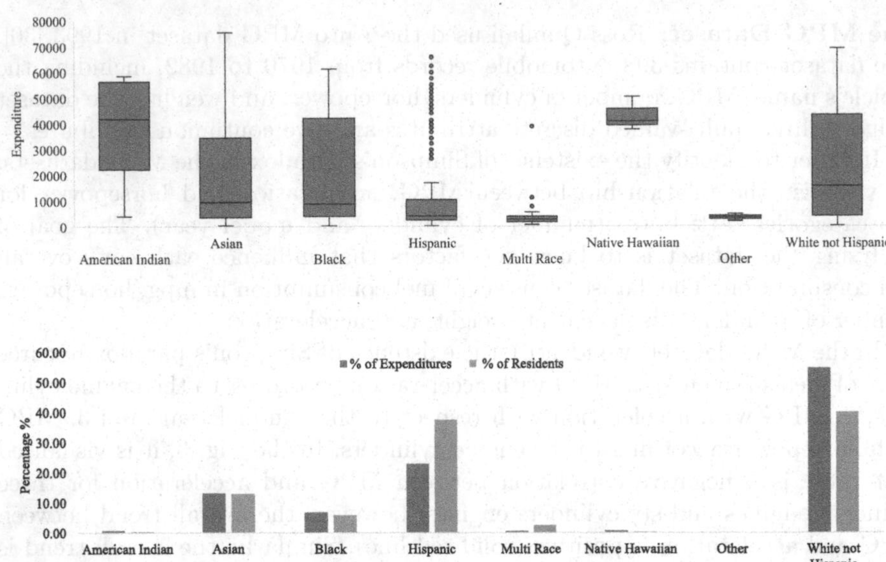

Fig. 4. Distribution of expenditure as per the ethnic groups.

5.1 The Dataset

We use the same dataset to analyse the original claims. The dataset is publicly available at [1]. The dataset mainly consists various information of one thousand disabled residents (DRs) under six important variables (ID, age group, gender, expenditures, ethnicity). Each DR has a unique identification, i.e., "ID". The state department uses AGE to decide the financial needs and other essential needs of DRs. The age groups of the residents are divided into six age groups. (0–5 years old, 6–12 years old, 13–17 years old, 18–21 years old, 22–50 years old, and 51 years old). These groups are based on the amount of financial assistance required at each stage of age. E.g., The 0–5 age group (preschool age) has the fewest needs and thus requires the least funding.

The"Expenditures" variable represents the annual expenditures made by the state to support each resident and their family. Information about the expenditures, the number of residents and their percentage as per ethnicity is given in Table 2. The expenditures include all the expenses, including psychological services, medical fees, transportation and housing costs such as rent (especially for adult residents). As far as the case is concerned, "ethnicity" is the most important demographic variable in the dataset. The dataset includes eight ethnic groups.

As demonstrated in Fig. 4, the population difference between the Hispanic and the White non-Hispanic people is significantly less. However, there is a big difference between the distribution of assistance to the Hispanic and the White non-Hispanic group. Therefore, these two populations are selected for the case study for further investigation.

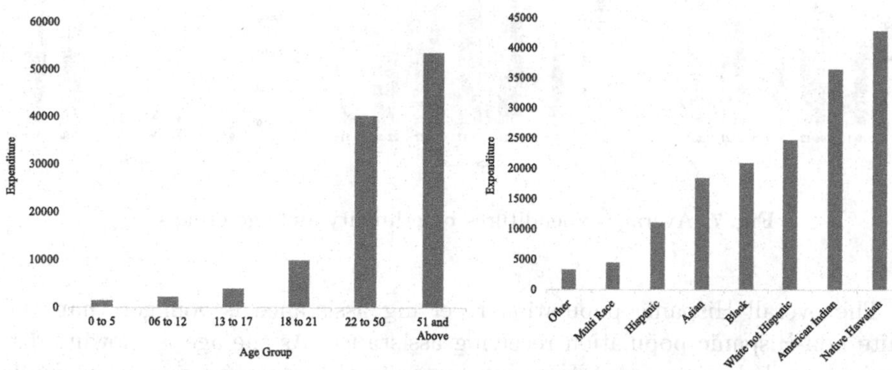

Fig. 5. 1. Average expenditure by age group, 2. Average expenditure by ethnicity.

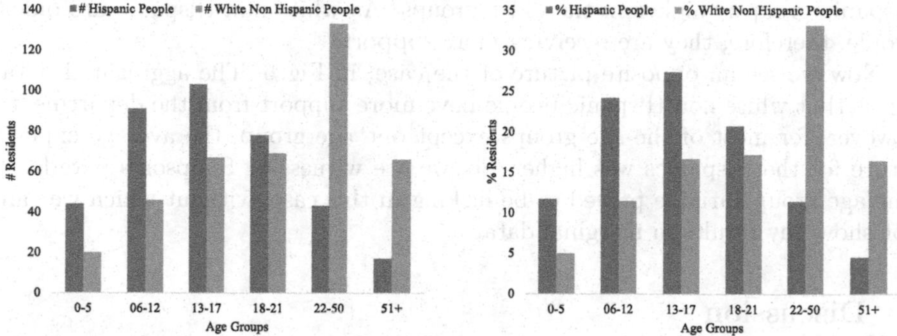

Fig. 6. 1. Hispanic and white non Hispanic residents with their age groups, 2. Percentage of Hispanic and white non Hispanic residents according the age groups.

5.2 Data Analysis

We begin the data analysis by comparing the total amount of expenditure in relation to different ethnic groups. As per the bar chart given in Fig. 5, It is clear that the average expenditure on Hispanic residents is significantly lower than the White non-Hispanic residents. Moreover, the analysis of average expenditure by the age groups shows that the average expenditure was very high for the older age groups. As per Fig. 5, it is also a clear case of age discrimination. However, age is not considered a factor for the discrimination because older people are eligible to get higher expenditures (Fig. 7).

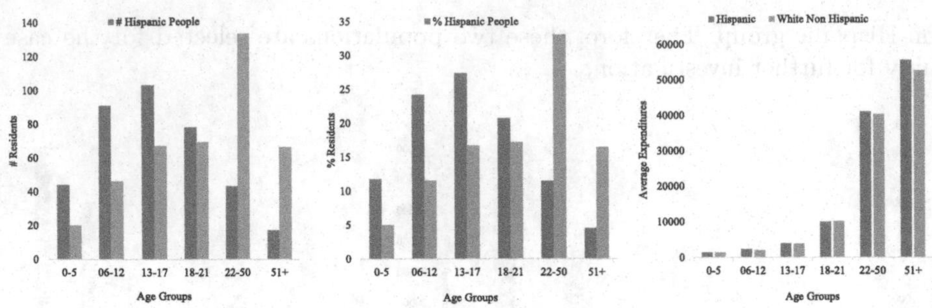

Fig. 7. Average expenditures by ethnicity and age groups.

The overall Hispanic population receiving assistance is younger than the white non-Hispanic population receiving assistance. As the age is showing discriminatory behaviour, therefore, we compare the average amount of funds received by the two observed ethnic groups as per their age groups in Fig. 6. It is clear that the number of beneficiaries from the Hispanic group is higher in the lower age groups, while the number of beneficiaries from the white non-Hispanic group is higher in the older groups. As white non-Hispanic are older people, therefore, they are receiving more support.

Now we see an opposite picture of the case, in Fig. 6. The aggregated data shows that white non-Hispanic people have more support from the department; however, for most of the age groups except one age group, the average expenditure for the Hispanics was higher. So, we are witnessing Simpson's paradox!. The age group variable proved to be lurking in this case, without which we can not show any results in marginal data.

6 Discussion

The existence of statistical paradoxes in benchmark datasets and in real-life case studies provides a direction to understand the causality in decision making. We noticed that most machine learning and deep learning algorithms focus only on identifying correlations rather than identifying the real or causal relationships between data items. Therefore, understanding and evaluating causality is an important term to be discussed in big data, Data Science, AI and ML.

7 Conclusion

Handling statistical paradoxes is a complex challenge in AI, ML and Big Data. Different paradoxes state the possibilities of errors in the outcomes of automatic data analysis conducted for AI, Ml and big data based applications. In this paper, we discussed the existence of Berkson's paradox and demonstrate the existence of Simpson's paradox and in two real datasets. Statistical paradoxes in data reflect the importance of probabilities and causal inference and seek a manual inspection of datasets. We argue that if confounding effects are not properly addressed in datasets, outcomes of an data analysis can be completely opposite. However, with the right tools and data analysis, a good analyst or data scientist can handle it in a better way. The statistical paradoxes confirm essential statistical evaluation for datasets and demonstrate the importance of human experts in the loop to examine and query Big datasets.

References

1. California Department of Developmental Services CDDS expenditures. https://kaggle.com/wduckett/californiaddsexpenditures
2. Agrawal, R., Srikant, R.: Fast algorithms for mining association rules in large databases. In: Proceedings of VLDB 1994 - The 20th International Conference on Very Large Data Bases, pp. 487–499. Morgan Kaufmann (1994)
3. Alipourfard, N., Fennell, P.G., Lerman, K.: Can you trust the trend? Discovering Simpson's paradoxes in social data. In: Proceedings of the Eleventh ACM International Conference on Web Search and Data Mining, WSDM 2018, pp. 19–27. Association for Computing Machinery, New York (2018). https://doi.org/10.1145/3159652.3159684
4. Alipourfard, N., Fennell, P.G., Lerman, K.: Using Simpson's paradox to discover interesting patterns in behavioral data. In: Proceedings of the Twelfth International AAAI Conference on Web and Social Media. AAAI Publications (2018)
5. Berkson, J.: Limitations of the application of fourfold table analysis to hospital data. Biometrics Bull. **2**(3), 47–53 (1946). http://www.jstor.org/stable/3002000
6. Blyth, C.R.: On Simpson's paradox and the sure-thing principle. J. Am. Stat. Assoc. **67**(338), 364–366 (1972)
7. Breiman, L.: Random forests. Mach. Learn. **45**(1), 5–32 (2001)
8. Cattell, R.B.: P-technique factorization and the determination of individual dynamic structure. J. Clin. Psychol. (1952)
9. Commission, E., Centre, J.R., Wenzl, T.: Smoking and COVID-19: a review of studies suggesting a protective effect of smoking against COVID-19. Publications Office (2020). https://doi.org/10.2760/564217
10. Conger, A.J.: A revised definition for suppressor variables: a guide to their identification and interpretation. Educ. Psychol. Measur. **34**(1), 35–46 (1974)
11. Cortes, C., Vapnik, V.: Support-vector networks. Mach. Learn. **20**(3), 273–297 (1995)
12. Dawid, A.P.: Conditional independence in statistical theory. J. Roy. Stat. Soc.: Ser. B (Methodol.) **41**(1), 1–15 (1979). https://doi.org/10.1111/j.2517-6161.1979.tb01052.x
13. Draheim, D.: DEXA'2019 keynote presentation: future perspectives of association rule mining based on partial conditionalization, Linz, Austria, 28th August 2019. https://doi.org/10.13140/RG.2.2.17763.48163
14. Draheim, D.: Future perspectives of association rule mining based on partial conditionalization. In: Hartmann, S., Küng, J., Chakravarthy, S., Anderst-Kotsis, G., A Min Tjoa, Khalil, I. (eds.) Database and Expert Systems Applications. LNCS, vol. 11706, p. xvi. Springer, Heidelberg (2019) (2019)
15. Fisher, R.A.: The use of multiple measurement in taxonomic problems. Ann. Eugen. **7**(2), 179–188 (1936). https://doi.org/10.1111/j.1469-1809.1936.tb02137.x

16. Fisher, R.A.: III. The influence of rainfall on the yield of wheat at rothamsted. Philos. Trans. R. Soc. London Ser. B Containing Papers Biological Character **213**(402–410), 89–142 (1925)
17. Freitas, A.A., McGarry, K.J., Correa, E.S.: Integrating Bayesian networks and Simpson's paradox in data mining. In: Texts in Philosophy. College Publications (2007)
18. Griffith, G.J., et al.: Collider bias undermines our understanding of COVID-19 disease risk and severity. Nat. Commun. **11**(1), 5749 (2020). https://doi.org/10.1038/s41467-020-19478-2
19. Kaushik, M., Sharma, R., Peious, S.A., Draheim, D.: Impact-driven discretization of numerical factors: case of two- and three-partitioning. In: Srirama, S.N., Lin, J.C.-W., Bhatnagar, R., Agarwal, S., Reddy, P.K. (eds.) BDA 2021. LNCS, vol. 13147, pp. 244–260. Springer, Cham (2021). https://doi.org/10.1007/978-3-030-93620-4_18
20. Kaushik, M., Sharma, R., Peious, S.A., Shahin, M., Ben Yahia, S., Draheim, D.: On the potential of numerical association rule mining. In: Dang, T.K., Küng, J., Takizawa, M., Chung, T.M. (eds.) FDSE 2020. CCIS, vol. 1306, pp. 3–20. Springer, Singapore (2020). https://doi.org/10.1007/978-981-33-4370-2_1
21. Kaushik, M., Sharma, R., Peious, S.A., Shahin, M., Yahia, S.B., Draheim, D.: A systematic assessment of numerical association rule mining methods. SN Comput. Sci. **2**(5), 1–13 (2021). https://doi.org/10.1007/s42979-021-00725-2
22. Kievit, R., Frankenhuis, W., Waldorp, L., Borsboom, D.: Simpson's paradox in psychological science: a practical guide. Front. Psychol. **4**, 513 (2013). https://doi.org/10.3389/fpsyg.2013.00513
23. Kim, Y.: The 9 pitfalls of data science. Am. Stat. **74**(3), 307–307 (2020). https://doi.org/10.1080/00031305.2020.1790216
24. King, G., Roberts, M.: EI: A (n R) program for ecological inference. Harvard University (2012)
25. Ma, H.Y., Lin, D.K.J.: Effect of Simpson's paradox on market basket analysis. J. Chin. Stat. Assoc. **42**(2), 209–221 (2004). https://doi.org/10.29973/JCSA.200406.0007
26. MacKinnon, D.P., Fairchild, A.J., Fritz, M.S.: Mediation analysis. Annu. Rev. Psychol. **58**(1), 593–614 (2007). https://doi.org/10.1146/annurev.psych.58.110405.085542. pMID: 16968208
27. Pearl, J.: Causal inference without counterfactuals: comment. J. Am. Stat. Assoc. **95**(450), 428–431 (2000)
28. Pearl, J.: Understanding Simpson's paradox. SSRN Electron. J. **68** (2013). https://doi.org/10.2139/ssrn.2343788
29. Pearson Karl, L.A., Leslie, B.M.: Genetic (reproductive) selection: inheritance of fertility in man, and of fecundity in thoroughbred racehorses. Philos. Trans. R. Soc. Lond. Ser. A **192**, 257–330 (1899)
30. Quinlan, J.: Combining instance-based and model-based learning. In: Machine Learning Proceedings 1993, pp. 236–243. Elsevier (1993). https://doi.org/10.1016/B978-1-55860-307-3.50037-X
31. Robinson, W.S.: Ecological correlations and the behavior of individuals. Am. Sociol. Rev. **15**(3), 351–357 (1950)
32. Rosenbaum, P.R., Rubin, D.B.: The central role of the propensity score in observational studies for causal effects. Biometrika **70**(1), 41–55 (1983)
33. Simpson, E.H.: The interpretation of interaction in contingency tables. J. Roy. Stat. Soc.: Ser. B (Methodol.) **13**(2), 238–241 (1951)

34. Srikant, R., Agrawal, R.: Mining quantitative association rules in large relational tables. In: Proceedings of the 1996 ACM SIGMOD International Conference on Management of Data, pp. 1–12 (1996)
35. Taylor, S.A., Mickel, A.E.: Simpson's paradox: a data set and discrimination case study exercise. J. Stat. Educ. **22**(1), 8 (2014). https://doi.org/10.1080/10691898.2014.11889697
36. Tu, Y.K., Gunnell, D., Gilthorpe, M.S.: Simpson's paradox, lord's paradox, and suppression effects are the same phenomenon-the reversal paradox. Emerg. Themes Epidemiol. **5**(1), 1–9 (2008)
37. Von Kugelgen, J., Gresele, L., Scholkopf, B.: Simpson's paradox in COVID-19 case fatality rates: a mediation analysis of age-related causal effects. IEEE Trans. Artif. Intell. **2**(1), 18–27 (2021). https://doi.org/10.1109/tai.2021.3073088
38. Yule, G.U.: Notes on the theory of association of attributes in statistics. Biometrika **2**(2), 121–134 (1903)

Localized Metric Learning for Large Multi-class Extremely Imbalanced Face Database

Seba Susan(✉) and Ashu Kaushik

Department of Information Technology, Delhi Technological University, Delhi 110042, India
seba_406@yahoo.in

Abstract. Metric learning serves to mitigate, to a great extent, the class-imbalance problem associated with large multi-class image databases. However, the computational complexity associated with metric learning increases when the number of classes is very large. In this paper, a novel localized metric learning scheme is proposed for a large multi-class extremely imbalanced face database with an imbalance ratio as high as 265:1. The Histogram of Gradient (HOG) features are extracted from each facial image and these are given as input for metric learning. The proposed scheme involves confining the metric learning process to local subspaces that have similar class populations. The training dataset is divided into smaller subsets based on the class populations such that the class imbalance ratio within a local group does not exceed 2:1. The locally learnt distance metrics are then, one by one, used to transform the entire input space. The nearest neighbor of the test sample, in the training set, is noted for each transformation. A comparison amongst all transformations for the closest nearest neighbor in the training set establishes the class of the test sample. Experiments are conducted on the highly imbalanced benchmark Labeled Faces in the Wild (LFW) dataset containing 1680 classes of celebrity faces. All classes are retained for the experimentation including those minority classes having just two samples. The proposed localized metric learning scheme outperforms the state of the art for face classification from large multi-class extremely imbalanced face databases.

Keywords: Extremely imbalanced dataset · Face recognition · Subspace-learning · Multi-class · Metric learning · Large margin nearest neighbor

1 Introduction

An uneven class distribution in multimedia data is a matter of concern to machine learning researchers. The reason is the bias induced while training the classification model that mis-classifies most of the minority classes whose samples are scant as compared to that of the majority classes. The class imbalance problem, as it is generally called, is a well-known highlighted issue in data mining [1]. The conventional solutions mostly revolve around resampling strategies [2] and cost-sensitive learning [3]. Resampling involves data manipulation by oversampling the minority class, undersampling the majority class, or hybrid sampling approaches [4]. Cost-sensitive learning, on the other hand, modifies

U. K. Rage et al. (Eds.): DASFAA 2022 Workshops, LNCS 13248, pp. 64–78, 2022.
https://doi.org/10.1007/978-3-031-11217-1_5

the learning process such that it becomes more sensitive to the cost function pertaining to the under-represented minority class [5]. However, these solutions, that are popular in data mining, were not found easy to adapt for multimedia datasets, mainly due to the large scale of the datasets involved, the large number of classes, and the high imbalance ratio between the majority and the minority classes [6]. Oversampling the minority class created replicas that further increased the scale of the dataset, while undersampling led to loss of discriminative information. Experiments to generate new minority samples were conducted in [7] that involved affine transformations of samples from the under-represented minority class. Data augmentation using Generative Adversarial Networks (GAN) is yet another technique to correct class imbalance by creating synthetic images for the minority class [8]. Creation of synthetic samples further increases the computational overhead associated with a large dataset, with parallel computing and high-performance computing machines usually being involved [9].

Most of the solutions to the class-imbalance problem in literature, including those discussed above, are meant for binary classification problems i.e. for datasets having a single majority class and a single minority class. All these algorithms work microscopically with majority class samples and/or minority class samples as the chief particles of interest, and focus on either eradicating the redundant majority samples or replicating the minority samples. The severity of the class imbalance problem is based on the ratio of the majority class population to minority class population that is defined as the imbalance ratio (IR). IRs of 50:1, 100:1 or higher are considered as indicators of extremely imbalanced datasets. In cases where the dataset was multi-class, researchers transformed the problem into a two-class one by segregating the classes into two clusters based on the relative similarity between class populations, for the purpose of testing the utility of their sampling algorithms [10]. Figure 1 illustrates the binarization of several popular multi-class datasets in various works in literature.

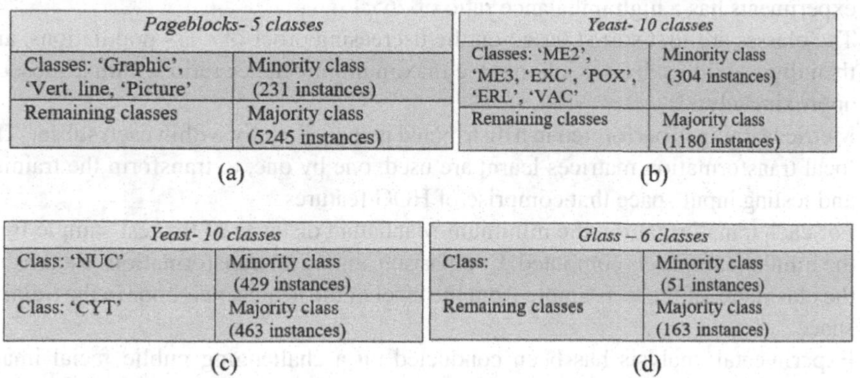

Pageblocks- 5 classes	
Classes: 'Graphic', 'Vert. line, 'Picture'	Minority class (231 instances)
Remaining classes	Majority class (5245 instances)

(a)

Yeast- 10 classes	
Classes: 'ME2', 'ME3, 'EXC', 'POX', 'ERL', 'VAC'	Minority class (304 instances)
Remaining classes	Majority class (1180 instances)

(b)

Yeast- 10 classes	
Class: 'NUC'	Minority class (429 instances)
Class: 'CYT'	Majority class (463 instances)

(c)

Glass – 6 classes	
Class:	Minority class (51 instances)
Remaining classes	Majority class (163 instances)

(d)

Fig. 1. Binarization of multi-class datasets in different works in literature: (a) *Pageblocks* by Barua et al. (2014) [11] (b) *Yeast* by Barua et al. (2014) [11] (c) *Yeast* by Tao et al. (2019) [12] (d) *Glass* by Liu et al. (2020) [13]

Apart from serving the purpose of testing and trying out new resampling strategies, the binarization of multiple classes destroys the original class information and renders the existing solutions useless for classifying the original multi-class dataset. Adaptation of existing binarized classification solutions to large multi-class multimedia datasets where the number of classes is too high and the imbalance condition is severe, is thus highly impractical. In our paper, we propose a novel solution based on distance metric learning for tackling the class-imbalance in large multi-class image datasets. Distance metric learning [14] induces data space transformation that results in an improved classification for a dataset with imbalanced class-distribution. The chief advantage over resampling strategies is the prevention of loss of information by undersampling, and the creation of redundancy and computational overhead by oversampling. Though this idea has been explored before by Feng et al. (2018) [15], Wang et al. (2018) [16], Wang et al. (2021) [17] and Susan and Kumar (2019) [10], it has not been investigated, to the best of our knowledge, for larger datasets having more than thousand classes and for a multi-class scenario with severe class-imbalance problem. The reason is the computational complexity of metric learning that attains impractical limits when the number of classes are extremely high (>1000). Specifically, we focus on the large multi-class Labeled Faces in the Wild (LFW) face database [25] containing 1680 classes, with a highly uneven class population profile ranging from 530 to 2. The researchers in [35] constrain metric learning to selected samples of the top-186 classes of LFW. However, in the current work, all the classes of the LFW dataset are involved in the metric learning process. To minimize the computational complexity, the low-dimensional HOG features are used to prove the effectiveness of the proposed localized metric learning. The contributions of our work are summarized as follows:

1. A novel localized metric learning scheme is introduced for handling severe class-imbalance in large multi-class face datasets; the LFW face dataset used in our experiments has a high imbalance ratio of 265:1.
2. The classes are first sorted based on the decreasing order of class populations, and then divided into subsets such that the maximum imbalance ratio within a subset is approximately 2:1.
3. Metric learning is performed in a distributed manner, locally within each subset. The local transformation matrices learnt are used, one by one, to transform the training and testing input space that comprise of HOG features.
4. For each transformation, the minimum Manhattan distance of the test sample from the training samples is computed. Comparison among all transformations establishes the class label of the test sample from the label of the nearest neighbor in the training space.
5. Experimental analysis has been conducted on a challenging public facial image database having 1680 classes with all the minority classes retained, and results are compared with the state of the art.
6. Our classification scores are observed to be the highest among all existing methods in terms of accuracy, AUC score and F1-score, for a very challenging train-test split of 50:50 with 2-fold cross-validation.

The organization of this paper is as follows. Section 2 reviews the metric learning concepts and the limitations of its direct application to large multi-class datasets. Section 3 describes the proposed localized metric learning scheme and Sect. 4 describes the experimental setup and analyzes the results. The final conclusions are drawn in Sect. 5.

2 Metric Learning as an Antidote for the Class Imbalance Problem

Distance Metric Learning (ML) is the process of learning a distance metric from labeled training samples for efficient classification [14]. The process involves learning a transformation matrix that brings similar points closer and pushes dissimilar points farther apart. The result is reduced intra-class differences and increased inter-class differences. Without depleting or adding new samples, metric learning achieves improved classification for imbalanced datasets. For best results, ML is used along with distance or similarity classifiers in the classification module [18]. Weinberger et al. proposed the Large Margin Nearest Neighbor (LMNN) metric learning [19, 20] that learns a Mahalanobis distance metric for k-Nearest Neighbor (kNN) classification by semidefinite programming. We use LMNN for metric learning and data space transformation in our experiments, hence we start with a brief discussion on its procedure.

Consider an input space of n training samples $\boldsymbol{X} = \{\vec{x}_i\}, i = 1, 2, \ldots, n$. The corresponding class labels are denoted by $\{y_i\}, i = 1, 2, \ldots, n$. The objective of LMNN is to learn a transformation matrix $\boldsymbol{T}$ that transforms the input space such that the k nearest neighbors of a training sample mandatorily belongs to the same class. The squared distance between two samples $\boldsymbol{x_i}$ and $\boldsymbol{x_j}$ in the linearly transformed input space is computed as

$$D = \left\| T(x_i - x_j) \right\|^2 \tag{1}$$

The authors in [19, 20] have proposed $k = 3$ to be a suitable choice, and this is adopted for our implementation of LMNN in this work. For classes with samples fewer than 3, k was reduced accordingly. The cost function of LMNN imposes a penalty on large distances between a training sample and its k nearest neighbors belonging to the same class, and another penalty on small distances between the training sample and training samples belonging to other classes. The latter penalty is responsible for the term "large margin nearest neighbor" since it ensures that a large margin is maintained between samples belonging to different classes. The Euclidean distance between samples in the transformed space is proved to be equivalent to the Mahalanobis distance between the samples in the original input space [20]. The newly transformed input space $\boldsymbol{X_{new}}$ yields improved performance for distance- or similarity-based classifiers such as kNN. The transformed input space is defined, in pure mathematical terms, by the matrix multiplication of the original input space with the transposed transformation matrix as shown in Eq. (2).

$$X_{new} = X \times T^t \tag{2}$$

In case of imbalanced class-distribution for a real-world dataset, the majority samples outnumber the minority population, generating a decision bias towards the majority class.

The role of metric learning in reducing the class-imbalance was explored in the works of Feng et al. (2018) [15], Wang et al. (2018) [16], Wang et al. (2021) [17] and Susan and Kumar (2019) [10]. The idea was to transform the data space and concentrate the class populations to confined spaces leading to better classification results. These methods, notably, marked a departure from the conventional solution of resampling that relied on depletion of majority samples and/or replication of minority samples for balancing the imbalanced dataset. Sampling was integrated with ML in a hybrid scheme in [21], where the distance metric was learnt from pruned datasets.

The computational complexity of metric learning constrains its use for large multi-class datasets. The real issue is the computation of the second penalizing term in the cost function that computes distance of a sample with all other samples of other classes. Hence, most of the imbalance treatment solutions mentioned above are for binary classes only with even multi-class datasets being recategorized to binary classes, as illustrated in Fig. 1. The limitations of the existing applications of ML to correct class-imbalance are summarized as follows.

1. The existing ML applications for class-imbalance treatment are limited, based on the size of the dataset and the number of classes (chosen to be two for most applications).
2. The reason is the large number of computations required to calculate inter-class distances and intra-class distances between every pair of training samples, for computing the cost function in metric learning.
3. Even if multi-class computations are somehow managed by high performance computing machines, the resulting transformation matrix may not be altogether accurate in the presence of so many classes due to spatial limitations in adjusting so many classes.
4. The transformation matrix $\boldsymbol{T}$ being a $d \times d$ matrix, the efficiency of transformation for the multi-class scenario is limited by the size of d.
5. Metric learning does not take into account the non-linear relationships between samples and class labels, instances of which are present in most of the real-world datasets.

Due to the above listed issues, ML has been rarely applied to correct class-imbalance in large multi-class multimedia databases. Some of the few works that did address this problem did so under computational constraints, and in no case, to the best of our knowledge, did the number of classes exceed 1000. In [22], the majority class was undersampled to reduce the population, followed by ML. In [23], metric learning was used to improve road traffic sign detection from a LIDAR equipped vehicle, with 112 sign categories affected by the class-imbalance problem; 20 to 80 samples were selected from each category to form a prototype set to ease the ML computations. Deep metric learning was used in [24], in an imbalanced class scenario, to augment the results of uncorrelated cost-sensitive multi-set learning (UCML) that identifies discriminative features for classification. Here, balanced subsets of *Majority class: Minority class* were created using Generative Adversarial Network (GAN), and the distance metric was learnt within each subset using an overall loss function incorporating the individual loss functions of all subsets. The datasets were converted to two-class datasets for the experiments.

In our work, we seek to address the above limitations, for an extremely imbalanced large multi-class publicly available dataset LFW containing 1680 classes of celebrity facial images, for which conventional ML is rendered impractical. Further details of our proposed technique are presented in Sect. 3.

3 Localized Metric Learning – The Proposed Approach

Learning local metrics is considered computationally economical and efficient for large multi-class datasets rather than computing a global metric. This is the hypothesis on which we construct our idea. The concept was noted in [31] where a local similarity metric was computed for Locally Embedded Clustering (LEC), based on kd-tree partitions specific to certain localities. The objective function of SERAPH metric learning algorithm was decentralized in [32] to learn local metrics from a partitioned geographical region comprising of a finite number of nodes. With an aim to achieve parallel computation, the data in [33] is partitioned into medium-sized subsets, and results of local discriminative metric learning are aggregated into a global solution.

In our paper, we propose a localized metric learning scheme for the classification of facial images from large multi-class extremely imbalanced face databases containing more than thousand classes with numerous minority classes. Our technique, that is devoid of any sampling procedure, relies entirely on locally learnt transformation matrices. The local group formation, here, is based on the relative class sizes. LMNN is applied separately to each of the subsets, and the distance metric that is learnt from each local subspace is used, by turns, to transform the entire training and test space. For each test sample, the closest training sample in the transformed data space is determined in each case. A comparison among all transformations indicates the closest neighbor in the training space, the label of this training sample is assigned as the class label of the test sample. In summary, the work done in our paper is distinct in the following ways.

1. Our method partitions the training space into discrete subspaces based on relative class sizes and performs metric learning locally within each subspace. The task at hand is the classification of severely imbalanced large datasets having numerous classes exceeding 1000. We include all the minority classes of LFW face dataset that have at least two samples, of which one sample is applied for training and the other for testing, in a 50:50 split. To the best of our knowledge, very few works on LFW have included all the 1680 classes.
2. The locally learnt transformation matrices (7 in number for LFW) are used, in turns, to transform the entire input space including both training and test spaces.
3. The nearest (training) neighbor of the tests sample in the seven-times transformed training space provides the class label of the test sample. Therefore, a seven-fold chance is provided for each test sample to find its closest match in the training space as opposed to a single chance in the case of a global metric, the computation of which, for such a large database, is practically infeasible.
4. Our method boosts the classification accuracies, especially of the minority classes that have very few samples in the training space.

The steps are explained in greater detail in further sub-sections.

3.1 Division of Dataset Into Subsets

The first task is the division of the dataset into smaller subsets for the purpose of localized metric learning in which local transformation matrices are learnt from each subset. Figure 2 shows the division of the dataset into seven groups or subsets for the LFW dataset. The classes are divided such that the maximum imbalance ratio (IR) within a group is limited to approximately 2:1. This ensures that the metric learnt from a subset is not heavily biased towards one class, as is the situation with the LFW dataset, when considered as a whole. As observed from Fig. 2, *subset 1* is a majority class subset and *subset 7* contains the minority classes with least number of samples (= 2 samples in each class). The distance metric is now learnt locally within each subset using the principles of LMNN explained in Eq. (1) and Eq. (2).

subset 1	*subset 2*	*subset 3*	*subset 4*	*subset 5*	*subset 6*	*subset 7*
Class size: [530, 236] IR=2.24:1	Class size: [144, 71] IR=2.02:1	Class size: [60, 30] IR=2:1	Class size: [29, 14] IR=2.07:1	Class size: [13, 6] IR=2.16:1	Class size: [5, 3] IR=1.67:1	Class size: 2 IR=1:1

Fig. 2. Division of LFW dataset into local subsets named as *subset 1, subset 2,.....,subset 7*

3.2 Localized Metric Learning

Within each training subset, LMNN distance metric learning, whose details were discussed in Sect. 2, is locally applied. The overall process flow is shown illustrated in Fig. 3.

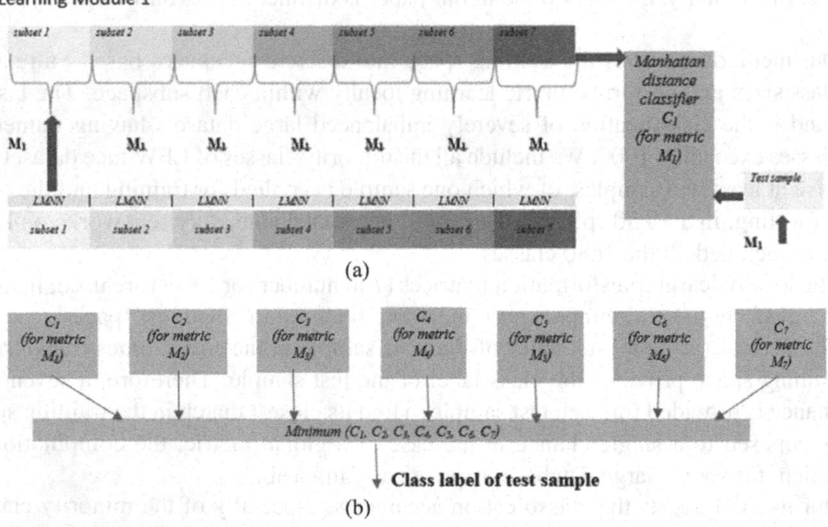

Fig. 3. Proposed scheme for localized metric learning (a) Local LMNN application and data space transformation using M_1 (b) Determining the closest nearest neighbor of the test sample in the training space using the *Minimum* (·) function over all possible transformations $M_1, M_2, \ldots, M_7$

The Mahalanobis distance metric locally learnt from each subset is applied to transform the complete training and testing data, as shown in Fig. 3(a). The figure shows Learning Module 1 that learns metric $\boldsymbol{M_1}$ from *subset 1*, and then uses it to transform the entire input space. The other Learning Modules numbered from 2 to 7 learn similar metrics from subsets 2 to 7, respectively. The process of determining the closest neighbor of the test sample in the training space is illustrated in Fig. 3(b). The minimum Manhattan distance of the test sample from its nearest neighbor in the training space, under all seven transformations, indicates the class of the test sample. The computations are detailed below.

The metric $\boldsymbol{M_1}$ learnt from *subset 1* is used to transform the entire training space and also the test sample. The transformation metric in (2) is set equal to $\boldsymbol{M_1}$ for *subset 1.*

$$T = M_1 \forall subset1 \tag{3}$$

The Manhattan or city-block distance of the transformed test sample from the training data transformed by $\boldsymbol{M_1}$ is computed to determine the nearest neighbor in the transformed training space. The city-block distance between two input vectors is the sum of the absolute differences between feature values. The class of the test sample and the corresponding minimum distance are given by Eqs. (4) and (5) respectively.

$$Class(test^{(M_1)}) = \underset{\forall l}{\arg\min} \sum_j \left| X_{new}^{(j)}\left(subset1^{(l)}\right) - X_{new}^{(j)}(test) \right| \tag{4}$$

$$dist(test^{(M_1)}) = \min_{\forall l} \sum_j \left| X_{new}^{(j)}\left(subset1^{(l)}\right) - X_{new}^{(j)}(test) \right| \tag{5}$$

where, *subset 1* has *l* training samples and each training sample has *j* features.

Similarly, the transformation matrices $\boldsymbol{M_2},\ldots,\boldsymbol{M_7}$ are derived locally from the training subsets: *subset 2,...subset 7*, respectively, and are used to transform the entire input space, one by one. The class of the test sample is determined by finding the minimum Manhattan distance of the test sample from all the training samples in the seven-times transformed training space. This is summarized in Eqs. (6) and (7).

$$c = \underset{\forall i}{\arg\min}\, dist(test^{(M_i)}) \tag{6}$$

$$Class(test) = Class(test^{(M_c)}) \tag{7}$$

3.3 Algorithm

The universal features considered are the Histogram of Gradient (HOG) features [26] since we wish to have a fair comparison among the classification methodologies without altering the feature space. It is understood that HOG features can be easily substituted by any of the other popular image features for the application. The algorithm for the application of localized metric learning on an extremely imbalanced face database is given below.

Algorithm Localized metric learning for extremely imbalanced face database

Input: *Training set*, *Test* sample
Output: Class label of *Test* sample
Find the 1784-dimensional HOG features for each facial image
Divide *Training set* into seven subsets based on similar class populations
FOR each subset i, DO
  Learn the LMNN distance metric $\boldsymbol{M_i}$
  Transform the entire input space using $\boldsymbol{M_i}$ as per Eq (2)
  Find the minimum distance of the transformed *Test* sample from the transformed training space to determine its nearest neighbor
REPEAT for i=1, 2, …., 7
Find the nearest neighbor of the *Test* sample in the seven-times transformed training space
Assign the class of the nearest neighbor to the *Test* sample as per Eq. (7)

4 Results and Discussions

The experiments were performed on an Intel dual core processor with graphics card, clocked at 2.8 GHz. The software platform is Python 3.7. The proposed approach was implemented as per the procedure outlined in Sect. 3. The dataset for our experiments is the benchmark Labeled Faces in the Wild (LFW) dataset comprising of cropped facial images of celebrities [25]. Only classes with at least two samples were considered since at least one sample from a class is required for training purpose. The parameter k for LMNN was set to 3 for the subsets 1 to 5, while it was set to 2 for subset 6, and 1 for subset 7 that contained only 2 samples per class. For developing a challenging classification problem to test the robustness of our algorithm, 50% of the images in a celebrity class were assigned for training and the remaining 50% images for testing. Two-fold cross-validation (V, CV) was performed by swapping the train and test sets. Most of the current works on LFW consider subsets of LFW that contain majority classes only for efficient classification and for presenting decent scores. Very few works include the 2-sample classes (minority classes) for the experimentation, despite of their dominant presence, since they are mostly misclassified and cause the AUC and F1-scores to drop. A total of 1680 classes of celebrities were shortlisted contributing to 4857 samples in training and 4307 samples in testing. This dataset has severe class-imbalance issues, with the class population ranging from 530 (George Bush) to 2 (Michel Duclos), leading to an imbalance ratio as high as 265:1. The number of classes with the minimum number of samples (2 Nos.) is 779 out of 1680 while the number of classes having a single digit class population is 1522 out of 1680, underlining the severity of the class imbalance problem. The unevenness in class-distribution can be observed in the population statistics of LFW compiled in Fig. 4. In this scenario, the classification accuracy of a large-population class would be high and that of a low-population class would be poor due to a decision bias that favors the largely populated class with more than sufficient classes. The problem is aggravated by the presence of noise and outliers that are common characteristics of all classes in real-world datasets.

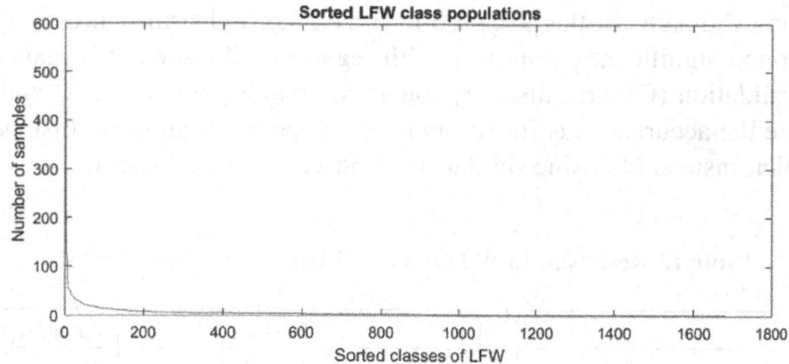

Fig. 4. LFW Class distribution sorted in decreasing order of class population ranging from 530 to 2

Metric learning transforms the data space within each local subset for efficient class separation. A demonstration of the same is shown for *subset 1* in Fig. 5 where the two classes of George Bush (label 533) versus Colin Powell (label 310) are shown for Feature $1 = 1762$ column of HOG feature and Feature $2 = 1763$ column of HOG feature.

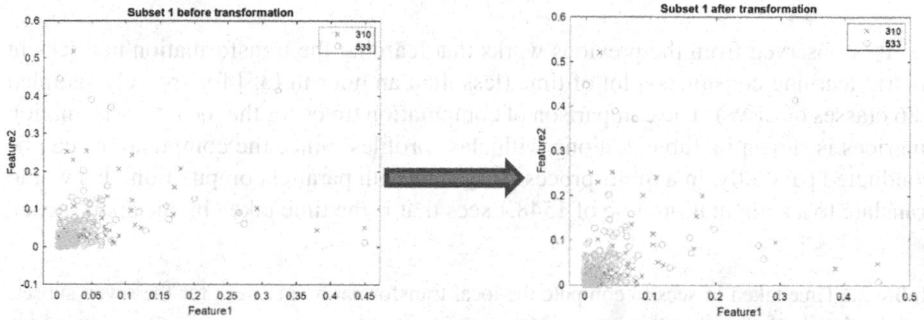

Fig. 5. Effect of local transformation on subset 1 comprising of class labels 533 (George Bush) and 310 (Colin Powell)

We compare our results with that of similar works that have tackled classification of large image datasets: HOG features with Cosine similarity measure by Chen et al. [28], HOG features with the Mahalanobis similarity metric by Fetaya and Ullman [29], HOG features with the Euclidean distance by Bhele and Mankar [30], HOG features with metric learning with majority classes [35]. The universal features considered, in all cases, are the Histogram of Oriented Gradients (HOG) features [26] since we wish to have a judicious comparison among the classification methodologies without altering the feature space. It is understood that HOG features can be easily substituted by any of the other popular image features for the application. Table 1 shows the results of our experiments in terms of AUC scores, accuracy (%) and F1-score (macro). AUC is the Area Under Curve score derived from the Receiver Operating Characteristic (ROC) curve. As observed, the proposed method outperforms the existing methods by a considerable margin. As compared to [35] which is metric learning with majority classes, the Validation (V) results of the proposed localized metric learning involving all 1680 classes are found significantly improved with regard to all three performance metrics. The Cross-Validation (CV) results were found comparable for the AUC and F1-score metrics, while the accuracy was found improved when the Manhattan distance is used as the classifier, instead of cosine similarity, as observed from Table 1.

Table 1. Results of LFW face recognition using HOG features

Method	AUC		F1-score		Accuracy	
	V	*CV*	*V*	*CV*	*V*	*CV*
Cosine similarity [28]	0.544	0.541	0.078	0.076	21.5%	19.1%
Manhattan [29]	0.55	0.544	0.087	0.082	22.96%	21.1%
Euclidean [30]	0.544	0.541	0.078	0.076	21.5%	19.1%
Metric learning with majority classes [35]	0.556	0.554	0.1006	0.097	26.8%	24.6%
Ours (Cosine similarity)	0.558	0.551	0.1023	0.095	26.07%	23.65%
Ours (Manhattan distance)	0.5625	0.551	0.1124	0.0962	27.72%	24.87%

It is observed from the previous works that learning the transformation matrices in metric learning consumes a lot of time (less than an hour in [35] for sparsely sampled 186 classes of LFW). The comparison of computation times for the local transformation matrices is shown in Table 2, along with class profiles. Since the computations can be conducted parallelly, in a multi-processor system with parallel computation, this would translate to a computation time of 1548.1 secs that is the time taken by the sixth subset.

Table 2. Time taken in secs to compute the local transformation matrices for the seven subsets and the details of classes within each subset

Subset	M_1	M_2	M_3	M_4	M_5	M_6	M_7
Time (secs)	55.96	67.75	154.94	276.05	692.07	1548.1	151.34
Number of classes	2	5	27	72	205	589	779
Class size	[530, 236]	[144, 71]	[60, 30]	[29, 14]	[13, 6]	[5, 3]	[2]

Figure 6 shows the improvement in performance using the proposed localized metric learning scheme. The results are shown for some of the minority classes with very low-class populations (ranging from 2 to 10) and some of the majority classes with high populations (ranging from 530 to 55) in Fig. 6(a) and Fig. 6(b), respectively. Our method thus improves classification scores for both the majority classes and the minority classes, especially for the minority classes as evident from the significant improvement in classification accuracies in Fig. 6(a). Parallelization of the computations using a multi-GPU system is the next stage of our work.

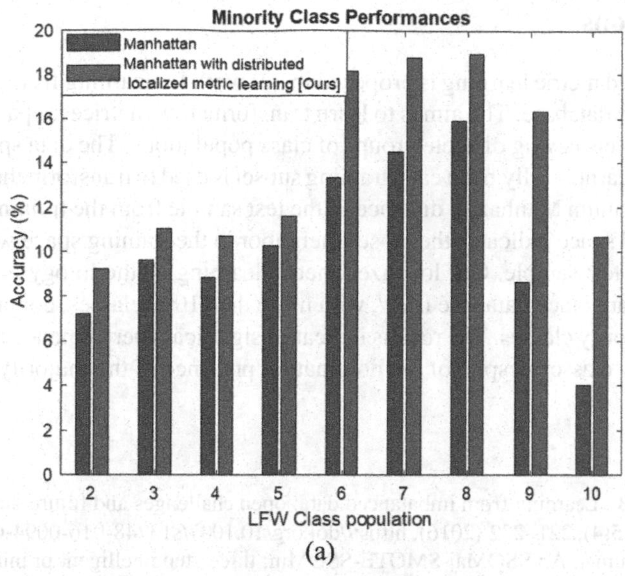

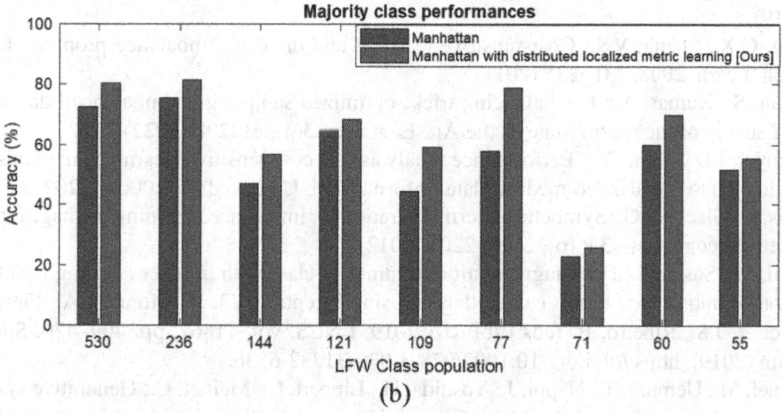

Fig. 6. (a, b) Minority and majority class performances using the Manhattan distance classifier, with and without the proposed metric learning scheme

Overall, in our work, metric learning is introduced as an efficient solution for learning from large multi-class imbalanced datasets. This is achieved by localizing the metric learning to training subspaces having similar class populations. Our approach, notably, improved the classification performance of minority classes (Fig. 6(b)) that are usually discarded by the current deep learning techniques [34] due to inadequate training examples in the LFW minority classes to train deep neural networks.

5 Conclusions

A novel localized metric learning is proposed in our work for learning from an extremely imbalanced face database. The aim is to learn transformation matrices separately for different class subsets having discrete groups of class populations. The data space transformation metric learnt locally from each training subset is used to transform the entire input space. The minimum Manhattan distance of the test sample from the training samples in the transformed space indicates the closest neighbor in the training space whose label is assigned to the test sample. Our localized metric learning methodology is applied successfully on a large face database LFW, with more than 1000 classes, containing a large quantity of minority classes. The results indicate a significant performance improvement for the minority classes despite of the dominating presence of the majority classes.

References

1. Krawczyk, B.: Learning from imbalanced data: open challenges and future directions. Prog. Artif. Intell. **5**(4), 221–232 (2016). https://doi.org/10.1007/s13748-016-0094-0
2. Susan, S., Kumar, A.: SSOMaj-SMOTE-SSOMin: three-step intelligent pruning of majority and minority samples for learning from imbalanced datasets. Appl. Soft Comput. **78**, 141–149 (2019)
3. Ling, C.X., Sheng, V.S.: Cost-sensitive learning and the class imbalance problem. Encycl. Mach. Learn. **2008**, 231–235 (2011)
4. Susan, S., Kumar, A.: The balancing trick: optimized sampling of imbalanced datasets—a brief survey of the recent State of the Art. Eng. Rep. **3**(4), e12298 (2021)
5. Mienye, I.D., Sun, Y.: Performance analysis of cost-sensitive learning methods with application to imbalanced medical data. Inform. Med. Unlocked **25**, 100690 (2021)
6. Piras, L., Giacinto, G.: Synthetic pattern generation for imbalanced learning in image retrieval. Pattern Recogn. Lett. **33**(16), 2198–2205 (2012)
7. Saini, M., Susan, S.: Data augmentation of minority class with transfer learning for classification of imbalanced breast cancer dataset using inception-V3. In: Morales, A., Fierrez, J., Sánchez, J.S., Ribeiro, B. (eds.) IbPRIA 2019. LNCS, vol. 11867, pp. 409–420. Springer, Cham (2019). https://doi.org/10.1007/978-3-030-31332-6_36
8. Rezaei, M., Uemura, T., Näppi, J., Yoshida, H., Lippert, C., Meinel, C.: Generative synthetic adversarial network for internal bias correction and handling class imbalance problem in medical image diagnosis. In: Medical Imaging 2020: Computer-Aided Diagnosis, vol. 11314, p. 113140E. International Society for Optics and Photonics (2020)
9. Rezaei, M., Yang, H., Meinel, C.: Recurrent generative adversarial network for learning imbalanced medical image semantic segmentation. Multimedia Tools Appl. **79**(21–22), 15329–15348 (2019). https://doi.org/10.1007/s11042-019-7305-1
10. Susan, S., Kumar, A.: DST-ML-EkNN: data space transformation with metric learning and elite k-nearest neighbor cluster formation for classification of imbalanced datasets. In: Chiplunkar, N.N., Fukao, T. (eds.) Advances in Artificial Intelligence and Data Engineering. AISC, vol. 1133, pp. 319–328. Springer, Singapore (2021). https://doi.org/10.1007/978-981-15-3514-7_26
11. Sukarna Barua, M., Islam, M., Yao, X., Murase, K.: MWMOTE--majority weighted minority oversampling technique for imbalanced data set learning. IEEE Trans. Knowl. Data Eng. **26**(2), 405–425 (2014). https://doi.org/10.1109/TKDE.2012.232
12. Tao, X., et al.: Real-value negative selection over-sampling for imbalanced data set learning. Expert Syst. Appl. **129**, 118–134 (2019)

13. Liu, T., Zhu, X., Pedrycz, W., Li, Z.: A design of information granule-based under-sampling method in imbalanced data classification. Soft. Comput. **24**(22), 17333–17347 (2020). https://doi.org/10.1007/s00500-020-05023-2
14. Moutafis, P., Leng, M., Kakadiaris, I.A.: An overview and empirical comparison of distance metric learning methods. IEEE Trans. Cybern. **47**(3), 612–625 (2016)
15. Feng, L., Wang, H., Jin, B., Li, H., Xue, M., Wang, L.: Learning a distance metric by balancing kl-divergence for imbalanced datasets. IEEE Trans. Syst. Man Cybern. Syst. **49**(12), 2384–2395 (2018)
16. Wang, N., Zhao, X., Jiang, Y., Gao, Y.: Iterative metric learning for imbalance data classification. In: Proceedings of the 27th International Joint Conference on Artificial Intelligence, pp. 2805–2811 (2018)
17. Wang, C., Xin, C., Zili, X.: A novel deep metric learning model for imbalanced fault diagnosis and toward open-set classification. Knowl.-Based Syst. **220**, 106925 (2021)
18. Kulis, B.: Metric learning: a survey. Found. Trends Mach. Learn. **5**(4), 287–364 (2012)
19. Weinberger, K.Q., Blitzer, J., Saul, L.K.: Distance metric learning for large margin nearest neighbor classification. In: Advances in Neural Information Processing Systems, pp. 1473–1480 (2006)
20. Weinberger, K.Q., Saul, L.K.: Distance metric learning for large margin nearest neighbor classification. J. Mach. Learn. Res. **10**(2) (2009)
21. Susan, S., Kumar, A.: Learning data space transformation matrix from pruned imbalanced datasets for nearest neighbor classification. In: 2019 IEEE 21st International Conference on High Performance Computing and Communications; IEEE 17th International Conference on Smart City; IEEE 5th International Conference on Data Science and Systems (HPCC/SmartCity/DSS), pp. 2831–2838. IEEE (2019)
22. Ghanavati, M., Wong, R.K., Chen, F., Wang, Y., Perng, C.-S.: An effective integrated method for learning big imbalanced data. In: 2014 IEEE International Congress on Big Data, pp. 691–698. IEEE (2014)
23. Tan, M., Wang, B., Zhaohui, W., Wang, J., Pan, G.: Weakly supervised metric learning for traffic sign recognition in a LIDAR-equipped vehicle. IEEE Trans. Intell. Transp. Syst. **17**(5), 1415–1427 (2016)
24. Jing, X.-Y., et al.: Multiset feature learning for highly imbalanced data classification. IEEE Trans. Pattern Anal. Mach. Intell. **43**(1), 139–156 (2019)
25. Huang, G.B., Mattar, M., Berg, T., Learned-Miller, E.: Labeled faces in the wild: a database for studying face recognition in unconstrained environments (2008)
26. Dalal, N., Triggs, B.: Histograms of oriented gradients for human detection. In: 2005 IEEE Computer Society Conference on Computer Vision and Pattern Recognition (CVPR 2005), vol. 1, pp. 886–893. IEEE (2005)
27. Dadi, H.S., Pillutla, G.K.M.: Improved face recognition rate using HOG features and SVM classifier. IOSR J. Electron. Commun. Eng. **11**(04), 34–44 (2016)
28. Chen, D., Cao, X., Wen, F., Sun, J.: Blessing of dimensionality: high-dimensional feature and its efficient compression for face verification. In: Proceedings of the IEEE Conference on Computer Vision and Pattern Recognition, pp. 3025–3032 (2013)
29. Abuzneid, M.A., Mahmood, A.: Enhanced human face recognition using LBPH descriptor, multi-KNN, and back-propagation neural network. IEEE Access **6**, 20641–20651 (2018)
30. Bhele, S.G., Mankar, V.H.: Recognition of faces using discriminative features of LBP and HOG descriptor in varying environment. In: 2015 International Conference on Computational Intelligence and Communication Networks (CICN), pp. 426–432. IEEE (2015)
31. Fu, Y., Li, Z., Huang, T.S., Katsaggelos, A.K.: Locally adaptive subspace and similarity metric learning for visual data clustering and retrieval. Comput. Vis. Image Underst. **110**(3), 390–402 (2008)

32. Shen, P., Xin, D., Li, C.: Distributed semi-supervised metric learning. IEEE Access **4**, 8558–8571 (2016)
33. Li, J., Lin, X., Rui, X., Rui, Y., Tao, D.: A distributed approach toward discriminative distance metric learning. IEEE Trans. Neural Netw. Learn. Syst. **26**(9), 2111–2122 (2014)
34. Taigman, Y., Yang, M., Ranzato, M.A., Wolf, L.: Deepface: closing the gap to human-level performance in face verification. In: Proceedings of the IEEE Conference on Computer Vision and Pattern Recognition, pp. 1701–1708 (2014)
35. Susan, S., Kaushik, A.: Weakly supervised metric learning with majority classes for large imbalanced image dataset. In: Proceedings of the 2020 the 4th International Conference on Big Data and Internet of Things, pp. 16–19 (2020)

Top-*k* Dominating Queries on Incremental Datasets

Jimmy Ming-Tai Wu[1(✉)], Ke Wang[1], and Jerry Chun-Wei Lin[2]

[1] College of Computer Science and Engineering, Shandong University of Science and Technology, Qingdao, China
wmt@wmt35.idv.tw

[2] Department of Computer Science, Electrical Engineering and Mathematical Sciences, Western Norway University of Applied Sciences, Bergen, Norway
jerrylin@ieee.org

Abstract. Top-k dominance (TKD) query for incomplete datasets is a popular preference query for incomplete data, which analyzes the dominance relationships among objects in a dataset by a dominance method to reveal the top-k most valuable information in the dataset. At present, in-depth research has been conducted on this topic, and efficient query algorithms based on various pruning strategies have been proposed, as well as optimization algorithms based on a distributed computing framework for processing large-scale datasets. With the advent of the information age, data update iterations are accelerated, and in the face of dynamically updated data, the traditional TKD query algorithm based on static data can no longer meet our needs, and an efficient algorithm based on the dynamically updated data set environment is needed. In this paper, we conduct an in-depth study on the TKD query problem for dynamically updated incomplete datasets, and propose a dynamic update parallel algorithm based on MapReduce framework. The algorithm utilizes the query results of historical datasets, avoids the repeated analysis of the dominant relationships between historical objects, optimizes the computation process, reduces the space occupation, and proves through experiments that the dynamic update algorithm has more obvious advantages compared with the traditional algorithm.

Keywords: Incomplete datasets · Top-k dominance query · Static datasets · Dynamic update · Parallel compute

1 Introduction

In the era of big data, data integrity among data quality issues has attracted a lot of attention. The data integrity problem may be the problem of missing data due to data hiding, transmission signal loss or other reasons. Incomplete datasets are common in real life, such as MovieLens, a movie evaluation system

Supported by Shandong Provincial Natural Science Foundation (ZR201911150391).

U. K. Rage et al. (Eds.): DASFAA 2022 Workshops, LNCS 13248, pp. 79–88, 2022.
https://doi.org/10.1007/978-3-031-11217-1_6

with several classic works, each movie can be represented by a tuple, and the user's evaluation of it as an element in the tuple, because there is no guarantee that every user of the evaluation has seen all the movies in the system, so there must be some missing elements in the movie tuple. Due to the rise of incomplete datasets, the research on incomplete datasets has gradually become hot, including the research on incomplete relational database models [2,8,11]; The research on skyline queries based on incomplete data [7,12,15], where [12] firstly proposed the idea of skyline queries on incomplete datasets and designed a skyline query algorithm (ISkyline) for incomplete datasets; An in-depth study of the indexing problem on incomplete datasets [3,17], in which two retrieval strategies for incomplete high-dimensional datasets are proposed and compared with exhaustive search, verifying that both retrieval strategies are efficient; And a study of the top-k query problem on incomplete datasets [9,22], the top-k query problem on asynchronous incomplete data streams is studied in the literature [9] and an efficient query algorithm based on an object pruning strategy is proposed.

TKD query is an emerging preference query that is based on top-k query and skyline query. TKD query avoids the setting of scoring function compared to top-k query and can determine the number of results returned by the query compared to skyline query. TKD query was originally proposed by Papadias *et al.* [18] and proposed a branch-and-bound skyline (BBS) algorithm based on TKD queries to solve data mining problems on complete datasets. Next, Yiu *et al.* [27] proposed a TKD query algorithm for multidimensional complete datasets, in which a novel data structure aR-tree was designed for efficient data traversal, and the algorithm was experimentally proven to be effective. Tiakas *et al.* [23] investigated the use of an asymptotic approach to TKD query was investigated and various pruning strategies were devised and verified that the algorithm possesses better performance. With the rise of incomplete datasets, the study of TKD queries has been extended to incomplete datasets [1,6,16,26], which includes the study of pruning strategies in performing TKD queries and how to implement TKD queries for large-scale incomplete datasets.

With the development of information technology, the update of data sets becomes frequent. How to efficiently perform data mining on dynamically updated data is a popular topic of much attention at present. The simplest way to solve this problem is to re-execute the traditional algorithm once for the updated dataset, but this solution not only does not make full use of the query results of the historical dataset, but also generates a large number of repeated calculation problems, which wastes a lot of time and space and causes untimely data query. Through in-depth research on this topic, a large number of related algorithms are proposed, including incremental association rule mining processing algorithms [4,5,10,13,20,24,25] and incremental sequence model mining algorithms [14,19], which dynamically update the dataset mining algorithm utilizes the original query results, reduces the number of scans of the database, and improves query efficiency. Compared to dynamically updated complete datasets, the study of TKD queries for dynamically updated incomplete datasets is more difficult, because the problem of uncertain missing data in the dataset has to be considered on top of dealing with dynamically complete datasets.

Based on the consideration of the above problems, this paper proposes a TKD query algorithm for incremental update of incomplete data based on MapReduce architecture, which divides the analysis and calculation of incremental update dataset into two parts, the update of historical object dominant scores and the calculation of new object dominant scores, the method makes full use of the query results of historical dataset, avoids the repeated calculation of data, and achieves an efficient query processing. Moreover, the algorithm is based on MapReduce parallel architecture, which decomposes the complex analysis and calculation tasks into multiple subtasks assigned to different nodes for parallel calculation, and can realize the analysis and processing of large-scale incomplete datasets.

2 Literature Review

This section describes top-k dominated queries for incomplete datasets and related work on dynamic incremental database data mining.

2.1 Top-k Dominance Query

Miao *et al.* [16] first started their research on the TKD query problem for incomplete data and proposed various algorithms (ESB, UBB, BIG, and IBIG) to solve the problem. ESB algorithm proposes the concept of bucket in order to apply the dominant transferability to TKD queries of incomplete data sets. It stores the objects in buckets according to the missing dimensions, and then prunes the objects in the buckets using the concept of k-skyband. UBB algorithm analyzes the dominance score of object in each dimension and uses the minimum dominance number as the dominance upper bound of the object and filters the objects in the dataset by this dominance upper bound first, avoiding a large number of unnecessary computational processes. BIG introduces the bitmap used for querying complete datasets to incomplete data sets, and uses the bitmap index to obtain the dominance score of objects by fast bit-by-bit calculation, which has a greater improvement in time consumption compared with the first two algorithms, but the algorithm will occupy a lot of memory and disk space due to the bitmap index and the storage of object sets. IBIG is an optimization of BIG algorithm, which effectively solves the problem of storage space occupation of BIG algorithm through bitmap compression technology and chunking strategy.

Ezatpoor *et al.* [6] found that most of the existing algorithms are effective for TKD queries on small incomplete datasets, but when the data size becomes large, the query task becomes difficult, and even traditional single machine query algorithms may fail the task due to lack of memory or disk space. Ezatpoor *et al.* performed the above problem in depth and proposed an algorithm based on MapReduce distributed framework (MRBIG). MRBIG is based on bitmap indexing, which decomposes some of the tasks originally executed on a single machine into multiple subtasks and distributes them to different nodes for parallel computation, thus realizing the analytical computation of large-scale data sets.

Wu *et al.* [26] proposed two high-performance algorithms (EHBIG, IEHBIG) based on MapReduce architecture to improve the efficiency of TKD queries for large-scale incomplete datasets. EHBIG proposes the concept of maximum domination number of objects according to the domination relationship and designs an efficient pruning strategy to reduce the computational effort. However, EHBIG completes the query through MapReduce iterations, and when there are thousands of objects in the dataset, EHBIG may have to iterate through thousands of MapReduce tasks, which generates a large amount of resource consumption. IEHBIG algorithm computes the dominance scores of all objects in the dataset by one MapReduce task, avoiding the problem of resource wastage due to MapReduce iterations, but the algorithm is not designed with a pruning strategy and requires more memory space. The two algorithms proposed in the article have their own advantages and disadvantages, and need to be chosen in conjunction with reality.

2.2 Dynamic Update of Data Mining

Cheung *et al.* [4] firstly studied the maintenance of association rules in incremental datasets and proposed the association rule mining algorithm (FUP) for incremental datasets, which is based on the logic of the idea of Apriori algorithm to determine whether to rescan the historical datasets by mining the added datasets. FUP algorithm considers the association rule maintenance problem of incremental datasets, but does not consider the existence of deletion of old things in the update of datasets in the actual environment, so the FUP_2 algorithm [5] is proposed to deal with the association rule maintenance problem of dynamic datasets in the case of deletion of old transactions.

Hong *et al.* [10] first proposed the concept of pre-large itemsets, which are filtered by two given thresholds, similar to the concept of buffers. When the number of new transactions is within the calculated threshold, only the support numbers of frequent and pre-large itemsets need to be updated, reducing the amount of computation, and when the cumulative number of new transactions exceeds the calculated maximum threshold, the original database is rescanned again. This algorithm can effectively handle the situation where there are fewer transactions in the historical data set and more transactions in the new data set, which is more suitable for handling real-life cases.

Saleti *et al.* [21] studied sequence pattern mining in incremental update datasets and proposed a parallel algorithm based on MapReduce architecture to achieve efficient sequence pattern mining in large-scale incremental datasets in response to the trend of big data. Wu*et al.* [25] applied the concept of pre-large to find high average-utility patterns in incremental update data sets, and proposed APHAUI algorithm, which achieves efficient querying of high average-utility patterns by setting two upper bounds of pub and lead-pub.

3 Query Base Preparation

In this section, the dominance relation, object score, and top-k dominance query for incomplete datasets are introduced. First, Table 1 gives a sample movie rating dataset consisting of five movies by four users, and the rating of movie m_1 is shown by the tuple (3,4,-,2), where the rating value of the third dimension is missing.

Table 1. Recommendation system dataset example.

ID	Movie name	Audience ratings			
		a_1	a_2	a_3	a_4
m_1	The Lion King (1994)	3	4	-	2
m_2	Forrest Gump (1994)	4	-	-	3
m_3	The Blind Side (2009)	-	-	2	-
m_4	The Martian (2015)	-	4	3	2
m_5	Zootopia (2016)	3	-	2	3

Definition 1. ***A dominance relation between incomplete objects***
Objects p and q in the incomplete dataset S, comparable dimension IDs are recorded in the set C_{pq} *and* C_{pq} $!= \emptyset$. *p dominates q (record as* $p \prec q$*) if it satisfies* $\forall\ d_i \in C_{pq}$, $p[d_i] \leq q[d_i]$ *and* $\exists\ d_j \in C_{pq}$, $p[d_j] < q[d_j]$.

Definition 2. ***Object score***
For an object p in the incomplete dataset S, the number of objects in the dataset S that p dominates is recorded as the score of p, denoted as Score(p), according to the dominance relationship in Definition 1.

Definition 3. ***Top-k domination query***
Calculate the score of each object in the dataset S and return the top-k objects with the highest scores, denoted as S_k.

4 Algorithm Description

This section presents the incremental update TKD query algorithm for the incomplete dataset proposed in the paper. The definition of object dominance relationship shows that when two objects are analyzed for dominance relationship, the objects are compared in terms of values by dimension. So the algorithm first processes the storage format of the added data and stores the multidimensional incomplete data in HBase, a distributed file storage database, by dimension to facilitate subsequent comparisons. After the preprocessing of the storage format for the new data set, the dimension values saved in the historical data

Table 2. Recommendation system new data example.

ID	Movie name	Audience ratings			
		a_1	a_2	a_3	a_4
m_6	La La Land (2016)	4	4	2	3
m_7	Uncle Drew (2018)	4	3	-	2

set in HBase are updated. For example, when the historical dataset shown in Table 1 welcomes the addition of new data, as shown in Table 2, the dataset in Table 2 is first preprocessed and then the information saved in HBase is updated, and the data in HBase after the update is shown in Table 3 (the missing data are represented by 0).

Table 3. HBase storage data example.

Row key	Column family:Qualifier	Column value
d_1	info:1	3, 4, 0, 0, 3 + 4, 4
d_2	info:2	4, 0, 0, 4, 0 + 4, 3
d_3	info:3	0, 0, 2, 3, 2 + 2, 0
d_4	info:4	2, 3, 0, 2, 3 + 3, 2

After updating the information in HBase, the next step is to use the MapRedcue architecture to calculate the object domination number, which is divided into two parts, namely, the update of the domination score of historical objects and the calculation of the domination score of new objects, which can be subdivided into the domination relationship of historical objects to new objects, the domination relationship of new objects to historical objects, and the domination relationship between new objects. The three parts are shown in Fig. 1. Compared with the traditional algorithm for static data sets, the incremental update algorithm updates the dominance scores of historical objects using the query results of historical data sets, which avoids the repeated calculation of dominance relationships among historical objects and improves the query performance.

The next step is to introduce the incremental update algorithm in detail by example. Map stage reads the column with RowKey 1 in HBase, and analyzes the dominance of the first dimension by this column, firstly, the evaluation value of the historical object is compared with the evaluation value of the new object in turn, and the bit vectors $[Q_1]$ and $[T_1]$ are obtained. When the evaluation value of the new object is missing or not better than the evaluation value of the historical object, it is recorded as 1, otherwise it is recorded as 0, and the comparison result is recorded by the bit vector $[Q_1]$; When the evaluation value of the new object is missing or equal to the evaluation value of the historical object, it is recorded as 1, otherwise it is recorded as 0, and it is expressed by

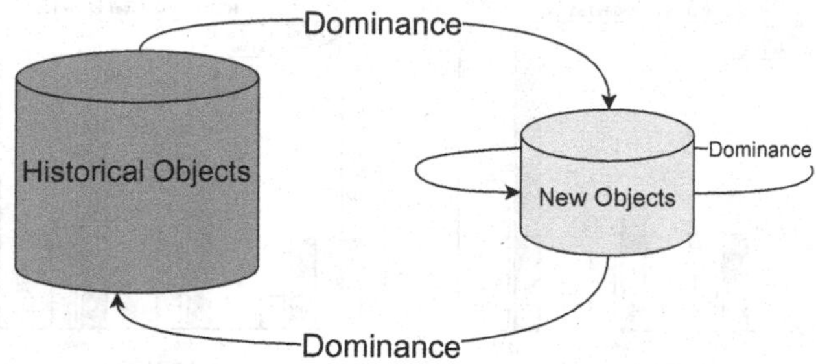

Fig. 1. Incremental update TKD query algorithm calculation logic for incomplete datasets.

the bit vector $[T_1]$. Then, the evaluation value of the new object is compared with all the evaluation values in the dimension in turn, and the bit vectors $[Q_1]$ and $[T_1]$ of the new object are calculated by the above-mentioned comparison method to obtain the dominance of the new object over all the objects in the dimension. Next, the remaining dimensions are compared in turn. Reduce stage takes the intersection of $[Q_i]$ and $[T_i]$ of each object to obtain the bit vectors $[Q]$ and $[T]$ of the objects, and then obtains the set Q and T of the objects. From the above analysis, it is known that the set Q of object p contains objects that are dominated by p, objects that cannot be compared with p, and objects that have no dominance relationship with p, while the set T of object p contains objects that cannot be compared with p and objects that do not have a dominant relationship with p. Therefore, we can obtain $Score(p) = |Q - T|$. For example, the bit vector $[Q_1] = 11$, $[Q_2] = 10$, $[Q_4] = 11$, $[T_1] = 00$, $[T_2] = 10$, and $[T_4] = 01$ for object M_1, and the bit vector $[Q] = 10$, $[T] = 00$, set $Q = [M_6]$, set $T = \emptyset$, so the dominance of M_1 over the new object is M_1 dominates M_6, and finally the incremental update algorithm obtains $Score(M_1) = Score_{History} + Score_{New} = 3$.

5 Experiment and Analysis

Through the introduction of the algorithm in Sect. 4, it can be analyzed that compared with the traditional algorithm based on static data, this algorithm reduces a lot of calculation process and improves the query efficiency. In this section, experiments are conducted in a Hadoop cluster built by five PCs and one switch to compare the query performance of the algorithm. The operating system of the PC is Ubuntu 18.10 and the Hadoop Framework version is 2.8.5. The experimental results are shown in Fig. 2 below. Through the experimental results, it can be found that the query performance of the algorithm proposed in this paper is significantly better than the traditional algorithm on the TKD query of dynamic incremental incomplete data sets.

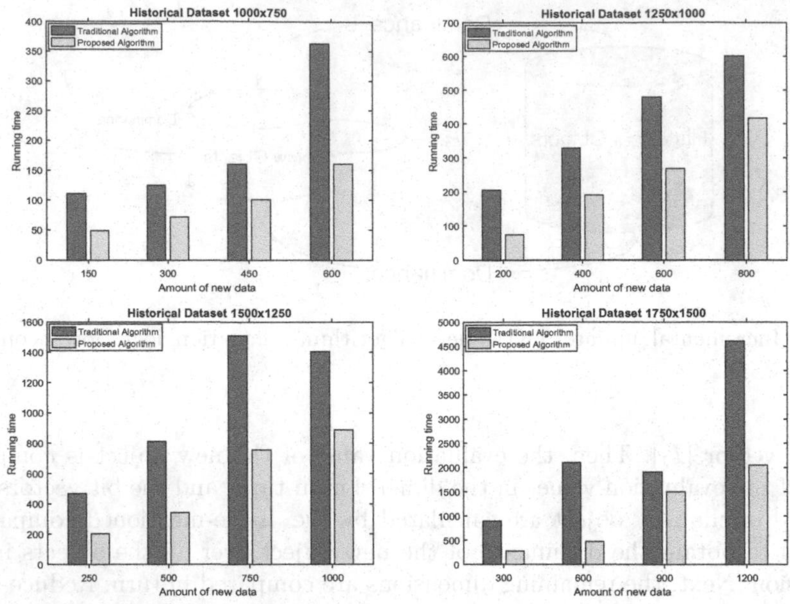

Fig. 2. Runtime under different database sizes.

6 Conclusion

In this paper, we propose an incremental update algorithm based on incomplete datasets, which makes full use of the query results of previous historical datasets when calculating the dominance relationships between historical objects and new objects, avoiding the repeated analysis of dominance relationships between historical objects and reducing a large number of complex calculations. The algorithm is implemented by MapReduce, which decomposes complex analysis and computation tasks into multiple subtasks distributed to different nodes in the cluster for parallel execution, and demonstrates through experimental results that the algorithm achieves efficient query processing of dynamically updated incomplete datasets.

In addition, in the actual production activities, the operations on the database are not only the addition of data, but also the deletion and modification operations are very common and important. So, we will explore these research directions more deeply next.

References

1. Amagata, D., Sasaki, Y., Hara, T., Nishio, S.: Efficient processing of top-k dominating queries in distributed environments. World Wide Web **19**(4), 545–577 (2015). https://doi.org/10.1007/s11280-015-0340-6

2. Antova, L., Koch, C., Olteanu, D.: From complete to incomplete information and back. In: Proceedings of the 2007 ACM SIGMOD International Conference on Management of Data, pp. 713–724 (2007)
3. Canahuate, G., Gibas, M., Ferhatosmanoglu, H.: Indexing incomplete databases. In: Ioannidis, Y., et al. (eds.) EDBT 2006. LNCS, vol. 3896, pp. 884–901. Springer, Heidelberg (2006). https://doi.org/10.1007/11687238_52
4. Cheung, D.W., Han, D.-W., Ng, V.T., Wong, C.Y.: Maintenance of discovered association rules in large databases: an incremental updating technique. In: Proceedings of the Twelfth International Conference on Data Engineering, pp. 106–114. IEEE (1996)
5. Cheung, D.W., Lee, S.D., Kao, B.: A general incremental technique for maintaining discovered association rules. In: Database Systems For Advanced Applications 1997, pp. 185–194. World Scientific (1997)
6. Ezatpoor, P., Zhan, J., Wu, J.M.-T., Chiu, C.: Finding top-k dominance on incomplete big data using mapreduce framework. IEEE Access **6**, 7872–7887 (2018)
7. Gao, Y., Miao, X., Cui, H., Chen, G., Li, Q.: Processing k-skyband, constrained skyline, and group-by skyline queries on incomplete data. Expert Syst. App. **41**(10), 4959–4974 (2014)
8. Green, T.J., Tannen, V.: Models for incomplete and probabilistic information. In: Grust, T., et al. (eds.) EDBT 2006. LNCS, vol. 4254, pp. 278–296. Springer, Heidelberg (2006). https://doi.org/10.1007/11896548_24
9. Haghani, P., Michel, S., Aberer, K.: Evaluating top-k queries over incomplete data streams. In: Proceedings of the 18th ACM conference on Information and Knowledge Management, pp. 877–886 (2009)
10. Hong, T.-P., Wang, C.-Y., Tao, Y.-H.: A new incremental data mining algorithm using pre-large itemsets. Intell. Data Anal. **5**(2), 111–129 (2001)
11. Imieliński, T., Jr, W.L.: Incomplete information in relational databases. In Readings in Artificial Intelligence and Databases, pp. 342–360. Elsevier (1989)
12. Khalefa, M.E., Mokbel, M.F., Levandoski, J.J.: Skyline query processing for incomplete data. In: 2008 IEEE 24th International Conference on Data Engineering, pp. 556–565. IEEE (2008)
13. Lee, C.-H., Lin, C.-R., Chen, M.-S.: Sliding-window filtering: an efficient algorithm for incremental mining. In: Proceedings of the Tenth International Conference on Information and Knowledge Management, pp. 263–270 (2001)
14. Lin, M.-Y., Lee, S.-Y.: Incremental update on sequential patterns in large databases. In: Proceedings of Tenth IEEE International Conference on Tools with Artificial Intelligence (Cat. No. 98CH36294), pp. 24–31. IEEE (1998)
15. Lofi, C., Maarry, K.E., Balke, W.T.: Skyline queries in crowd-enabled databases. In: Proceedings of the 16th International Conference on Extending Database Technology, pp. 465–476 (2013)
16. Miao, X., Gao, Y., Zheng, B., Chen, G., Cui, H.: Top-k dominating queries on incomplete data. IEEE Trans. Knowl. Data Eng. **28**(1), 252–266 (2015)
17. Ooi, B.C., Goh, C.H., Tan, K.L.: Fast high-dimensional data search in incomplete databases. In: VLDB, pp. 357–367 (1998)
18. Papadias, D., Tao, Y., Fu, G., Seeger, B.: Progressive skyline computation in database systems. ACM Trans. Database Syst. (TODS) **30**(1), 41–82 (2005)
19. Parthasarathy, S., Zaki, M.J., Ogihara, M., Dwarkadas, S.: Incremental and interactive sequence mining. In: Proceedings of the Eighth International Conference on Information and Knowledge Management, pp. 251–258 (1999)
20. Pudi, V., Haritsa, J.R.: Quantifying the utility of the past in mining large databases. Inf. Syst. **25**(5), 323–343 (2000)

21. Saleti, S., Subramanyam, R.: A mapreduce solution for incremental mining of sequential patterns from big data. Expert Syst. App. **133**, 109–125 (2019)
22. Soliman, M.A., Ilyas, I.F., Ben-David, S.: Supporting ranking queries on uncertain and incomplete data. VLDB J. **19**(4), 477–501 (2010)
23. Tiakas, E., Papadopoulos, A.N., Manolopoulos, Y.: Progressive processing of subspace dominating queries. VLDB J. **20**(6), 921–948 (2011)
24. Wang, K.: Discovering patterns from large and dynamic sequential data. J. Intell. Inf. Syst. **9**(1), 33–56 (1997)
25. Wu, J.M.-T., Teng, Q., Lin, J.C.-W., Cheng, C.-F.: Incrementally updating the discovered high average-utility patterns with the pre-large concept. IEEE Access **8**, 66788–66798 (2020)
26. Wu, J.M.-T., Wei, M., Wu, M.-E., Tayeb, S.: Top-k dominating queries on incomplete large dataset. J. Supercomput. **78**, 1–22 (2021)
27. Yiu, M.L., Mamoulis, N.: Efficient processing of top-k dominating queries on multi-dimensional data. VLDB **7**, 483–494 (2007)

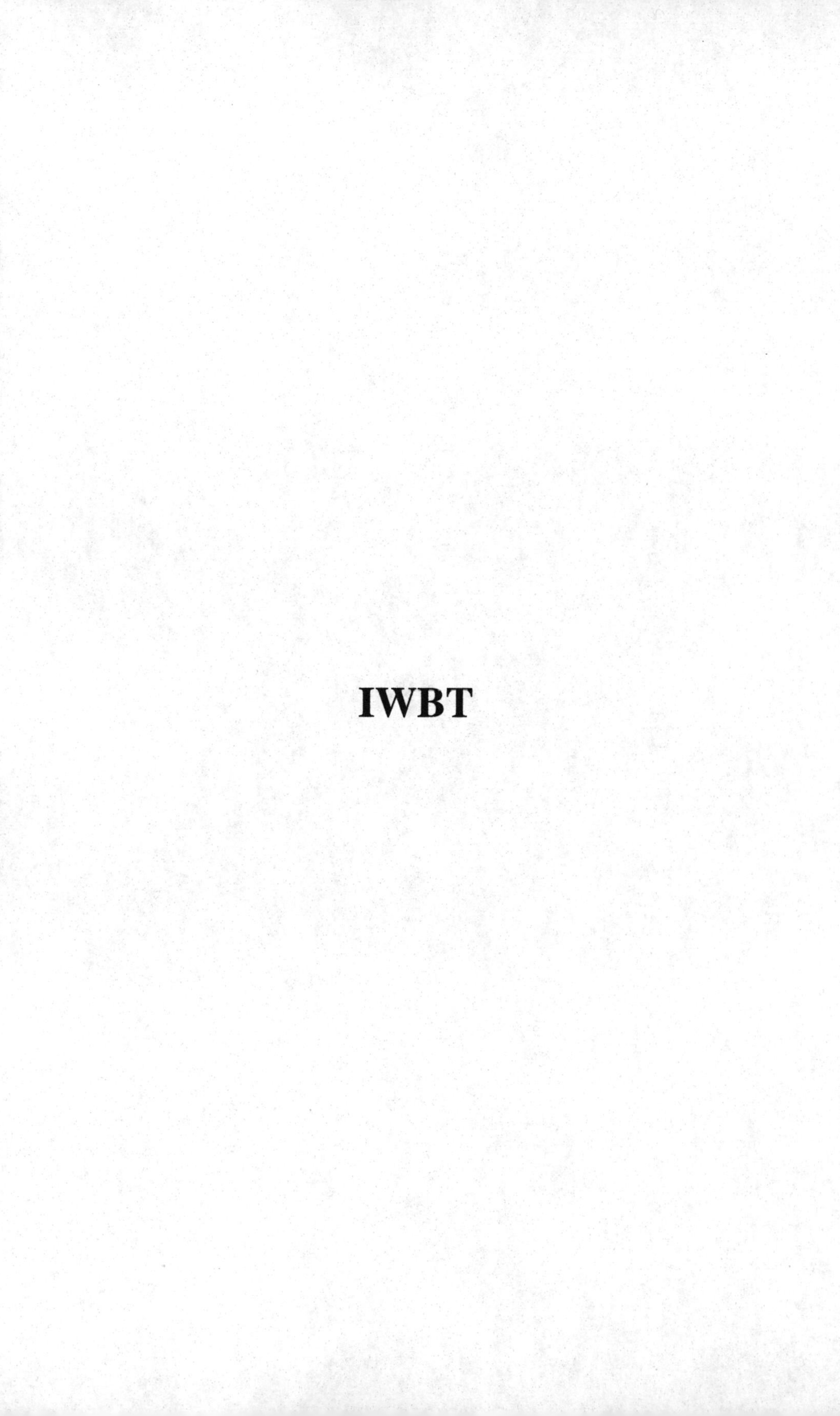
IWBT

Collaborative Blockchain Based Distributed Denial of Service Attack Mitigation Approach with IP Reputation System

Darshi Patel(✉) and Dhiren Patel

Department of Computer Science, Gujarat Vidyapith, Ahmedabad 380014, Gujarat, India
pdarshi94@gmail.com

Abstract. Due to the high accessibility of the internet and unsecured internet-connected devices, Distributed Denial of Service attacks (DDoS) become devastating among all cyber-attacks. Attack complexity and sophistication have been increasing day by day. Many mitigation techniques and methods are available through which the risk of DDoS attacks will be reduced at a significant level. It is ideal to implement distributed solutions to mitigate DDoS attacks because of distributed nature of DDoS attacks. Many collaborators who share the same interest, participate in DDoS attack mitigation in a collaborative DDoS mitigation approach. But, creating a collaboration in a transparent, secured, and distributed way is still a challenge. Blockchain technology presents the ideology of decentralized, distributed, secured, and immutable ledger. Blockchain is an emerging technology to mitigate many cyber threats because of its fundamental characteristics like decentralization, immutability, authenticity, trust management, and integrity. In DDoS attacks mitigation, Blockchain technology is utilized in many different ways but, still in the initial phase. Due to all these characteristics of Blockchain, Blockchain become the ideal choice to create collaboration among peers. As well as it would be also useful to signalize the attack among various collaborators. Accuracy of shared IP addresses is equally important in the mitigation of DDoS Attacks and Botnet identification. IP Reputation systems are generally designed to get accurate information about IP addresses. In this paper, we present how blockchain is used collaboratively to mitigate DDoS attacks in various domains. This paper aim at providing a brief overview of DDoS mitigation strategies, the architecture of the proposed DDoS mitigation solution. This paper also discusses the integration of IP reputation scheme for Blacklisted IP addresses with Blockchain to build transparency and accuracy of data. Accurate IP addresses data would be helpful in not only DDoS Attack Mitigation but, in many attacks related to IP Addresses, identification of the Bots, Intrusion prevention system (IPS), and Intrusion detection system (IDS).

Keywords: Blockchain technology · DDoS attacks · Mitigation · IP reputation · Collaborative approach

U. K. Rage et al. (Eds.): DASFAA 2022 Workshops, LNCS 13248, pp. 91–103, 2022.
https://doi.org/10.1007/978-3-031-11217-1_7

1 Introduction

Distributed Denial of Service attacks is one of the most adverse threats which target the availability of system and services. In this attack, attackers often flood the system or system resources with a high volume of legitimate or illegitimate requests in the resulting system that becomes unavailable for their legitimate users [10]. DDoS attacks are larger and extended forms of DoS attacks. In DoS attacks, attack-traffic is originated from one particular source whereas multiple sources are used to attack the target system in DDoS attacks. The DDoS attack is not limited to the flood of the services or available bandwidth of the system but, it also consumes system memory and processing power. Huge financial losses, Reputation of the system, and loss of prospective customers are the adverse effects of DDoS attacks. And sometimes DDoS attack is used as a tool to spread the malware in the system. DDoS attacks themselves do not steal the information but, some DDoS attacks are used as a facade for other malicious activities like dismantling system firewalls and weakening their security code for future attacks and gain understanding about system's security policy while DDoS attacks target the system and whole system resources are occupied to counter the ongoing DDoS attacks.

Blocking traffic from malicious sources becomes one of the prominent solutions to counter DDoS attacks. It is also beneficial to the legitimate user by getting service on time without any delay and improves the overall performance of the system.

There are many solutions available to mitigate DDoS attacks. But, solutions are vary depending upon the different types of attacks. DDoS attacks can be classified as Volume-based Attacks, Protocol Attacks, and Application Layer Attacks. Bandwidth saturation mainly happens in volumetric or volume-based attacks. The intensity of attacks is measured in bits per second (Bps). Protocol attacks are performed by using the flow or vulnerability of existing communication protocols. And occupy the server resources and communication devices. This type of attack is measured in bits per second (Bps). In application-layer attacks, attackers flood the web-server or services by web requests which are measured in packets per second (Pps) [1, 10].

The rest of the paper has seven sections. Section 2 provides the introduction and fundamentals of Blockchain technology. Moreover, a detailed overview of DDoS mitigation approaches, techniques, and Reputation system are also covered in this section. Problem Description is discussed in Sect. 3. Section 4 details the related work of an existing system. And in Sect. 5 research questions are described. Proposed Architecture, integration of Reputation system with proposed Architecture, implementation strategy, and advantages of the proposed system are discussed in Sect. 6. The research paper is concluded in Sect. 7.

2 Theoretical Background

2.1 Blockchain Technology – Brief

The idea of Blockchain is not limited to a chain of Blocks and its significance is not limited to one type of Data structure. But, Blockchain is a distributed ledger technology that is cryptographically secured and maintains trust between multiple domains without any special need of central authority. Blockchain provides immutability, security, reliability and authenticity, integrity through cryptography and digital signature to the stored data. These security elements are never found in any other record-keeping system.

Blockchain is originally invented for the transfer of tokens in a peer-to-peer network. Unlike traditional system blockchain do not require any third party to manage trust between sender and receiver of the token. The work which is performed by the third party in the traditional system is being done with the help of consensus protocols.

Blockchain is different from other distributed ledgers because of five elements:

Block. Block is the smallest key unit of Blockchain which stores numbers of transactions in it and is interlinked with other Blocks which creates the chain of Blocks. The first Block of Blockchain is called the Genesis block. All the transactions and hash of the previous block are used to compute block hash which becomes a pointer to the newly generated block.

Transactions. Transactions are the data stored in the blockchain. After validating by the miner, the transaction is officially included in the block. Transactions are immutable after it is recorded into the blockchain, no deletion and no modification have happened. These records are accessible throughout without any authentication. This shows the beauty of blockchain technology.

Peer to Peer Network. Generally, a Blockchain network is a peer-to-peer network where no fixed protocol, topology. This peer-to-peer network usually operates on ad-hoc protocol and is implemented on ad-hoc topology. All the nodes of the network are considered equally and a new node can join at any time and leave the network at any time. All the node in the network has a full copy of Blockchain. So, it prevents blockchain from the single point of failure.

Mining. Mining is the process of generating a new block of blockchain by solving a computation puzzle. A miner is a person who performs all these tasks to mine a block and is rewarded by the pre-decided token. Miners validate the transactions and included them in the blocks. Mining decides the functioning and implementation of blockchain and is a core element of Blockchain.

Consensus. Unlike in traditional systems, Blockchain does not have any central authority that manages the level of trust between parties involved in the transaction. So, in blockchain this trust management work is implemented with the help of a consensus mechanism. The validity of any transaction or record is achieved through the consensus of all the miners. Proof of Work (PoW), Proof of Stack (PoS) are well-known examples of consensus algorithms. Block mining reward is also decided based on the consensus mechanism implemented on the blockchain.

One of the well-known applications of blockchain is crypto-currency. But, use cases and practical implementations of blockchain are not limited to it. Some areas like supply chain, energy, asset management, health care, banking, voting also utilize Blockchain to fulfill different purposes [4, 11].

2.2 Overview of DDoS Mitigation Approaches and Techniques

DDoS mitigation approach is categorized in Collaborate approach and Non-collaborative approach. These categories are formed based on actors involved in DDoS mitigation. Generally, all the DDoS mitigation techniques and strategies fall into these categories.

Collaborative DDoS Mitigation Approach. The concept of collaborative DDoS mitigation is coming from the nature of DDoS attacks. The intensity and traffic volume of DDoS attack is tremendous as well as attackers become sophisticated that use various attack techniques to make the attack undetectable. So, to counter the adverse effect of DDoS attacks collaboration between members of the same interest is required. In collaborative DDoS mitigation, participating networks and organizations provide joint effort in terms of attack prevention, detection, defense, and mitigation. Attack traffic is blocked near the origin of the attack due to collaborative efforts is become a major advantage of this approach. Secondly, Repetitive actions and efforts in DDoS attacks detection are avoided and the attack detection and mitigation process become fast and efficient. Finally, through attack signalization, all other participating members of collaboration become alert about ongoing DDoS attacks and modify their attack mitigation strategies accordingly.

Maintaining and establishing trust between collaborative parties is most important, which becomes the biggest challenge in the collaborative mitigation approach. Secondly, authenticity and trust in shared data are very difficult to develop. For that certificate authority is needed to give the certificate of validity and authenticity of shared data.

Collaboration communities play a significant role in DDoS attacks mitigation. These communications are created especially for attack information sharing, detection, and signalized attacks. The majority of these communities are private communities charging membership fees for their service.

Some fee-based collaboration communities are the Anti-Phishing Working Group (APWG), the Research and Education Networking (REN) Information Sharing and Analysis Center (REN-ISAC), and the Forum of Incident Response and Security Teams (FIRST).

The Advanced Cyber Defense Centre (ACDC), the GÉANT Task Force on Computer Security Incident Response Teams1 (TF-CSIRT), and the GÉANT Special Interest Group on Network Operations Centres2 (SIG-NOC) are some of the non-fee-based communities [18].

There are some implementations of collaborative mitigation in the internet world.

DOTS Protocol. DDoS Open Threat Signaling (DOTS) protocol is developed by Internet Engineering Task Force (IETF) for DDoS attacks information sharing and mitigation. DOTS is developing a standards-based approach related to DDoS detection, classification, traceback, and mitigation in the context of a larger collaborative system at the service provider level. DOTS requires client-server architecture to share attack information in centralized as well as distributed domains. DOTS protocol is less implemented as a solution because of architectural complexity [7].

Real-Time Inter-Network Defense (RID). RID is developed by the Managed Incident Lightweight Exchange (MILE). RID outlines a proactive inter-network communication method to facilitate sharing incident handling data while integrating existing detection, tracing, source identification, and mitigation mechanisms for a complete incident-handling solution [2]. It is an application layer protocol and uses an XML feature for encryption and data security [13].

Non-collaborative DDoS Mitigation Approach. An individual network or organization that implements an action plan or mitigation strategy without being part of any collaboration is called the non-collaborative approach. The organization is fully reliant on its strategies and resources for DDoS mitigation. It makes them better control over their security plan and mitigation policy. As well as mitigation is not general like a collaborative approach, application-specific mitigation could be implemented. Latency in terms of attack detection, defense, and mitigation is decreased in the non-collaborative approach, which is become a key factor in attack mitigation. Low latency improves system performance for end-users while the organization is targeted.

One of the major drawbacks of this strategy is the available bandwidth and network resources. DDoS attacks become devastating and the capacity and intensity of attacks are also increased with time. In that case, organizations have limited bandwidth and capacity to deal with future attacks. So, it becomes expensive as well as infeasible to purchase hardware equipment, and mitigation plans only to counter DDoS attacks for any small and medium-sized organization.

In a non-collaborative mitigation approach, defining the normal traffic state of a system by identification of traffic patterns is the fundamental step in DDoS mitigation and also required for threat detection and alert. It also includes separate human traffic from non-human traffic (bots). Comparison of signatures, examine different attributes of traffic like IP addresses, cookies, HTTP headers, Javascript footprints become useful to identify normal traffic and attack traffic patterns.

After attack traffic identification, the next phase is filtering. Generally, filtering is performed through IP reputation list or database, Blacklisted/Whitelisted IP addresses, tracing connections, deep packet inspection (DPI), and rate-limiting [6]. Networks that have a high capacity of filtering generally work as scrubbing centers or traffic scrubbing filters. Traffic addressed to potential target is generally rerouted through these scrubbing centers, malicious traffic discarded, and filtered or clean traffic is transferred to their original destination.

Mode of operation is another way to categorize DDoS attacks mitigation strategies. Manual mitigation is generally not used by an organization because DDoS attacks have become more sophisticated and automated. The majority of organizations prefer to use automated mitigation because it becomes faster and easy to implement [6].

Non-collaborative approach is categorized into three different categories:

Do It Yourself or In-premise. In do-it-yourself mitigation approach, mitigation equipment is generally placed in-between the internet and the organization's resources. This mitigation equipment may be a load balancer, firewall, packet filter which filters all the traffic to and from the internet and protects the organization's resources.

Outsource/External Services. DDoS mitigation service providers or cloud-based mitigation providers or CDN providers provide services to mitigate DDoS. They charge for their mitigation services based on the size of the attack or based on clean traffic. Two Operating models of External services are On-demand and Always-on.

In On-demand, mitigation is provided on the demand of victims of DDoS by mitigation service provider immediately at the time of the attack. The initial level of attack detection is done at the victim's end and requested to mitigate. The mitigation service provider redirects the traffic and injects BGP routes to filter the traffic and block the attacker's origin. And send the clean traffic to the victim.

The mitigation service provider serves the mitigation service and traffic filtering service permanent basis in Always-On model. Mitigation service providers advertise the client IP address space on behalf of a client and provide services using shared delivery infrastructure like content delivery network (CDN). Client traffic is always going through mitigation providers' networks for getting filtered traffic. Always-on approach is used with a combination of CDN to improve website performance.

Hybrid. In a hybrid approach, Do it yourself and the mitigation service provider are work together. Hybrid approach protects against large-scale attacks without added additional cost [5].

2.3 IP Reputation Scheme

IP Reputation platforms are generally responsible to generate IP reputation score of an individual IP address. And this reputation score is used to identify fraud and malicious activity based on IP addresses. Use of IP reputation is not limited to finding out any malicious domain using reputation score. But, to identify the host of fraud websites, to filter the spam mail or message, to identify the Bots and filter the DDoS attack traffic from normal traffic based on this reputation score. Many applications use IP reputation systems as input like mail servers, IPS, IDS, and DDoS attack mitigation systems. Generally blacklisted IP addresses are categorized as blacklisted based on their Low reputation score.

These IP Reputation datasets are generally maintained by the organization's internal staff which is available on paid or free basis. Security vendors or government organizations rely upon private feeds which are generally paid. And public feeds or datasets are generally used by an individual user to check the reputation of a small set of IP addresses on a free basis. VirusTotal, IPVoid, BrightCloud, MXToolbox, AbuseIPDB are some examples of public threat intelligence services or public reputation systems [16].

3 Problem Description

Existing schemes which use blockchain as a medium to share attack information to all the participants of the collaboration can be implemented easily on any network infrastructure without the help of any special requirements. The main purpose behind designing such types of schemes is to bring transparency, security in attack information sharing.

But, the main concern is that, the accuracy of the attack information which has been shared among all the peers in collaboration. Without authenticity and accuracy of attack information, all collaborators get wrong information about the attack. And as result, they change their network infrastructure security policy according to this shared information. So, legitimate user traffic gets blocked before it reaches to destination.

Sharing whitelisted and blacklisted IP addresses among multiple domains through blockchain and smart contracts brings robustness and decentralization. But, if attack information is unauthenticated and false then individual domains block or allow traffic based on wrong data which becomes very dangerous for that autonomous system's security even it would hack the domain and take control of the whole system through DDoS attack or other network attacks.

The authenticity of the source which provides whitelisted and blacklisted IP addresses is a major concern. And data which are shared with the help of smart contracts are also unauthenticated and invalid. If an attacker compromises one node of a peer-to-peer network, and implement the smart contracts which contain invalid IP addresses, using these unauthenticated and invalid contracts, all other peers of the network change their traffic policy and network setting, which is dangerous for the valid user also because they also neglected due to wrong information sharing.

The existing IP Reputation System faces main four operational challenges while assigning Reputation score to IP addresses are High-management cost, increased False-positive rate, high consumption time, and availability of data sources that provide IP Reputation to the IP addresses [20].

4 Related Work

There are several related works, concepts that help in developing this algorithm, for brevity, we only discuss only some of them. DOTS protocol coveredboth intra-organization and inter-organization communications. DOTS requires servers and clients organized in both centralized and distributed architectures to advertise black or whitelisted addresses [7].

Cochain-SC [9] proposes Intra-domain and Inter-domain DDoS mitigation solutions through Blockchain. Cochain-SC presents a complete DDoS mitigation solution, by including mitigation techniques of both domains. In Intra-domain mitigation, Software-defined networking (SDN) is used for attack detection and blocking attack traffic within the network. Inter-domain attack mitigation is performed through collaboration between domains using Blockchain [9].

BloSS [15] presents effective mitigation of DDoS attacks with the help of Blockchain and SDN. BloSS mainly provides its solution in the context of cooperative defense by sharing attack information and tracking reputation in a fully distributed manner. It also used different techniques and approaches to encourage cooperative defense between various Autonomous Systems (AS). Not only provide attack signaling in a decentralized way but, also have an incentive scheme for service providers.

Research [14], proved that it might be possible to discover IP addresses related to a financial botnet, by combining the information coming from well-known IP blacklists providers. Set of metrics that help to calculate the reputation score of the collected

IP address. However, Intelligence sharing may not be facilitated due to less system configuration and constraint (Table 1).

Table 1. Comparative analysis of existing DDoS mitigation solutions

Existing solutions	Collaboration medium	Network	Detection/mitigation	IP reputation system
[15]	Blockchain	AS	No	No
[9]	Blockchain	AS	Yes	No
[8]	Blockchain		Yes	No
[7]	Client-Server architecture	AS	Yes	No
[17]	Blockchain	Private Network	Yes	No
[19]	Client-Server architecture (C-to-C protocol)	AS	No	No
[21]	Blockchain	Security operation Centers (SOCs)	No	No

5 Research Questions

Many research gaps and challenges are present in existing DDoS Mitigation schemes not only in single but also in multiple domains. The expected contribution of this work is categorized as below: 1). To form a collaboration between single as well as multiple domains, 2). DDoS attack detection and mitigation (at specific network end), 3). Implement IP Reputation Scheme for all the blacklisted IP addresses.

Present research work which utilized as blockchain to whitelisted and blacklisted IP address sharing, only share the whitelisted and blacklisted IP addresses which are found by the firewall or existing DDoS detection software or intrusion prevention and detection software. So, the authenticity of these IP addresses is generally depending on the network's security policy and Access Control List (ACL),which becomes biased towards the authenticity of IP address for another domain. For Example, Blacklisted IP for one domain may be a genuine IP address for another domain. So, to prevent this problem, one general or universal scheme is required, which would only find the reputation score for the IP address. And based on the reputation score, a particular domain decides which addresses are accepted or considered as a blacklisted IP address for their network.

- How proposed system minimizes the False-positive rate of DDoS attack detection?
- Can the proposed system decrease the management cost of the IP reputation system?

- Does the proposed system helps to decrease the severity of a DDoS attack?
- How proposed system provides transparency, security, and authenticity to share attack information among all the collaborators without the help of any specialized infrastructure?
- How proposed system assigns IP Reputation score to detected IP addresses from Firewall, Intrusion Prevention, and detection system?

6 Approach and Next Step

Proposed system use Blockchain to mitigate DDoS attack on single as well as multiple domains. The proposed system uses blockchain technology to transfer whitelisted and blacklisted IP addresses in transparent manner. But not only transfer IP addresses but along with the reputation score of particular IP through the smart contracts. IP reputation score helps individual domains to decide which addresses are considered as blacklisted or whitelisted based on the reputation score of the IP address. Due to the transparent, secure, and authenticated nature of blockchain, blockchain become a medium to share attack information with all the collaborative domains in transparent, secured, and without any help of certificate authority.

6.1 Proposed Architecture

Proposed Architecture of DDoS mitigation using Blockchain would work with single as well as multiple domains. Figure 1 shows the proposed blockchain based Architecture for DDoS Attack mitigation.

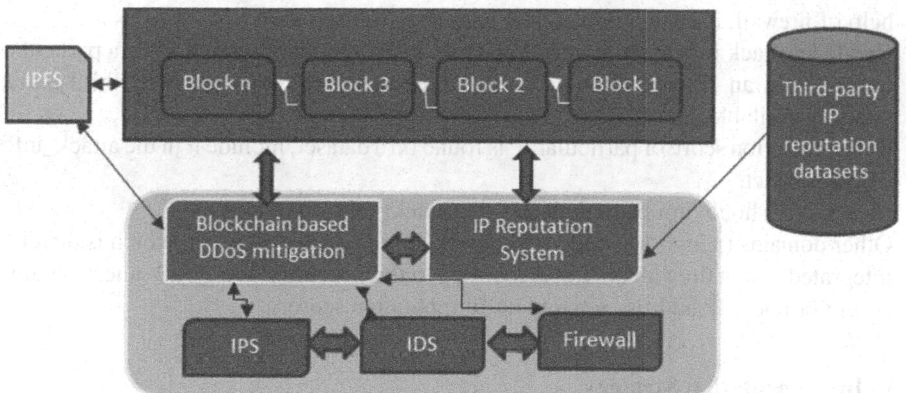

Fig.1. Proposed system architecture

This Proposed Architecture for DDoS attack Mitigation is mainly divided into two parts: 1). Blockchain layer and 2). Internal layer.

The Blockchain layer will work as a collaborative medium to share blacklisted and whitelisted IP address sharing in a transparent and secured and authenticated way. In this proposed system, Polygon [12] Ethereum compatible blockchain solution and smart contracts will be used for that purpose. All the blacklisted IP addresses would store on Interplanetary File System (IPFS) [3], which would also integrate with Blockchain to store the hash of IPFS stored data.

The internal layer resides at the individual network end, where three components named IPS, IDS, and Firewall would use to filter incoming and outgoing traffic and identify the blacklisted IP addresses. Blockchain based DDoS mitigation module would work as an integrator between IPS, IDS, Firewall, and Blockchain.

Third-party IP reputation system provides reputation score to blacklisted IP addresses which would perform by IP reputation system module. All these modules are linked together for efficient transmission of data between all the modules.

6.2 Integration of IP Reputation System with Proposed Architecture

Third-party IP reputation providers or publicly available IP reputation databases provide reputation score of IP addresses which would be found by IPS, IDS, and firewall. This mapping would be performed at the IP reputation system module. The final blacklisted IP addresses along with the reputation score would be inserted in the smart contract and stored on the blockchain. Through this smart contract, all other collaborative domains get blacklisted IP addresses.

The following steps describe the process of integration of IP reputation system with Proposed Architecture.

- Get a Reputation score of IP addresses through third-party organizations and publicly available datasets.
- Find the blacklisted and whitelisted IP addresses at individual network end with the help of firewall, intrusion prevention, and detection software.
- Create an attack info smart contract and check whether the detected IP of a particular network has an IP reputation score in the existing reputation dataset (third party, publicly available datasets).
- If the reputation score of particular IP is found on a dataset, include it in the attack_info smart contract.
- Else this IP should not be included in the attack_info smart contract.
- Other domains (network) would find the attack_info smart contract, which is directly integrated with a firewall or IPS. According to their security policy, all other domains accept or reject IP addresses based on IP reputation score.

6.3 Implementation Strategy

Proposed Architecture collaboratively mitigates DDoS attack using Blockchain along with IP Reputation system. For building collaboration between different collaborators, Blockchain and smart contracts may be used which provide transparency, authenticity, and security to the data. Third-party IP reputation datasets become a source to get the IP reputation data of a given set of IP addresses.

Blockchain. Polygon [12], Ethereum compatible layer 2 solutions may be used to bring transparency, authenticity, and security of the Attack information sharing among various collaborators. Blockchain development platform, Ethereum charges higher Gas fees, and scalability also becomes an issue. So, Polygon is the suitable solution that helps to create Ethereum compatible solutions for collaboration among various collaborators. IPFS would be used for storing IP addresses along with their IP reputation score in a decentralized manner.

Software-Defined Networking. Unlike traditional network architecture, Software-defined Networking decouples data plane from the control plane. Due to this separation, management and monitoring of the network become easy and efficient. In this proposed system, SDN would be used for detection and blocking the attack traffic near the source of an attack. IDS, firewall, IPS would be implemented through the SDN in the proposed system.

Third-Party IP Reputation Datasets. Various publicly available IP reputation Datasets and databases would be available as a resource provider to decide the IP reputation score of blacklisted IP addresses. According to the IP reputation score, other collaborator domains decide policy for their ACL.

6.4 Advantages of Proposed System Over Existing Implementation

IP reputation systems decide the actual reputation of IP addresses using publicly available datasets. The IP reputation system would give the IP reputation to all the blacklisted IP addresses so that collaborators' domains decide their ACL based on this IP reputation score, it would be helpful to lower the false positive rate of attack detection. In existing implementations, blacklisted IP addresses are directly shared using smart contracts, any validation and accuracy mechanism is not implemented. But in the proposed solution, validation and accuracy of IP address data would be implemented through an IP reputation system.

In the proposed system, IP reputation data would be shared in a decentralized, transparent, and secured way using blockchain. Polygon, Ethereum compatible solution would be decreased the Gas cost and transaction fees. As well as, all the IP address data along with their reputation score would be stored on the IPFS. Management cost [20] of the proposed system would remain lower as compared with other mediums like client-server architecture.

Accurate and early detection of DDoS attacks near the source of attack helps to prevent the risk of DDoS attacks. Attack information sharing among all the collaborators, other collaborators would change their ACL policy based on attack information. This would decrease the severity of ongoing DDoS attacks by blocking attack traffic near the source of an attack.

Existing collaborative solutions like DOTS rely on specialized client-server architecture for attack information sharing among other collaborators. As well as change the existing infrastructure to implement this DOTS solution, which would become a costly and unfeasible solution for various collaborators. Whereas implementing a blockchain based reputation system, would be transparent, decentralized, secured, and compatible with existing infrastructure.

In the proposed solution, the IP reputation system would assign the IP reputation score to the detected IP addresses which would be detected by IPS, IDS, and firewall. There would be an algorithm that integrates IP addresses to their IP reputation score calculated by third-party IP reputation databases.

7 Conclusion

DDoS mitigation using Blockchain is a novel way to mitigate DDoS attacks because DDoS attacks are distributed in nature. Blockchain is a ledger management technology in a distributed environment. Features like reliability, security, immutability of blockchain prove it as an appropriate solution for mitigating any cyber-attacks especially, DDoS attacks. Some mitigation solutions are designed with the help of Blockchain in the market. Blockchain is also utilized as a medium of attack information sharing in distributed and secured manner. A comprehensive review of existing DDoS mitigation solutions using Blockchain is discussed in this paper. Additionally, Blockchain with the combination of other technologies such as SDN, machine learning proves as an efficient solution to tackle DDoS attacks. This proposed architecture would share blacklisted IP addresses with the help of Blockchain technology in a transparent, secured way. Authenticity of Blacklisted IP addresses would be calculated by the Reputation system. And Reputation system would be also helpful to decide whether the particular IP address would allow or not in the network. A Blockchain-based mitigation solution would also create transparency in the collaboration process. However, a lot of work remains in these domains. There are immense capabilities of Blockchain to serve as a complete solution in terms of detection, prevention of attacks in the initial phase. However, Scalability and storage are two challenges, which require attention while utilizing Blockchain as an attack mitigation solution.

References

1. Ahamed, J.N., Iyengar, N.: A review on Distributed Denial of Service (DDoS) mitigation techniques in cloud computing environment. IJSIA **10**, 277–294 (2016) https://doi.org/10.14257/ijsia.2016.10.8.24
2. Aujla, G.S., Singh, M., Bose, A., Kumar, N., Han, G., Buyya, R.: BlockSDN: blockchain-as-a-service for software defined networking in smart city applications. IEEE Network **34**, 83–91 (2020). https://doi.org/10.1109/MNET.001.1900151
3. Benet, J.: IPFS - Content Addressed, Versioned, P2P File System. arXiv:1407.3561 [cs]. (2014)
4. Bodkhe, U., et al.: Blockchain for industry 4.0: a comprehensive review. IEEE Access. **8**, 79764–79800 (2020). https://doi.org/10.1109/ACCESS.2020.2988579
5. DDoS Mitigation Fundamentals. https://www.first.org/education/trainings. Accessed 26 Jun 2021
6. DDoS mitigation (2021). https://en.wikipedia.org/w/index.php?title=DDoS_mitigation&oldid=1023844160

7. draft-nishizuka-dots-inter-domain-mechanism-02 - Inter-organization cooperative DDoS protection mechanism. https://datatracker.ietf.org/doc/draft-nishizuka-dots-inter-domain-mechanism/. Accessed 22 Jul 2021
8. El Houda, Z.A., Hafid, A., Khoukhi, L.: Co-IoT: A collaborative DDoS mitigation scheme in IoT environment based on blockchain Using SDN. In: 2019 IEEE Global Communications Conference (GLOBECOM), pp. 1–6 (2019). https://doi.org/10.1109/GLOBECOM38437.2019.9013542
9. Houda, Z.A.E., Hafid, A.S., Khoukhi, L.: Cochain-SC: an intra- and inter-domain DDoS mitigation scheme based on blockchain using SDN and smart contract. IEEE Access. **7**, 98893–98907 (2019). https://doi.org/10.1109/ACCESS.2019.2930715
10. Mahjabin, T., Xiao, Y., Sun, G., Jiang, W.: A survey of distributed denial-of-service attack, prevention, and mitigation techniques. Int. J. Distrib. Sens. Netw. **13**, 1550147717741463 (2017). https://doi.org/10.1177/1550147717741463
11. Narayanan, A., Bonneau, J., Felten, E., Miller, A., Goldfeder, S.: Bitcoin and Cryptocurrency Technologies: A Comprehensive Introduction. Princeton University Press, Princeton (2016)
12. polygon.technology: Polygon-Ethereum's Internet of Blockchains. https://polygon.technology/lightpaper-polygon.pdf. Accessed 25 Jan 2022
13. Real-time Inter-network Defense (RID). https://datatracker.ietf.org/doc/html/rfc6545, Accessed 23 Jun 2021
14. Riccardi, M., Oro, D., Luna, J., Cremonini, M., Vilanova, M.: A framework for financial botnet analysis. In: 2010 eCrime Researchers Summit, pp. 1–7. IEEE, Dallas (2010). https://doi.org/10.1109/ecrime.2010.5706697
15. Rodrigues, B., Scheid, E., Killer, C., Franco, M., Stiller, B.: Blockchain Signaling System (BloSS): cooperative signaling of distributed denial-of-service attacks. J. Netw. Syst. Manage. **28**(4), 953–989 (2020). https://doi.org/10.1007/s10922-020-09559-4
16. Shewale, S., Deshpande, L.: Analytical Study on IP Reputation Services & Automation for Traditional Method. IJARCCE. **10**, 4 (2021)
17. Spathoulas, G., Giachoudis, N., Damiris, G.-P., Theodoridis, G.: Collaborative blockchain-based detection of distributed denial of service attacks based on internet of things botnets. Future Internet. **11**, 226 (2019). https://doi.org/10.3390/fi11110226
18. Steinberger, J.: Distributed DDoS Defense (2018). https://doi.org/10.3990/1.9789036545815
19. Hameed, S.: SDN based collaborative scheme for mitigation of DDoS Attacks. Future Internet. **10**, 23 (2018). https://doi.org/10.3390/fi10030023
20. Usman, N., et al.: Intelligent dynamic malware detection using machine learning in IP reputation for forensics data analytics. Futur. Gener. Comput. Syst. **118**, 124–141 (2021). https://doi.org/10.1016/j.future.2021.01.004
21. Yeh, L.-Y., Lu, P.J., Huang, S.-H., Huang, J.-L.: SOChain: a privacy-preserving DDoS data exchange service over SOC consortium blockchain. IEEE Trans. Eng. Manage. **67**, 1487–1500 (2020). https://doi.org/10.1109/TEM.2020.2976113

Model-Driven Development of Distributed Ledger Applications

Piero Fraternali, Sergio Luis Herrera Gonzalez(✉), Matteo Frigerio, and Mattia Righetti

Dipartimento di Elettronica, Informazione e Bioingegneria, Politecnico di Milano, Piazza Leonardo da Vinci 32, 20133 Milan, Italy
sergioluis.herrera@polimi.it

Abstract. The Distributed Ledger Technology (DLT) is one of the most durable results of virtual currencies, which goes beyond the financial sector and impacts business applications in general. Developers can empower their solutions with DLT capabilities to attain such benefits as decentralization, transparency, non-repudiability of actions and security and immutability of data assets, to the price of integrating a distributed ledger framework into their software architecture. Model-Driven Development (MDD) is the discipline that advocates the use of abstract models and of code generation to reduce the application development and integration effort by delegating repetitive coding to an automated model-to-code transformation engine. In this paper, we explore the suitability of MDD to support the development of hybrid applications that integrate centralized database and distributed ledger architectures and describe a prototypical tool capable of generating the implementation artefacts starting from a high level model of the application and of its architecture.

Keywords: Blockchain · Distributed ledger · MDD · IFML

1 Introduction

Distributed Ledger Technology (DLT), popularized by the advent of virtual currencies, is having great impact also on applications outside the financial sector, such as those for the insurance industry, for the public administration, for NFT trading and more. The common trait of the business sectors in which DLT holds the greatest potential is the need of sharing data and transactions within a decentralized distributed network in a transparent yet secure way. Most of the time DLT functionalities must be integrated within a traditional application architecture. For example, an insurance business may start introducing the DLT technology in the claim payment workflow, while retaining a more traditional database-driven centralized architecture in the other processes. As the employment of DLT in businesses matures one can expect that companies will implement a migration strategy to progressively port, totally or in part, their business processes to this new architecture. This phenomenon resembles the transition to B2B application architectures in the nineties, when companies started migrating their business processes

U. K. Rage et al. (Eds.): DASFAA 2022 Workshops, LNCS 13248, pp. 104–119, 2022.
https://doi.org/10.1007/978-3-031-11217-1_8

to the Web in the wake of the success of B2C applications. Integrating centralized database-driven and decentralized ledger-driven architectures for developing hybrid applications poses new design and implementation challenges. At the conceptual level the boundary must be defined between the data and the operations that reside in either of the two platforms. At the physical level, suitable interfaces must be implemented between data and transactions in the distributed ledger and in the database. The development of hybrid DLT and database driven applications and business processes requires adequate methodologies and tools to reduce effort and cost, enforce uniform design patterns across projects, and ease the migration of workflows from one architecture to the other. One such methodology is Model-Driven Development (MDD), defined as the software engineering discipline that advocates the use of *models* and of *model transformations* as key ingredients of software development [14]. Abstraction is the most important aspect of MDD, which enables developers to create and validate a high level design of their application and introduce implementation-level architecture details at a later stage of the realization process. Implementation and architecture details are introduced via model transformations, which iteratively refine the high-level initial model, eventually getting to the final executable solution [2]. Nowadays MDD is applied in practice by the so-called low-code software development platforms[1], which exploit Platform Independent Modeling languages and code generators to automate the production of e.g., mobile, Web and enterprise applications. In this paper, we investigate the application of MDD to the development of hybrid applications that mix features of both centralized database-driven and decentralized ledger-driven architectures. The contributions of the paper can be summarized as follows:

- We introduce the class of hybrid DLT/DB applications, which consists of enterprise applications that implement data and processes on both distributed ledger and centralized database infrastructures.
- We describe the MDD process of hybrid DLT/DB applications in terms of inputs, activities and outputs. Specifically, we start from a development process scheme conceived for data intensive multi tier applications and discuss its extensions with activities that cope with the DLT requirements.
- We propose to model the design of hybrid DLT/DB solutions with a simple extension of the Domain Model and of the Interface and Action Models. For the sake of illustration, we express such models with OMG's UML Class Diagrams or Entity-Relationship Diagrams, OMG's Interaction Flow Modeling Language (IFML) diagrams, and an abstract action language for IFML [11]. The extension caters for the requirements posed by the integration of DLT into a traditional database-driven application development. The designer simply annotates the Domain Model entities and relationships to specify the platforms where such primitives are materialized. In this way, the operations of the Action Model that affect domain objects can infer the platform(s) in which the operations are executed. Similarly, the components of the Interface Model implicitly derive the content to publish in the interface from the appropriate data source.

[1] Examples of low code platforms are WebRatio, Mendix and Outsystems.

- We illustrate a prototypical version of a code generator mapping the high-level specifications of a hybrid DLT/DB application into the executable code for the Java EE architecture integrated with a popular DLT framework (Hyperledger Fabric). The code generator takes in input the Domain, Interface and Action Models and produces a fully functional multi-organization and multi-role hybrid DLT/DB application with Web/mobile GUIs.
- We showcase the MDD approach in the realization of the blueprint application of Hyperledger Fabric, a financial certificate trading solution.
- We discuss the integration of the developed modeller and code generator into the WebRatio[2] commercial low-code platform.

1.1 Running Example

To illustrate the MDD of hybrid DLT/DB application, we exploit throughout the paper an exemplary application built on top of the blueprint Papernet network introduced the Hyperledger Fabric tutorial[3]. Papernet is a commercial paper network that allows participants to issue, trade and redeem commercial papers. The hybrid DLT/DB application will permit the authorized personnel of business companies to create commercial papers and share them in a DLT network, where they can be purchased and redeemed by the employees of financial trading companies. The hybrid nature of the application stems from the necessity to exploit a traditional centralized database architecture for storing the employee data, implementing role-based access control, recording in the company's own books both the details of the commercial certificates and other relevant accounting and internal auditing data (e.g., the certificate's creator). To such needs the DLT requirements add up: the operations on the papers must be implemented so as to ensure transparency, accountability and non-repudiability of operations.

2 Background

2.1 Distributed Ledger Technology and Hybrid DLT/DB Applications

Distributed ledger technology (DLT) is an approach for sharing data across a distributed network of participants with the guarantee of immutability of transactions. DLT evolved from the Peer-to-Peer (P2P), file sharing and blockchain technologies. In a P2P network the peers are connected computer systems and the assets are shared directly without a central server. In 2008 the Bitcoin virtual currency applied the P2P paradigm to financial assets [10]. The underlying *blockchain* data sharing technology opened the way to other P2P asset management frameworks, yielding to DLT as a general-purpose architecture. The blockchain is a specific type of DLT that uses cryptographic and algorithmic methods to create and

[2] https://www.webratio.com.

[3] https://hyperledger-fabric.readthedocs.io/en/release-1.4/tutorial/commercial_paper.html.

verify a continuously growing append-only chain of transaction blocks constituting the ledger. Additions are initiated by a node that creates a new block of data, e.g., containing several transaction records. Information about a new data block is shared across the network and all participants collectively determine the block validity according to a predefined algorithmic validation method. Only after validation all participants add the new block to their respective ledgers. With DLT no single entity in the network can amend past data and approve additions. An attacker willing to corrupt the ledger must gain control over the majority of the nodes.

The **Smart contracts** [15] are code packages deployed and executed in the nodes of a DLT network. They are programs that run, control or document relevant actions in the network. The source of a smart contract is stored in the blockchain, allowing any interested party to inspect its code and current state and verify its functionality; in this way, also the operational semantics of a smart contract cannot be changed without the consensus of the network participants. Smart contracts are replicated on all the network nodes; when a smart contract executes on a node, others can verify the result and the operations performed by the smart contract are recorded in the blockchain permanently.

A **Hybrid DLT/DB application** is a distributed application that: 1) has a client-server multi-tier architecture; 2) manages persistent data; 3) involves transactions that update data both in traditional database storage and in distributed ledger storage; 4) requires DLT-enabled transparent sharing and non-repudiability of operations; 5) exposes its functionality to the end-users through one or more client-side (web or mobile) interfaces; 6) optionally exposes its functionality to other applications through APIs (e.g., as Restful services).

The **State of a hybrid DLT/DB application** consists of two elements: the ledger state i.e., an immutable log of transactions and the world state, i.e., a database with business objects managed by application transactions. In a hybrid DLT/DB application the world state can be further distinguished into internal and external. This leads to a tripartite notion of the state of a hybrid DLT/DB application, in which each level has its specific update semantics:

- The *external world state* is the state of the objects in databases external to the network. It is updated by external operations and transactions, whose effect is not automatically recorded and shared in the ledger.
- The *internal world state* is the state of the objects recorded in the network data store. It is updated by smart contracts, whose effect is automatically recorded and shared in the ledger.
- The *ledger (or blockchain) state* is the state of the distributed ledger containing the log of the smart contract operations. It cannot be updated explicitly but only by the system as an effect of the smart contract execution. In other terms, the ledger state is read-only for the application business logic and can only used by applications to visualize smart contract execution history.

Permissionless vs Permissioned DLT. DLT systems can be *permissionless* or *permissioned.* In permissionless systems such as those underlying the virtual currencies, the participants can join or leave the network at will, without being authorized by any entity. There is no central owner, and identical copies of the ledger are distributed to all network participants. In permissioned DLT, members are pre-selected by someone, in general, an owner or an administrator of the ledger, who controls network access and sets the participation rules. Permissioned DLT systems have been conceived to support the use of the technology in business contexts in which the sharing of data and operations are constrained to a set of entities that satisfy the network access rules.

DLT Frameworks for Application Development. The adaptation of DLT to general-purpose applications has led to the advent of software frameworks supporting the integration of DLT functions in business applications. *Hyperledger Fabric*[4] is an open-source enterprise-grade permissioned DLT platform with features such as participant identifiability, high transaction throughput, low transaction confirmation latency, and transaction privacy and confidentiality control. The architecture separates the transaction processing workflow into three different stages. The smart contracts, also called chaincode, comprise the system distributed logic for processing and agreement. The transaction ordering module and the transaction validation and commitment module implement the serialization and persistence of operations. Such separation reduces the number of verification levels, mitigates the network bottlenecks, improves network scalability and the overall performances. When a participant submits a transaction proposal, the network peers need to endorse it. When a majority of peers have agreed, an ordering service creates a block of transactions to be validated. After validation, the transactions are committed to the ledger. Since only confirming instructions, such as signatures and read/write sets, are sent across the network, the scalability and performance of the network are enhanced. The plug-in and component-based architecture of the platform also simplifies the reuse of existing features and the integration of custom modules.

2.2 MDD with the Interaction Flow Modeling Language

The Interaction Flow Modeling Language (IFML) [11] is an OMG standard for the platform-independent specification of interactive applications that allows developers to describe the organization of the interface, the content to be displayed, and the effect produced by the user interaction or by system events. The business logic of the actions activated by the user interaction can be modelled with any behavioral language, e.g., with UML sequence diagrams or with IFML extensions for action modelling [3].

Interface Structure. The core IFML element for describing the interface structure is the *ViewElement*, specialized into *ViewContainer* and *ViewComponent*. ViewContainers denote the modules that comprise the interface content. They can include ViewComponents which represent the actual content elements. Figure 3 shows the IFML specification of one of the interfaces of the Papernet application. The front-end comprises two ViewContainers: *Home* and *Issue*

[4] https://www.hyperledger.org/use/fabric.

Paper, which in turn comprise the ViewComponents that specify their content. Different types of ViewComponents can be used to describe alternative content patterns. The basic ones are *Detail* ViewComponents that denote the publication of a single object, *List* ViewComponents that denote the publication of multiple objects, and *Form* ViewComponents that denote an input form. Depending on their type, ViewComponents can output parameters: a form has output parameters corresponding to the submitted values and a List ViewComponent has an output parameter that identifies the selected item(s). IFML shows the abstract source from which ViewComponents derive their content with a *DataBinding* element, which references an object class of the application Domain Model. The object(s) bound to a ViewComponent can be constrained by a *selector condition*, which can be parametric. For example, in Fig. 3 the *Paper Details* ViewComponent displays the data of a *Paper* object. The selector condition `[creator = ?]` on the ViewComponent expresses a required input parameter corresponding to the identifier of the object to display. The parameter input-output dependencies between components are specified with *Flows*.

Events, Navigation Flows and Data Flows. ViewContainers and ViewComponents can be associated with *Events*, denoted as circles, to express that they support the user interaction. For example, a List ViewComponent can be associated with an Event for selecting one or more items, and a Form ViewComponent with an Event for input submission. The effect of an Event is represented by a *NavigationFlow*, denoted by an arrow, which connects the Event to the ViewElement/Action affected/triggered by it. IFML specifies (also implicitly) the input, output, and parameter passing from the source to the target of the NavigationFlow. For example, in Fig. 3 the NavigationFlow from the *Issued papers* List ViewComponent to the *PaperDetails* Detail ViewComponent denotes that the user can interact with the list by selecting one item. Such an event determines the (re)computation of the content of the *Paper Detail* ViewComponent based on the identifier of the *Paper* object selected from the list. Input-output dependencies between ViewComponents can also be specified independently of interaction events, using *DataFlows* denoted as dashed arrows.

Actions. The above mentioned list selection Event expresses a user interaction which has the sole effect of updating the interface content. Events can also specify the triggering of business logic, executed prior to updating the state of the user interface. The IFML *Action* construct is represented by a hexagon symbol (see Fig. 3) and denotes an invoked program treated as a black box, possibly exposing input and output parameters. The effect of an Event firing an Action and the possible parameter passing rules are represented by a NavigationFlow connecting the Event to the Action and possibly by DataFlows incoming to the Action from other IFML elements. The execution of the Action may trigger another Action, cause a change in the state of the interface and produce output parameters; this is denoted by termination events associated with the Action and connected by a NavigationFlow to the Action executed after it or to the ViewElement affected by it. In Fig. 3 the outgoing NavigationFlow of the *Create Paper* Action is connected to the *Paper Details* ViewComponent and is associated by default with the output parameter of the Action (this is the identifier of the Paper object created by the Action, as shown Fig. 4). When the Action completes, the *Home* ViewContainer is displayed and the output parameter of the *Create Paper* Action determines the object shown in the *Paper Details* ViewComponent, i.e., the newly created Paper object.

3 Development of Hybrid DLT/DB Applications

We adopt an enterprise application integration perspective whereby the specific DLT requirements add up to the requirements typical of enterprise applications. For this reason, we start from a development process scheme typical of multi-tier data-intensive applications and extend such a workflow with the inputs, activities and outputs needed to address the integration of DLT requirements.

The development process of data-driven enterprise applications encompasses the major phases of requirements specifications, design, implementation and maintenance/evolution. In this paper, we do not address the modalities in which such major concerns are addressed in a practical software life cycle (e.g., in a SCRUM agile development method) but rather focus on the input, output and tasks that characterize each concern to show how DLT requirements impact.

3.1 Requirement Specification

Requirements specification collects and formalizes the essential knowledge about the application domain and expected functions. The input is the set of business requirements that motivate the application development and all the available information on the technical, organizational, and managerial context. The output is a document specifying what the application must do. In a traditional database-driven application the specifications typically comprises the identification of the user roles, the use cases of each role with pre-conditions, workflow and post-conditions, and a dictionary of the essential data. When DLT requirements come into play, the requirements specifications should also address the data and transaction sharing requirements. This can be done by identifying the organizations participating to the network to which the user roles belong and making explicit the operations of the use case workflows whose execution should be tracked in the network.

Running Example. Two types of organizations can join the Papernet network: *Issuers* are companies that create commercial papers to fund their operations, and *Traders* are financial organizations that transact such certificates as a form of investment. The application role models comprise *Issuer employees* and *Trader employees* who interact on behalf of their respective organizations. The relevant use cases comprise the *Issue paper* use case by the issuer employees and the *Buy Paper* and *Redeem Paper* use cases by the trader employees. The data dictionary comprises as entities the organizations, the organizations' roles, the users, the commercial papers and the issuing, buying and redeeming operations. The application publishes three user interfaces: one public interface for logging in, one protected interface for issuer employees and one protected interface for trader employees. Figure 1 shows a simplified and partial excerpt of the requirements specifications of the Papernet application.

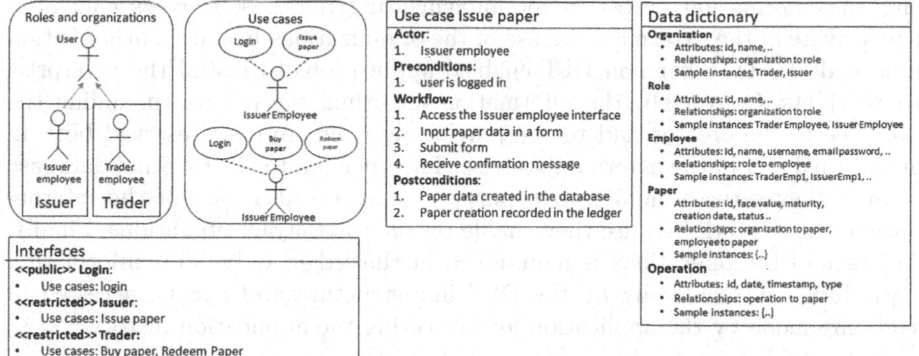

Fig. 1. The simplified requirements specifications of the Papernet application

3.2 Data Design

Data design is the activity that takes in input the data dictionary and the use cases and creates the domain model of the application, using such notations as UML class diagrams or Entity-Relationship diagrams. In the design of the domain model of a hybrid DLT/DB application entities and relationships can be stereotyped with the three storage modes (`external`, `internal`, `ledger`) to show the level at which the information resides in the application state. The entities and relationships defined at the ledger level are implicitly read-only. The default storage mode is defined to be `external`, because it is assumed that only a minority of critical data will be stored in the internal database.

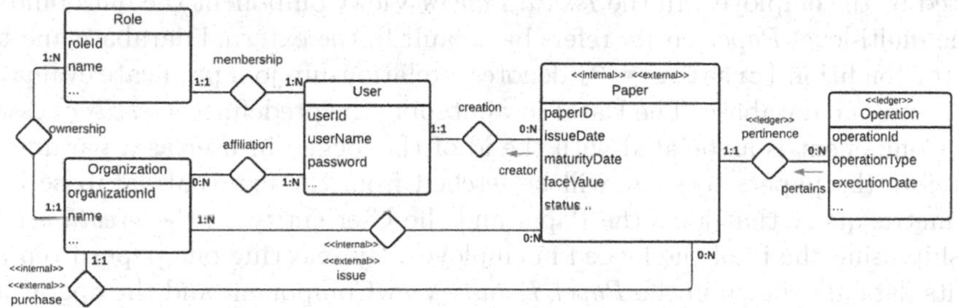

Fig. 2. Domain model of the PaperNet application

Running Example. The domain model of the Papernet application shown in Fig. 2 contains five entities (Organization, User, Role, Paper, and Operation) and seven relationships (ownership, membership, affiliation, creation, issue, purchase and pertinence). The diagram shows which features are stored in the external database only (User, Role, Organization, membership, ownership, affiliation), which ones both in the internal and in the external database (Paper, issue and purchase) and which ones in the ledger only (Operation and pertinence). Note that the information necessary for implementing role based access control is kept private in the external database of the organizations because authorization is needed also by other non DLT-enabled applications to install the enterprise access rights. Conversely, the information regarding the papers (including the reference to the creator and to the purchasing company) is preserved both in the internal and in the external database. They pertain to the external database because they serve administrative purposes and are also part of the internal state of the network because they enable the smart contract operations. Finally, the trace of the operations is maintained in the ledger only. This information is produced automatically by the DLT infrastructure and can be accessed in read-only mode by the application for inspecting the application history.

3.3 Interface Design

Interface design maps the use cases into the IFML model of the GUIs that support the users' workflows. Each interface can be public or restricted to one or more roles and contains the ViewComponents, Flows and abstract Actions needed to support the associated use case(s). To disambiguate the case in which an entity or relationship belongs to multiple state levels, a default rule is introduced for the data binding of a ViewComponent. If no stereotype is used `external` is assumed. Otherwise `internal` or `ledger` can be specified.

Running Example. The Papernet application comprises one public and two restricted interfaces, as shown in Fig. 1. The Issuer employee can access the interfaces shown in Fig. 3. In the public interface a Form ViewComponent and a Login Action let the employee access the restricted interface. The restricted interface comprises a *Home* ViewContainer with a list of the commercial papers created by the employee. In the *Issued Papers* ViewComponent the data binding to the multi-level *Paper* entity refers by default to the external database and the selector condition `[creator = ?]` denotes a relationship join predicate evaluated in the external database. The DataFlow outgoing the predefined *GetUser* session ViewComponent is associated with the id of the logged in user as a parameter. Therefore the papers to show will be fetched from the external database by a parametric query that joins the Paper and the User entity on the *creation* relationship using the id of the logged in employee. By selecting one paper from the list, its data are shown in the *Paper Details* ViewComponent and the operations executed on it are displayed in the *Operations* ViewComponent. Note that in a real application, the *Paper Details* ViewComponent may show administrative attributes of the commercial paper that are pertinent to the business but not maintained in the network, which motivates the decision of binding the ViewComponent to the external database. A DataFlow between the *Paper Details* and the *Operations* ViewComponents expresses the input-output dependency between the two elements. In the *Operations* ViewComponent the data binding to the ledger *Operation* entity refers to this data layer and thus the selector condition `[pertains = ?]` also refers to a relationship predicate over the ledger. Therefore the operations to show will be fetched by retrieving from the ledger the records related to the displayed paper. The interface also contains an *Issue Paper* ViewContainer, with a Form ViewComponent for inputting the data of a new paper. The Form comprises the fields corresponding to the attributes of the Paper entity (not shown in the IFML graphic notation). The submission of the *Issue* form (denoted by the event and NavigationFlow associated with this ViewComponent) triggers the execution of the *Create Paper* Action, whose internal workflow is specified in the action model of Fig. 4.

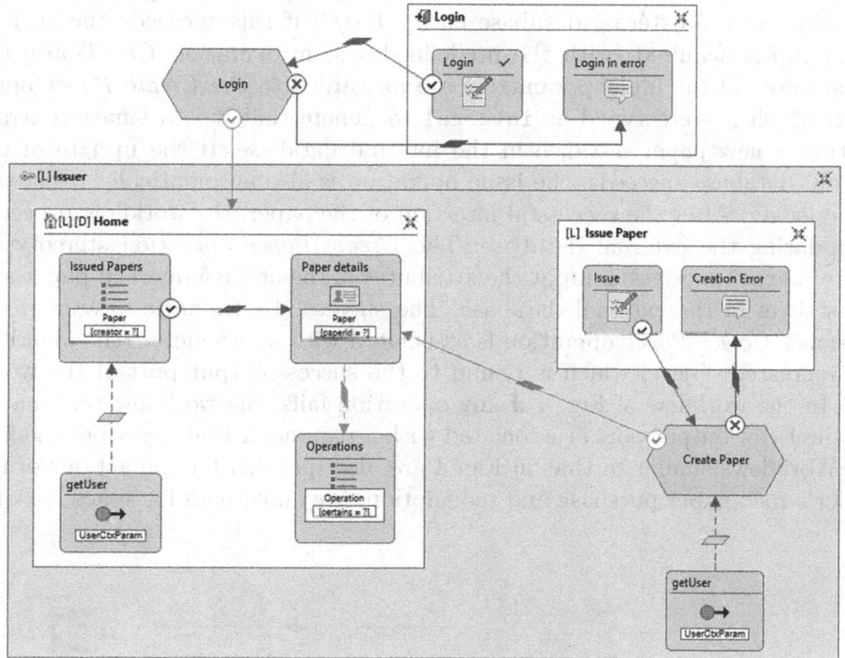

Fig. 3. IFML model of the public Login interface and of the restricted Issuer interface. The specification of the Create Paper abstract action is provided in Fig. 4.

3.4 Operation Design

Operation design maps each abstract Action into a detailed workflow. We express the workflow with the IFML action language extension of the WebRatio low code platform[5]. The action language comprises predefined operations, such as CRUD primitives bound to the entities and relationships of the domain model, system and session operations and more. When a CRUD operation affects a multi-level

[5] https://www.webratio.com.

entity or relationship, the `internal` and `external` stereotypes can be associated with its data binding to disambiguate the data source. We assume `external` as the default. Given that applications have read-only access to the ledger, CRUD operations bound to ledger entities and relationships are not meaningful.

Running Example. The Papernet application comprises four operations (login, issue, buy, redeem) which are modelled as abstract Actions in the interfaces of the respective actor. Figure 4 shows the operation model of the *Create paper* abstract Action specified in the Issuer's interface (Fig. 3). The workflow has an input port that specifies the parameters consumed by the Action. They are the creator's id, the face value and the maturity date (such parameters match the inputs to the abstract Action of Fig. 3). The Action workflow starts by verifying the creator's credentials (with the *Verify User Identity* predefined read operation on the external database entity *User*); if this succeeds the current timestamp is acquired (with the predefined system operation *Get Time*); the timestamp and the input parameters are forwarded to the *Create Paper* operation, which is stereotyped as `internal` to denote that it is a smart contract creating a new paper instance in the internal database. If the update of the internal database succeeds, the issue operation is also automatically registered to the ledger. After the successful insertion of the paper, the workflow proceeds by updating the external database. The *Create Paper* operation stereotyped with `external` receives in input the attributes of the new commercial paper and stores them in the external database. The successful termination event of the `external` *Create Paper* operation is associated with a parameter (the identifier of the created object) which is bound to the success output port of the workflow. In the workflow of Fig. 4 if any operation fails, the workflow terminates, and the Error output port is associated with a parameter that describes the failure. Workflows similar to that in Fig. 4 are also specified for the actions of the Trader's role: paper purchase and redemption. We omit them for space reasons.

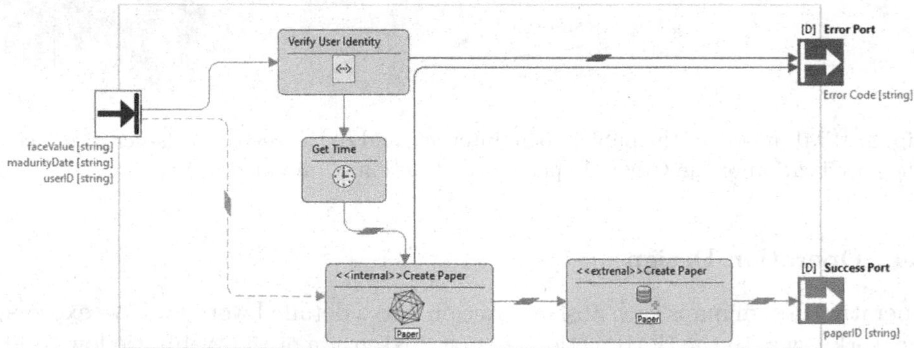

Fig. 4. Operation model of the Create paper abstract action

3.5 Architecture Design

Architecture design describes at a high level the runtime infrastructure onto which the application is deployed. We assume a fixed architecture pattern consisting of a Java Enterprise back-end connected to one or more relational databases and a permissioned DLT network implemented with Hyperledger Fabric.

Running Example. Under the above mentioned assumptions, architecture design boils down to specifying the configuration parameters of the architecture of the specific application. These amount to the URLs and permissions to access the database by each organization and the network channels, peers and policies for each organization and for the channel. In the running example, we install a single *Trading* channel with two peers for each organization. Figure 5 represents the network model designed with the MDD tool described in Sect. 4.

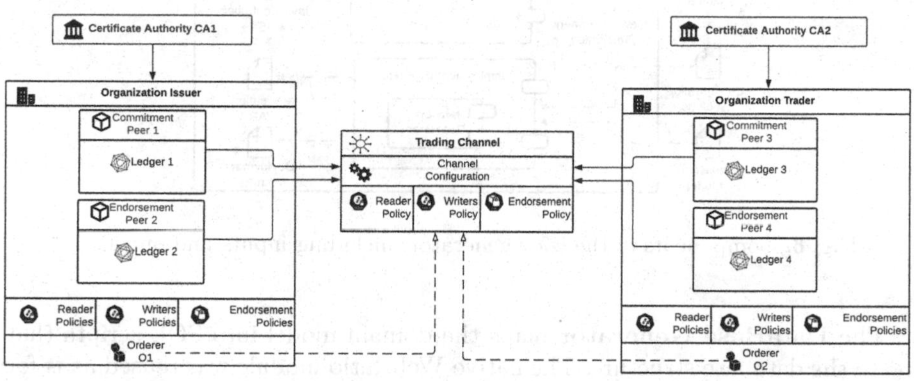

Fig. 5. Network model

4 Implementation

To implement the described MDD approach, we extended the code generator of the WebRatio low code platform, which supports Entity-Relationship data modelling and IFML user interface modelling. The WebRatio code generator is template-based and enables developers to refine the implementation of the predefined components and to add new components and templates. We customized the behaviour of the data-driven ViewComponents to support the binding to internal database entities and relationships and added the action templates for CRUD operations bound to the internal database so as to map them to smart contracts. Furthermore, a completely new model editor and a deployment code generator were developed to map the network model into the artefacts required to install the network. Figure 6 illustrates the architecture, inputs and outputs of the MDD code generator of hybrid DLT/DB applications.

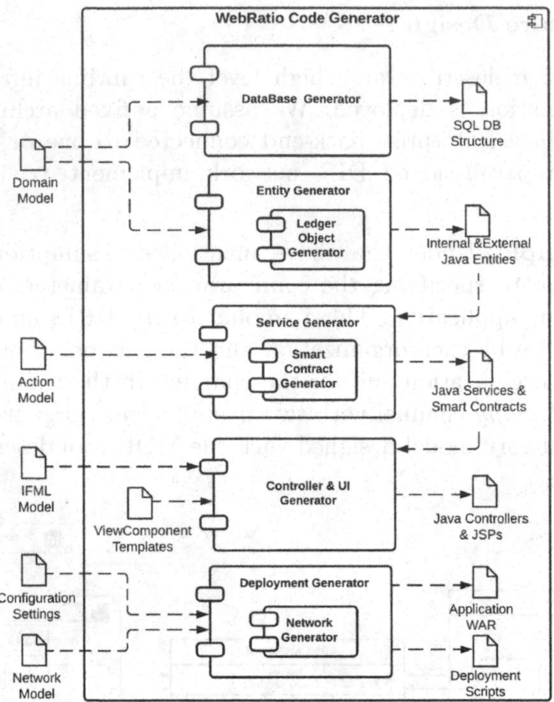

Fig. 6. Components of the code generator, including inputs and outputs

The **DataBase Generator** maps the domain model into SQL scripts that create the database structure. The native WebRatio module was reused as-is for mapping the external entities. No code generation is necessary for the Hyperledger Fabric store because all internal entities map to standard collections, and ledger entities map to JSON objects extracted via API calls.

The **Entity Generator** maps the domain model into Hibenate classes. The module has been extended with a *Ledger Object Generator* to map the internal database entities onto Java classes extending the Hyperledger Fabric Framework and the ledger entities onto JSON objects extracted from the blockchain.

The **Service Generator** takes in input the action model and the domain entities and outputs service classes that implement the operations and the control logic of the action workflows. It exploits code templates and generates the service API and the implementation code for the predefined system and CRUD operations. We extended the module by adding a *Smart Contract Generator*, which produces the smart contracts for managing the internal objects.

The **Controller and UI generator** takes in input the IFML models, the service interfaces and the predefined ViewComponent templates and outputs Java controllers and GUI views. We extended the module by adding code templates for the ViewComponents bound to the internal and ledger entities.

The **Deployment generator** takes in input the configurations defined in the architecture model and creates a WAR package containing the generated code of the Web/mobile application. We extended the module by adding a *Network generator* that produces the artefacts for installing the network.

To model the network during architecture design, we defined a domain specific language with the Eclipse Modeling Framework (EMF)[6]. We identified the network entities, attributes, and relationships, represented them in an Ecore meta-model and built a graphical editor that supports the creation of network models. The Network generator, built using Acceleo[7], takes in input the network model and produces the following files for instantiating and starting the network automatically:

- configtx.yaml: contains the information to configure the channel(s);
- base.yaml: Docker-Compose file used to define a basic peer container configuration; it provides the standard actions for all the peer setup.
- dockercompose.yaml: Docker-Compose file used to define and run the Hyperledger Fabric containers that make up the network;
- dockercomposeca.yaml: Docker-Compose file used to define and run the Fabric-CA containers.
- create_channel.sh: script used to create a channel and to join the peers to it;
- deploy_chaincode.sh: script used to deploy a smart contract to a channel according to the Fabric chaincode life cycle.

5 Related Work

Efforts have been done to apply MDD to DTL applications. In [12] the author propose a method to model the behaviour of Etherium smart contracts using Entity-Relationship Diagrams, UML class diagrams, and BPMN process models. Marchesi et al. [8] discuss a design method for blockchain applications covering the definition of system goals, use cases, data structures, implementation of smart contracts and integration and testing. UML stereotypes are defined to account for DLT concepts. In [7,16] Business Process Models and data registry models are used as input for a code generator that creates smart contracts executable in the blockchain network. Corradini et al. [4] implement a similar approach for multi-party business processes interacting with multiple blockchain networks, by including a choreography model in which the target of each task is defined. Similar approaches for the generation of smart contracts can be found in [6] where enhanced state diagrams are used to represent the entities life-cycle as a set of states and transitions and a code generator takes the diagram as input and creates a smart contract for each transition. The FSolidM tool [9] employs a similar technique but uses finite state machines to represent states and transitions. In [5] the authors propose B-MERODE, an MDD methodology for blockchain applications that represents the application in 5 layers: Domain, Permission, Core Information System Services, Information System Services, and Business Process. The models

[6] https://www.eclipse.org/modeling/emf/.

[7] https://www.eclipse.org/acceleo/.

can be created using UML Class diagrams, finite state machines, BPM or tables. The method includes also compliance checks. However, no code generation is provided. The same authors propose the MDE4BBIS framework [13], an extension to B-MERODE that defines the transformations for generating smart contracts from activity models. CEPchain [1] is a tool to connect a blockchain to a Complex Event Process (CEP) Engine; the author use domain-specific languages to model the smart contract behaviour, the CEP domain, and the events that trigger processes. The models are translated into smart contracts and into event patterns code deployed on the network and on the CEP, allowing the CEP engine to trigger transactions on the blockchain when event pattern conditions are met. Most of the revised works focus on defining software engineering artefacts for modeling blockchain applications and helping developers in the design process. Some of them include code generation, but only focused on the generation of smart contracts, leaving the rest of the application development and network deployment to be performed manually. Our approach provides an MDD method for designing hybrid DLT/DB solutions and a platform that generates the complete code for the deployment of the network, the installation of the peers and of the smart contracts and a fully functional web/mobile application for interacting with the network and the ledger, internal and external data.

6 Conclusion

In this paper, we have described a novel MDD method and toolchain for supporting the development of so-called hybrid DLT/DB applications. The method has been implemented by extending a commercial MDD platform with features specific to DLT applications. The current implementation maps the models into a fully functional Hyperledger Fabric network, the smart contracts realizing the network operations and a web/mobile GUI for interacting with both database and ledger content. Future work will focus on improving the generated code for advanced non-functional requirements such as scalability and atomicity of transactions spanning the internal and external database and on testing the usability of the method and tools with developers of real-world applications.

Acknowledgement. This work has been supported by the European Union's Horizon 2020 project PRECEPT, under grant agreement No. 958284.

References

1. Boubeta-Puig, J., Rosa-Bilbao, J., Mendling, J.: CEPchain: a graphical model-driven solution for integrating complex event processing and blockchain. Exp. Syst. Appl. **184**, 115578 (2021). https://doi.org/10.1016/j.eswa.2021.115578, https://www.sciencedirect.com/science/article/pii/S0957417421009805
2. Brambilla, M., Cabot, J., Wimmer, M.: Model-Driven Software Engineering in Practice, 2nd edn. Synthesis Lectures on Software Engineering, Morgan & Claypool Publishers (2017). https://doi.org/10.2200/S00751ED2V01Y201701SWE004, https://doi.org/10.2200/S00751ED2V01Y201701SWE004

3. Brambilla, M., Fraternali, P.: Interaction Flow Modeling Language: Model-driven UI Engineering of Web and Mobile Apps with IFML. Morgan Kaufmann, Waltham (2014)
4. Corradini, F., et al.: Model-driven engineering for multi-party business processes on multiple Blockchains. Blockchain Res. Appl. **2**(3), 100018 (2021). https://doi.org/10.1016/j.bcra.2021.100018, https://www.sciencedirect.com/science/article/pii/S2096720921000130
5. Amaral de Sousa, Victor, Burnay, Corentin, Snoeck, Monique: B-MERODE: a model-driven engineering and artifact-centric approach to generate blockchain-based information systems. In: Dustdar, Schahram, Yu, Eric, Salinesi, Camille, Rieu, Dominique, Pant, Vik (eds.) CAiSE 2020. LNCS, vol. 12127, pp. 117–133. Springer, Cham (2020). https://doi.org/10.1007/978-3-030-49435-3_8
6. Garamvölgyi, P., Kocsis, I., Gehl, B., Klenik, A.: Towards model-driven engineering of smart contracts for cyber-physical systems. In: 2018 48th Annual IEEE/IFIP International Conference on Dependable Systems and Networks Workshops (DSN-W), pp. 134–139 (2018). https://doi.org/10.1109/DSN-W.2018.00052
7. Lu, Q., et al.: Integrated model-driven engineering of Blockchain applications for business processes and asset management. Softw. Pract. Exp. **51**(5), 1059–1079 (2021)
8. Marchesi, M., Marchesi, L., Tonelli, R.: An agile software engineering method to design Blockchain applications. In: Proceedings of the 14th Central and Eastern European Software Engineering Conference Russia, pp. 1–8 (2018)
9. Mavridou, A., Laszka, A.: Designing secure Ethereum smart contracts: a finite state machine based approach. CoRR abs/1711.09327 (2017), http://arxiv.org/abs/1711.09327
10. Nakamoto, S.: Bitcoin: a peer-to-peer electronic cash system. Decentral. Bus. Rev. 21260 (2008)
11. OMG: Interaction flow modeling language (IFML), version 1.0 (2015). http://www.omg.org/spec/IFML/1.0/
12. Rocha, H., Ducasse, S.: Preliminary steps towards modeling blockchain oriented software. In: 2018 IEEE/ACM 1st International Workshop on Emerging Trends in Software Engineering for Blockchain (WETSEB), pp. 52–57. IEEE (2018)
13. de Sousa, V.A., Burnay, C.: MDE4BBIS: a framework to incorporate model-driven engineering in the development of blockchain-based information systems. In: 2021 Third International Conference on Blockchain Computing and Applications (BCCA), pp. 195–200. IEEE (2021)
14. Stahl, T., Völter, M.: Model-Driven Software Development: Technology, Engineering Management. Wiley, Chichester (2006)
15. Szabo, N.: Formalizing and securing relationships on public networks. First Monday **2** (1997)
16. Tran, A.B., Lu, Q., Weber, I.: Lorikeet: A model-driven engineering tool for blockchain-based business process execution and asset management. BPM (Dissertation/Demos/Industry), pp. 56–60 (2018)

Towards a Blockchain Solution for Customs Duty-Related Fraud

Christopher G. Harris(✉)

School of Mathematical Sciences, University of Northern Colorado, Greeley, CO 80639, USA
christopher.harris@unco.edu

Abstract. Customs duties are meant to control the flow of goods across international borders and represent a significant source of government revenue. In many parts of the world, the enforcement of customs duties is labor-intensive and fraught with inefficiencies, fraud, and corruption, leading to a substantial loss of revenue. We propose a blockchain model built on Hyperledger Fabric that can reduce some of these hindrances and pay for itself quickly through more efficient transactions and a reduction in opportunities for fraud. We illustrate how permissionless blockchain models are a better fit for international trade and show how ordered transactions in a blockchain can help reduce fraud related to avoiding duties on the import and export of goods. Last, we conduct some benchmarking on the scalability of our proposed model and examine some of the effects it may have on performance.

Keywords: Hyperledger fabric · Permissioned blockchain · Federated blockchain · International trade · VAT gap · Customs fraud · Transparency · Provenance · Supply chains · Customs procedures

1 Introduction

Global supply chains are networks that can span across multiple continents and countries for sourcing and supplying goods and services. These networks have been identified as clear beneficiaries of new distributed ledger technologies such as blockchain and smart contracts based on the impact these new technologies might deliver.

When goods are moved internationally through a global supply chain, customs duties are typically required when a good enter a country. These customs duties are designed to protect each country's interests by controlling the flow of goods across international borders, particularly the import and export of restricted goods. This flow is typically controlled by a duty - a tax or tariff - which is charged when goods enter or exit a country through a customs port of entry. Customs duty is usually based on the value of goods or is based on a product's weight, dimensions, or other specific criteria and may vary by country of origin and product. In 2017, Value-Added Tax (VAT) accounted for 62.3% of consumption tax revenues in the OECD on average, making it an important source of government revenue [1].

U. K. Rage et al. (Eds.): DASFAA 2022 Workshops, LNCS 13248, pp. 120–134, 2022.
https://doi.org/10.1007/978-3-031-11217-1_9

Besides bribery of officials and the intentional delay of imports, fraud constitutes a significant threat to the customs process. Customs duty fraud, also known as "the VAT gap," is the difference between the expected revenue and the actual revenue collected. Each year the loss to countries due to fraud and lack of information is staggering – for example, in 2020 in the EU Member states alone, customs duty fraud is estimated at €164 billion ($185.7 billion) [2]. This is due to tax fraud, tax evasion, tax avoidance, and inadequate tax collection systems. Not surprisingly, this makes any technology that can reduce these losses very impactful. Customs duty fraud is commonly accomplished using various techniques: by underreporting the quantity or weight of goods being imported, forging the bills of lading by misidentifying the product or good, or misreporting the category the goods from the Harmonized Schedule (HS). HS codes, a standardized numerical method of classifying traded products used by customs authorities in over 200 countries around the world, are an industry classification system used to determine duties on 98% of international goods [3]. Therefore, an immutable record of an item's HS category, quantity, and description as it is exported from one country and imported into another can assist in reducing fraud.

Another form of fraud involves misrepresenting the provenance of an imported good. Provenance – the place of origin or earliest known history of an item – is an important consideration for determining customs duties. Sometimes the customs offices of countries that cannot detect provenance provide "worst-case" tariffs. For instance, the US government cited circumvention as a justification for applying steel and aluminum tariffs to all countries, including countries that do not pose a national security threat. It is steel from China, not Canada, that was the stated target of the tariffs, but Chinese producers could circumvent these taxes and reach the US market through Canada to evade the tariffs. As a result, tariffs on these goods were applied to all countries; however, blockchain technology could allow shippers and national customs agencies to verify the provenance of goods and broad trade restrictions and tariffs would not be required, making them more effective and saving end consumers money.

A blockchain serves as open distributed ledgers hosted by numerous decentralized devices called nodes. Blockchain transactions between two or more parties (regardless of whether the parties are trusted or untrusted) can be recorded in a manner that is both verifiable and immutable, reducing the risk of fraud. When transactions involving the customs documents occur, additional blocks are created to record this information, then the ledger is updated with these blocks, and the updated ledger is subsequently synchronized across each node.

In this paper, we describe how a blockchain can provide a robust solution to substantially reduce customs duty-related fraud. We propose a Hyperledger Fabric architecture and benchmark it to evaluate its performance in a customs-based environment. We also provide examples to demonstrate how this proposed framework can tackle the inefficiencies and fraud related to an inability to establish provenance and the misrepresentation of imported goods.

2 Related Literature

Much of the research in blockchains and smart contracts has focused on they can revolutionize global supply chains. For example, Saberi, Kouhizadeh, Sarkis, and Shen looked at how blockchains can affect supply chain management efforts [4]. Chang, Iakovou, and Shi looked at the challenges and opportunities of blockchain approaches in [5], indicating that customs and government officials can overcome the obstacles that have plagued the International Trade Data System' (ITDS) - also known as 'Single Window' - efforts by riding the wave of private investment in blockchain. An excellent survey of blockchain use in global supply chain solutions is provided in [6].

Researchers have also realized the value of blockchain in customs enforcement. In [7], Okazaki illustrates the potential for blockchain, indicating that through blockchain customs offices would be able to collect the necessary data in an accurate and timely way, envisions a future where customs officials can use the blockchain can clear goods, coordinate with tax authorities for revenue compliance, and will help to better combat illicit financial flows. Others have investigated how VAT fraud in the EU can be solved by blockchain, illustrating the failures of other pre-blockchain attempts at mitigating fraud, such as the VAT Information Exchange System (VIES) [8]. Another solution, Digital Invoice Customs Exchange (DICE) improves upon VIES but is based on a centralized (instead of a distributed) ledger such as blockchain. The inherent weaknesses of a centralized ledger would be resolved by blockchain, including a single point of failure, the potential for corruption, and lack of data insecurity (due to the ability to easily change information in one location with little oversight), and the inability to capture relevant transactional data.

In [9], McDaniel and Norberg reveal that blockchain could discourage corruption by simplifying procedures and reducing the number of government offices and officials involved in each transaction. Moreover, a blockchain network would allow for disintermediation in trade by requiring total, decentralized consensus within the network. This would help to ensure that no single party could be cheated by another without consequence.

Another benefit is that the real-time auditability of blockchain allows users to see when and where disputes arise and exactly what the discrepancies are. Figure 1 shows the stakeholders involved with blockchain implementations, many of whom would be involved in fraudulent activities, which introduces additional complexities to any tracking solution.

Blockchains are used to prove the provenance of goods, tracing their supply chain journey from the source to the end purchaser. For example, Fiji and New Zealand piloted the use of blockchain to track tuna, showing when and where it was caught [10] to verify to consumers that the tuna provided was sustainably caught.

There have been few articles that demonstrate blockchain-related implementations specifically to aid in customs-related tasks. In [11], the authors implement a privacy-preserving system for freight declaration. In [12], the authors focus on illegal exports and propose a blockchain solution to track potentially illegal goods exiting the country. In [13], a focus on cross-validating customs forms is examined using the blockchain, with the authors demonstrating a reduction in the customs clearance time and proportion of goods needed for sampling.

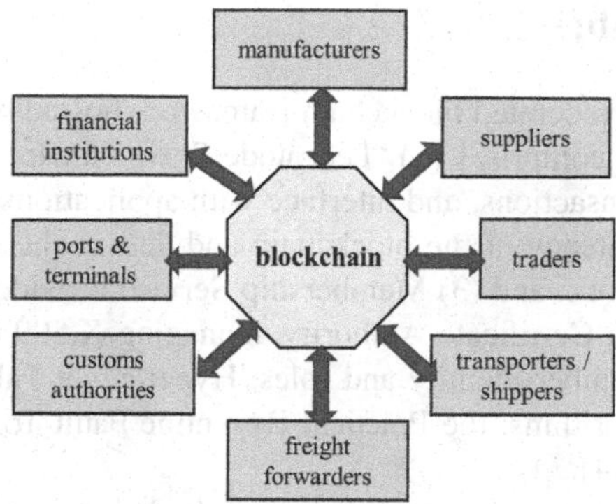

Fig. 1. Stakeholders involved with an import/export solution to reduce customs fraud

3 Types of Blockchains for Customs Enforcement

All blockchain platforms are distributed ledgers, are immutable, and make use of consensus mechanisms; however, there are several blockchain paradigms. Most distributed blockchain systems, such as Ethereum and Bitcoin are permissionless, therefore, any node can participate in the consensus process – a process wherein transactions are ordered and bundled into blocks. Permissionless systems, therefore, rely on probabilistic consensus algorithms (e.g., Proof of Work, Proof of Stake), which eventually guarantee ledger consistency with a high degree of probability; however, permissionless systems are still vulnerable to ledger divergence (also known as a ledger "fork"), where different participants in the network have a different view of the accepted transaction order.

While permissionless blockchains follow Nakamoto's original design for a blockchain to serve as an open, free, public system [4], a permissioned blockchain is effectively the opposite. Permissioned blockchains, such as Hyperledger Fabric and Ripple, are closed ecosystems where participation needs to be approved by a central authority. Transactions on permissioned blockchains, also called private blockchains, are validated only by approved members of that blockchain. As a result, participants in a permissioned blockchain can dictate the network's structure and consensus policies and control who sees transactional information.

Because consensus can be done more efficiently, permissioned blockchains deliver better performance. Permissioned blockchains also offer better governance structures, are more customizable, provide access controls, and are more scalable. Drawbacks of permissioned blockchains are that they must rely on the security of its members, can be more easily censored or regulated, are less transparent to the public, and are less anonymous – which have limited impact on the actors involved with international trade such as the banking sector, shippers and freight forwarders, buyers, and sellers.

A third type is federated blockchains. Although they share similar scalability and privacy protection level with a permissioned blockchain, their main difference is that a set of nodes, named leader nodes, is selected instead of a single entity to verify the transaction processes. This enables a partially decentralized design where leader nodes can grant permissions to other users.

4 Hyperledger Fabric

Hyperledger Fabric is a federated blockchain framework hosted by The Linux Foundation. A Fabric Network comprises (1) "Peer nodes", which execute chaincode, access ledger data, endorse transactions, and interface with applications; (2) "Orderer nodes" which ensure the consistency of the blockchain and deliver the endorsed transactions to the peers of the network; and (3) Membership Service Providers (MSPs), each generally implemented as a Certificate Authority, managing X.509 certificates which are used to authenticate member identity and roles. Hyperledger Fabric allows for use of different consensus algorithms; the Practical Byzantine Fault Tolerance (PBFT) is the most commonly used [14].

Fabric-based applications that want to update the ledger are involved in a process with three phases that ensures all peers in a blockchain network keep their ledgers consistent with each other. One mechanism featured by Fabric is a kind of a node called an orderer (or "ordering node") that performs transaction ordering. Working with other contributing nodes, these form an ordering service. Because Fabric's design relies on deterministic (instead of probabilistic) consensus algorithms, any block that is validated is guaranteed to be final and correct. Once a transaction has been written to a block, its position in the ledger is immutably assured. Fabric ledgers cannot fork the way they do in many other distributed blockchains.

4.1 Transaction Overview

Fabric-based applications that wish to update the ledger are involved in a three-phase process that ensures all peers in a blockchain network keep their ledgers consistent with each other. First, a client application sends a transaction proposal to a subset of peers that will invoke a smart contract to produce a proposed ledger update and then endorse the results.

Endorsement is a mechanism in Fabric to check the validity of a transaction (see [15] for an overview and evaluation of various Fabric endorsement policies). The endorsing peers do not apply the proposed update to their copy of the ledger since transactions need to be ordered across multiple nodes; instead, the endorsing peers return a proposal response to the client application. The endorsed transaction proposals will ultimately be ordered into blocks in phase two, and then distributed to all peers for final validation and commitment in phase three.

Figure 2 illustrates the process of how a participant (i.e., client) invokes a transaction request through the client application while Fig. 3 provides a swim lane diagram of how these nodes interrelate. A transaction request could, for example, be from a supplier uploading the customs declaration information so that it can be tracked by the other entities shown in Fig. 1.

1. The client application broadcasts the transaction invocation request to the endorser peer.
2. The endorser peer checks the Certificate Authority details and other information necessary to validate the transaction. It then executes the chaincode and returns the endorsement responses to the client. The endorser peer sends transaction approval

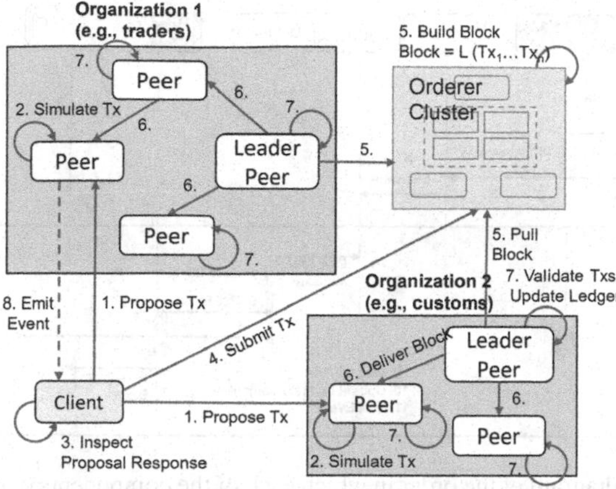

Fig. 2. An illustration of the eight steps used in our proposed Fabric model for ordering and executing chaincode-based transactions.

or rejection as part of the endorsement response. If the customs declaration already exists in the system, for example, the transaction will be rejected as we would not permit two transactions for the same set of goods

3. The client evaluates the response from the endorser peer. This is typically done through a software application.
4. If approved, the client now sends the approved transaction to the orderer peer to be properly ordered and be included in a block.
5. The orderer node includes the transaction (comprised of key-value pairs) into a block and forwards the block to the anchor nodes of different member organizations of the Fabric network. The leader peers deliver the block to the other peers within the organization.
6. Anchor nodes then broadcast the block to the other peers inside their organization. This allows consistency within the organization.
7. These individual peers then update their local ledger with the latest block. Thus, synchronization occurs across the entire network. Every node in the system now has the updated information about the transaction.
8. The event is then emitted back to the client indicating success or failure. Thus, it provides a message to the calling application regarding the success of the customs declaration upload to the block.

4.2 Ensuring Transaction Order

An essential part of Fabric is the ordering service. This is one of Fabric's strengths and is ideal for industries where the ordering of transactions needs to be ensured. This may indicate the provenance of goods in a shipment, from raw components to finished products. Special nodes called *orderers* receive transactions from many different application

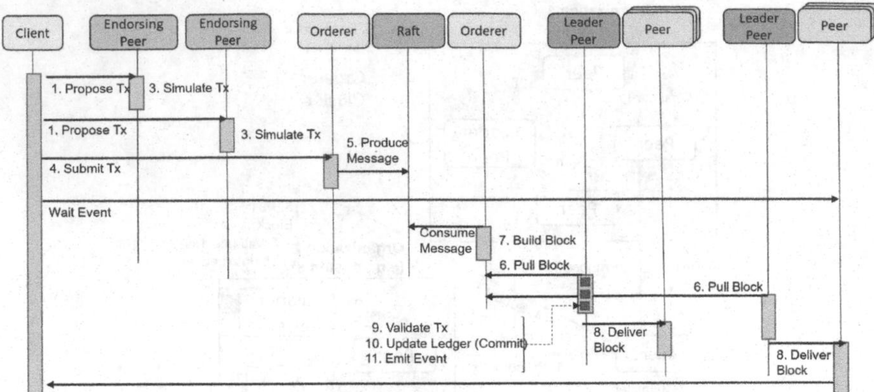

Fig. 3. A swim lane diagram of the order in which each of the components of our proposed fabric model interacts to write a transaction to the block.

clients concurrently. These orderers work together to collectively form the ordering service with the role to arrange batches of submitted transactions into a precise sequence and package them into blocks, which become part of the blockchain [16].

The most prominent ordering services in Fabric are Solo, Raft, and Kafka. Solo involves only a single ordering node and Kafka is a more challenging ordering service to implement while maintaining few benefits over a Raft-based implementation. Kafka and Zookeeper are not designed to be run across large networks but run in a tight group of hosts, limiting a Kafka cluster to a single organization, and rendering it a poor choice for our model.

Orderer peers maintain the central communication channel for Fabric. Thus, the orderer peer is responsible for ensuring a consistent ledger state across the network. An orderer peer creates the block and delivers that to all the peers. In addition to promoting finality, separating the endorsement of chaincode execution from ordering gives Fabric advantages in performance and scalability, eliminating bottlenecks that can occur when the same nodes perform both execution and ordering.

In addition to ordering transactions, orderers also maintain the list of organizations that are permitted to create channels. This list is known as the consortium, and it is maintained in the configuration of the orderer system channel. They also enforce basic access control for channels, restricting who can read and write data to them, and who can configure them. Because of their central role in access control, configuration transactions are processed by the orderer, to ensure that the requestor has the proper administrative rights. If so, the orderer validates the update request against the existing configuration, generates a new configuration transaction, and packages it into a block that is relayed to all peers on the channel. The peers then process these configuration transactions to verify that the modifications approved by the orderer do indeed satisfy the policies defined in the channel. For instance, this ensures that there is one declaration form for a shipper for a set of goods. When it is checked by a customs official in the destination country, a record of the customs official performing the task is stored in the block, which can help detect customs officials receiving bribes for accepting fraudulent declaration forms.

It can also record the provenance of goods so the opportunity to circumvent tariffs is reduced. Also, given the heterogeneous environments for international trade, including situations where a node (e.g., a customs officer at a remote frontier) may be unavailable for extended periods, each of these properties can ensure the integrity of the information written to and read from the block despite these conditions.

5 Detecting and Mitigating Customs Duty Fraud Using Fabric: Three Scenarios

The incentives for fraudsters involved with international trade to avoid customs duties are significant; many perceive it as a victimless crime [17]. Moreover, even as technology becomes more sophisticated, the chance of customs fraud detection remains low [18]. Here we examine three scenarios of customs duty fraud and how the Fabric architecture (and blockchains in general) can increase the chance of fraud detection.

5.1 Underreporting the Cargo Weight or Quantity

This is the most prevalent type of customs duty fraud. Underreporting fraud may be "lost" goods – when a trader underreports cargo and is then able to steal the difference. Or it could be achieved when shippers misrepresent the amount or quality of shipped goods to appear to comply with customs requirements or to pay lower import duties [9].

The proposed Fabric solution is secure by design, and each transaction is uploaded to the chain if and only if it is agreed upon by the other peers. Therefore, it is nearly impossible to edit past transactions without the other peers in the network approving them. And a record of the trader or shipper providing the information is written to the blockchain, providing a shipment history. Figure 4 shows an illustration of how these transactions are written and read from the blockchain. A trader or shipper with one falsified transaction would be investigated for all other transactions, making it easy to detect patterns of fraud.

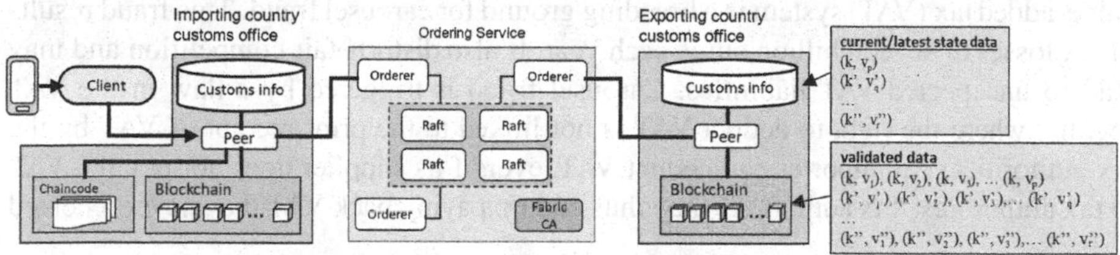

Fig. 4. An illustration of customs forms validity check of customs information between the importing country's customs office and the exporting country's customs office. Each organization maintains a copy of the documents issued when the customs form was generated by the trader in the exporting country. The validated data is stored in the blockchain upon successful execution of the Chaincode terms and each organization is provided an immutable copy of the transaction. Thus, the customs document can be validated before importing country's customs office approves the goods and the correct duty amount can be charged.

There are other intangible benefits to this solution as well. The international shipping industry carries 90% of the world's trade in goods and still largely relies on paper documentation [19]. Since any blockchain can simplify the flow of information, it would discourage corruption by simplifying procedures and reducing the number of government offices and officials involved in each transaction. In turn, this reduces the opportunities for bribery and corruption in port.

5.2 Relying on a Bill of Lading for Customs Audits

At a port of entry, customs officials rely on a bill of lading to determine which arriving containers to audit. A bill of lading is an aggregation of information about the goods in a single shipment of containers coming into a port. It is sent to the relevant customs agency at least 24 h before the ship arrives in port [11]; however, there are several shortcomings of this current system:

- The bill of lading does not tell the customs agency through which other ports a container of goods may have traversed, making circumventing difficult to detect.
- The bill of lading is an aggregation of the list of goods, so it is not the source of data
- The bill of lading is only required to be received 24 h before a ship's arrival.

Using the blockchain model discussed here, all information about each good (not an aggregate) would be presented in real-time, allowing for customs officials to make a better determination of which containers to audit. Also, since all information is written to the blockchain, a complete history, including intermediate ports, could easily be identified.

5.3 Misrepresenting the Country of Origin: Shipping Goods to an Intermediate Country to Avoid Import Tariffs

Misrepresenting the country of origin is one known ploy to avoid customs duties. Along with this is another type of fraud: carousel fraud. The current European Union (EU) value-added tax (VAT) system is a breeding ground for carousel fraud. This fraud results in tax losses of several billion euros each year. It also distorts fair competition and may lead to unexpected VAT liabilities. Carousel fraud is triggered by a flaw in the VAT regime, where the right to deduct VAT is not linked to the prior receipt of VAT by the tax authorities. An importer can deduct VAT, even if its supplier does not pay the VAT to tax authorities. Tax authorities may thus end up paying back VAT they never received [20].

Blockchain makes it possible to record transactions and chronological events across supply chains in an irreversible manner and gives supply chain partners access to this transaction history, which is essential to establish a good's provenance. Therefore, customs agents can be aware if a product has components from a country upon which tariffs are imposed. Combining provenance and HS code functionality, allows customs to trace the source and track subsequent modifications of products, allowing the tariffs on goods to be targeted and reducing the potential. Likewise, for detecting carousel fraud, customs agents can work with tax authorities and check the status of VAT paid on a good before allowing a deduction.

Four types of assurances can be provided from the blockchain: origin assurance, authenticity assurance, custody assurance, and integrity assurance [21]. Blockchain technologies such as Fabric deliver these assurances by providing traceability, certifiability, trackability, and verifiability of product information, respectively, as the good traverses the global supply chain. In addition to the above assurances, the blockchain may also be used to improve the detection of illicit trade flows and deter illegitimate efforts to circumvent trade rules [9]. Such applications could aid customs and law enforcement in facilitating the flow of legitimate trade.

6 Benchmarking Scalability

We wish to examine the feasibility of our prescribed model's performance. This scalability benchmark will be discussed in much more detail in a future article, but we briefly describe it here. Unfortunately, the performance of both permissionless and permissioned/federated blockchain systems lags far behind that of a typical database. The primary bottleneck is the consensus protocol used to ensure consistency among nodes. PBFT does not scale to many nodes [22].

Another source of inefficiency is the order-execute transaction model, in which transactions are first ordered into blocks, then they are executed sequentially by every node. This model, which occurs in Hyperledger Fabric v0.6, is inefficient since there is no concurrency in transaction execution [15]. However, version v1.1 of Hyperledger Fabric and later, a new transaction model called the execute-order-validate model is used. Mimicking the optimistic concurrency control mechanisms in advanced database systems, this execute-order-validate model consists of three phases described in Sect. 4. In the first phase, transactions are executed (or simulated) speculatively. This simulation does not affect the global state of the ledger. In the second phase, they are ordered and grouped into blocks. In the third phase, called validation or commit, they are checked for conflicts between the order and the execution results. Finally, non-conflicting transactions are committed to the ledger.

By allowing parallel transaction execution, Fabric v1.1 can potentially achieve higher transaction throughputs than systems that execute transactions sequentially. However, it introduces an extra communication phase compared to the previous Fabric v0.6 order-execute model, thus incurring more overhead.

Fabric consists of three types of ordering services, namely Solo, Kafka, and Raft. Solo does not have a consensus algorithm and is only used during the development of blockchain networks while Kafka and Raft are crash fault tolerant. We evaluate the throughput and latency of Fabric v1.1 in a local cluster of up to 32 nodes, running with Raft ordering service. Our work differs from Blockbench [22], which benchmarks an earlier version of Fabric with an order-execute transaction model. Other benchmarking studies such as [23] did not consider the effect of scaling of the cluster on performance. Similarly, other studies (e.g., [24, 25]) fix the size of the ordering cluster and use fewer than 10 nodes. In contrast, we examine the impact of scaling ordering cluster on the overall performance, using up to 32 nodes in a cluster, which is likely to occur when customs offices, importers, and exporters interact.

In [26], Surjandari et. al. found Raft was superior to Kafka in terms of success rate and throughput rate when conducting invoke transactions due to its simpler framework Raft employs the Raft consenter which was created directly from ordering service nodes whereas Kafka requires Kafka brokers and an elaborate Zookeeper Ensemble. Indeed, most other performance benchmarking studies (e.g. [27]) examine Kafka, whereas our focus here is on benchmarking Raft.

Although Hyperledger Caliper [28] is the official benchmarking tool for Fabric, it offers little support and documentation on how to benchmark a real distributed Fabric ordering service such as Kafka or Raft. Most of the documentation and scripts are considering the Solo orderer and a single client. To accommodate benchmarking at scale, we use a set of scripts to start up a Raft-based Fabric network across cluster nodes, launch additional benchmarking tools, and benchmark performance across distributed clients.

6.1 Benchmarking Setup

A Hyperledger Fabric network topology is described by the number of endorsing peers N, number of clients, C, with mutual transaction send rate, T, number of Fabric orderers, O, number of Raft consenters, R. All the peers belong to different organizations and serve as endorsing peers, P. As with [27], the block size is set to 100 transactions/block, and the timeout is set to 2s, and we set the number of clients, C, to 1.

We use Smallbank [29], a widely used OLTP database benchmark workload tester. Simulating typical customs brokerage transfer scenario and a large class of transactional workloads allow us to test Fabric at scale. These experiments were run on a 48-node commodity cluster. Each server node has an Intel Xeon E5–1603 @ 2.80 GHz, 16 GB of RAM, a 1 TB hard drive, and a Gigabit Ethernet card. Our nodes are running Ubuntu 20.04.3 (Focal Fossa).

Overall, as the number of Fabric orderers increases, system throughput performance significantly degrades due to increased communication overhead – in Fig. 5(a)) we increase the number of orderers from $O = 4$ to $O = 12$. In our subsequent experiments, we fix the number of Fabric orderers to $O = 4$ and vary the number of Raft consenters, R, to assess the impact on scaling Fabric.

At the networking level, we find that the client, peers, Fabric orders, and Raft consenters exhibit the highest traffic both with sending and receiving. The high traffic of the Fabric orderer is explained by its double role, as receiver of requests from the client and as a dispatcher of requests to Raft.

Next, we fix the request rate and investigate the impact of increasing the number of peers. We fix the total request rate to $C * T = 300$ and $C * T = 400$, which represent the points before and after saturation occurs. We vary the number of Raft consenters from $R = 4$ to $R = 16$ in multiples of 4. We increase the number of peers to $P = 32$. The average throughput and latency of each experimental setting are shown in Fig. 5(b) and Fig. 5(c), respectively.

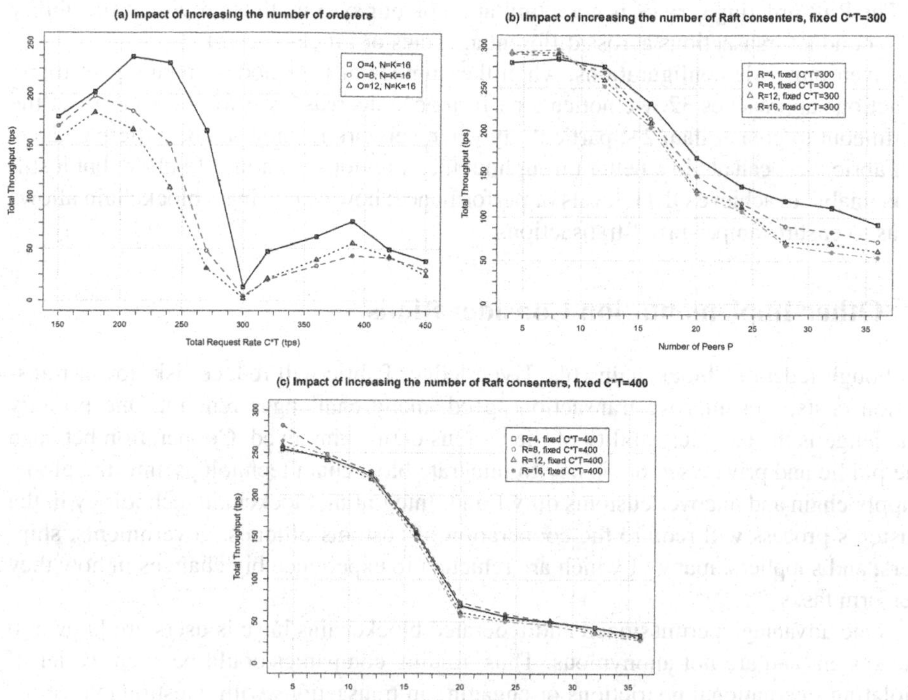

Fig. 5. Changes in performance as (a) we increase the number of orderers from O = 4 to 12; we increase the number of Raft consenters, R, from 4 to 16 and we increase the number of peers, P, from 4 to 32 for (b) C * T = 300 and (c) C * T = 400.

6.2 Performance Benchmarking Results and Analysis

Increasing the number of peers strongly degrades the system's throughput and limits scalability. We examine the system logs and observe that scaling the number of peers produces significant overhead in both the endorsement and ordering phases of the transaction. As the number of endorsing peers grows, each client must wait for a larger set of endorsements from a larger set of peers to prepare the endorser transaction. From the logs, we observe that the clients return numerous timeout errors while collecting the endorsements from peers and, accordingly, discard those transaction proposals. Consequently, clients send endorser transactions to the orderers at a decreased rate due to dropped transactions and the time overhead for collecting endorsements. Unlike our findings when scaling the peers, we find that scaling the Raft consenters does not impact the throughput pattern or scalability of the system, which is similar to what [27] found performing a similar benchmark with the Kafka brokers.

Our proposed model would be limited by the number of transactional peers used in a system, limiting the size of the overall customs offices and importers/exporters involved in transactions. In a typical customs system as proposed here, few of the peers would be involved with the transactional activity of ordering; while ordering is a necessity, we can effectively limit this to make it practical. We note the following observations:

- The Raft ordering service is not a limitation on our system, thus providing the ability to conduct transactions across different agencies or offices (peers)
- Given our node configurations, when the number of peer nodes engaging in transactions approaches 32, we notice a performance decrease; we would need to set the timeout to greater than 2 s, particularly when network latency can be anticipated.
- Fabric v.1.1 can attain a better throughput than previous versions of Fabric, but it still is unable to achieve OLTP levels of performance; however, using a blockchain allows us to ensure tamper-proof transactions.

7 Other Implementation Considerations

Although federated blockchains like Hyperledger Fabric will reduce risk, lower transaction costs, and improve transaction speed, many challenges remain. One primary challenge is the technical ability of the various entities involved. Cooperation between the public and private sectors is key to integrate blockchain technology into the global supply chain and uncover customs duty fraud. Integrating blockchain technology in the customs process will require the cooperation of customs officials, governments, shippers, and suppliers, many of which are reluctant to experience big changes in how they perform tasks.

One advantage permissioned and federated blockchains have is users are known to the system and are not anonymous. Thus, trading companies could be even warier of violating international restrictions or engaging in transactions with questionable actors in the importing and exporting processes.

8 Conclusion

We have proposed a blockchain solution to help combat customs duty fraud. Many of the tasks currently completed on paper or using electronic documents could be integrated into a blockchain. For networks where user anonymity is unnecessary - such as customs authorities and others in the global supply chain - federated blockchains such as Fabric can provide assurances of accurate ordered transactions that can scale as countries adopt its use. Our proposed system, built on Fabric, maintains the advantages of a permissioned distributed ledger, such as improved security, immutability, visibility and traceability, and the individual control of networks and channels.

We have also illustrated three different scenarios involving customs duty fraud and have demonstrated how this proposed system can remediate this risk, which in turn increases revenue and reduces corruption. However, our system is not without its challenges. We understand that implementing a system is a large undertaking and would be assisted by having a killer app as a front end that can allow each of the tasks a customs officer is likely to perform to be accomplished quickly and more easily.

We have investigated the performance of the number of nodes, orderers, and Raft consenters using Smallbank and a modified version of Hyperledger Caliper as benchmarking tools. Although increasing the number of orderers degrades the system beyond; also, an increase in the number of peers negatively affects system throughput and limits scalability as we approach 32 nodes, while additional stress on the Raft ordering service does not. We note while the benefits of our proposed model provide for tamper-proof transactions, which is ideal in a customs environment to fight fraud, it is not meant to, nor does it need to, be transaction intensive.

Blockchain technology is still a nascent technology, with its vast potential understood but only limited endeavors into solving real-world problems. In future work, we plan to investigate several smaller, more targeted aspects of how Fabric can assist with specific tasks of fraudulent behavior and benchmark their performance. We also plan to introduce front-end applications to make the reading and writing of transactions much easier to accomplish.

References

1. Tax Foundation. New European commission report: vat gap. https://taxfoundation.org/vat-gap-eu-europe-2020/
2. European Commission. Tax and customs union. https://ec.europa.eu/taxation_customs/vat-gap_en
3. Pierce, J.R., Schott, P.K.: A concordance between ten-digit US harmonized system codes and SIC/NAICS product classes and industries. J. Econ. Soc. Meas. **37**(1–2), 61–96 (2012)
4. Saberi, S., Kouhizadeh, M., Sarkis, J., Shen, L.: Blockchain technology and its relationships to sustainable supply chain management. Int. J. Prod. Res. **57**(7), 2117–2135 (2019)
5. Chang, Y., Iakovou, E., Shi, W.: Blockchain in global supply chains and cross border trade: a critical synthesis of the state-of-the-art, challenges and opportunities. Int. J. Prod. Res. **58**(7), 2082–2099 (2020)
6. Juma, H., Shaalan, K., Kamel, I.: A survey on using blockchain in trade supply chain solutions. IEEE Access **7**, 184115–184132 (2019)
7. Okazaki, Y.: Unveiling the potential of blockchain for customs. In: WCO Research Paper No 45 (2018)
8. Ainsworth, R.T., Shact, A.: Blockchain (distributed ledger technology) solves VAT fraud. In: BU. School of Law, Law, and Economics Research Paper, pp. 16–41 (2016)
9. McDaniel, C.A., Norberg, H.C.: Can blockchain technology facilitate international trade? In: Mercatus Research Paper (2019)
10. Visser, C., Hanich, Q.: How blockchain is strengthening tuna traceability to combat illegal fishing The Conversation, pp. 1–4 (2018)
11. Juma, H., Shaalan, K., Kamel, I.: Customs-based blockchain solution for exportation protection. In: Shen, H., Sang, Y. (eds.) PAAP 2019. CCIS, vol. 1163, pp. 405–416. Springer, Singapore (2020). https://doi.org/10.1007/978-981-15-2767-8_36
12. Vos, D., et al.: DEFenD: a secure and privacy-preserving decentralized system for freight declaration (2018). arXiv preprint arXiv:1803.09257
13. Wu, B., Li, Y., Liang, X.: A smart contract-based risk warning blockchain symbiotic system for cross-border products. In: Chao, K.-M., Jiang, L., Hussain, O.K., Ma, S.-P., Fei, X. (eds.) ICEBE 2019. LNDECT, vol. 41, pp. 274–289. Springer, Cham (2020). https://doi.org/10.1007/978-3-030-34986-8_20
14. De Angelis, S.: Assessing security and performances of consensus algorithms for permissioned blockchains (2018). arXiv preprint arXiv:1805.03490
15. Soelman, M., Andrikopoulos, V., Pérez, J.A., Theodosiadis, V., Goense, K., Rutjes, A.: Hyperledger fabric: evaluating endorsement policy strategies in supply chains. In: 2020 IEEE International Conference on Decentralized Applications and Infrastructures (DAPPS), pp. 145–152. IEEE (2020)

16. Harris, C.G.: Using Hyperledger fabric to reduce fraud in international trade. In: 2021 IEEE International Conference on Artificial Intelligence and Blockchain Technology (AIBT). IEEE (2021)
17. Widdowson, D.: Bordering on corruption: an analysis of corrupt customs practices that impact the trading community. World Customs J. **7**(2), 11–21 (2013)
18. Ferreira, C., Engelschalk, M., Mayville, W.: The challenge of combating corruption in customs administrations. Many Faces Corruption **11**, 367–386 (2007)
19. IBM and Maersk: The paper trail of a shipping container. Infographic (2017). https://www-01.ibm.com/common/ssi/cgi-bin/ssialias?htmlfid=XI912347USEN
20. Wolf, R.: VAT carousel fraud: a European problem from a Dutch perspective. ntertax **39**(1), 30–31 (2011)
21. Montecchi, M., Plangger, K., Etter, M.: It's real, trust me! Establishing supply chain provenance using blockchain. Bus. Horiz. **62**(3), 283–293 (2019)
22. Dinh, T.T.A., Wang, J., Chen, G., Liu, R., Ooi, B.C., Tan, K.L.: Blockbench: a framework for analyzing private blockchains. In: Proceedings of the 2017 ACM International Conference on Management of Data, pp. 1085–1100 (2017)
23. Androulaki, E., et al.: Hyperledger fabric: a distributed operating system for permissioned blockchains. In: Proceedings of the thirteenth EuroSys Conference, pp. 1–15 (2018)
24. Baliga, A., Solanki, N., Verekar, S., Pednekar, A., Kamat, P., Chatterjee, S.: Performance characterization of hyperledger fabric. In: 2018 Crypto Valley Conference on Blockchain Technology (CVCBT), pp. 65–74. IEEE (2018)
25. Thakkar, P., Nathan, S., Viswanathan, B.: Performance benchmarking and optimizing hyperledger fabric blockchain platform. In: 2018 IEEE 26th International Symposium on Modeling, Analysis, and Simulation of Computer and Telecommunication Systems (MASCOTS), pp. 264–276. IEEE (2018)
26. Yusuf, H., Surjandari, I.: Comparison of performance between Kafka and raft as ordering service nodes implementation in Hyperledger fabric. Int. J. Adv. Sci. Technol. **29**(7s), 3549–3554 (2020)
27. Nguyen, M.Q., Loghin, D., Dinh, T.T.A.: Understanding the scalability of Hyperledger Fabric (2021). arXiv preprint arXiv:2107.09886
28. Hyperledger Caliper. https://www.hyperledger.org/projects/caliper
29. Cahill, M.J., Röhm, U., Fekete, A.D.: Serializable isolation for snapshot databases. ACM Trans. Database Syst. (TODS) **34**(4), 1–42 (2009)

Securing Cookies/Sessions Through Non-fungible Tokens

Kaushal Shah[1](✉), Uday Khokhariya[1], Nidhay Pancholi[1], Shambhavi Kumar[1], and Keyur Parmar[2]

[1] Pandit Deendayal Energy University (PDEU), Gandhinagar, India
shah.kaushal.a@gmail.com

[2] Sardar Vallabhbhai National Institute of Technology, Surat, India
keyur@coed.svnit.ac.in

Abstract. Cookies are used as authentication tokens after successfully validating users by web applications. As they are stored on the client's side, it makes them vulnerable to hijacking, stealing, and unauthorized distribution. There are methods in the literature that are developed to protect cookies. However, cookie stealing and impersonating is still a widely adopted practice. Session cookies are used so that the user does not have to log in again and again. When an attacker accesses these cookies, he/she can join the user's active session as well; this phenomenon is called cookie hijacking. Here, we are proposing a model using the concept of blockchain, non-fungible tokens and smart contracts, which prevent the attacker from performing unauthorized tasks even when an individual gets access to the user's session cookies. The web server uses the unique identification address of the user to generate a session ticket which would represent the ownership of the verified user. Whenever, a request is made, it gets authenticated by the blockchain; thereby making the cookie verification decentralized. This method ultimately aims to prevent unauthorized users from performing tasks through a user's active session, which will decrease identity stealing and imitation through cookies.

Keywords: Non-Fungible Token (NFT) · Cookies · Session hijacking · Blockchain · Smart contracts

1 Introduction

In recent years, there has been much development in Information Security. However, lot of identity stealing still takes place [1]. There has been rapid growth in the processing power available and therefore, protecting data has become even more challenging. Some of the technologies being employed to tackle this are Firewalls, Two Factor Authentication, HTTPS. However, no system is perfect, and for every use case, a different set of information needs to be protected, and hence the technologies keep developing. When the Internet is within reach of almost anyone on Earth, identity impersonation has become a common problem [2]. The attacker poses as someone they are not to perform tasks that

U. K. Rage et al. (Eds.): DASFAA 2022 Workshops, LNCS 13248, pp. 135–146, 2022.
https://doi.org/10.1007/978-3-031-11217-1_10

he/she is not authorized to perform. One way to pose as an authorized user and interact with the web servers is through cookie hijacking [3].

The HTTP is a stateless protocol in which the web browsers send the requests, and the servers respond. All requests are treated as independent requests; hence no state is retained [4], and to include the functionality of "state," cookies are used. Cookies are stored on the client-side, and hence this makes the cookies vulnerable to attacks. Cookie hijacking is when a malicious user somehow gets the cookie of another user, poses as the authorized user and utilizes them for accessing vital information [5]. Many websites use the functionality of 'Remember Me', so that the users do not have to log in again through their credentials whenever they visit the website from their device. The necessary information to perform this function is stored in the session cookie present on the user's device. The attackers often steal these session cookies and utilize them to enter into the user's active session directly [6]. A way of eliminating session hijacking is by tracking the number of active requests and changes in address of the requesting user.

We propose a model including blockchain, non-fungible tokens (NFT), and smart contracts, which denies the attackers the opportunity to steal the users' cookies and join an active user session. Blockchain is a shared, immutable ledger that stores vital information [7]. Transactions that are approved are stored in blocks that contain the cryptographic hash of its previous block. Any manipulation in the transactions after a particular transaction has been approved will cause the block's hash to change, which would be enough to identify malicious activity. These blockchains store the NFTs, which contain a unique identification key, distinguishing them from additional tokens and other metadata [8]. Smart contracts will be used to detect whether the user is authorized and has subsequent permissions to make a valid request. A smart contract is a piece of code that is executed when a specific set of conditions are met. The parties in the contract agree to the set codes that are electronically programmed and respond to certain encoded conditions [9].

2 Background

With the internet becoming an integral part of people's lives, privacy has become a significant concern as discussed in [10–12]. With the advancement of web tracking techniques, it has become more challenging for ordinary users to secure their systems from peeking eyes. Nowadays, almost every website asks for authentication before exploring further and uses cookies to make the interaction smoother [6]. By signing in or taking the cookies from a web app, users' personal information is no longer secured. These websites require information like name, age, earning, likes, dislikes, preferences, and many more. The websites are in the race of making money and establishing their presence. For this, they try to attract the user by showing something that fascinates them.

Once end-users accept the cookies, the websites can track the pages they visit and record their activity, leading to privacy breaching activities [13]. Some applications have premium features available. Hackers are grabbing these cookies from premium users and making them publicly available. This gives the access of account to unauthorized users without paying for a subscription, generating losses for the service providers. Therefore, we see the cookie hijacking is an issue for ordinary users and cause trouble for service providers and established firms like Netflix, Amazon Prime, and Grammarly.

3 Literature Review

The newly budding blockchain technology has been implemented in several sectors, and now it is being incorporated to protect the data which is not to be shared. A lot of research work has been done on the blockchain [14–16]. Here firstly, we will understand the basics of blockchain and cookies.

3.1 Blockchain

Blockchain is an emerging technology that is currently being explored widely. Several blocks or ledgers are put together to constitute a blockchain. A group of people creates their separate ledgers or blocks. These ledgers are rigid, and a cryptographic hash key connects each ledger. These keys are strenuous to decode, and they make sure that the authenticity of the transaction is maintained [17]. They first validate the entries, safeguard them and also preserve the previous ones. When an online transaction is initiated, a third party carries out the transaction, which cannot be manipulated. The breach can occur if this third party has any trust issues, and blockchain helps in safeguarding it.

The blockchain ledgers are protected using cryptographic techniques. When a transaction is initiated, a digital signature is required as the public key on the receiver's end, and this digital signature is signed by the sender using his private key [16]. The authentication of each transaction is checked, and then each transaction is stored in the public ledger. These ledgers are to be stored properly. They should be arranged in a proper order which is ensured by the use of blockchain. If we try to define blockchain in simpler words, it is a special database that keeps track of all the transactions or operations taking place in the system. A block or ledger has two main components: the header and the body. The header has all the essential information like block id, date of creation, and links to the next or previous block. The body contains all the information related to the transaction [7]. One of the most widely adopted conception of the blockchain is bitcoin [16].

3.2 Cookies

With the advancement in internet technology, it has become nearly impossible to maintain privacy [18]. A few Social Media websites like Facebook have become a significant part of the lives of the people. These websites have currently procured 80% of users' browsing history [19]. These techniques of tracking information have also evolved with time. The stored cookies often act as bridges and make it easier for the tracers to study the patterns and interests of the user.

Cookies are texts in the form of a string transferred from the website stored on the user's device. They help the computer remember some basic information about the user [6]. The cookies stored include the user's latest browsing history, the forms he has filled, or the websites he has visited. A cookie can be designed according to the willingness of the manufacturer. They can last on the user's system for a time ranging from a few minutes to forever. However, cookies are not illegal or useless assets and are required in some cases like e-commerce websites. These websites use cookies and store the user-selected items in the shopping cart. The information about the user and his selections is stored using cookies.

Cookies are easily accessible and are often used by a few websites to understand users' browsing patterns or habits [2]. This practice rings an alarming concern regarding the privacy risk that the users are being exposed to [20]. It has been clearly stated by the user advocates, marketers, and policymakers that the cookies are resulting in a breach of privacy. The summary of the existing literature is shown in Table 1.

Table 1. Summary of the existing work.

Reference no.	Observations
[4]	• Cookie hijacking is causing serious privacy issues • A proper model has to be developed to stop the cloning of cookies and impersonating other users
[6]	• The paper proposes replacing standard cookies with one-time cookies • Their model generates a new cookie for each request, making cookie stealing obsolete to perform other tasks
[20]	• Cookies are being cloned by third parties, which is hindering the privacy of the users • The cookie uses and disclosures have increased from 2000 to 2007 • The web environment has increased the threat and the need to develop a method to ensure their safety
[21]	• Introduction to tracking NFT using Hyperledger Fabric program • Demonstrates an implementation of blockchain technique in most scalable format • The paper has elaborated the errors that occur when several databases from different data servers are combined at one place
[22]	• The blockchain system along with NFT can be easily used to replace any paper information • The end-to-end encryption would provide efficient security and speed up the authentication of users • The IdChain concept can be implemented to encourage use of encrypted smart cards
[23]	• Consumer has the obsolete right on the information that is being stored online • Several laws are being implemented to control the computer crimes
[24]	• E-commerce websites are the ones indulged with extensive use of cookies
[25]	• The paper analyzed the exposure of sensitive user information like bank details can be forged if we accept cookies for a web page • The data travels in clear text and black hat hackers can have access to it pretty easily • The current proposed solutions have complex implementation and there are loop holes when deployed with real-world traffic
[26]	• The paper proposes a decentralized method for validating licenses through cryptocurrency and blockchain concepts • It also shows how end-to-end encryption can be maintained
[27]	• The web pages are highly modified on the way from service provider to web browser • Page rewriting software can introduce some loopholes into websites before they reach the target web browser • Web tripwires are introduced to detect any change in the web page

4 The Proposed Model

The need of the hour is to identify the adversaries and individuals trying to enter the network by exploiting the vulnerabilities of cookies and interdict them from breaking into the system. We propose a scalable model as an extension of the current cookie system model by decentralizing the authentication procedure and making the overall system more transparent.

4.1 Properties

We have recognized some properties to secure the cookies which are publicly verifiable and robust to changes. These properties form the basis of the model:

Integrity. Implementation of the model should not compromise user's data and should be significant in maintaining the state in the network.

Independent Authentication. Anyone can verify the cookies even if they are not a part of the network, which establishes trust and makes the system unbiased.

Addressing Thefts. The proposed mechanism should be efficient in identifying and blocking unauthorized users from accessing the content.

Reliability. The system must be reliable as the verification is decentralized and distributed in the blockchain, saving the server time and resources.

Versatility. The proposed mechanism should not disrupt the existing user experience. It should be adaptable to current devices and easily adjustable.

Transparency. The network's working should be transparent, which allows the system to be monitored or trace back in cases of suspicion.

4.2 Preliminaries

Non-Fungible Token (NFT). It is a unique token identifying a digital asset on the blockchain which is irreplaceable and immutable [28]. We use NFTs to represent a user who is a part of the network to establish a transparent, verifiable ecosystem.

Elliptic Curve Digital Signature Algorithm (ECDSA). It is the modified version of the Digital Signature Algorithm (DSA) which uses Elliptic Curve Cryptography (ECC) which allows authentication and identification of the signee of a hash without unlocking the message or revealing the private key [29]. This method has been utilized in our model for the validation of cookies signed by the server.

Cryptographically Hashed Message (CHM). The message is encoded using known cryptography methods to pass it safely to the networks without revealing the secured data in the transmission process. The salted version of nonce and expiration time is used in the creation of CHM.

Session Ticket. The extension of standard cookies can be stored and passed in the network to establish trust among different entities [30]. The combination of CHM and the digital signature of the server is referred to as a session ticket.

Unique Identifier (UID). It is a unique identification address representing a particular user on the network and establishing a distinctive identity. We are using UID as a part of the digital signature signed by the server on the session ticket.

4.3 Model Description

Each account that is a part of the network would be represented by a Non-Fungible Token (NFT) stored on the blockchain and linked to the user's wallet with its corresponding public and private keys. The concept of asymmetric cryptography would be utilized to authenticate the user reaching the server for access grant. The working flow of authenticating a user into the network is shown in Fig. 1.

The user sends the address of the owned NFT to the server in order to make a claim. The server fetches the request and verifies the claim by confirming that the requested NFT is a part of the system or not. The authentication procedure only begins once it is confirmed that the claimed NFT was issued by the network, which helps in discarding the false claims. If the claim is approved as valid, it fetches the public key of the owner's address and encodes a nonce number using that public key which only its corresponding private key can decode. This method ensures that the user can always transfer the ownership of the NFT to a third party without requesting the system's permission, as the server always takes the current owner's public key.

Requesting user needs to decrypt the cryptographic hash and send the result back to the server, verifying its integrity from the hash table. If the value stored in the hash table and the one obtained from the user matches, the user is granted access of the network, or else it is denied. When the user enters the network, the server sends the required cookies. HTTP being a stateless protocol, cookies are used as authentication tokens by the server to verify the user's identity using a session id generated only after successful login. However, to validate the user's authenticity using cookies, confirmation from the participating nodes in the blockchain is proposed to decentralize the verification process. The server establishes trust on the blockchain, which further ensures that the user can be trusted or not.

Upon proven authentication of the device into the network, a nonce and expiration time of the session are generated and stored along with the UID of the device in the server's hashed database. All UID values in the database show that those are the users who have been verified and are currently a part of the network. The only column of UID is shared with the blockchain nodes while nonce and expiration time columns are kept hidden. If a user logs out from the network, the corresponding UID gets removed from the shared database, maintaining order and integrity. The permission to perform read and write operations would only be granted to the server, whereas nodes can only read the UID values available in the hashed database to perform the validation checks.

The generated nonce is salted with an expiration time, and the obtained result is encrypted using the server's public key. The resultant hash is referred to as CHM and

2 Claim Verification
The server validates the claim by checking whether the stored NFT is a part of the network before authenticating the user.
4 Decryption
The user unlocks the message using owned private key and sends the decrypted result back to the server.
6 Session Ticket
The server generates a session ticket and attaches it with cookies which proves the user's identity in the network.
1 NFT Claim
User claims the ownership of an NFT stored on the blockchain and sends its address to the server for verification.
3 Hash Generation
If the claim gets verified, the server fetches the public key of NFT's owner and encodes a nonce value with the obtained public key.
5 Confirmation
The server confirms the obtained nonce value with the one already stored in the hash table.

Fig. 1. Authentication of the user into the network.

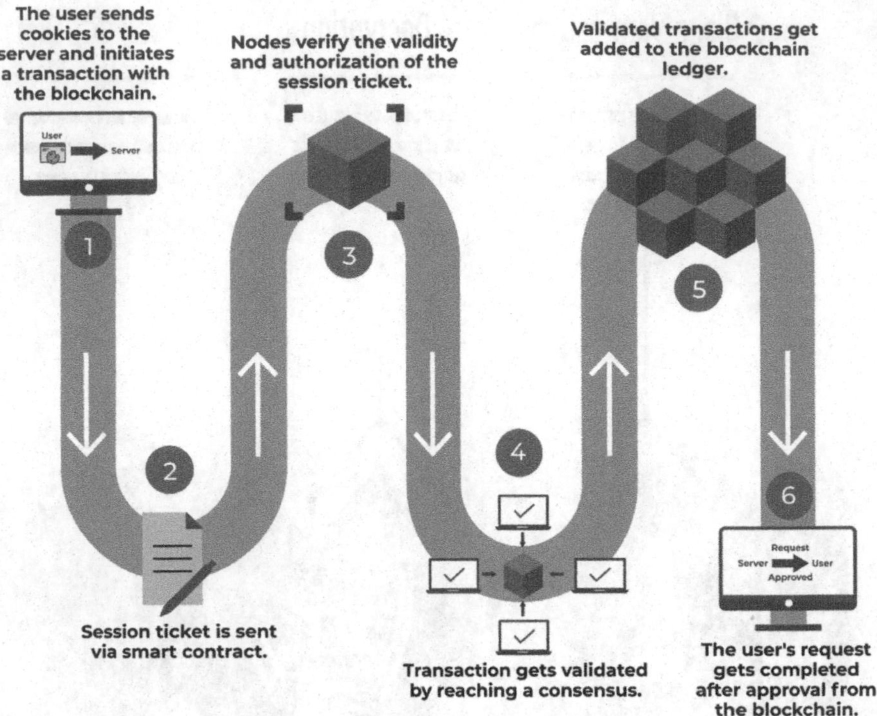
The user sends cookies to the server and initiates a transaction with the blockchain.
User
Server
Nodes verify the validity and authorization of the session ticket.
Validated transactions get added to the blockchain ledger.
1
2
3
4
5
6
Session ticket is sent via smart contract.
Transaction gets validated by reaching a consensus.
Request
Server
User
Approved
The user's request gets completed after approval from the blockchain.

Fig. 2. Validation of cookies in the blockchain through verification of the session ticket.

signed using ECDSA, where the UID of the user is used as the random number required for generating the signature. CHM, along with signature, acts as the session ticket for the user, which would be attached with the cookie, and its encryption ensures that it cannot be modified and nonce does not get compromised.

Whenever a user's web app requests a new page, cookies are sent to the server, and a transaction proceeds through smart contract. During the execution of the smart contract, nodes receive the session ticket and UID from the user. Since nodes have access to the shared hashed database of the server, they confirm whether the session has expired or not by checking whether the obtained UID exists. If the session has not expired, they validate that the session ticket passed by the user has been signed by the server or not. Signature validation of session ticket uses ECDSA, where random number should match to user's fetched UID and public key should belong to the server. Here, instead of explicitly providing the server's public key, it is fetched using its NFT address to allow multiple servers into the network and prevent the signing of same CHM from different servers. It also helps in fending off any ambiguity in ownership and solving the double-spending problem of the same NFT.

After all successful checks, the transaction would be validated, and on reaching a consensus, it would be added to the blockchain. The working flow of the validation of cookies in the blockchain is shown in Fig. 2. The server keeps a pointer on the blockchain, and as new transactions get added, they get stored in the server's database. Whenever the user's sent cookie gets verified through the transaction obtained from the blockchain, the sent request is completed as the authenticity of the user has already been established. If the server suspects a false claim, it can always trace back the chain of transactions, since they are stored on the blockchain.

5 Security Analysis and Discussion

In our model, cookies are sent to the server as well as the blockchain where server doesn't need to verify the user, but instead validation checks are performed by the blockchain nodes. Due to decentralization of validation checks, there is an addition of time requirement for the request fulfillment but it also adds a layer of security which prevents unauthorized access to a user's session. Offsetting of time requirements can be ensured by separating mainnet using a bridge and establishing a sidechain where all the transactions keep happening. When a threshold of transactions has happened, they get pushed to the main chain which ensures credibility of every transaction happening on the sidechain.

Table 2. Comparison of Cookie model, One time cookie and Proposed model.

S. no.	Topic	Cookie model	One time cookie	Proposed model
1	Cookie transfer protocols	HTTPS	HTTPS	HTTPS + Blockchain Consensus Protocol
2	Replication of requests	Not secured	Not secured	Secured
3	Validation procedure	Centralized	Centralized	Decentralized
4	User authentication from cookies	Dependent	Independent	Independent
5	Traceability of session hijacker	Mutable	Mutable	Immutable
6	Cookie modification during request	Not required	Required for each individual request	Not required
7	Functioning of verification procedure	Non-transparent	Non-transparent	Transparent
8	Cookie substitution attack	Vulnerable	Invulnerable until session secret gets compromised	Invulnerable until majority network gets compromised

When an attacker succeeds in getting access to the session cookies of an authorized user, he/she would not be able to authenticate them because he/she/it wouldn't be able to replicate the UID. An alternative to this scenario can be modifying the CHM, which cannot happen without compromising the server's private key and access to nonce stored in the hashed database. Even after modification of CHM, it needs to be added in the hashed database for which attacker needs to learn corresponding user's private key and carry out the login procedure. This leads to the establishment of the fact that majority of the network needs to be compromised for an attacker to hijack another user's session. However, if this happens, it can always be traced back to faulty transactions as long as the server does not get compromised.

The proposed model accounts for lower overhead, lower storage requirement and higher latency in comparison to the current implementations of cookies and sessions. Server only needs to perform read operations on the blockchain to verify a user back into the network which leads to reduced overhead on server as well. The latency of the network is heavily impacted due to decentralization process as each request needs to be authenticated by blockchain nodes and added to the public ledger for server to keep a track of authenticated users. Network latency requires significant improvement in blockchain network by reducing difficulty level of mining transactions, lesser transactions in a block and distribution of different transactions to nodes by making decentralized groups among the nodes as well. The comparison of the proposed model with the existing models is shown in Table 2.

There is also a requirement of remuneration of the blockchain nodes for the mining process which leads to higher investments along the user's side to make a valid request. If the server decides to waive off the gas fees for the network participants, it can lead to request spam attack on the network. The addition of gas fees makes the model more expensive in comparison to current cookie model being employed for verification.

The model has some added advantages as it combines the properties of blockchain, non-fungible tokens (NFTs), and smart contracts. NFTs allow the transfer of ownership without the involvement of the server as an intermediary. Blockchain ensures immutability and robustness in the verification process. The model ensures transparency in the network and throughout the login process. It also prevents cookie substitution during the transmission process and avoids leading to privacy breaches. Overhead on the server also gets reduced due to the distributed login validation process.

6 Conclusions and Future Works

Identity impersonation is a common threat through cookie cloning. The attacker manages to access a user's session cookie and joins the active session to perform tasks that the attacker is unauthorized to perform is called session hijacking. The model seeks to prevent the attacker from joining an active user session through blockchain, non-fungible tokens, and smart contracts. Instead of the webserver validating the login, the validation task is decentralized among the blockchain nodes and performed by the miners, which achieves the purpose of curbing session hijacking. To validate false claims, more than half of the blockchain nodes are required to be compromised. Cookie validation process can be enhanced by increasing the computing power of participating nodes in the blockchain. The proposed system can be improved further by decreasing the response time of the blockchain. The network can be scaled by increasing the participating nodes in the blockchain but it can also add up to the transaction delays in mining process as vote of confidence needed by the network gets raised. The proposed model achieves its purpose of securing sessions even after the access of user cookies, along with it being compatible, transparent, decentralized, and immutable but needs significant improvement in reducing network latency to account for industry level implementations. Further research upon sidechain validation process and lower gas fee requirement for request fulfilment can increase the overall performance of the system to improve its adaptability.

References

1. Ahmad, A., Maynard, S.B., Shanks, G.: A case analysis of information systems and security incident responses. Int. J. Inf. Manage. **35**(6), 717–723 (2015)
2. Sipior, J.C., Ward, B.T., Mendoza, R.A.: Online privacy concerns associated with cookies, flash cookies, and web beacons. J. Internet Commer. **10**(1), 1–16 (2011)
3. Sivakorn, S., Polakis, I., Keromytis, A.D.: The cracked cookie jar: HTTP cookie hijacking and the exposure of private information. In: 2016 IEEE Symposium on Security and Privacy (SP), pp. 724–742. IEEE (2016)
4. Putthacharoen, R., Bunyatnoparat, P.: Protecting cookies from cross site script attacks using dynamic cookies rewriting technique. In:13th International Conference on Advanced Communication Technology (ICACT2011), pp. 1090–1094. IEEE 92011)
5. Singh, T.: Prevention of session hijacking using token and session id reset approach. Int. J. Inf. Technol. **12**, 781–788 (2020)
6. Dacosta, I., Chakradeo, S., Ahamad, M., Traynor, P.: One-time cookies: preventing session hijacking attacks with stateless authentication tokens. ACM Trans. Internet Technol. (TOIT) **12**(1), 1–24 (2012)
7. Kiviat, T.I.: Beyond bitcoin: Issues in regulating blockchain transactions'. Duke Law J. **65**, 569 (2015)
8. Regner, F., Schweizer, A., Urbach, N.: NFTs in practice – non-fungible tokens as core component of a blockchain-based event ticketing application. In: Proceedings of the Fortieth International Conference on Information Systems, Munich, Germany, pp. 1–17 (2019)
9. Alkhajeh, A.: Blockchain and Smart Contracts: the Need for Better Education. Rochester Institute of Technology (2020)
10. Shah, K.A., Jinwala, D.C.: Privacy preserving, verifiable and resilient data aggregation in grid-based networks. Comput. J. **61**(4), 614–628 (2018)
11. Shah, K., Jinwala, D.: Privacy preserving secure expansive aggregation with malicious node identification in linear wireless sensor networks. Front. Comput. Sci. **15**(6), 1–9 (2021). https://doi.org/10.1007/s11704-021-9460-6
12. Shah, K., Patel, D.: Exploring the access control policies of web-based social network. In: Kumar, A., Paprzycki, M., Gunjan, V.K. (eds.) ICDSMLA 2019. LNEE, vol. 601, pp. 1614–1622. Springer, Singapore (2020). https://doi.org/10.1007/978-981-15-1420-3_168
13. Helling, B.: Web-site sensitivity to privacy concerns: collecting personally identifiable information and passing persistent cookies. First Monday (1998)
14. Lepore, C., Ceria, M., Visconti, A., Rao, U.P., Shah, K.A., Zuanolini, L.: A survey on blockchain consensus with a performance comparison of PoW, PoS and pure PoS. Mathematics **8**(10), 1782 (2020)
15. Folk-Sullivan, B.: Feasibility Study of the Usage of Blockchain Technology in Online Privacy Protection (2018)
16. Monrat, A.A., Schelén, O., Andersson, K.: A survey of blockchain from the perspectives of applications, challenges, and opportunities. IEEE Access **7**, 117134–117151 (2019)
17. Shrier, D., Weige, W., Pentland, A.: Blockchain & infrastructure (identity, data security). Mass. Inst. Technol. Connect. Sci. **1**(3), 1–19 (2016)

18. Cahn, A., Alfeld, S., Barford, P., Muthukrishnan, S.: An empirical study of web cookies. In: Proceedings of the 25th International Conference on World Wide Web, pp. 891–901 (2016)
19. Libert, T.: Exposing the hidden web: An analysis of third-party HTTP requests on 1 million websites. *arXiv preprint* arXiv:1511.00619(2015)
20. Miyazaki, A.D.: Online privacy and the disclosure of cookie use: effects on consumer trust and anticipated patronage. J. Public Policy Mark. **27**(1), 19–33 (2008)
21. Bal, M., Ner, C.: NFTracer: a Non-Fungible token tracking proof-of-concept using Hyperledger Fabric. arXiv preprint arXiv:1905.04795 (2019)
22. Talamo, E., Pennacchi, A.: IdToken: a new decentralized approach to digital identity. Open Identity Summit 2020 (2020)
23. Jones, M.L.: Cookies: a legacy of controversy. Internet Histories **4**(1), 87–104 (2020)
24. Park, J.S., Sandhu, R.: Secure cookies on the web. In: IEEE Internet Computing, vol. 4, issue number 4, pp. 36–44, July – August 2000. https://doi.org/10.1109/4236.865085
25. Sivakorn, S., Keromytis, A.D., Polakis, J.: That's the way the cookie crumbles: evaluating HTTPS enforcing mechanisms. In: Proceedings of the 2016 ACM on Workshop on Privacy in the Electronic Society, pp. 71–81 (2016)
26. Herbert, J., Litchfield, A.: A novel method for decentralised peer-to-peer software license validation using cryptocurrency blockchain technology. In: Proceedings of the 38th Australasian Computer Science Conference (ACSC 2015), vol. 27, pp. 27–25 (2015)
27. Reis, C., Gribble, S.D., Kohno, T., Weaver, N.C.: Detecting in-flight page changes with web tripwires. In: NSDI, vol. 8, pp. 31–44 (2008)
28. Sghaier Omar, A., Basir, O.: Capability-based non-fungible tokens approach for a decentralized AAA framework in IoT. In: Choo, K.-K., Dehghantanha, A., Parizi, R.M. (eds.) Blockchain Cybersecurity, Trust and Privacy. AIS, vol. 79, pp. 7–31. Springer, Cham (2020). https://doi.org/10.1007/978-3-030-38181-3_2
29. Khalique, A., Singh, K., Sood, S.: Implementation of elliptic curve digital signature algorithm. Int. J. Comput. Appl. **2**(2), 21–27 (2010)
30. Gutzmann, K.: Access control and session management in the HTTP environment. IEEE Internet Comput. **5**(1), 26–35 (2001)

Chinese Spelling Error Detection and Correction Based on Knowledge Graph

Ximin Sun[1], Jing Zhou[1(✉)], Shuai Wang[2], Huichao Li[2], Jiangkai Jia[2], and Jiazheng Zhu[3]

[1] State Grid Electronic Commerce Co., Ltd./State Grid Financial Technology Group,Beijing, China
{sunximin,zhoujing}@sgec.sgcc.com.cn

[2] State Grid Ecommerce Technology Co., Ltd.,Beijing, China
{wangshuai1,lihuichao,jiajiangkai}@sgec.sgcc.com.cn

[3] College of Intelligence and Computing, Tianjin University, Peiyang Park Campus, Tianjin, China
jiazhengzhu@tju.edu.cn

Abstract. Spelling error correction is a task in which errors in a natural language sentence can be detected and corrected. In this paper, we consider Chinese spelling error correction (CSC) for generality. A previous state-of-the-art method for this task connects a detection network with a correction network based on BERT by soft masking. This method does solve the problem that BERT has the insufficient capability to detect the position of errors. However, we find that it still lacks sufficient inference ability and world knowledge by analyzing its results. To solve this issue, we propose a novel correction approach based on knowledge graphs (KGs), which queries triples from KGs and injects them into the sentences as domain knowledge. Moreover, we leverage MLM as correction to improve the inference ability of BERT and adopt a denoising filter to increase the accuracy of results. Experimental results on the SIGHAN dataset verify that the performance of our approach is better than state-of-the-art methods.

Keywords: Spelling error correction · Knowledge graph · BERT

1 Introduction

Spelling error correction is a significant task in NLP with many real-world applications, such as optical character recognition [4,12] and speech recognition [6]. Its purpose is to correct spelling errors at the word or text level [1–3,12–15]. Our primary objective is Chinese spelling error correction(CSC) at the character level in this paper, which is challenging to complete because it needs human-level language understanding ability. Moreover, most methods for other languages like English are incapable of being validly used for the Chinese language because

U. K. Rage et al. (Eds.): DASFAA 2022 Workshops, LNCS 13248, pp. 149–159, 2022.
https://doi.org/10.1007/978-3-031-11217-1_11

there is little in common between Chinese and other languages. The unique characteristic of Chinese language, which is no delimiters between words, makes the analysis of the syntax and semantics of Chinese characters primarily rely on its contextual information.

In the area of CSC or more generally spelling errors, many solutions have been proposed. The popular methods are primarily divided into two categories. One is to use conventional machine learning or rule learning, and the other is to use deep neural networks. For example, [12]employed the traditional machine learning method, a comprehensive CSC framework consisting of three components: error detection, candidate generation, and final candidate selection. [14] proposed a pointer network, which employs a duplication mechanism converting the original sentence into a new one in which the spelling errors are corrected.

More recently, the pretrained language model such as BERT [6] and XLNET [23] has been successfully applied to a variety of language understanding tasks, including CSC. [13] selects the characters to be corrected(including the correct characters) in the corresponding position from the list of candidate characters at each sentence position based on BERT. [15] proposed a neural network structure for CSC consisting of one network for detecting errors and another network for correcting errors in the corresponding positions. In the correction stage, the character-level BERT is employed to predict the character with the highest probability among the candidate characters at each position in the given sentence.

Table 1. Examples of Chinese spelling errors

Wrong: “她通过努力拿到了冠军, 心里十分高心”
Correct: “她通过努力拿到了冠军, 心里十分高兴”
Wrong: “芜湖新闻报道有一男子落入青戈江”
Correct: “芜湖新闻报道有一男子落入青弋江”

These methods are powerful, but the analysis of the failure cases in their experiments demonstrates that the accuracy and precision of correction on the CSC task can be further improved. On the one side, the model requires some inference ability. For instance, shown in Table 1, in the sentence “她通过努力拿到了冠军, 心里很高心”, the last character ‘心’ should be corrected as ‘兴’ because the word is intended to express the meaning of ‘happy’ in this scene. On the other side, world knowledge is also essential for the model to correct spelling errors, such as ‘青弋江’, which is a river mistakenly written as ‘青戈江’. Based on our observation, about 39% of the uncorrected results are due to lack of world knowledge, 45% to lack inference ability, and 16% to some other reasons. We believe that this issue is general and emerge great trouble for employing the pipeline.

In order to solve the above problem, we firstly utilize knowledge graphs (KGs) to enhance the error correction ability of the model, which can facilitate the model to obtain world knowledge and make valid inferences with relevant knowledge. Due to the MLM(masked language model) method used by BERT, the model tends not to correct errors. Therefore, modifying MLM as correction [8] is adopted to decrease the difference between the pre-training and fine-tuning stages by using similar words for the masking purpose. Further, we filter the output of the model to improve the positive discrimination by providing higher confidence for the more similar words to the original, which is more precise than the ones proposed by [3,5]. The filter is efficient in terms of obtaining the goal of obtaining the highest accuracy and minimum damage recall rate.

We conducted experiments to compare our method with several baselines. The dataset SIGHAN is used as the benchmark dataset in our experiments. Experimental results show that our method achieves higher performances to the baselines including previous state-of-the-art models on the benchmark dataset in terms of all types of measures.

2 Model

The framework of our approach is shown in Fig. 1. Based on the intuition of knowledge conducting, we propose a character-level error correction model based on knowledge fusion for learning world knowledge and inference ability. Firstly, we define a sentence of n words as $S_1 = \{x_0, x_1, x_2, ..., x_n\}$, and the output sentence is $S_2 = \{y_0, y_1, y_2, ..., y_n\}$, where the wrong words in S_1 are substituted by the corresponding correct words to generate S_2. The overall model for the task can be regarded as a mapping function: $f : X \to Y$.

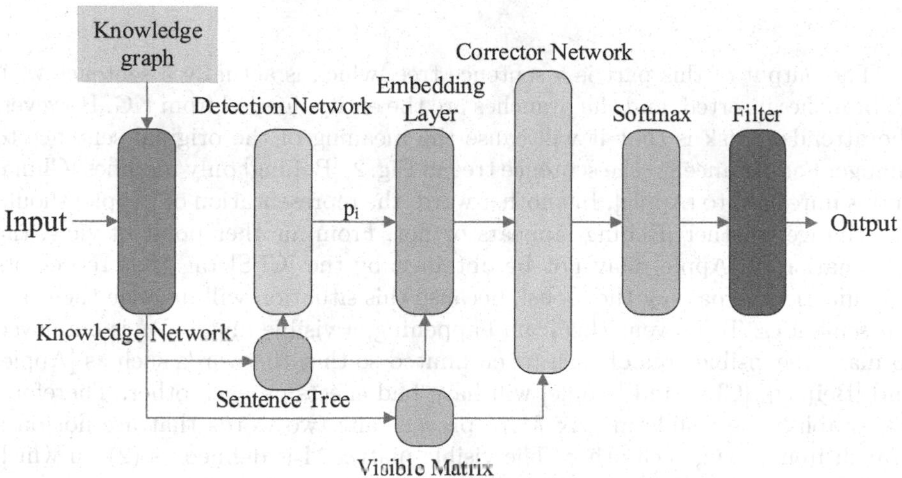

Fig. 1. Overview of our model.

2.1 Knowledge Network

In this part, we inject triples in KGs into the sentences as world knowledge and introduce a visible matrix [9] to overcome heterogeneous embedding space and knowledge noise issues. Specifically, given a KG K and a sentence $s = \{x_0, x_1, x_2, ..., x_n\}$, the network outputs a sentence tree t and a visible matrix M. First, the sentence s is divided into several words in which the entity names are extracted to query their corresponding triples E from the KG.

$$E = \{(x_i, r_{i0}, x_{i0}), ..., (x_i, r_{ik}, x_{ik})\} \tag{1}$$

The triples E is injected into s in the form of a sentence tree t by concatenating the corresponding position with its triples. The depth of a sentence tree is fixed to 1 to limit its complexity because a sentence tree is likely to have various branches to query their corresponding triples, which results in the entity based on the knowledge graph in triples to derive branches iteratively.

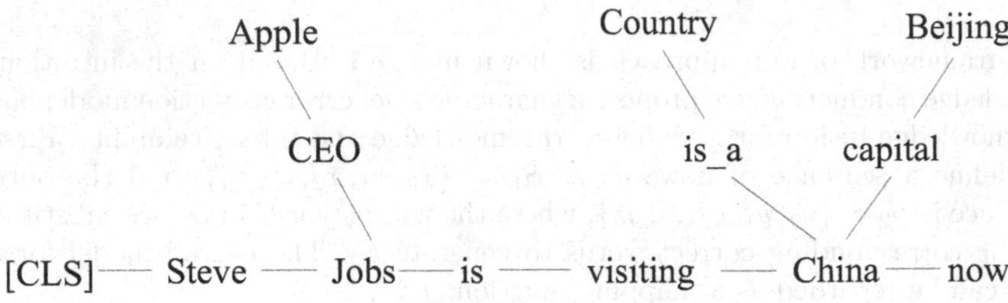

Fig. 2. An example of a sentence tree.

The output of this part is a sentence tree, which is actually a sentence with all branches inserted, and the branches are the entity queried from KG. However, the attendant risk is that it will cause the meaning of the original sentence to change. For instance, in the sentence tree in Fig. 2, [Beijing] only modifies [China] and is unrelated to [Apple]. In another word, the representation of [Apple] should not change whether [Beijing] appears or not. From another point of view, the information of [Apple] may not be obtained by the [CLS] tag that represents classification, bypassing the [Jobs], because this situation will damage the original semantics. To prevent this from happening, a visible matrix M is employed to make the visible area of each token limited so that the words such as [Apple] and [Beijing], [CLS] and [Apple] will have bad effects on each other. Therefore, we establish the visible matrix M to prevent any two words that are not on a branch from seeing each other. The visible matrix M is defined as (2), in which the first line means that x_i and x_j are in the same branch, and the second means that they are not.

$$M_{ij} = \begin{cases} 0 & x_i \ominus x_j \\ -\infty & x_i \oslash x_j \end{cases} \tag{2}$$

2.2 Detection and Correction Network

The Detection network is used to predict the probabilities p_i of errors for every character. In this work, we realize it as Bi-GRU where the probabilities p_i is defined as

$$p_i^d = \sigma(Wh_i + b) \tag{3}$$

where σ denotes the sigmoid function, h_i denotes the hidden state of Bi-GRU, W and b are parameters. Moreover, the hidden state is defined as

$$\overrightarrow{h_i} = GRU(\overrightarrow{h}_{i-1}, e_i) \tag{4}$$

$$\overleftarrow{h_i} = GRU(\overleftarrow{h}_{i+1}, e_i) \tag{5}$$

$$h_i = [\overrightarrow{h_i}, \overleftarrow{h_i}] \tag{6}$$

The correction network based on MLM as correction BERT employs similar words for masking instead of using [MASK]. The input is the sequence of the output of the embedding layer. The result is a sequence of characters. It predicts the probabilities of every word in the entire sequence, which should be corrected as another character in the word space. The probability is defined as

$$p_i^c = softmax(Wh_i^c + b)[j] \tag{7}$$

where softmax is the softmax function, $h_i^{'}$ denotes the hidden state, W and b are parameters.

To connect the two networks, we adopt a soft way to pass the former output to the latter input in the embedding layer. The embedding layer is to transform the sentence obtained from the knowledge network into an embedding representation based on BERT. In this way, the input of the correction network is the weighted sum of token embedding and mask embedding:

$$e_i' = p_i * e_{mask} + (1 - p_i) * e_{i-token} \tag{8}$$

where $e_{i-token}$ is the i-token embedding and e_{mask} is the mask embedding.

2.3 Filter

The final filter utilizes the feature of Chinese character similarity to improve accuracy. In detail, we adopt ideographic description sequence (IDS) based on the open dataset - Kanji Database Project[1] to obtain vision and phonological representations for all CJK Unified Ideographs in all CJK languages. To compute the visual and phonological similarity of two characters, we calculate one minus normalized Levenshtein edit distance between their corresponding representations.

1 http://kanji-database.sourceforge.net/

	IDS	MC	CC	JO	K	V	V-sim	P-sim
干 (dry)	⿱一⿻一丨	gan4	gon1	kan	kan	can	0.835	0.279
于 (at)	⿱一⿻一亅	yu2	jyu1	vu	wu	ku		
用(use)	⿵⿰丨𠃌⿻⿱一一丨	yong4	jung6	you	yong	dung	0.889	0.093
甩(throw)	⿵⿰丨𠃌⿻⿱一一乚	shuai3	lat1	shutsu	shutg	null		

Fig. 3. Examples of the computation of character similarities in terms of vision and pronunciation. The IDS stands for the method of calculating visual similarity(V-sim). And MC, CC, JO, K, and V represent the pronunciation of each country, namely China, Japan, Korea, and Vietnam.

There are two advantages to this: one is the similarity value ranging from 0 to 1, which is convenient for filtering; the other is that if two pairs of candidates have the same edit distance, we prefer that the similarity of candidates with greater complexity is relatively higher as shown in Fig. 3. Moreover, contextual confidence and character similarity are both considered in our method to improve the positive discrimination of the results by filtering the output of the correction network, which means that the word with higher similarity and confidence is more likely to select out.

2.4 Loss

The whole learning of our model is driven by optimizing two objective functions, which are detecting and correcting errors respectively. Their definitions are

$$L_d = -\sum_{i=1}^{n} \log p_i^d \tag{9}$$

$$L_c = -\sum_{i=1}^{n} \log p_i^c \tag{10}$$

where L_d is the loss of detection, and L_c is the loss of correction. The final objective function of the model is the linear sum of the two functions in training.

$$L = L_d + \lambda L_c \tag{11}$$

3 Experiments and Evaluation

3.1 Datasets and Baseline

Our method is evaluated on a well-known benchmark for CSC, the SIGHAN dataset. For training, validation, and testing, we split the datasets randomly by 8:1:1. In terms of knowledge graphs, we employ two Chinese KGs, CN-DBpedia [10] and HowNet [11]. CN-DBpedia is a general domain structured encyclopedia Knowledge graph proposed and maintained by the Fudan University, containing millions of entities and relationships. It is mainly for in-depth mining of Chinese encyclopedia websites with a single data source. HowNet is a large-scale knowledge graph in which most Chinese characters are labeled with sememes and the relationship between concepts and the attributes of concepts as the basic content is revealed.

We employ several methods which performed well before as baselines for comparison, including HanSpeller++ [12], FASPell [13], Confusionset [14], and Soft-Masked BERT [15]. FASPell proposed a new paradigm by utilizing transformers to denoise and a confidence-similarity decoder to filter candidate characters. HanSpeller++ is a unified framework. It first generates a set of possible candidates for the input sentence, and employs a filter to re-rank the candidates and select the best result as the new sentence. Confusionset adopted a pointer network and duplicate mechanism to generate characters. Soft-Masked BERT proposed a neural architecture for CSC consisting of two networks for error detection and correction.

3.2 Evaluation

In our experiments, we train our model with a batch size of 16 and learning rate is 0.0001 and utilized sentence-level accuracy, precision, recall, and F1 score for evaluation. For our model, the accuracy is defined as

$$Accuracy(f;D) = \frac{1}{n}\sum_{i=1}^{n}(f(x_i) = label_i) \tag{12}$$

And the precision, recall, and F1 are defined as

$$Precision = \frac{TP}{TP+FP} \tag{13}$$

$$Recall = \frac{TP}{TP+FN} \tag{14}$$

$$F1 = 2\cdot\frac{Pre.\cdot Rec.}{Pre.+Rec.} \tag{15}$$

Table 2. Comparisons with diferent methods on CSC dataset.

Test set	Method	Result			
		Acc.	Prec.	Rec.	F1.
SIGHAN	HanSpeller++(2015)	69.2	79.7	51.5	62.5
	FASPell (2019)	73.7	66.6	59.1	62.6
	Confusionset(2019)	-	71.5	59.5	64.9
	BERT	76.6	65.9	64.0	64.9
	Soft-masked BERT(2020)	77.4	66.7	66.2	66.4
	Ours (HowNet)	79.8	83.6	68.8	75.5
	Ours (CN-DBpedia)	80.1	85.8	71.4	77.9

3.3 Results

As shown in Table 2, our method outperforms the baselines on the SIGHAN dataset. Especially in the F1 score, we obtain a considerable improvement of 11.5% compared to the best result of the previous methods. It means that our model indeed gains more vital error detection and correction capabilities. The encyclopedic KG(CN-DBpedia) performs better than the encyclopedic KG(HowNet) in our method, which means that the model actually has a demand for world knowledge. We also get the highest recall score among the baseline on the SIGHAN dataset. The 71% recall rate means more than 71% errors will be found and corrected.

We also carry out ablation study on our model. As shown in Table 3, we remove MLM as correction (Mac), filter, and knowledge network, respectively. The results show that the three major components of our model, Mac, Filter, and KG are necessary for achieving high performance.

Table 3. Ablation study of our method on SIGHAN.

Method	**result**			
	Acc.	Prec.	Rec.	F1.
Ours	80.1	85.8	71.4	77.9
Ours-Mac	79.6	85.4	70.5	77.1
Ours-filter	79.5	84.3	71.6	76.2
Ours-KG	79.0	84.6	70.1	75.5

4 Related Work

A large amount of studies have been conducted on Chinese spelling error correction(CSC), which is crucial for natural language applications, such as optical character recognition [4,12], speech recognition [6], and search [16]. Most of these studies for CSC can be divided into two broad types. One is that a fixed combination of similar Chinese characters is used as a general example of candidate characters. The filter selects the best candidate words to replace the specified sentence. This simple design has two main bottlenecks which is insufficient CSC resources [3,18] and underutilization character similarity respectively [5,13]. The negative impact of the two bottlenecks has been unsuccessfully mitigated.

The other type employs language models and rules [8,9,14,17], which followed the comprehensive CSC framework consisting of three components or treated it as a sequence labeling task to employ CRF or pretrained language model [12,13]. BERT (Bidirectional Encoder Representations from Transformers) [7] is a pre-trained model with Transformer encoder as its architecture, which has achieved great success in NLP studies, such as those in the GLUE challenge [19]. It underlines that the conventional language model with only one direction or the shallow splicing of two normal language models for pre-training is no longer used. However, a new masked language model (MLM) is utilized to generate deep bidirectional language representation. Later, they further proposed an upgraded version of BERT, which mainly changes the training sample generation strategy of the initial pre-training stage, called whole word masking (wwm). The technique is employed to optimize the original masking in the MLM target. The original word segmentation method randomly selects WordPiece [20], while in the wwm setting, the WordPiece subwords of a whole word and other tokens of the same word are all masked [21]. Recently, other language representation models have also been proposed, such as GPT-3 [22], XLNET [23], Roberta [24], ALBERT [25], and MASS [26].

5 Conclusion

In this paper, we have proposed a novel correction model as a way of efficiently leveraging external knowledge to improve Chinese spelling error correction. A critical advantage of our approach is that our model could capture adequate world knowledge and reinforce inference ability. Extensive empirical evaluations and ablation study demonstrate that our proposed way is beneficial and promising.

References

1. Yu, J., Li, Z.: Chinese spelling error detection and correction based on language model, pronunciation, and shape. In: 3rd CIPS-SIGHAN Joint Conference on Chinese Language Processing, pp. 220–223. Association for Computational Linguistics, Wuhan, China (2014)
2. Yu, L.-C., Lee, L.-H., Tseng, Y.-H., Chen, H.-H.: Overview of SIGHAN 2014 bake-off for Chinese spelling check. In: 3rd CIPS-SIGHAN Joint Conference on Chinese Language Processing, pp. 126–132. Association for Computational Linguistics, Wuhan, China (2014)

3. Wang, D., Song, Y., Li, J., Han, J., Zhang, H.: A hybrid approach to automatic corpus generation for Chinese spelling check. In: 23rd Conference on Empirical Methods in Natural Language Processing, pp. 2517–2527. Association for Computational Linguistics, Brussels, Belgium (2018)
4. Afli, H., Qiu, Z., Way, A., Sheridan, P.: Using SMT for OCR error correction of historical texts. In: 10th International Conference on Language Resources and Evaluation, pp. 962–966. European Language Resources Association, Portorož, Slovenia (2016)
5. Liu, C.-L., Lai, M.-H., Chuang, Y.-H., Lee, C.-Y.: Visually and phonologically similar characters in incorrect simplified Chinese words. In: 23rd International Conference on Computational Linguistics, pp. 739–747. Chinese Information Processing Society of China, Beijing, China (2010)
6. Guo, J., Sainath, T.N., Weiss, R.J.: A spelling correction model for end-to-end speech recognition. In: 43rd International Conference on Acoustics, Speech and Signal Processing, pp. 5651–5655. IEEE, Brighton, United Kingdom (2019)
7. Devlin, J., Chang, M.-W., Lee, K., Toutanova, K.: BERT: pre-training of deep bidirectional transformers for language understanding. In: 23rd Conference of the North American Chapter of the Association for Computational Linguistics, pp. 4171–4186. Association for Computational Linguistics, Minneapolis, MN, USA (2019)
8. Cui, Y., Che, W., Liu, T., Qin, B., Wang, S., Hu, G.: Revisiting pre-trained models for Chinese natural language processing. In: 25th Findings of Empirical Methods in Natural Language Processing, pp. 657–668. Association for Computational Linguistics (2020)
9. Liu, W., et al.: K-BERT: enabling language representation with knowledge graph. In: 34th AAAI Conference on Artificial Intelligence, pp. 2901–2908. AAAI Press, New York (2020)
10. Xu, B., et al.: CN-DBpedia: a never-ending Chinese knowledge extraction system. In: 30th International Conference on Industrial Engineering and Other Applications of Applied Intelligent Systems, pp. 428–438. Springer, Arras, France (2017)
11. Dong, Z., Dong, Q., Hao, C.: HowNet and its computation of meaning. In: 23rd International Conference on Computational Linguistics: Demonstrations, pp. 53–56. Association for Computational Linguistics, Beijing, China (2010)
12. Xiong, J., Zhang, Q., Zhang, S., Hou, J., Cheng, X.: HANSpeller: a unified framework for Chinese spelling correction. Int. J. Comput. Linguist. Chinese Lang. Process. **20**(1), 38–45 (2015)
13. Hong, Y., Yu, X., He, N., Liu, N., Liu, J.: FASPell: a fast, adaptable, simple, powerful Chinese spell checker based on DAE-decoder paradigm. In: 5th Workshop on Noisy User-generated Text, pp. 160–169. Association for Computational Linguistics, Hong Kong, China (2019)
14. Wang, D., Tay, Y., Li, Z.: Confusionset-guided pointer networks for Chinese spelling check. In: 57th Conference of the Association for Computational Linguistics, pp. 5780–5785. Association for Computational Linguistics, Florence, Italy (2019)
15. Zhang, S., Huang, H., Liu, J., Li, H.: Spelling error correction with soft-masked BERT. In: 58th Annual Meeting of the Association for Computational Linguistics, pp. 882–890. Association for Computational Linguistics, Online (2020)
16. Gao, J., Li, X., Micol, D., Quirk, C., Sun, X.: A large scale ranker-based system for search query spelling correction. In: 23rd International Conference on Computational Linguistics, pp. 358–366. Tsinghua University Press, Beijing, China (2010)

17. Tseng, Y.-H., Lee, L.-H., Chang, L.-P., Chen, H.-H.: Introduction to sighan 2015 bake-off for Chinese spelling check. In: 8th SIGHAN Workshop on Chinese Language Processing, pp. 32–37. Association for Computational Linguistics, Beijing, China (2015)
18. Zhao, H., Cai, D., Xin, Y., Wang, Y., Jia, Z.: A hybrid model for Chinese spelling check. ACM Trans. Asian Low Resour. Lang. Inf. Process. **16**(3), 21:1–21:22 (2017)
19. Wang, A., Singh, A., Michael, J., Hill, F., Levy, O., Bowman, S.R.: GLUE: a multi-task benchmark and analysis platform for natural language understanding. In: 23rd Workshop: Analyzing and Interpreting Neural Networks for NLP, pp. 353–355. Association for Computational Linguistics, Brussels, Belgium (2018)
20. Wu, Y., et al.: Google's Neural Machine Translation System: Bridging the Gap between Human and Machine Translation (2016). http://arxiv.org/abs/1609.08144
21. Cui, Y., et al.: Pre-training with whole word masking for Chinese BERT. http://arxiv.org/abs/1906.08101 (2019)
22. Brown, T.B., et al.: Language models are few-shot learners. In: 34rd Annual Conference on Neural Information Processing Systems. Association for Computational Linguistics, Vancouver, virtual (2020)
23. Yang, Z., Dai, Z., Yang, Y., Carbonell, J.G., Salakhutdinov, R., Le, Q.V.: XLNet: generalized autoregressive pretraining for language understanding. In: 33rd Annual Conference on Neural Information Processing Systems, pp. 5754–5764. Association for Computational Linguistics, Vancouver, BC, Canada (2019)
24. Liu, Y., et al.: RoBERTa: A Robustly Optimized BERT Pretraining Approach. arXiv preprint arXiv:1907.11692 (2019)
25. Lan, Z., Chen, M., Goodman, S., Gimpel, K., Sharma, P., Soricut, R.: ALBERT: a lite BERT for self-supervised learning of language representations. In: 8th International Conference on Learning Representations. OpenReview.net, Addis Ababa, Ethiopia (2020)
26. Song, K., Tan, X., Qin, T., Lu, J., Liu, T.-Y.: MASS: masked sequence to sequence pre-training for language generation. In: 36th International Conference on Machine Learning, pp. 5926–5936. PMLR, California, USA (2019)

Construction and Application of Event Logic Graph: A Survey

Bin Zhang[1], Ximin Sun[1(✉)], Xiaoming Li[2], Dan Liu[2], Shuai Wang[2], and Jiangkai Jia[2]

[1] State Grid Electronic Commerce Co., Ltd./State Grid Financial Technology Group, Beijing, China
{zhangbin,sunximin}@sgec.sgcc.com.cn
[2] State Grid Ecommerce Technology Co., Ltd., Beijing, China
wangshuai1@sgec.sgcc.com.cn

Abstract. Since being proposed in 2017, event logic graph has attracted more and more researchers' attention and has been gradually applied in various fields such as finance, health care, transportation, information, politics, etc. Unlike the traditional knowledge graph describing static entities and their attributes and relationships, the event logic graph describes the evolution rules and patterns between events. The construction of event logic graphs is significant for understanding human behavior patterns and mining event evolution rules. The survey first systematically combs the work of constructing an event logic graph, including event extraction and event relationship extraction methods. Secondly, the typical application of the event logic graph is explained. Finally, the challenges of event logic graph construction are analyzed, and future research trends are prospected.

Keywords: Event logic graph · Event extraction · Relation extraction

1 Introduction

Human activities are often related to events, and events are one of the core concepts of human society. The evolution and development of events have their basic principles, which makes events often occur in sequence. For example, the sentence "After arriving at the train station, he bought a train ticket and then waited in the waiting hall" shows the sequence of events: "arrive at the train station" → "buy a train ticket" → "waiting for a train". This is a typical pat-tern in scenes where people buy train tickets. There-fore, studying the evolutionary laws and patterns of events is valuable for understanding human behavior and social development laws. However, the nodes of the traditional knowledge graph represent static entities, and the edges represent the attributes of entities and the relationships between entities, which ignores the event logic in the real world and lacks the description of dynamic attributes. Therefore, in actual application, the traditional knowledge graph is limited to answering the question of what is

U. K. Rage et al. (Eds.): DASFAA 2022 Workshops, LNCS 13248, pp. 160–174, 2022.
https://doi.org/10.1007/978-3-031-11217-1_12

what, and it is weak in solving the problems of how to do it, why to do it, and how to do it next.

To solve the problems of the above-mentioned traditional knowledge graph, the concept of event logic graph came into being. The concept of event logic graph was proposed by Liu Ting at the China Computer Conference (CNCC) in October 2017. An event logic graph is defined as an event logic knowledge base in the form of a directed graph, which reveals the laws and patterns of evolution be-tween events. As shown in Fig. 1, the nodes in the event logic graph are events, and the edges between nodes describe the relationship between events generally including four types: succession relationship, causality relationship, conditional relationship, and upper-lower relationship. The events in the event logic graph are defined as triples, where P represents the action, S represents the subject of the action, and O represents the object of the action. It should be noted that the events in the graph must be abstract, generalizable, and semantically complete [18]. Abstraction and generalizability mean that the event logic graph does not care about the specific participants of the event and the place and time of the occurrence. For example, "watching a movie" can be used as an event node in the event logic graph, and it doesn't matter who watched the movie and where. The semantic integrity of an event means that people need to understand the meaning of the event without any doubt. For example, (eat, meal), (watch, movie) are valid event tuples, and (eat, what), (go, somewhere) are incomplete, illegal event tuples with ambiguous semantics. The succession relationship in the event logic graph refers to the partial order relationship in which two events occur one after another in time. Causality means that in two events, the occurrence of the previous event is the cause of the latter. The conditional relationship refers to the condition that the last event is the occurrence of the next event. The subordinate relationship refers to the "is-a" relationship, including the nominal and verb subordinate relationships.

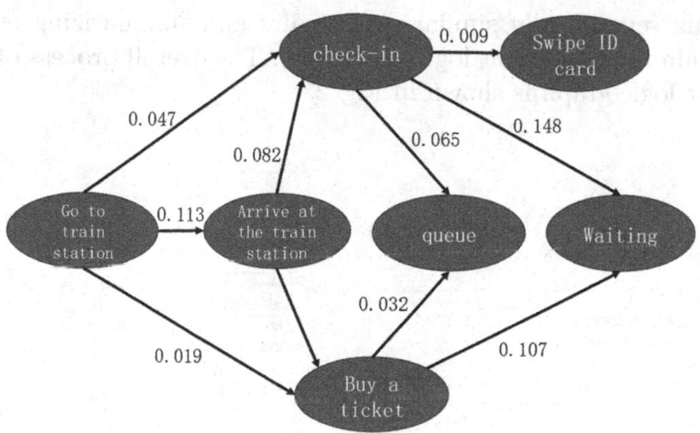

Fig. 1. An example of an event logic graph.

The main differences between the event logic graph and the traditional knowledge graph are: 1) The research object of the traditional knowledge graph is the static nominal entity, and the attributes of the entity and the relationship between the entities, while the research object of the event logic graph is the dynamic predicate events and events 2) The relationship between entities in the knowledge graph is definite, while the relations between events in the event logic graph are mostly uncertain relations expressed by probability; 3) The goal of the knowledge graph is to realize the interconnection of everything. The goal of event logic graph is to build a full logic library. But from the perspective of organizational form, both knowledge graphs and event logic graphs store knowledge in the form of directed cyclic graphs.

The construction and reasoning of the event logic graph are essential for understanding human behavior patterns and discovering the law of event evolution. Therefore, the event logic graph has received more and more attention from researchers and has gradually been applied to finance [50,54,62], medical [65], transportation [21], information [20,55,63], and politics [29,30], and many other fields. However, there is no comprehensive work on the construction technology of the event logic graph, so this article systematically expounds on the key technologies and applications of the event logic graph, and analyzes the opportunities and challenges that the event logic graph currently faces.

2 Event Extraction

As the concept of event logic graphs is relatively new, the current research on event logic graphs mostly focuses on the construction technology of event logic graphs. The construction process of the event logic graph mainly includes: 1) extracting events from the original data; 2) identifying the succession, causality, condition or subordinate relation-ship between event pairs; 3) identifying the direction of the causality and the succession relationship; 4) Combine event pairs by connecting semantically similar event pairs and summarizing each specific event to obtain the final event logic graph [18]. The overall process of constructing the event logic graph is shown in Fig. 2.

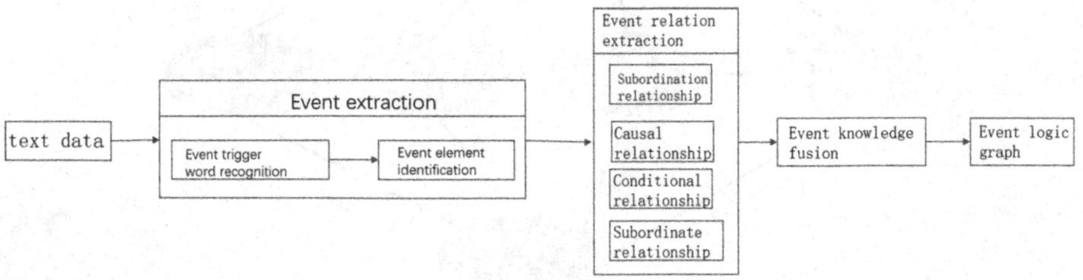

Fig. 2. The basic process of event graph construction.

Event extraction aims to detect the existence of predicate events from unstructured text data, and if so, extract event-related information, such as the "5W1H" of the event (i.e., who, when, where, what, why, and How). The event extraction in the construction of the event logic graph requires the event to be abstract, generalizable, and semantically complete. Among them, judging whether the event is semantically complete needs to be combined with the application scenario of the event logic graph. When the application scenario is unknown, the structure of the event is generally specified in the form of (subject, action, object), which requires the final output (subject, action, object) form of the event. The method of event extraction in the process of event logic graph construction can be mainly divided into three categories: methods based on pattern matching, methods based on machine learning and methods based on deep learning.

2.1 Methods Based on Pattern Matching

The event extraction method based on pattern matching requires a predefined language pattern, and then text information is matched according to the language pattern. The language model can be constructed from the original text or the annotated text, but both require professional knowledge. In the online extraction phase, if the event matches a predefined template, then the event and its parameters will be extracted. The earliest event extraction system based on pattern matching is AutoSlog proposed by Riloff et al. [52], a system for extracting specific areas of terrorist events. Inspired by the AutoSlog system, researchers have proposed pattern- matching-based event extraction systems for different fields, including biomedical event extraction systems [6,7,9], financial event extraction systems [2,5], etc.

In addition to manually constructing event patterns, researchers have proposed weakly supervised or enumerated methods that automatically acquire more patterns only using little preclassified seed patterns or training data. Based on the previous AutoSlog system, Riloff et al. [53] developed the AutoSlog event extraction system to achieve automatic mode construction. Based on the thirteen language modes defined in the AutoSlog system, it uses a syntax analyzer CIRCUS to obtain new event modes from unlabeled corpus. Take the sentence "World trade center was bombed by terrorists." as an example. We can obtain the subject "world trade center", the verb phrase "was bombed" and the preposition phrase "by terrorists" from syntactic analysis. Combining the "$\langle subject \rangle$ *passive-verb*" and "*passive–verb prep* $\langle np \rangle$" in the language model, we can get the hidden event modes "$\langle x \rangle$ *was bombed*" and "*bombed by* $\langle y \rangle$". Whether to use the new language model depends on the statistical scores in the domain-related and domain-independent documents. In addition, researchers have also proposed some methods to learn new patterns from seed patterns by designing ma-chine learning algorithms. The NEXUS system developed by Piskoriski et al. [49]. learns candidate event patterns from a small group of annotated corpora through a machine learning algorithm based on entropy maximization. Cao et al. [49] proposed a pattern technique to import frequent patterns from an external corpus through active learning.

The advantage of the event extraction method based on pattern matching is simple and efficient. Using expert knowledge to construct event patterns can achieve higher accuracy with lower computational complexity and computational resources, so it has been widely used in the industry. However, due to the need to pre-define language patterns, the pattern matching method is only suitable for event extraction in specific domains. David et al. [1] proposed an event extraction in a specific domain due to the need to define language patterns in advance. Moreover, pre-defining language models that rely on expert knowledge requires high labor costs.

2.2 Methods Based on Machine Learning

The event extraction method based on machine learning transforms event category and event element recognition into classification problems, learns a classifier from training data, and uses the classifier to extract events from new text. The key to event extraction in machine learning methods is to select features and then construct a classifier based on the selected features. Event extraction based on machine learning can be divided into four sub-tasks: 1) Trigger detection, that is, determines whether the event exists. If it exists, then detects the corresponding event trigger in the text. 2) Trigger or event type recognition, that is, use a classifier to trigger Or event classification. 3) Parameter detection, that is, which parameter to detect which entity, time and value, etc. 4) Parameter role recognition, that is, classifies the parameter's role according to the recognized event type. Its overall structure is shown in Fig. 3.

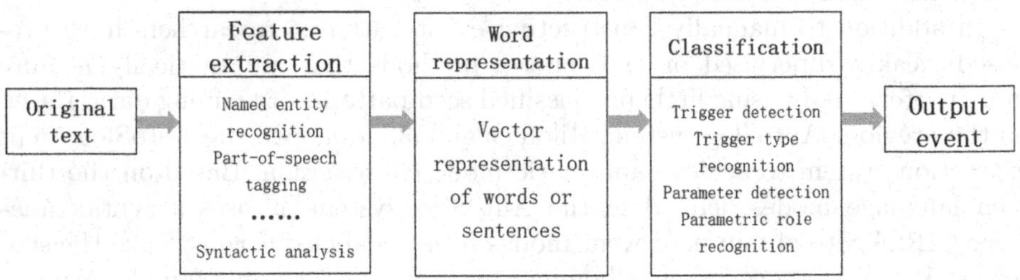

Fig. 3. Event extraction based on machine learning.

Pipeline processing framework that includes two consecutive classifiers: the first classifier is called TiMBL, which uses the nearest neighbor learning algorithm to detect triggers; the second classifier is called MegaM, which uses maximum entropy Learner to identify parameters. In order to train the TiBML classifier, vocabulary features, WordNet features, context features, dependent features, and related entity features are used; trigger words, event types, entity mentions, entity types, and dependent paths between trigger words and entities are used to Train the MegaM classifier. Some other pipeline classifiers are trained using different types of features. For example, Chieu and Ng [14] added single-character and double-character features, and used maximum entropy for classification. In the field of biomedicine, more domain-specific features and expertise are used for classifier training, such as frequency features, label features, path features, etc. [22,41,42].

The advantage of the pipeline classification model is that each sub-task is independent of the other, the division of labor is clear, and the efficiency is high. But the independence between subtasks is also a disadvantage of the pipeline model. The downstream task cannot influence the decision of its previous classifier and cannot make good use of the interdependence of different subtasks. On the other hand, pipeline classification models may encounter error propagation problems, where errors in upstream classifiers may propagate to downstream classifiers and may reduce their performance. In contrast, useful information from one subtask in joint learning can be forwarded to the next subtask, or backward to the previous subtask, so valuable information can be obtained from the interaction between tasks [64]. Li et al. [33] proposed a joint model for trigger detection and event type recognition, using an inference framework based on integer logic programming (ILP) to integrate two types of classifiers. In particular, they proposed two trigger detection models: one is based on conditional random fields, and the other is based on maximum entropy. Li et al. [34] also used a similar method to train a joint model for argument determination and role recognition at the next level. Chen et al. [10] trained two SVM-based joint classifiers for one stage of subtasks but with more language features. Li et al. [36] expressed event extraction as a structured learning problem and proposed a joint extraction algorithm that integrates local and global features into a structured perceptron model to predict event triggers and independent variables simultaneously. Judea and Strube [26] observed that event extraction is structurally the same as frame semantic analysis. Therefore, they optimized and retrained the frame semantic analysis system SEMAFOR [15] for structural event extraction.

The advantage of event extraction based on machine learning is that it does not have strict requirements on the form and content of the corpus, but it also has obvious shortcomings. One is that feature extraction will cause error propagation. Existing tools such as Stanford CoreNLP have disadvantages such as low accuracy and weak generalization ability. Errors will propagate along the task chain and ultimately affect the effect of event extraction. On the other hand, the event extraction method based on machine learning relies on a large amount of labeled corpus, and insufficient data will cause serious data sparse problems.

2.3 Methods Based on Deep Learning

In recent years, deep learning technology has been intensively studied and applied to various classification tasks. This technology uses multi-layer connected artificial neural networks to construct artificial neural networks. In an artificial neural network, each layer can learn to transform its lower layer input into a more abstract and composite representation, and then input it to its own higher layer until its output features are used in the highest level of classification.

Compared with classic machine learning techniques, using deep learning for event extraction helps greatly reduce the difficulty of feature engineering. Nguyen and Grishman et al. [45] used convolutional neural networks (CNN) for event extraction for the first time by identifying trigger words and their event types. Chen et al. [12] proposed a dynamic multi-pooling convolutional neural network (DMCNN), which evaluates each part of a sentence by extracting a dynamic multi-pooling layer of vocabulary-level and sentence-level features. There are other works to improve the CNN model structure for event extraction [8,27,32,73]. Researchers have also proposed some models based on Recurrent Neural Network (RNN) to take advantage of the interdependence of words by inputting words according to the order of the words in the sentence (forward or backward) [11,19,44,67]. In addition, in some work, the syntactic dependency between words is also used to expand the RNN structure for event extraction [56,57,72]. Recently, some scholars have applied graph neural network (GNN) to event extraction [40,46,51]. The key to event extraction with GNN is to construct a graph for the words in the text. Rao et al. [51] used a semantic analysis technique called Abstract Meaning Representation (AMR) [3] to standardize many vocabularies and syntactic changes in the text, and output a directed acyclic graph to capture the "who is right" in the text. The concept of "who did what". The above three basic types of neural network architectures have their own advantages and disadvantages when they are used to capture different features, relationships, and dependencies in text for event extraction. Many researchers have proposed hybrid neural network models that combine different neural networks to combine the advantages of each model [23,24,28,38,69,70]. In addition, the attention mechanism is also combined with various neural network models for event extraction tasks [17,39,61,66,71].

Compared with pattern matching methods and traditional machine learning methods, the event extraction method based on deep learning has significantly improved the recognition effect. Through the learning of a large amount of supervised data, the model also has strong generalization ability. The disadvantage is that the quality of labeling is more demanding. High. Due to the high cost of data annotation, the current event extraction methods of machine learning and deep learning still cannot completely replace the method of pattern matching in the construction of the event logic graph. The method based on pattern matching is still the mainstream event extraction method in the construction of event logic graph in the industry.

3 Event Relation Extraction

The event relationship extraction in the event logic graph depends on the domain and task, and it mainly includes four types: succession relationship, causality relationship, conditional relationship, and subordinate relationship [64]. Different relationship extraction models need to be constructed for other relationships of events, divided into two types: feature learning-based methods and deep learning-based methods. The method based on feature learning transforms relationship recognition into classification problems through feature learning such as vocabulary, semantics, grammar, and kernel functions, and uses traditional machine learning methods such as support vector machines, decision trees, and Bayes classifiers for classification. For example, some researchers have introduced temporal relationships between feature learning events such as parts of speech and syntactic tree paths [4,31,47,59]. However, the method based on feature learning relies on complex feature engineering, and the error of feature extraction will propagate to the relational reasoning stage. In addition, the method based on features mainly depends on a specific domain, and the scalability is weak.

At present, the most widely used method in event relationship recognition is the method based on deep learning. Some works use long and short-term memory neural networks (LSTM) to extract the sequence features between words in the text for the recognition of the temporal relationship of events [13,37,48, 58] and causality [25]. Other works use convolutional neural networks (CNN) to extract semantic features and combine prior human knowledge to capture language clues of causality for event relationship extraction [16,35,43]. Although the deep neural network model has better performance in event relationship extraction, the current problem with this method is that it requires a large amount of annotation data to learn and high quality of data annotation.

In general, the current research progress of event relationship extraction is relatively slow. This is mainly due to the following difficulties faced by event relationship extraction: 1) Relationship extraction depends on domains, and relationships defined in different domains are usually challenging to reuse; 2) Relationship extraction The data set of is relatively scarce; 3) The semantic boundary between relations is not clear, that is, the classification of specific relations in the labeling process is prone to divergence, resulting in lower quality of the data set.

4 Applications

Due to the ability of describing dynamic event relationship in the real world, event logic graph has a very broad application prospect and makes up for the shortage of traditional knowledge graph. By deducing the cause-effect and cascade relationships between events, event logic graph enables people to discover hot events faster, better understand the status, cause and effect of many related events, and help people to grasp the potential impact of events more accurately, predict the future development trend of events and make timely responses. At the same time, event logic graph plays an increasingly important role in various industries such as investment research, media creation, public opinion monitoring, credit risk control, etc. In this chapter, we will briefly introduce the three application directions of event logic graph, namely, detection of hot events, analysis of event context, and prediction of future events [68].

4.1 Detect Hot Events

Event logic graph can be applied to the detection of hot events. Because of the sudden and changeable nature of hot events, it is difficult to capture them in time by traditional methods. The relevant analysis technology based on event logic graph can be used to monitor and analyze online public opinion and capture hot events in real time and quickly. By further refining hot events into industry hotspots, regional hotspots, and other specific hotspots, hotspots can be presented from multiple perspectives by using event logic graph related technology. At the same time, for media workers or related enterprises, the timely detection of hot events means that the detected hot events can be pushed to users in time to get more hot news at the first time.

In the field of consumer intent recognition and recommendation systems, based on the event logic graph, it is possible to find certain salient events that may become consumer intent, and these salient events can trigger a series of subsequent consumer events, such as the event of "going on a trip" that may result in "buying a ticket, the event of "going on a trip" may lead to "buying a ticket", "booking a hotel" and other consumption events. Identifying these events and the possibility of the chain of these events depending on the direction of the event development in the event logic graph will help recommend more reasonable personalized products to potential consumers [60].

4.2 Analyze the Event Lineage

Event logic graph are also often applied to the analysis of event sequences. Since the nodes of a fact graph represent events, and the edges represent the probability of occurrence from one event to another, a fact graph can be used to correlate the causes and consequences of an event by using the relationships between the recorded events to form a chain of events and present it to the user. When users search for an event they are interested in, they can also get the cause and result of the event, so that they can clearly understand the development of the event, which greatly improves the efficiency of knowledge retrieval and provides more convenience to human beings.

In the field of intelligent QA, event logic graph is also expected to be widely used. Unlike traditional static knowledge graph that can only answer common sense questions such as time, place, and people, if an event occurs in a conversation, event logic graph can capture the customer's needs based on the dynamic change of the event state, and thus be able to include the preconditions and subsequent events of the event in the reply. With this as the support, the intelligent QA system combined with the event logic graph can further give more intelligent and reasonable responses.

4.3 Predict Future Events

Another important scenario for event logic graph is future prediction. By analyzing the development of historical events, event logic graph plays an important role in predicting possible future events in various industries. Using event logic graph to predict future events mainly consists of two steps: 1) for a new event, find the most similar event to the new event by calculating its similarity to each event in the event logic graph; 2) predict the development direction of the new event based on the most similar event in the abstract event logic graph.

In terms of public opinion early warning, event logic graph can predict possible future events by the events that have occurred at the present stage and make a certain degree of prediction on the development process of the events. Once an event evolves adversely or has an uncontrollable trend, the monitoring system can respond in time and send an early warning to the relevant parts, enabling the relevant departments to intervene in the first place to prevent greater losses. In the financial field, how to predict future events has always been a concern. The application of event logic graph can grasp the dynamics of the industry based on historical events and predict the trend of industry development. Event logic graph can accurately grasp the development trend of the market, so as to quickly make corresponding adjustments in response to market changes and avoid risks for market regulators and investors in general. The prediction of future events has a very important role in real life, and the application of event logic graph in predicting the direction of future events will be more extensive, playing an important role in prediction and early warning, providing a strong guarantee for timely risk avoidance and thus creating great social value.

5 Summary and Prospect

As a unique knowledge graph, the event logic graph has aroused the interest of more and more researchers. The event logic graph describes and portrays the dynamic evolution law between events by constructing a directed graph with events as nodes and relationships between events as edges. The event logic graph aligns with people's cognition of the sequence, causality, and other relations of events in human society. It is bound to have broader application scenarios. However, although more and more researchers are working to improve the construction effect of the event logic graph, the existing construction methods still have a lot of space for improvement. In terms of event extraction, current research is still limited to independent texts, and the research on event extraction across documents is still few and immature. Cross-document and cross-language event extraction research will be more extensive. In the aspect of event relationship recognition, enhancing the model's cross-domain generalization ability is an urgent problem to be solved. The data can be expanded by aligning the knowledge base with the unstructured text to improve the scalability and generalization ability of the model.

References

1. Ahn, D.: The stages of event extraction. In: Proceedings of the Workshop on Annotating and Reasoning About Time and Events, pp. 1–8 (2006)

2. Arendarenko, E., Kakkonen, T.: Ontology-based information and event extraction for business intelligence. In: Ramsay, A., Agre, G. (eds.) AIMSA 2012. LNCS (LNAI), vol. 7557, pp. 89–102. Springer, Heidelberg (2012). https://doi.org/10.1007/978-3-642-33185-5_10
3. Banarescu, L., et al.: Abstract meaning representation for sembanking. In: Proceedings of the 7th Linguistic Annotation Workshop and Interoperability with Discourse, pp. 178–186 (2013)
4. Bethard, S., Savova, G., Chen, W.-T., Derczynski, L., Pustejovsky, J., Verhagen, M.: SemEval-2016 task 12: clinical tempeval. In: Proceedings of the 10th International Workshop on Semantic Evaluation (SemEval-2016), pp. 1052–1062 (2016)
5. Borsje, J., Hogenboom, F., Frasincar, F.: Semi-automatic financial events discovery based on lexico-semantic patterns. Int. J. Web Eng. Technol. **6**(2), 115–140 (2010)
6. Bui, Q.-C., Campos, D., van Mulligen, E., Kors, J.: A fast rule-based approach for biomedical event extraction. In: Proceedings of the BioNLP Shared Task 2013 Workshop, pp. 104–108 (2013)
7. Bui, Q.-C., Sloot, P.M.A.: A robust approach to extract biomedical events from literature. Bioinformatics **28**(20), 2654–2661 (2012)
8. Burel, G., Saif, H., Fernandez, M., Alani, H.: On semantics and deep learning for event detection in crisis situations (2017)
9. Casillas, A., De Ilarraza, A.D., Gojenola, K., Oronoz, M., Rigau, G.: Using Kybots for extracting events in biomedical texts. In: Proceedings of BioNLP Shared Task 2011 Workshop, pp. 138–142 (2011)
10. Chen, C., Ng, V.: Joint modeling for Chinese event extraction with rich linguistic features. In: Proceedings of COLING 2012, pp. 529–544 (2012)
11. Chen, Y., Liu, S., He, S., Liu, K., Zhao, J.: Event extraction via bidirectional long short-term memory tensor neural networks. In: Sun, M., Huang, X., Lin, H., Liu, Z., Liu, Y. (eds.) CCL/NLP-NABD-2016. LNCS (LNAI), vol. 10035, pp. 190–203. Springer, Cham (2016). https://doi.org/10.1007/978-3-319-47674-2_17
12. Chen, Y., Xu, L., Liu, K., Zeng, D., Zhao, J.: Event extraction via dynamic multi-pooling convolutional neural networks. In: Proceedings of the 53rd Annual Meeting of the Association for Computational Linguistics and the 7th International Joint Conference on Natural Language Processing (Volume 1: Long Papers), pp. 167–176 (2015)
13. Cheng, F., Miyao, Y.: Classifying temporal relations by bidirectional LSTM over dependency paths. In: Proceedings of the 55th Annual Meeting of the Association for Computational Linguistics (Volume 2: Short Papers), pp. 1–6 (2017)
14. Chieu, H.L., Ng, H.T.: A maximum entropy approach to information extraction from semi-structured and free text. In: AAAI/IAAI 2002, pp. 786–791 (2002)
15. Das, D., Schneider, N., Chen, D., Smith, N.A.: Probabilistic frame-semantic parsing. In: Human Language Technologies: The 2010 Annual Conference of the North American Chapter of the Association for Computational Linguistics, pp. 948–956 (2010)
16. De Silva, T.N., Zhibo, X., Rui, Z., Kezhi, M.: Causal relation identification using convolutional neural networks and knowledge based features. Int. J. Comput. Syst. Eng. **11**(6), 696–701 (2017)
17. Ding, R., Li, Z.: Event extraction with deep contextualized word representation and multi-attention layer. In: Gan, G., Li, B., Li, X., Wang, S. (eds.) ADMA 2018. LNCS (LNAI), vol. 11323, pp. 189–201. Springer, Cham (2018). https://doi.org/10.1007/978-3-030-05090-0_17
18. Ding, X., Li, Z., Liu, T., Liao, K.: ELG: an event logic graph. arXiv preprint arXiv:1907.08015 (2019)

19. Ghaeini, R., Fern, X.Z., Huang, L., Tadepalli, P.: Event nugget detection with forward-backward recurrent neural networks. arXiv preprint arXiv:1802.05672 (2018)
20. Hu, H.: Research on the construction and application of causal graph for hot topics. Master's thesis, Qingdao University (2020)
21. Zhu, H.: Research on causality of aviation safety accident based on Event Evolutionary Graph. Ph.D. thesis, Civil Aviation University of China (2019)
22. Hakala, K., Van Landeghem, S., Salakoski, T., Van de Peer, Y., Ginter, F.: EVEX in ST'13: application of a large-scale text mining resource to event extraction and network construction. In: Proceedings of the BioNLP Shared Task 2013 Workshop, pp. 26–34 (2013)
23. Hong, Y., Zhou, W., Zhang, J., Zhou, G., Zhu, Q.: Self-regulation: employing a generative adversarial network to improve event detection. In: Proceedings of the 56th Annual Meeting of the Association for Computational Linguistics (Volume 1: Long Papers), pp. 515–526 (2018)
24. Ji, Y., Wang, J., Li, S., Li, Y., Lin, S., Li, X.: An anomaly event detection method based on GNN algorithm for multi-data sources. In: Proceedings of the 3rd ACM International Symposium on Blockchain and Secure Critical Infrastructure, pp. 91–96 (2021)
25. Jinghang, X., Wanli, Z., Shining, L., Ying, W.: Causal relation extraction based on graph attention networks. J. Comput. Res. Dev. **57**(1), 159 (2020)
26. Judea, A., Strube, M.: Event extraction as frame-semantic parsing. In: Proceedings of the Fourth Joint Conference on Lexical and Computational Semantics, pp. 159–164 (2015)
27. Kodelja, D., Besançon, R., Ferret, O.: Exploiting a more global context for event detection through bootstrapping. In: Azzopardi, L., Stein, B., Fuhr, N., Mayr, P., Hauff, C., Hiemstra, D. (eds.) ECIR 2019. LNCS, vol. 11437, pp. 763–770. Springer, Cham (2019). https://doi.org/10.1007/978-3-030-15712-8_51
28. Kuila, A., Chandra Bussa, S., Sarkar, S.: A neural network based event extraction system for Indian languages. In: FIRE (Working Notes), pp. 291–301 (2018)
29. Bai, L.: The construction of the eventic graph for the political field. Ph.D. thesis (2020). J. Chin. Inf. Process
30. Bai, L.: Event evolution graph construction in political field. Ph.D. thesis, University of International Relations (2020)
31. Leeuwenberg, A., Moens, M.F.: Structured learning for temporal relation extraction from clinical records. In: Proceedings of the 15th Conference of the European Chapter of the Association for Computational Linguistics: Volume 1, Long Papers, pp. 1150–1158 (2017)
32. Li, L., Liu, Y., Qin, M.: Extracting biomedical events with parallel multi-pooling convolutional neural networks. IEEE/ACM Trans. Comput. Biol. Bioinf. **17**(2), 599–607 (2018)
33. Li, P., Zhu, Q., Diao, H., Zhou, G.: Joint modeling of trigger identification and event type determination in Chinese event extraction. In: Proceedings of COLING 2012, pp. 1635–1652 (2012)
34. Li, P., Zhu, Q., Zhou, G.: Joint modeling of argument identification and role determination in Chinese event extraction with discourse-level information. In: Twenty-Third International Joint Conference on Artificial Intelligence (2013)
35. Li, P., Mao, K.: Knowledge-oriented convolutional neural network for causal relation extraction from natural language texts. Expert Syst. Appl. **115**, 512–523 (2019)

36. Li, Q., Ji, H., Huang, L.: Joint event extraction via structured prediction with global features. In: Proceedings of the 51st Annual Meeting of the Association for Computational Linguistics (Volume 1: Long Papers), pp. 73–82 (2013)
37. Lim, C.-G., Choi, H.-J.: LSTM-based model for extracting temporal relations from Korean text. In: 2018 IEEE International Conference on Big Data and Smart Computing (BigComp), pp. 666–668. IEEE (2018)
38. Liu, J., Chen, Y., Liu, K.: Exploiting the ground-truth: an adversarial imitation based knowledge distillation approach for event detection. In: Proceedings of the AAAI Conference on Artificial Intelligence, vol. 33, pp. 6754–6761 (2019)
39. Liu, S., Chen, Y., Liu, K., Zhao, J.: Exploiting argument information to improve event detection via supervised attention mechanisms. In: Proceedings of the 55th Annual Meeting of the Association for Computational Linguistics (Volume 1: Long Papers), pp. 1789–1798 (2017)
40. Liu, X., Luo, Z., Huang, H.: Jointly multiple events extraction via attention-based graph information aggregation. arXiv preprint arXiv:1809.09078 (2018)
41. Majumder, A., Ekbal, A., Naskar, S.K.: Biomolecular event extraction using a stacked generalization based classifier. In: Proceedings of the 13th International Conference on Natural Language Processing, pp. 55–64 (2016)
42. Martinez, D., Baldwin, T.: Word sense disambiguation for event trigger word detection in biomedicine. BMC Bioinform. **12**, 1–8 (2011). https://doi.org/10.1186/1471-2105-12-S2-S4
43. Mekuriaw, M.: Automatic causal relation extraction for amharic language texts using CNN. Ph.D. thesis (2020)
44. Nguyen, T.H., Cho, K., Grishman, R.: Joint event extraction via recurrent neural networks. In: Proceedings of the 2016 Conference of the North American Chapter of the Association for Computational Linguistics: Human Language Technologies, pp. 300–309 (2016)
45. Nguyen, T.H., Grishman, R.: Event detection and domain adaptation with convolutional neural networks. In: Proceedings of the 53rd Annual Meeting of the Association for Computational Linguistics and the 7th International Joint Conference on Natural Language Processing (Volume 2: Short Papers), pp. 365–371 (2015)
46. Nguyen, T.H., Grishman, R.: Graph convolutional networks with argument-aware pooling for event detection. In: Thirty-Second AAAI Conference on Artificial Intelligence (2018)
47. Ning, Q., Feng, Z., Wu, H., Roth, D.: Joint reasoning for temporal and causal relations. arXiv preprint arXiv:1906.04941 (2019)
48. Ning, Q., Subramanian, S., Roth, D.: An improved neural baseline for temporal relation extraction. arXiv preprint arXiv:1909.00429 (2019)
49. Piskorski, J., Tanev, H., Atkinson, M., van der Goot, E., Zavarella, V.: Online news event extraction for global crisis surveillance. In: Nguyen, N.T. (ed.) Transactions on Computational Collective Intelligence V. LNCS, vol. 6910, pp. 182–212. Springer, Heidelberg (2011). https://doi.org/10.1007/978-3-642-24016-4_10
50. Shi, Q.: Research on the key technology of consumer intention recognition and prediction based on event logic graph, pp. 1–z. Harbin Institute of Technology (2020)
51. Rao, S., Marcu, D., Knight, K., Daumé III, H.: Biomedical event extraction using abstract meaning representation. In: BioNLP 2017, pp. 126–135 (2017)
52. Riloff, E., et al.: Automatically constructing a dictionary for information extraction tasks. In: AAAI, vol. 1, p. 2-1. Citeseer (1993)

53. Riloff, E., Shoen, J.: Automatically acquiring conceptual patterns without an annotated corpus. In: Third Workshop on Very Large Corpora (1995)
54. Chen, S.: Study on the method of constructing the event logic graph of housing price changes. Master's thesis, Harbin Institute of Technology (2020)
55. Zhang, S., Wang, L., Lou, G.: Research on network public opinion analysis and judgment system based on knowledge graph. J. Mod. Inf. (2021)
56. Sha, L., Qian, F., Chang, B., Sui, Z.: Jointly extracting event triggers and arguments by dependency-bridge RNN and tensor-based argument interaction. In: Thirty-Second AAAI Conference on Artificial Intelligence (2018)
57. Tai, K.S., Socher, R., Manning, C.D.: Improved semantic representations from tree-structured long short-term memory networks. arXiv preprint arXiv:1503.00075 (2015)
58. Tourille, J., Ferret, O., Neveol, A., Tannier, X.: Neural architecture for temporal relation extraction: a Bi-LSTM approach for detecting narrative containers. In: Proceedings of the 55th Annual Meeting of the Association for Computational Linguistics (Volume 2: Short Papers), pp. 224–230 (2017)
59. Verhagen, M., Pustejovsky, J.: Temporal processing with the TARSQI toolkit. In: COLING 2008: Companion Volume: Demonstrations, pp. 189–192 (2008)
60. Xiang, W.: Reviews on event knowledge graph construction techniques and application. Comput. Modernization **1**(10), 10–16 (2020)
61. Wu, W., Zhu, X., Tao, J., Li, P.: Event detection via recurrent neural network and argument prediction. In: Zhang, M., Ng, V., Zhao, D., Li, S., Zan, H. (eds.) NLPCC 2018. LNCS (LNAI), vol. 11109, pp. 235–245. Springer, Cham (2018). https://doi.org/10.1007/978-3-319-99501-4_20
62. Ding, X.: Research on social media-based market sentiment prediction method, pp. 1–z. Harbin Institute of Technology (2016)
63. Shan, X., Pang, S., Liu, X., Yang, J.: Research on internet public opinion event prediction method based on event evolution graph. Inf. Stud. Theory Appl. **43**(10), 165 (2020)
64. Xiang, W., Wang, B.: A survey of event extraction from text. IEEE Access **7**, 173111–173137 (2019)
65. Tian, Y., Li, X.: Analysis on the evolution path of COVID-19 network public opinion based on the event evolutionary graph. Inf. Stud. Theory Appl. **44**(3), 76–83 (2021)
66. Yadav, S., Ramteke, P., Ekbal, A., Saha, S., Bhattacharyya, P.: Exploring disorder-aware attention for clinical event extraction. ACM Trans. Multimed. Comput. Commun. Appl. (TOMM) **16**(1s), 1–21 (2020)
67. Yan, S., Wong, K.-C.: Context awareness and embedding for biomedical event extraction. Bioinformatics **36**(2), 637–643 (2020)
68. Hu, Z., Jin, X., Chen, J., Huang, G., et al.: Construction, reasoning and applications of event graphs. Big Data Res. **7**(3), 80–96 (2021)
69. Zeng, Y., Luo, B., Feng, Y., Zhao, D.: WIP event detection system at TAC KBP 2016 event nugget track. In: TAC (2016)
70. Zeng, Y., Yang, H., Feng, Y., Wang, Z., Zhao, D.: A convolution BiLSTM neural network model for Chinese event extraction. In: Lin, C.-Y., Xue, N., Zhao, D., Huang, X., Feng, Y. (eds.) ICCPOL/NLPCC 2016. LNCS (LNAI), vol. 10102, pp. 275–287. Springer, Cham (2016). https://doi.org/10.1007/978-3-319-50496-4_23
71. Zhang, J., Zhou, W., Hong, Yu., Yao, J., Zhang, M.: Using entity relation to improve event detection via attention mechanism. In: Zhang, M., Ng, V., Zhao, D., Li, S., Zan, H. (eds.) NLPCC 2018. LNCS (LNAI), vol. 11108, pp. 171–183. Springer, Cham (2018). https://doi.org/10.1007/978-3-319-99495-6_15

72. Zhang, W., Ding, X., Liu, T.: Learning target-dependent sentence representations for Chinese event detection. In: Zhang, S., Liu, T.-Y., Li, X., Guo, J., Li, C. (eds.) CCIR 2018. LNCS, vol. 11168, pp. 251–262. Springer, Cham (2018). https://doi.org/10.1007/978-3-030-01012-6_20
73. Zhang, Z., Xu, W., Chen, Q.: Joint event extraction based on skip-window convolutional neural networks. In: Lin, C.-Y., Xue, N., Zhao, D., Huang, X., Feng, Y. (eds.) ICCPOL/NLPCC -2016. LNCS (LNAI), vol. 10102, pp. 324–334. Springer, Cham (2016). https://doi.org/10.1007/978-3-319-50496-4_27

Enhancing Low-Resource Languages Question Answering with Syntactic Graph

Linjuan Wu(✉), Jiazheng Zhu, Xiaowang Zhang, Zhiqiang Zhuang, and ZhiYong Feng

College of Intelligence and Computing, Tianjin University, Tianjin 300350, China
{wulinjuan1997,jiazhengzhu,xiaowangzhang,zhuang,zyfeng}@tju.edu.cn

Abstract. Multilingual pre-trained language models (PLMs) facilitate zero-shot cross-lingual transfer from rich-resource languages to low-resource languages in extractive question answering (QA) tasks. However, during fine-tuning on the QA task, the syntactic information of languages in multilingual PLMs is not always preserved or even is forgotten, which may influence the detection of answer spans for low-resource languages. In this paper, we propose an auxiliary task to predict syntactic graphs to enhance syntax information in the fine-tuning stage of the QA task to improve the answer span detection of low-resource. The syntactic graph includes Part-of-Speech (POS) information and syntax tree information without dependency parse label. We convert the syntactic graph prediction task into two subtasks to adapt the sequence input of PLMs: POS tags prediction task and syntax tree prediction task (including depth prediction of a word and distance prediction of two words). Moreover, to improve the alignment between languages, we parallel train the source language and target languages syntactic graph prediction task. Extensive experiments on three multilingual QA datasets show the effectiveness of our proposed approach.

Keywords: Question answering · Low-resource language · Transfer learning · Syntactic graph

1 Introduction

Question Answering (QA) plays a critical role in assessing the ability of machines in natural language understanding. The field has made rapid advances in recent years, especially in the span extractive QA task of English (SQuAD v1.1 [1]), which has even exceeded human performance in the method based on pre-trained language models (PLMs) [2]. Due to insufficient annotated corpora in low-resource languages, extractive question answering is primarily exclusive to English. With the introduction of multilingual language models (e.g. multilingual BERT [2], XLM [3], XLM-Roberta [4], and [5]) and multilingual evaluation datasets (e.g. MLQA [6], XQuAD [7], and Tydi QA [8]), QA tasks are encouraged to expand from English to low-resource languages by cross-lingual transfer.

U. K. Rage et al. (Eds.): DASFAA 2022 Workshops, LNCS 13248, pp. 175–188, 2022.
https://doi.org/10.1007/978-3-031-11217-1_13

Table 1. The percentages of answer spans that respect syntactic constituent boundaries in four multilingual QA datasets in both English and Chinese.

	XQuAD	MLQA	TyDi QA-GoldP
English	89.08%	90.11%	89.12%
Chinese	88.05%	87.57%	-

For extractive QA, given a passage c and a question q, it requires models to extract a consecutive span from c as the answer to q. For low-resource languages, the cross-lingual transfer method based on multilingual PLMs could roughly locate the correct spans but still produce partially correct answers [6]. To address this issue, [9] introduces language-specific phrase boundary supervision in the fine-tuning stage to enhance answer boundary detection. [10] points out that the issue of partially correct answers requires consideration of syntactic structure, semantic coherence, and adequacy.

An intuitive assumption here is that most answer spans respect syntactic constituent boundaries (i.e. consisting of one or several complete syntactic constituents or phrases). On three multilingual QA evaluation datasets, we use Stanford CoreNLP[1] to collect syntax parse trees and calculate the percentages of ground-truth answers that respect syntactic constituent boundaries. As shown in Table 1, over 87% of answer spans respect syntactic constituent boundaries. Therefore, we focus on the impact of the syntactic information in the multilingual pre-trained representation on the low-resource languages QA.

The structural probe [11,12] finds that the multilingual BERT (mBERT) encoded the syntax tree of a sentence in embedding. It probes the syntax tree by looking for the linear transformation of the representation. By this structural probe, [13] analyze the evolution of the embedded syntax trees of BERT along the fine-tuning process on six different tasks, including QA. It shows that the syntactic information initially encoded in BERT is preserved during fine-tuning on the English QA task. Following [13], we explored the evolution of the embedded syntax tree of six languages in mBERT for the QA task. We found that syntactic information in mBERT was forgotten in three languages and preserved in the other three languages (detail in Sect. 2). It suggests that the syntactic information languages in multilingual PLMs are not always preserved during fine-tuning on the QA, even are forgotten.

Since understanding the syntax of the language may help optimize the language modeling objective, we propose an auxiliary syntactic graph prediction task to enhance syntactic information for low-resource languages to improve the answer span detection. Similar to [14], the syntactic graph includes word-level and sentence-level syntax: Part-of-Speech (POS) information and syntax tree information (without label about dependency relation compare to the syntactic dependency parse tree), as shown in Fig. 1(b). However, the syntactic graph is structural, and texts of QA and input of PLM are sequence; combining the

[1] https://stanfordnlp.github.io/CoreNLP/.

two types of information is difficult. Inspired by [11], we convert the syntactic graph prediction task into two subtasks that can be performed on the sequence: POS tags prediction and syntax tree prediction (including depth prediction of a word and distance prediction of two words in the syntax tree). We collect labeled parallel corpus to learn syntactic knowledge of English and low-resource target languages at the same time, as shown in Fig. 1(c). With the help of English as the hub language, our model improves the alignment between languages. Together with the auxiliary task trained on the syntactically labeled corpus, we fine-tune the multilingual PLMs on the rich-resource source language (i.e. English) and zero-shot transfer to the target language QA.

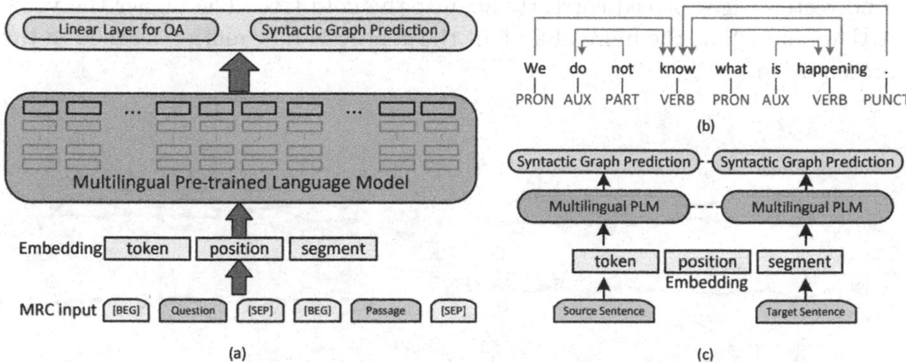

Fig. 1. Diagram of our training procedure for low-resource languages QA based on PLM. (a) The overall framework of our model for extractive QA tasks. (b) The illustration of the syntactic graph. The purple one-way arrow line represents the syntax tree information; the orange line and orange label represent the POS information. (c) The flow of syntactic graph prediction task. The input sentence is a parallel sentence pair, and the output is the prediction result of the syntactic graph information corresponding to the sentence.

To summarise, our main contributions are as follows.

- We propose an auxiliary syntactic graph prediction task on top of multilingual PLMs that enhances syntactic information to improve the accuracy of the answer span detection for low-resource QA.
- To adapt to the input of the sequence model and improve the alignment between languages, we convert the syntactic graph prediction task into two sequence-level subtasks and parallel train on the source and target languages.
- Different languages have different syntaxes, but not all low-resource languages have syntactic graph corpus. In further experiments, our model improves the performance of the low-resource target languages that without training datasets to learn syntactic information.

Experiments on three multilingual QA datasets (XQuAD, MLQA, and TyDi QA) with 17 different low-resource target languages demonstrate that our model can significantly improve performance over strong baselines.

2 Syntactic Information Evolvement in mBERT

We explored how syntactic information of different languages in mBERT evolves during the fine-tuning process on the QA task by the structural probe [11]. The probe can evaluate how well syntax trees are embedded in the representation space by a linear transformation, including two evaluations: tree distance evaluation and tree depth evaluation. We trained the probe for tree distance evaluation, in which squared L2 distance encodes the distance between words in the syntax tree. The evaluate metrics are the Undirected Unlabeled Attachment Score (UUAS) and the Spearman correlation between actual and predicted distances in sentences with lengths between 5 and 50 (DSpr). UUAS is the percent of undirected edges placed correctly against the gold tree. The higher the value of UUAS and DSpr, the more abundant the syntactic information learned in the representation.

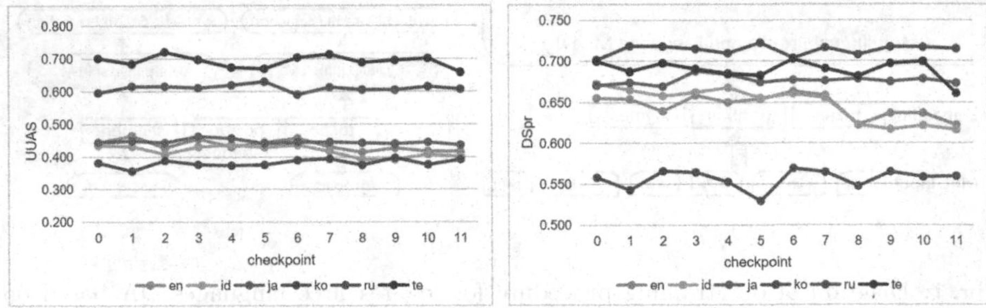

Fig. 2. The results (UUAS and DSpr) of the syntactic information evolution experiment on QA task based on mBERT fine-tuned only on English dataset.

About the experimental setting, we fine-tune the mBERT on the QA task for two epochs with a learning rate of 5e−5 and batch size of 24, saving every 1000 batch as one checkpoint. [12] show that the 7th layer representation of mBERT has richer syntactic information for various languages. We train six probes on the 7th layer embedding of the different checkpoint models generated along the fine-tuning process for English (en), Indonesia (id), Japanese (ja), Russian (ru), Korean (ko), and Telugu (te). We repeat training these probes five times with different random seeds and averaged them.

The results are shown in Fig. 2, English (en), Indonesia (id), and Telugu (te) have a fluctuation on UUAS and DSpr, and to fell finally along with the fine-tuning. Japanese (ja) shows a stable performance along during fine-tuning. Russian (ru) and Korean (ko) fluctuate to a certain extent and finally show an insignificant increase in UUAS and DSpr. The results have shown that, along with the fine-tuning of the QA task, the syntactic information initially encoded in the mBERT is forgotten in English, Indonesia, and Telugu. In Japanese, Russian, and Korean, that syntactic information is preserved and not significantly enhanced. It suggests that the syntactic information languages in multilingual PLMs are preserved or forgotten during fine-tuning on the QA.

3 Related Work

[15] investigate cross-lingual transfer capability of multilingual BERT (mBERT) on extractive QA tasks and find that zero-shot learning based on PLM is feasible, even between distant languages, such as English and Chinese. Various approaches have been proposed on top of multilingual MRC based on PLMs [9,16–19].

[9] is a work most relevant to our goal of enhancing syntactic information for multilingual QA. It presents two additional tasks: mixMRC and LAKM, introducing phrase boundary supervision into the fine-tuning stage. The mixMRC uses machine translation to translate English QA data into multilingual training data. LAKM presents an additional phrase corpus to guide the masking task of the pre-trained model. The two additional tasks may suffer from low translation quality and resource limits. Therefore, unlike this multilingual QA study, we mainly consider enhancing the syntax information already mastered by multilingual PLMs to improve multilingual QA by zero-shot transfer.

4 Method

In this section, we first introduce the overall training procedure and then describe the syntactic graph and the auxiliary syntactic graph prediction task.

The overview of our training procedure is shown in Fig. 1. Our method is built on top of multilingual pre-trained models (mBERT and XLM). We concatenate texts with special tokens [BEG] and [SEP] as the input sequence of our model and transform word embedding into contextually-encoded token representations using PLM. Finally, this contextual representation is used for the extractive QA task in the source language (i.e., English) and the syntactic graph prediction task.

We only fine-tune the PLMs on the English corpus of the extractive QA task and evaluate them in other low-resource languages. This zero-shot transfer method can make our model free from the dilemma of the scarcity of target language training data.

In the following, we introduce our utilized syntactic graph, prediction task, and training strategy. This syntax task will jointly train with our primary task to boost low-resource language QA performance.

4.1 Syntactic Graph

To enhance the syntactic information of the multilingual PLM, we want to incorporate the syntactic structure of low-resource languages into the pre-trained representations. The syntactic structure can be divided into word level, phrase level, and sentence level. For the extractive QA task, the model read a passage and related question, and detect the words that can be the answer boundary's start or end in the passage. Most answer spans respect syntactic constituent boundaries, as the statistic shown in Table 1. Therefore, we combine word-level POS information and sentence-level syntax tree information to construct a syntactic graph. The syntax tree is a syntactic dependency parse tree without the dependency label.

The illustration can be seen in Fig. 1(b). Firstly, POS information is the word-level syntax tag that integrates the type and function of a word. Our syntactic graph incorporates POS information by generating a neighbor node for each word. A syntax tree describes the grammatical relations that hold among words. We incorporate this information into the syntactic graph by adding directed edges between the word nodes.

4.2 Syntactic Graph Prediction Task

The syntactic graph shows the syntactic information we use, which is structural. However, the input of the pre-training model is a sequence, so we divide the task of syntactic graph prediction into two subtasks to predict Part-of-Speech tags information and syntax trees, respectively. The prediction of the syntax tree includes the depth prediction of a single word in the tree and the distance prediction of two words in the tree.

Part-of-Speech Tags Prediction. POS tags prediction is a sequence labeling task, which can be regarded as a multi-class classification problem for each word in a sentence. Hence, we define POS loss as a cross-entropy style loss as follows:

$$\begin{aligned} L_{POS} &= \sum_i \{ \sum_{j=1}^{n} -\mathrm{I}(y_i == j) \log[\mathrm{softmax}(g(e_i))_j] \} \\ &= \sum_i [-\log \mathrm{softmax}(g(e_i))_{y_i}] \end{aligned} \tag{1}$$

where $g(\cdot)$ is a linear layer with input e_i that from the multilingual PLM for the ith token in the input sentence. $\mathrm{softmax}(\cdot)_j$ estimates the probability of POS tag j, n is the number of different POS tags. y_i is the truth POS tag of ith word in the sentence. $\mathrm{I}(\cdot)$ is the judgment function, which indicates that when the bracket equation is true, take 1; otherwise, it is 0.

Syntax Tree Prediction. Inspired by [11], we formulate syntactic parsing from pre-trained word representations as two independent tasks: depth prediction of a word and distance prediction of two words in the parse tree. Given a matrix $B \in \mathbb{R}^{k \times m}$ as a linear transformation, the losses of these two subtasks are defined as:

$$L_{depth} = \sum_i (\|w_i\| - \|Be_i\|_2^2), \tag{2}$$

$$L_{distance} = \sum_{i,j} |d_T(w_i, w_j) - d_B(e_i, e_j)| \tag{3}$$

where $\|w_i\|$ is the parse depth of a word defined as the number of edges from the root of the parse tree to w_i, and $\|Bh_i\|_2$ is the tree depth L2 norm of the vector space under the linear transformation. $d_T(w_i, w_j)$ is the number of edges in the path between the ith and jth word in the parse tree T. As for $d_B(e_i, e_j)$, it can be defined as the squared L_2 distance after transformation by B:

$$d_B(e_i, e_j) = (B(e_i - e_j))^T (B(e_i - e_j)) \tag{4}$$

To induce parse trees, we minimize the summation of the above two losses L_{depth} and $L_{distance}$, and the loss of syntax tree prediction (STP) task is defined as:

$$L_{STP} = L_{depth} + L_{distance} \tag{5}$$

Finally, our training objectives are defined as follows:

$$L = \alpha L_{POS} + L_{STP} \tag{6}$$

where the α is the weight of POS loss, which is a hyperparameter.

As shown in Fig. 1(c), our training of syntactic graph prediction task has two parallel flows for the source language and low-resource target languages.

5 Experiment

5.1 Datasets and Baseline Models

To verify the effectiveness of our model for low-resource languages QA tasks, we conducted experiments on three multilingual question answering benchmarks:

MLQA [6] consists of over 5K extractive MRC instances in 7 languages: English (en), Arabic (ar), German (de), Spanish (es), Hindi (hi), Vietnamese (vi) and Chinese (zh). MLQA is also highly parallel, with QA instances parallel between 4 different languages on average.

XQuAD [7] consists of a subset of 240 paragraphs and 1190 question-answer pairs from the development set of SQuAD v1.1 together with their professional translations into ten languages: Spanish (es), German (de), Greek (el), Russian (ru), Turkish (tr), Arabic (ar), Vietnamese (vi), Thai (th), Chinese (zh), Hindi (hi), and Romanian (ro).

TyDi QA-GoldP is the gold passage task in TyDi QA [8] covers 9 typologically diverse languages: Arabic (ar), Bengali (bg), English (en), Finnish (fi), Indonesian (id), Korean (ko), Russian (ru), Swahili (sw), Telugu (te). It is more challenging as questions have been written without seeing the answers, leading to 3 and 2 times less lexical overlap than XQuAD and MLQA, respectively [21].

Our model implementation is based on two multilingual pre-trained models: **mBERT** and **XLM-100** [3]. They built on the transformer model, pretrain with mask language model task on Wikipedia corpus, but XLM-100 has more parameters and a more extensive shared vocabulary than mBERT. Our models based on two pre-trained models are named mBERT_QA_ours and XLM_QA_ours.

5.2 Setup and Evaluation Metric

For the syntactic graph prediction task, we collected approximately 26,000+ labeled parallel sentence pairs from the Universal Dependencies (UD 2.7) Corpus [22] as the training set. The training set covers 18 languages and overlapping with 10 target languages of three QA evaluation datasets: Arabic (ar), Chinese (zh), Finnish (fi), German (de), Hindi (hi), Indonesian (id), Korean (ko), Russian (ru), Spanish (es), Thai (th) and Turkish (tr). We used Universal POS tags and HEAD tags in UD 2.7 to compose our syntactic graph.

For our QA model and two baseline models, we fine-tuned them on the English SQuAD v1.1 [1] and evaluated them on the three multilingual QA benchmarks with 17 target languages. Among these 17 target languages, 10 languages have training data to learn syntactic graph information, and the other 7 target languages are implemented with the complete zero-shot transfer. We used a learning rate of 2e−5 to train and fine-tuned them for three epochs with a batch size of 32 based on mBERT. For XLM-100, we fine-tuned the baseline and our model for two epochs with a batch size of 16 and a learning rate of 3e–5. The hyperparameter α is set as 0.5.

The evaluation metrics for the extractive QA task are F1 and Exact Match (EM). Exact Match Score measures the percentage of predictions that exactly match any one of the ground truths. F1 score is used to measure the answer overlap between predictions and ground truths.

5.3 Experiment Results

We show the results of QA evaluation based on the mBERT and XLM-100 models on three datasets and show the results of the ablation experiment based on mBERT. The two models of the ablation experiment removed the POS loss and the syntax tree loss, respectively.

Results on MLQA. We first evaluated our method on the MLQA dataset in 6 languages, including the conditions of having training data for syntax tasks and zero-shot transfer setting. The results are shown in Table 2. Our method consistently outperforms two baselines in both two conditions. With training data of syntactic graph, our model based on XLM-100 outperforms the baseline by 9.5%, 6.4% in EM for Arabic (ar) and Spanish (es), respectively. Even under the zero-shot transfer setting, our model on Vietnamese (vi) also obtains 4.8% and 2.3% EM improvement over XLM-100 and mBERT, respectively. Compared with the experimental results of the existing method LAKM [9], we use a small training corpus, differing by 3 orders of magnitude on each language, and obtain performance close to it. The ablation experiments on MLQA show that both losses are essential, especially the syntax tree loss. When the syntax tree loss is removed, the effectiveness of our model on all 6 languages is dropped.

Table 2. EM and F1 score of 6 languages on MLQA. The left 5 languages have training data for syntactic graph prediction task while the Vietnamese (vi) is under zero-shot transferring. The underlined results are our best results in German and Spanish.

	MLQA (EM/F1)					
	ar	de	es	hi	zh	vi
XLM-100	27.0/62.8	43.5/71.3	42.7/73.8	29.3/56.4	30.1/53.7	37.4/65.0
XLM_QA_ours	**36.5/65.1**	**48.6/73.3**	**49.1/73.9**	**34.8/60.2**	**32.8/54.6**	**42.2/67.9**
mBERT	31.5/49.5	43.8/58.3	45.8/64.1	29.4/45.2	34.5/56.1	37.5/57.3
LAKM	-	**45.5/60.5**	**48.0/65.9**	-	-	-
mBERT_QA_ours	**32.7/50.0**	<u>44.2</u>/<u>58.6</u>	46.9/65.0	**31.9/46.9**	35.0/55.7	**39.8/59.3**
w/o L_{POS}	28.0/45.9	43.9/58.2	<u>47.8</u>/<u>65.4</u>	30.9/46.6	**36.2/56.7**	38.5/57.9
w/o L_{STP}	27.8/44.1	43.9/58.4	44.9/62.8	27.2/41.5	33.6/54.1	38.4/57.4

Results on XQuAD. We further evaluate our method on the XQuAD dataset and the results are shown in Table 3 and Table 4, which are under two different conditions. For the languages with training data of syntactic graph prediction task, our method based on XLM-100 and mBERT improved all of 7 languages. Especially, our method based on XLM-100 gets 5.4% and 9.6% improvement of EM score on Arabic (ar) and Thai (th) respectively. In the setting of completely zero-shot transfer, our model based on XLM-100 obtains a more significant improvement in all of the 4 languages. The ablation results in Table 4 show that our model with only syntax tree loss gets the best performance in the zero-shot transfer setting. It suggests that POS information is not helpful for zero-resource language QA on the XQuAD dataset, and even has a negative impact. During zero-shot transfer, POS information may interfere with the semantics of the target language. But the POS information of the language itself can really help the improvement of their QA tasks, as shown in the lower part of Table 3.

Table 3. EM and F1 score of 8 languages on the XQuAD dataset that have training data to learn more syntactic information.

	XQuAD (EM/F1)								
	ar	de	es	hi	ru	th	tr	zh	avg
XLM-100	35.6/72.4	53.8/80.9	54.6/81.0	39.9/64.9	54.0/79.5	10.3/27.0	42.0/72.4	42.7/65.4	41.7/67.9
XLM_QA_ours	**41.0/75.0**	**56.6/82.6**	**56.3/82.1**	**43.4/68.3**	**56.1/79.7**	**19.6/29.7**	**46.3/72.6**	**43.8/67.1**	**45.4/69.6**
mBERT	44.3/60.6	54.0/69.6	57.3/74.9	38.3/53.3	54.0/69.6	30.9/39.9	33.8/50.0	46.0/57.4	44.0/59.0
mBERT_QA_ours	**45.5/60.9**	**56.3**/70.9	57.5/**75.9**	**41.2/54.7**	**56.3/70.3**	**33.4/40.9**	34.0/50.8	47.2/56.9	**46.4/60.2**
w/o L_{POS}	40.0/56.7	56.0/**71.3**	**57.8**/75.5	40.8/55.3	54.6/69.0	32.0/40.6	**35.2/51.2**	**47.3/57.7**	45.4/59.7
w/o L_{STP}	39.7/54.5	55.7/71.0	54.5/72.4	35.7/49.6	53.3/68.8	30.3/37.8	30.3/37.8	45.5/54.0	43.7/55.7

Table 4. EM and F1 score of 3 languages on the XQuAD dataset under the zero-shot transfer setting.

	XQuAD (EM/F1)			
	el	ro	vi	avg
XLM-100	37.9/66.3	56.6/79.6	49.5/75.4	48.0/73.8
XLM_QA_ours	**47.2/70.9**	**59.7/81.0**	**52.9/75.8**	53.3/75.6
mBERT	**46.0**/61.1	58.3/72.5	46.1/65.9	50.1/66.5
mBERT_QA_ours	44.8/59.7	58.1/72.4	48.2/67.4	50.4/66.5
w/o L_{POS}	44.9/**61.3**	**59.0/72.7**	**49.2/68.1**	**51.0/67.4**
w/o L_{STP}	42.1/57.9	57.4/70.0	35.3/50.8	44.9/59.6

Table 5. EM and F1 score of 8 languages on the TyDi QA-GoldP dataset. The left 5 languages have training data for syntactic graph prediction task while the right 3 languages are under the zero-shot transfer method.

	TyDi QA-GoldP (EM/F1)							
	ar	fi	id	ko	ru	bg	sw	te
XLM-100	31.1/69.8	39.3/65.3	42.8/69.0	1.4/24.9	36.8/70.2	29.2/57.7	32.9/59.2	34.4/61.1
XLM_QA_ours	**39.7/72.4**	**42.2/65.6**	**48.0/75.4**	**4.0/29.1**	**37.4/72.4**	**39.8/62.8**	**35.6/61.1**	**40.5/62.8**
mBERT	43.8/59.5	44.0/56.9	45.3/59.8	41.7/49.8	41.4/64.4	39.8/54.9	32.3/50.0	39.0/48.2
mBERT_QA_ours	**45.3/60.5**	**44.8/58.4**	48.3/60.1	46.0/54.1	44.1/**64.5**	**49.6/60.8**	38.3/53.0	**42.2/51.4**
w/o L_{POS}	40.3/54.9	44.0/58.0	**49.0/62.1**	**46.4/55.1**	**44.2**/64.2	45.1/57.8	40.3/54.6	42.0/49.5
w/o L_{STP}	37.7/52.8	41.0/52.9	45.0/56.8	43.1/52.4	40.0/60.2	43.4/54.7	**42.9/55.4**	40.1/46.9

Results on TyDi QA-GoldP. The results of the TyDi QA-GoldP dataset across 8 typologically diverse languages are shown in Table 5. The EM and F1 in our model are superior to the two baselines in all of the 8 languages. Significantly, our method improves the EM in Arabic (ar), Indonesian (id), Bengali (bg), Telugu (te) by 8.6%, 5.2%, 10.6%, 6.1% over the XLM-100 baseline, respectively. Bengali (bg) and Telugu (te) are distant languages for English, without training corpus to learn syntactic information, but they significantly improve.

The evaluation results on three datasets powerful verify the effectiveness and generalization of our proposed method. The ablation results prove the effectiveness of each loss in the additional tasks.

6 Analysis

6.1 Why Use Parallel Sentence Pairs to Train Syntax Task?

In order to learn richer syntactic information, it is enough to use the labeled data only in the target language (this single-version model is named ours_single). Moreover, to align the target and source languages better, we use parallel sentence pairs to train the syntactic task to improve the cross-lingual effect. As the results on Tydi QA-GoldP datasets shown in Table 6, for languages having training data of syntax task, if only the target language sentence is used, there are fewer improvements than our model. However, the single-version model has a higher EM score on zero-resource languages in the TyDi QA-GoldP than our model. From the average improvement effect, our model still has better performance.

Table 6. EM and F1 score of 8 languages on the TyDi QA-GoldP dataset with different training strategies of the syntactic graph prediction task.

	TyDi QA-GoldP (EM/F1)								
	ar	fi	id	ko	ru	bg	sw	te	avg
mBERT	43.8/59.5	44.0/56.9	45.3/59.8	41.7/49.8	41.4/64.4	39.8/54.9	32.3/50.0	39.0/48.2	40.9/55.4
mBERT_QA_ours	**45.3/60.5**	**44.8/58.4**	**48.3/60.1**	**46.0/54.1**	**44.1/64.5**	49.6/60.8	**38.3/53.0**	42.2/**51.4**	**44.8/57.9**
ours_single	43.1/57.9	41.9/55.2	46.7/60.0	43.8/52.0	40.1/63.5	**52.2/62.5**	**38.3**/51.9	**42.3**/50.9	43.6/56.7

6.2 Why the Syntactic Graph Prediction Task Works?

Our method mainly aims to enhance syntactic information of pre-trained representations for low-resource languages to improve the detection of the answer span of QA tasks. Through the strategy of parallel training of source language and target language to learn syntactic information, we hope to strengthen alignment between languages and obtain a more universal multilingual representation. In order to examine (1) whether PLM representations encode richer syntactic information and (2) whether the syntactic information helps multilingual modeling, we conducted additional experiments and analyses.

Syntactic Information Evaluation. We continued the setting in Sect.2 to evaluate the syntactic information in mBERT representations during fine-tune process of our method. The last layer vector (i.e. 12th layer vector) is the input of our two tasks layer, so we experiment with the vector of the last layer in mBERT. As the results are shown in Fig. 3, the syntactic information initially encoded in

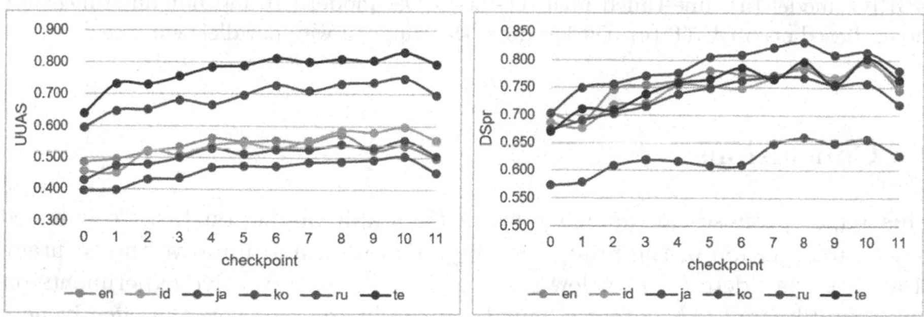

Fig. 3. The results (UUAS and DSpr from left to right) of the syntactic information evolution experiment on QA task with syntactic graph prediction task based on mBERT.

the 12th layer representation of mBERT is strengthened during fine-tune process of the QA task. It suggests that our syntactic graph prediction task enhances the syntactic information of pre-trained representation. Combined with the results of the QA benchmark experiment, it can be seen that the enhancement of syntactic information contributes to the improvement of QA tasks.

Visualization of Representations After Fine-Tuning. We randomly selected 100 parallel sentences from a 15-way parallel corpus [23] with 10,000 sentences. We visualize hidden representations of the last layer of the mBERT model, the last layer embedding of baseline QA model and our QA model based on mBERT, as shown in Fig. 4. An interesting phenomenon in Fig. 4(b) is that the representations of 15 languages (semantically equivalent to each other) show apparent language aggregation, which may reflect the limitation of its zero-shot transfer: the really distant and low-resource languages are challenging to transform following the source languages (English). However, It is clear to see that the last layer representations of our fine-tuned QA model based on mBERT make all of the parallel sentences in 15 languages closer to one another in space, blending language boundaries clearly seen from Fig. 4(a) and (b). On the other hand, it shows that our method can shorten the distance of language in the semantic space, thereby improving the performance of zero-shot transfer.

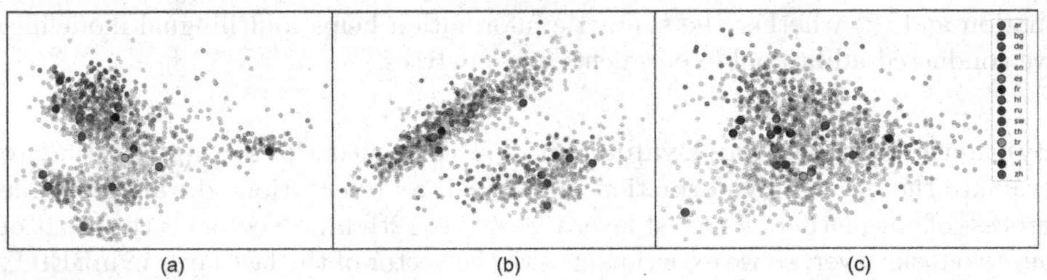

Fig. 4. PCA visualization of hidden representations from the last transformer layer of mBERT model (a), fine-tuned mBERT-based QA model (b) and our fine-tuned QA model based on mBERT (c). Darker dots: the same 15-way parallel sentence.

7 Conclusions

This paper proposes an auxiliary syntactic graph prediction task to enhance syntax information of the pre-trained representation and improve the accuracy of answer span detection for low-resource languages. Extensive experiments on three multilingual QA datasets have been conducted to prove the effectiveness of our proposed approach. Significantly, the zero-resource target languages without training datasets of syntax tasks have also been improved on the QA task, reflecting the generalization of our model. Further experimental analysis shows that our method effectively enriches the syntactic information of pre-training representations and strengthens the alignment between languages, which improves the performance on QA tasks for low-resource languages.

References

1. Rajpurkar, P., Zhang, J., Lopyrev, K., Liang, P.: SQuAD: 100, 000+ questions for machine comprehension of text. In: 21st International Conference on Empirical Methods in Natural Language Processing, pp. 2383–2392. Association for Computational Linguistics. Austin, Texas, USA (2016)
2. Devlin, J., Chang, M.-W., Lee, K., Toutanova, K.: BERT: pre-training of deep bidirectional transformers for language understanding. In: 23rd International Conference of the North American Chapter of the Association for Computational Linguistics: Human Language Technologies, pp. 4171–4186. Association for Computational Linguistics. Minneapolis, MN, USA (2019)
3. Conneau, A., Lample, G.: Cross-lingual language model pretraining. In: 33rd International Conference on Neural Information Processing Systems, pp. 7057–7067. Vancouver, BC, Canada (2019)
4. Conneau, A., et al.: Unsupervised cross-lingual representation learning at scale. In: 58th Conference of the Association for Computational Linguistics, pp. 8440–8451. Association for Computational Linguistics. Online (2020)
5. Artetxe, M., Schwenk, H.: Massively multilingual sentence embeddings for zero-shot cross-lingual transfer and beyond. Trans. Assoc. Comput. Linguist. **7**, 597–610 (2019)
6. Patrick, S.H., Lewis, B.O., Rinott, R., Riedel, S., Schwenk, H.: MLQA: evaluating cross-lingual extractive question answering. In: 58th Conference of the Association for Computational Linguistics, pp. 7315–7330 (2020)
7. Artetxe, M., Ruder, S., Yogatama, D.: On the cross-lingual transferability of monolingual representations. In: 58th Conference of the Association for Computational Linguistics, pp. 4623–4637 (2020)
8. Jonathan, H., et al.: TyDi QA: a Benchmark for information-seeking question answering in typologically diverse languages. Trans. Assoc. Comput. Linguist. **8**, 454–470 (2020)
9. Yuan, F., et al.: Enhancing answer boundary detection for multilingual machine reading comprehension. In: 58th Conference of the Association for Computational Linguistics, pp. 925–934. Association for Computational Linguistics (2020)
10. Gangi Reddy, R., et al.: Answer span correction in machine reading comprehension. In: 25th International Conference on Empirical Methods in Natural Language Processing, pp. 2496–2501. Association for Computational Linguistics (2020)
11. Hewitt, J., Manning, C.D.: A structural probe for finding syntax in word representations. In: 23rd International Conference of the North American Chapter of the Association for Computational Linguistics: Human Language Technologies, pp. 4129–4138. Association for Computational Linguistics. Minneapolis, MN, USA (2019)
12. Chi, E.A., Hewitt, J., Manning, C.D.: Finding universal grammatical relations in multilingual BERT. In: 58th Conference of the Association for Computational Linguistics, pp. 5564–5577. Association for Computational Linguistics (2020)

13. Pérez-Mayos, L., Carlini, R., Ballesteros, M., Wanner, L.: On the evolution of syntactic information encoded by BERT's contextualized representations. In: 16th Conference of the European Chapter of the Association for Computational Linguistics, pp. 2243–2258. Association for Computational Linguistics(2021)
14. Xu, K., Wu, L., Wang, Z., Yu, M., Chen, L., Sheinin, V.: Exploiting rich syntactic information for semantic parsing with graph-to-sequence model. In: 23th International Conference on Empirical Methods in Natural Language Processing, pp. 918–924. Association for Computational Linguistics. Brussels, Belgium (2018)
15. Hsu, T.-Y., Liu, C.-L., Lee, H.: Zero-shot reading comprehension by cross-lingual transfer learning with multi-lingual language representation model. In: 24th International Conference on Empirical Methods in Natural Language Processing and the 9th International Joint Conference on Natural Language Processing, pp. 5932–5939. Association for Computational Linguistics. Hong Kong, China (2019)
16. Liu, J., Shou, L., Pei, J., Gong, M., Yang, M., Jiang, D.: Cross-lingual machine reading comprehension. In: 24th International Conference on Empirical Methods in Natural Language Processing and the 9th International Joint Conference on Natural Language Processing, pp. 1586–1595. Association for Computational Linguistics. Hong Kong, China (2019)
17. Liu, J., Shou, L., Pei, J., Gong, M., Yang, M., Jiang, D.: Cross-lingual machine reading comprehension with language branch knowledge distillation. In: 28th International Conference on Computational Linguistics, pp. 2710–2721. International Committee on Computational Linguistics (2020)
18. Huang, W.-C., Huang, C., Lee, H.: Improving cross-lingual reading comprehension with self-training. arXiv preprint arXiv:2105.03627 (2021)
19. Gaochen Wu, Bin Xu, Yuxin Qin, Fei Kong, Bangchang Liu, Hongwen Zhao, Dejie Chang: Improving Low-resource Reading Comprehension via Cross-lingual Transposition Rethinking. arXiv preprint arXiv:2107.05002 (2021)
20. Jiao, J., Wang, S., Zhang, X., Wang, L., Feng, Z., Wang, J.: gMatch: knowledge base question answering via semantic matching. Knowl.-Based Syst. **228**, 107270 (2021)
21. Hu, J., Ruder, S., Siddhant, A., Neubig, G., Firat, O., Johnson, M.: XTREME: a massively multilingual multi-task benchmark for evaluating cross-lingual generalisation. In: 37th International Conference on Machine Learning, pp. 4411–4421. Proceedings of Machine Learning Research. Virtual Event (2020)
22. Zeman, D., et al.: Universal dependencies 2.7 (2020). http://hdl.handle.net/11234/570 1-3424
23. Conneau, A., et al.: XNLI: evaluating cross-lingual sentence representations. In: 23th International Conference on Empirical Methods in Natural Language Processing, pp. 2475–2485. Association for Computational Linguistics. Brussels, Belgium (2018)

Profile Consistency Discrimination

Jing Zhou[1], Ximin Sun[1(✉)], Shuai Wang[2], Jiangkai Jia[2], Huichao Li[2], Mingda Wang[2], and Shuyi Li[3]

[1] State Grid Electronic Commerce Co., Ltd./State Grid Financial Technology Group, Beijing, China
{zhujing,sunximin}@sgec.sgcc.com.cn
[2] State Grid Ecommerce Technology Group Co., Ltd., Beijing, China
{wangshuai1,jiajiangkai,lihuichao,wangmingda}@sgec.sgcc.com.cn
[3] College of Intelligence and Computing, Tianjin University, Tianjin, China
suyielee@tju.edu.cn

Abstract. Consistency discrimination between attribute information and generated response is vital to personality-based dialogue. On existing work, few works specially investigates the consistency. In this paper, we propose a feasible method to solve this problem. We combine the typical natural language inference model (ESIM) and natural language understanding model (Bert) to discriminate consistency. But ESIM will fail when the input is structured attribute information. To solve this, We introduce external knowledge to expand the attribute information. Additionally, we observed the characteristics in the dialogue and found that adding a keyword matching label to the generated response is effective. We experimented on KvPI dataset and analyze the impact of different data sizes on the model. Compared with traditional methods, our method overall achieved better results.

Keywords: Profile consistency · Personality-based dialogue · External knowledge · Natural language inference

1 Introduction

In open-domain dialogue systems, generating role consistent responses is a challenge. The data used on personality-based dialogue model is divided into three categorics: with implicit persona embedding [9], with explicit profiles [17], personal facts [13], and structured profile [24]. Since the Persona-Chat dataset [30] was proposed, many works are based on it. The research on role consistency has made great progress [8,10,11,22,23,25,27,29], but the development of role consistency dialogue is still limited duc to the lack of consistency discriminator.

Different from the previous method, some researchers have introduced consistency discriminator to assist dialogue generation. (Song et al. 2020) [23] recognizes inconsistent words through attention mechanism and replaces the words. (Song et al. 2020) [26] models role dialogue as reinforcement learning problem.

U. K. Rage et al. (Eds.): DASFAA 2022 Workshops, LNCS 13248, pp. 189–199, 2022.
https://doi.org/10.1007/978-3-031-11217-1_14

Table 1. The data format of KvPI

Key-value	Response	Label
Gender: Female Location: Beijing Constellation: Aquarius	I'm going to Shanghai tomorrow. Would you like to visit me in Beijing the day after tomorrow?	E
Gender: Male Location: Shanghai Constellation: Pisces	I am not Pisces.	C
Gender: Female Location: Guangdong, Shenzhen Constellation: Pisces	She is the goddess in my mind. She is the pride in my mind. No one can replace her.	E

The discrimination results are used as the feedback of the model. However, neither of these methods can be independent of the model, so it is difficult to transplant to other model or application.

Recently, (Song et al. 2020) [24] propose a model named KvBert to distinguish the consistency between the generated dialogue and role information. In addition to proposing KvBert, it also creates a human-annotated chinese data set with over 110K single-turn conversations, named KvPI. Each piece of data contains three representative domains(includes Gender, Location and Constellation) and a single-turn conversation. Some examples are shown at Table 1. Based on structured role profile, we propose a model or precisely a discriminator, which combines the typical natural language inference model (ESIM) and natural language understanding model (Bert) to discriminate consistency. Additionally, We found the defects of ESIM in dealing with structured information, and uses external knowledge to expand structured information. After observing and summarizing the dialogue data, we found that adding keyword matching labels can enhance semantic information of sentence and get better results.

Our contributions are three-fold:

- We propose a model or precisely a discriminator, which helps develop role dialogues from two aspects: dialogue generation and model evaluation.
- We combine ESIM and pre-trained Bert and overall achieve better result.
- In ESIM, when the premises are structural attribute information, it will fail. We propose that external knowledge base can be used to extend the premise and adding a consistency label to data can obtain better result.

2 Relate Work

2.1 Natural Language Inference

Understanding entailment and contradiction is the basis of understanding natural language. Natural language inference (NLI) tasks are mainly about premise and hypothesis. It is required to judge the relationship between premise and hypothesis. There are three kinds of relationships: Entailment, Neutral and Contradiction. It can be identified as a function: $f_{NLI}(p, h) \rightarrow \{E, N, C\}$. The output E, N and C represent entailment, neutral and contradiction.

After SNLI [2] proposed, many NLI tasks [12,15,20,21] base on it. ESIM [3] proposes a natural language inference model based on chain LSTM. It breaks the tradition that the natural language inference model is too complex in the past.

In recent years, (Welleck et al. 2019) [28] models the detection of dialogue consistency as a NLI task and propose DNLI dataset in the domain of persona-based dialogue. This data set provides a new way to solve the problem of consistency discrimination. However, in the existing NLI model, premises and hypotheses are sentences. When the premises are single word, these models will fail.

2.2 Role Consistency

There are few studies devoted to consistency discrimination. Most methods are part of the generation model [14,23,26,28].

At present, only KvBERT makes the discrimination of consistency independent of the generation model. It combines Bert and TreeLSTM to learn the consistency relationship between attribute information and response. It linearizes attribute information into a sequence and treating the responses as another sequence. The input of BERT embedding includes type embedding to distinguish different attribute information, inspired by tablebert [4]. On the tree-LSTM side, the profiles are parsed to predefined structure, so it requires to manually annotate the syntax structure tree. As mentioned above, three operations are performed to aggregate information from two embeddings. The aggregated embedding is followed up by a linear layer to obtain final representation.

But in the real world, a dialogue agent generates a huge number of sentences every day. Manual tagging can not guarantee real-time performance, so it is unreasonable to manually label every sentence to support the training of TreeLSTM. And it is not open source. Although it has achieved excellent results, it is difficult to reproduce and difficult to finish other tasks based on this model.

3 Consistency Discrimination

3.1 Problem Definition

Consistency errors can be interpret as two contradictory words that should not appear at the same sentence. For example, query: "Where do you live now? I want to have fun with you." and response: "I am living in Tianjin. Come on, I'm looking forward to it." with attribute information "Local: Beijing, Gender: Female, Constellation: Pisces". Obviously, the attribute information is contradictory to the generated answer.

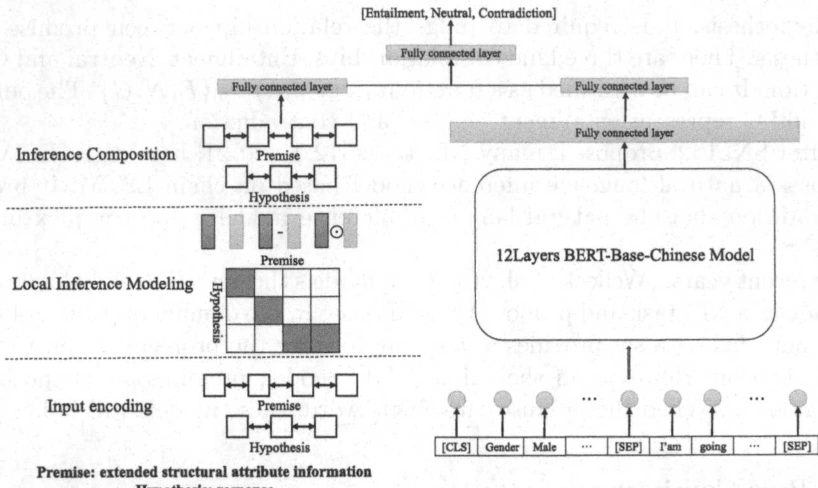

Fig. 1. The overall structure of our model. on the left of the model is ESIM model, whose inputs are extended attribute information and response, and Bert model on the right, whose inputs are the splicing of attributes and response. The output sequences on both sides are concatenated, and put into fully connected layer to obtain the consistency discrimination result. [CLS] is a special symbol added in front of every input example, and [SEP] is a special separator token (e.g. separating attribute information/response).

In this paper, We don't discuss logical contradiction. Our goal is to discriminate the consistency between structured role information and generated response. Generally, the dialogue query is less important in this work, because the dialogue query is always proposed by real human. It can be considered to satisfy persona consistency.

Formally, let $R = r_1, r_2, ..., r_m$ denote the set of response, and each $r_i \in R$ corresponds to a persona attribute p_j that satisfy the format of $\{k_1 : v_1, \ldots, k_n : v_n\}$, where k represent keys, and the v represent values. Our goal is to design a function that can figure out the consistency between r_i and p_j:

$$f(r_i, p_j) \in \{Entailed, Contradicted, Irrelevant\} \quad (1)$$

In every piece of data, r_i matchs p_j, which means only when $i = j$, it works. Simply, we let r, p denote a response and a persona attribute. l_r is the length of r and l_p is the length of p. In this paper, the premise refers to persona attribute p and the hypothesis refers to response r.

3.2 Consistency Discriminator

Our model combines pre-trained Bert model [5] and ESIM and hope that natural language understanding and natural language inference can complement each other. By introducing external common knowledge base, we solve the problem that it is unable to infer sentences from attribute value in ESIM. And we found that adding a keyword matching label to data can effectively improve the accuracy of consistency discrimination. The framework of the model is shown at Fig. 1.

Next, we will briefly introduce BERT and ESIM. Enhanced LSTM for natural language inference (ESIM) is a recently proposed text similarity calculation model. ESIM shows that well designing sequential inference models based on chain LSTMs can get better result in natural language inference. It is simple but efficient, and the training cost is faster than other natural language inference models. The structure of ESIM is shown in the left half of Fig. 1. It consists of three parts: input encoding, local inference modeling, inference composition. In this paper, we set the persona attribute as the input premise and the generated response as the hypothesis. However, ESIM can only learn the embedding of words contained in the data set, which is narrow sense. For natural language understanding, the broad sense of vocabulary is equally important. Therefore, it is feasible to combine ESIM and Bert. They are complementary.

When we have tried to use ESIM to judge the dialogue consistency directly, we found that it is difficult for ESIM to fully obtain the inference information. That is because this is a single-turn dialogue, and most responses are only related to one of the structural attribute information. Although all attribute information are concatenated together, the effect is similar to that of using a single related attribute word. Therefore, we pay attention to the characteristics of structural attribute information, which reminds us of the popular knowledge graphs. For a single word, it is very easy to find its related information. Expanding information through external common knowledge base can enrich attribute information. We chose CN-DBpedia[1] [1], a Chinese knowledge graph for expansion. Of course, using other external knowledge bases also work, such as zhishi.me[2]. Take the location of Fujian as an example in Table 2. The schematic diagram of expansion is shown in Fig. 2.

Table 2. Examples of external knowledge bases.

Head entity	Relation	Tail entity
Fujian	Another name	Haixi
Fujian	Provincial capital	Fuzhou
Fujian	Famous scenery	Gulangyu
Fujian	Famous scenery	Wuyi Mountain
Fujian	Dialect	Fuzhou dialect
Fujian	Dialect	hakka dialect
Fujian	Main city	Xiamen

[1] http://kw.fudan.edu.cn/cndbpedia/intro/.
[2] http://zhishi.me/.

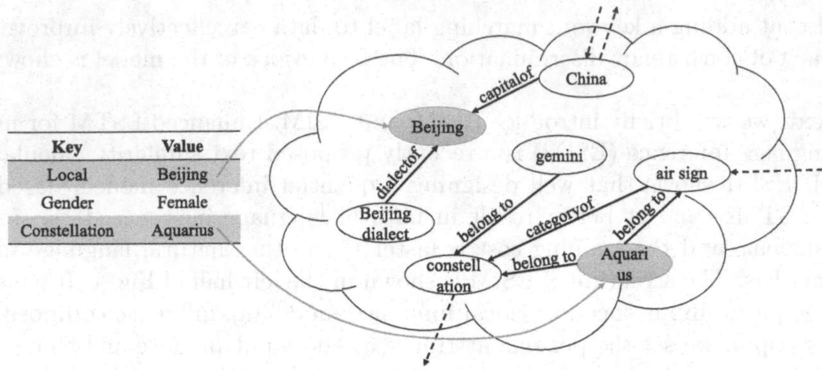

Fig. 2. This figure shows an example of structured attribute information extension. The data in the table is structured attribute information. On the right of the figure is an example in the common knowledge base.

For each piece of data, we directly obtain the topic of the conversation from the dataset. Then, the corresponding attribute value is determined according to the topic, and the attribute value is extracted. The extracted attribute value match the corresponding triples in CN-DBpedia. Next the matched results are spliced as one input of ESIM.

Additionally, We observe that in conversation data, as long as attributes appear in the response, it is more likely to be consistent. Take an example, key-value is {'constellation': 'Pisces', 'loc': 'Xi'an', 'gender': 'female'} and response is "Xi'an people are not addicted to playing mahjong.", which is consistent.

The value $v_i \in p$ will be matched to response r to get a matching result. We define that if $v_i \in r$, they are matched. If matched, we set it true. Otherwise, we set it false. It can be expressed with Eq. 2.

$$\mathrm{f_1(p,r)} = \begin{cases} \text{true, if matched} \\ \text{false, if not matched} \end{cases} \tag{2}$$

It seems easy to judge whether a response r contains attributes p, but for some attributes, such as "female", "male", they are rarely appear in conversations and keyword matching will be invalidated.

To resolve this, we expanded the vocabulary of gender. Specifically, we add some words, like "boy", "father", "sister", and use per-trained Bert model to predict gender. Thanks to the gender prediction, we can also get a preliminary result about inconsistency.

$$\mathrm{f_2(p,r)} = \begin{cases} \text{consistent, if keyword matched} \\ \text{inconsistent, if gender is opposite} \\ \text{neutral, if gender is unknow} \end{cases} \tag{3}$$

After keyword matching, matching result $\mathrm{f_2(p,r)}$ will be spliced at the end of the response sentence as a consistency label.

4 Experiments Setup

4.1 Implementation Details and Evaluations

To evaluate the performance of the proposed model, we carried out persona consistency experiments in a KvPI dataset. The pre-trained BERT-Base-Chinese model has 12-layer, 768 hidden units, 12 attention heads and 110 million parameters. For the ESIM, we set hidden size to 1024 and embedding size to 300, and output dimension to 300. The dimension of the final representation is 1068. The model is trained four epochs in total. The basic statistics of the KvPI dataset is shown on Table 3.

Table 3. key-value information

Domain	Entail	Contr	Irrelv	Train	Test
Gender	8270	6858	16201	28329	3000
Location	18468	17777	28759	69004	6000
Constell	6376	6365	9466	20207	2000
Total	33114	31000	54426	107540	11000

For evaluations, we used accuracy (acc) and F1 score to measure the overall performance. In addition, we want to analysis the robustness of the model. We extract 90%, 70%, 50% and 30% data from the training set to train the model, so as to compare the impact of different scale training sets on the model.

4.2 Baselines

In this experiment, we use various baseline models to discriminate the consistency between structured attribute information and generated response, and compare their performance. The baseline model used covers a variety of models, such as classification problems, natural language inference model, natural language understanding model and pre-trained model, etc.

BiLSTM. LSTM [6] can learn long-term dependence. BiLSTM [7] is composed of forward LSTM and backward LSTM. In sequence tagging task, BiLSTM can access to both past and future input features for a given time. BiLSTM can efficiently make use of the past features and the future features within a specific time frame, trained with forward propagation and back propagation.

SVM. SVM (Support vector machines) [16] is a classification model, usually used for binary classification problems. For multivariate classification, it is usually decomposed into multiple binary classification problems. This model was also applied in (Song et al. 2019) [24]. They named SVM with unigram features and bigram features as SVM+uni+bi and named SVM with extracted feature as SVM+uni+bi+overlap. We quote their results here.

Table 4. Evaluation results on the KvPI dataset.

Models	Accuracy	entail-f1	contr-f1	irrelv-f1	90%	70%	50%	30%
SVM+uni+bi	61.3 (14)	73.6 (18)	55.9 (5.5)	41.5 (3.5)	-	-	-	-
BiLSTM	65.3 (14)	70.2 (9.5)	60.9 (12.1)	61.2 (5.2)	-	-	-	-
SVM+uni+bi+overlap	68.7 (8.5)	76.2 (13)	65.1 (32)	50.3 (13)	-	-	-	-
ESIM	83.7 (0.8)	82.0 (4.1)	86.3 (1.7)	80.6 (1.7)	-	-	-	-
BERT	85.4 (1.6)	86.5 (2.2)	87.2 (0.8)	85.2 (2.2)	-	-	-	-
GPT	86.4 (0.5)	88.2 (0.9)	86.1 (2.2)	83.8 (2.2)	-	-	-	-
KvBERT	92.8 (1.7)	93.1 (1.2)	93.4 (2.5)	91.7 (2.6)	-	-	-	-
Our model	88.3 (0.8)	89.7 (2.7)	88.2 (1.4)	87.2 (1.3)	88.4	87.1	87.3	84.8

GPT. As a pre-trained transformers, GPT [18] have been shown effective for natural language understanding tasks. GPT as a baseline model allows us to better compare pre-trained models.

ESIM and Pre-trained BERT. Ablation study [19] is necessary for the two tower model. It is usually used in neural networks, especially in relatively complex neural networks. The main idea is to analyze the effectiveness of the network by deleting some parts of networks. Our model refers to ESIM and pre-trained BERT. In order to better reflect the validity of our model, so we set them as baseline models individually.

5 Result

The data in the table is the best result of the three runs, and the data in brackets is the variance of the three results. For each model, we use identical data sets, thus different results are solely due to different networks and methods. As shown in Table 4, our model achieves well performance on all metrics. It is better than other traditional method, except KvBERT.

From the experimental results of SVM model, we can see the difficulty of consistency discrimination task. This is a multi classification problem. Although the result of SVM with extracted feature is not satisfactory. Turn to BiLSTM, which is often used for natural language tasks. Its results are not good. The natural language inference model ESIM has higher accuracy than the previous models, which benefits from natural language inference to mining the relationship between attribute information and the response. It also proves that natural language inference is effective in this problem. GPT and Bert perform well as pre-trained models in this task. They are all outstanding in natural language understanding. Both our model and KvBERT combining natural language understanding and natural language inference, achieved the best results.

Table 5. Some case in experiment

Index	Key-value	Response	Domain	Iscorrect
1	Gender: female Location: Guangdong Constellation: Virgo	I still remember the words of Han people in Shenzhen	Location	Yes
2	Gender: female Location: None Constellation: Libra	So you are Pisces, and I am also.	Constell	Yes
3	Gender: female Location: foreign Constellation: Scorpio	If you go to Shanghai, will you run with them?	Location	Yes
4	Gender: female Location: Guangdong Constellation: Capricorn	I know it all in Dongguan.	Location	No
5	Gender: female Location: Beijing Constellation: Capricorn	Poof, no, no, no, you are not so carefree like me, you are a good girl	Gender	Yes
6	Gender: Male Location: Guangdong Constellation: Aries	Brother Hao got into trouble, I just ask casually. They say lions and Aries are good	Constell	Yes

Although our results were slightly inferior to KvBERT, but the training of KvBERT is complex, particularly for TreeLSTM. It not only needs complex training, but also needs additional annotation information for TreeLSTM. In contrast, our model is easy to implement and more extensible. Our model is easier to adapt to extended data sets according to specific tasks or adapt to other structural data sets.

In addition to comparing the accuracy of each model, we also analyze the impact of different scale training data on the model. With the decline of data scale, the accuracy of the model also shows a downward trend as a whole. But overall, the decline is not much. Although only using 30% of the training data, 84.8% accuracy can still be achieved.

6 Case Study

Some sample results are shown in Table 5. Most of the topics in the data are about location. This kind of dialogue with the theme of location is not difficult to distinguish. Because the place usually appears in the conversation, but unfortunately, the accuracy of the location in the data set will affect the discrimination of the model.

For example, in the case 4, the location is "Guangdong", and "Dongguan" was mentioned in the response. We know that "Dongguan" is a city in "Guangdong". But this is difficult for machines. The purpose of extending attribute information is to solve this problem. But obviously, this has not been solved. We further explored the reasons. We looked at the extended information about the attribute word "Guangdong". Then we found that the word "Dongguan" was not included in the extended information. It may not be summarized because "Dongguan" is neither the capital of "Guangdong" nor a major city.

But in case 1, the location is "Guangdong", and "Shenzhan" was mentioned in the response. "Shenzhen" appears in the extended information, so the model can distinguish correctly. Therefore, extended information can play a role for the model. However, it is difficult to ensure that the words in the conversation are included in the extended information. We believe that as long as the triples in the response are included in the knowledge base, the accuracy will be greatly improved. Such an experiment may be carried out in the follow-up work.

In addition, we note that it is difficult to distinguish gender, especially for short texts, which is the same for other models [24]. The short text contains too little information about gender, because people seldom emphasize their gender in dialogue. Constellations are relatively better.

7 Conclusion and Future Work

In this paper, we propose a consistency discriminator by combining ESIM and Bert and propose that external knowledge base can be used to extend the premise in ESIM and adding a consistency label to data can obtain better result.

We want to provide a solution to the problem of consistency in persona-based dialogue, aiming at structural attribute information. In the future, we will use graph network method to further fuse structure information and text information

References

1. Xu, B., Xu, Y., Liang, J., Xie, C., Liang, B., Cui, W., Xiao, Y.: CN-DBpedia: a never-ending chinese knowledge extraction system. In: Benferhat, S., Tabia, K., Ali, M. (eds.) IEA/AIE 2017. LNCS (LNAI), vol. 10351, pp. 428–438. Springer, Cham (2017). https://doi.org/10.1007/978-3-319-60045-1_44
2. Bowman, S.R., Angeli, G., Potts, C., Manning, C.D.: A large annotated corpus for learning natural language inference. In: EMNLP (2015)
3. Chen, Q., Zhu, X.D., Ling, Z., Wei, S., Jiang, H., Inkpen, D.: Enhanced LSTM for natural language inference. In: ACL (2017)
4. Chen, W., et al.: TabFact: a large-scale dataset for table-based fact verification. arXiv:abs/1909.02164 (2020)
5. Devlin, J., Chang, M.W., Lee, K., Toutanova, K.: BERT: pre-training of deep bidirectional transformers for language understanding. In: NAACL (2019)
6. Hochreiter, S., Schmidhuber, J.: Long short-term memory. Neural Comput. **9**, 1735–1780 (1997)
7. Huang, Z., Xu, W., Yu, K.: Bidirectional LSTM-CRF models for sequence taggin. arXiv:abs/1508.01991 (2015)
8. Kim, H., Kim, B., Kim, G.: Will I sound like me? improving persona consistency in dialogues through pragmatic self-consciousness. In: EMNLP (2020)

9. Li, J., Galley, M., Brockett, C., Spithourakis, G.P., Gao, J., Dolan, W.: A persona-based neural conversation model. arXiv:abs/1603.06155 (2016)
10. Lin, Z., Madotto, A., Wu, C.S., Fung, P.: Personalizing dialogue agents via meta-learning. In: ACL (2019)
11. Liu, Q., et al.: You impress me: Dialogue generation via mutual persona perception. In: ACL (2020)
12. Manning, C.D., MacCartney, B.: Natural language inference (2009)
13. Mazaré, P.E., Humeau, S., Raison, M., Bordes, A.: Training millions of personalized dialogue agents. In: EMNLP (2018)
14. Mesgar, M., Simpson, E., Gurevych, I.: Improving factual consistency between a response and persona facts. In: EACL (2021)
15. Parikh, A.P., Täckström, O., Das, D., Uszkoreit, J.: A decomposable attention model for natural language inference. In: EMNLP (2016)
16. Platt, J.: Sequential minimal optimization : a fast algorithm for training support vector machines. Microsoft Research Technical Report (1998)
17. Qian, Q., Huang, M., Zhao, H., Xu, J., Zhu, X.: Assigning personality/profile to a chatting machine for coherent conversation generation. In: IJCAI (2018)
18. Radford, A., Narasimhan, K.: Improving language understanding by generative pre-training (2018)
19. Ren, S., He, K., Girshick, R.B., Sun, J.: Faster R-CNN: towards real-time object detection with region proposal networks. IEEE Trans. Pattern Anal. Mach. Intell. **39**, 1137–1149 (2015)
20. Rocktäschel, T., Grefenstette, E., Hermann, K., Kociský, T., Blunsom, P.: Reasoning about entailment with neural attention. CoRR abs/1509.06664 (2016)
21. Sharma, A.: Sequential LSTM-based encoder for NLI (2017)
22. Song, H., Wang, Y., Zhang, K., Zhang, W., Liu, T.: BoB: BERT over BERT for training persona-based dialogue models from limited personalized data. In: ACL/IJCNLP (2021)
23. Song, H., Wang, Y., Zhang, W., Liu, X., Liu, T.: Generate, delete and rewrite: a three-stage framework for improving persona consistency of dialogue generation. In: ACL (2020)
24. Song, H., Wang, Y., Zhang, W., Zhao, Z., Liu, T., Liu, X.: Profile consistency identification for open-domain dialogue agents. In: EMNLP (2020)
25. Song, H., Zhang, W., Cui, Y., Wang, D., Liu, T.: Exploiting persona information for diverse generation of conversational responses. In: IJCAI (2019)
26. Song, H., Zhang, W., Hu, J., Liu, T.: Generating persona consistent dialogues by exploiting natural language inference. In: AAAI (2020)
27. Tigunova, A., Yates, A., Mirza, P., Weikum, G.: Charm: Inferring personal attributes from conversations. In: EMNLP (2020)
28. Welleck, S., Weston, J., Szlam, A.D., Cho, K.: Dialogue natural language inference. In: ACL (2019)
29. Wolf, T., Sanh, V., Chaumond, J., Delangue, C.: Transfertransfo: a transfer learning approach for neural network based conversational agents. arXiv:abs/1901.08149 (2019)
30. Zhang, S., Dinan, E., Urbanek, J., Szlam, A.D., Kiela, D., Weston, J.: Personalizing dialogue agents: I have a dog, do you have pets too? In: ACL (2018)

H-V: An Improved Coding Layout Based on Erasure Coded Storage System

Tiantong Mu[1,2], Ying Song[1,2,3(✉)], Mingjie Yang[1,2], Bo Wang[4], and Jiacheng Zhao[3]

[1] Beijing Key Laboratory of Internet Culture and Digital Dissemination Research, Beijing Information Science and Technology University, Beijing, China
songying@bistu.edu.cn

[2] Beijing Advanced Innovation Center for Materials Genome Engineering, Beijing Information Science and Technology University, Beijing, China

[3] State Key Laboratory of Computer Architecture, Institute of Computing Technology, Chinese Academy of Sciences, Beijing, China

[4] Software Engineering College, Zhengzhou University of Light Industry, Henan, China

Abstract. The failure of a single unreliable commodity components is very common in large-scale distributed storage systems. In order to ensure the reliability of data in large-scale distributed storage systems, a lot of studies have emerged one after another. Among them, Erasure Codes are widely used in actual storage systems, such as Hadoop Distributed File System (HDFS), to provide high fault tolerance with lower storage overhead. However, usually the recovery of erasure-coded storage system when encountering a node failure will result in severe cross-node and cross-rack bandwidth loss, which affects the efficiency of failure recovery and wastes additional resources. In this paper, we improve the erasure code storage strategy in Hadoop 3.x, propose $H^{RS(n,k)}$ - $V^{RS(n',k')}$ abbreviated as H-V, and add RS parity check inside the data nodes, effectively reduce cross-node and cross-rack data transmission during recovery, reduce the occupation of cross-rack bandwidth, and improve recovery efficiency. Theoretical analysis shows that compared with traditional RS erasure code storage, H-V can reduce the cross-node and cross-rack bandwidth of RS by at least 25% and respectively during data recovery. 62.5%; compared with D^3, H-V reduces the storage redundancy by up to 19.7% while reducing the cross-node and cross-rack bandwidth of D^3 by 25% and 12.5% during data recovery.

Keywords: Distributed storage system · Erasure coding · Data recovery

1 Introduction

Distributed storage systems are usually composed of many independent unreliable commodity components, and components failure are very common. In order

U. K. Rage et al. (Eds.): DASFAA 2022 Workshops, LNCS 13248, pp. 203–213, 2022.
https://doi.org/10.1007/978-3-031-11217-1_15

to ensure the high reliability and availability of data in such a distributed storage system, two common methods are to use multiple replicas and erasure codes to provide fault tolerance.

The multiple replicas method is to store data in a redundant way by copying multiple copies of the stored data. For example, Google File System (GFS) [1] and Hadoop Distributed File System (HDFS) [2]. By default, three copies of each data block are stored to tolerate dual component failures. The multi-copy form is easy to deploy and recover from failures, but the storage overhead is too large, and it is not suitable for systems with large data volumes and small disk space.

As an alternative, erasure codes provide fault tolerance close to multiple replications and lower storage overhead. This solution has been deployed in some distributed systems. For example, HDFS-RAID [3] deploys Reed-Solomon (RS) [4] code, which reduces storage redundancy from traditional 3x to 1.4x, saving more space. But using erasure codes, recovering a failed block requires multiple available blocks to be retrieved, which leads to high recovery costs. Although erasure codes improve storage efficiency, they significantly increase disk I/O and network bandwidth for failure recovery.

The large-scale distributed storage system is deployed on multiple independent data nodes, which are located in multiple racks, that is, multiple data nodes are placed on each rack to store the original data and parity data of the erasure coding system respectively. In order to maximize the data availability of distributed storage systems deployed using erasure codes, different blocks of erasure codes are stored in nodes in different racks. This data layout allows the system to tolerate a certain number of node failures and rack failures. However, this data block placement method will inevitably cause the repair of any faulty data block to retrieve available data blocks from other racks, which will occupy a large amount of cross-rack bandwidth. In a distributed storage system, the inner-rack bandwidth is considered sufficient, but the cross-rack bandwidth is not. Generally, the available cross-rack bandwidth of each node is only 1/20 to 1/5 of the internal rack traffic [5]. Therefore, cross-rack bandwidth is generally regarded as a scarce resource [6], and excessive cross-rack bandwidth will inevitably delay the recovery process.

In this paper, we propose an improved erasure code storage calculation method $H^{RS(n,k)}$ - $V^{RS(n',k')}$ abbreviated as H-V, which is suitable for distributed erasure code storage systems, such as HDFS. H-V adds parity calculation inside the node on the basis of traditional RS erasure code storage, also uses RS encoding, and stores the parity calculation result in the current node, although this will cause data redundancy, in case of node failure, parity blocks can be decoded from the node itself, so as to reduce the cross-rack bandwidth during a recovery.

2 Related Work

At present, there have been a lot of research achievements on the data recovery problem of HDFS, and there are many effective methods in different versions of Hadoop. During the widespread application of Hadoop 2.x, most recovery methods considered the frequency of data access, such as [7–9], among which literature [9] proposed an initiative based on popularity in the HDFS-RAID system, PP tracks popular data and immediately restores lost popular data when a Hadoop node fails. When these popular data are accessed, there is no need to spend time to recover the data, which significantly improves the efficiency of foreground work, reduces the data recovery time and the execution time of the foreground tasks, and minimizes the impact on other tasks in HDFS-RAID.

As Hadoop 3.x gradually dominates, there are more and more solutions to use erasure coding strategies and data layout changes. The existing data recovery work using erasure coding and data layout can be roughly divided into the following types: a new erasure coding scheme, a combination of multiple erasure codes coding methods, changes to the original data layout, etc.

A New Erasure Coding Scheme. Tai Zhou [10] used the BASIC framework and ZigZag decodable code design to implement a new storage code, and built it into a STORE system. In the data strip layout, the upper triangular matrix is arranged, and the lower triangular part is reflected and filled along the diagonal. When there is data loss, part of it can be recovered by diagonal copying, and part of it can be recovered by calculating the erasure code with BMBR code. Evaluated on a 21-nodes HDFS cluster, recovery bandwidth and disk I/O are almost minimized, and degraded read throughput is significantly improved. However, such a layout will result in repeated storage of some data blocks, resulting in a certain degree of redundancy. For single node failure, Runhui Li [11] constructed CORE's system, which enhanced the existing optimal regeneration code to support single and concurrent failures. CORE proposes two kinds of coded symbols to restore failed nodes by reconstructing virtual symbols and real symbols. CORE was implemented and evaluated on a Hadoop HDFS cluster test platform with up to 20 storage nodes, which proved that a prototype of CORE saves recovery bandwidth and improves recovery efficiency compared with the traditional recovery method based on erasure codes.

Combination of Multiple Erasure Codes. Currently, there are many types of erasure codes in use, such as RS, LRC, PC, MSR, etc. Each erasure codes have its own advantages and disadvantages. In order to make up for these shortcomings, some researchers use a combination of two erasure codes to improve the recovery efficiency of distributed storage systems. RS is the most commonly used, but recovery needs to read a large amount of data, LRC can reduce the recovery cost, but the storage efficiency is sacrificed. MSR code is designed to decrease the recovery cost with high storage efficiency, but its computation is too complex. To address this shortcoming, literature [12] propose an erasure code for multiple availability zones (called AZ-Code), which is a hybrid code by taking advantages of both MSR code and LRC codes. AZ-Code utilizes a specific MSR code as the local parity layout, and a typical RS code is used to generate the global parities. In this way, AZ-Code can keep low recovery cost with high reliability. There are also studies combining other erasure codes. Literature [13] introduced a new eras-ure code storage system HACFS, a new erasure-coded storage system that instead uses two different erasure codes and dynamically adapts to workload changes. It uses a fast code to optimize for recovery performance and a compact code to reduce the storage overhead. A novel conversion mechanism is used to efficiently upcode and downcode data blocks between fast and compact codes. Compared with the three popular single-code storage systems, the HACFS system always maintains low storage overhead and significantly improves recovery performance, effectively reducing the delay of degraded reads, recovery time, and disk/network bandwidth.

Change the Original Data Layout. For situations involving bandwidth occupancy between racks, Literature [14] defines Deterministic Data Distribution (D^3), placing multiple data blocks of the same strip on different nodes in the same rack to reduce the cross-rack traffic for single-node failure recovery, and propose an efficient recovery method based on D^3 to achieve load balancing of recovery traffic under single node failure. Not only between the nodes in the racks, but also between the racks to balance the recovery flow. If data nodes are distributed in a very large geographic area, that is, storage across data centers, the recovery performance of the code will be affected by more factors, including network and computing-related factors. Pablo [15] et al. proposed a code based on XOR, which supplemented the idea of parity copy and rack awareness, and expand it to MXOR (Multiple XOR), generate several vertical parity check blocks and horizontal parity check blocks for the data block, which can effectively deal with the failure of a node. Through the implementation and evaluation of the erasure code module in the XORBAS version of HDFS, the amount of data to be read during recovery is reduced and the recovery time is shortened.

3 $H^{RS(n,k)}$ - $V^{RS(n',k')}$ Encoding Method

RS(n, k) erasure code is a commonly used coding method in current erasure code storage systems. Given two positive integers, n and k, a (n, k) code encodes n data blocks into k additional parity blocks, such that any one of the (n + k) blocks can be reconstructed from any other k ones. All the (n + k) data/parity blocks form a stripe and (n + k) is called stripe size. Thus, a (n, k) code tolerates any k blocks being lost in a stripe, and it achieves the so-called maximum distance separable (MDS) property. Furthermore, RS codes satisfy the property of linearity. That is, given a (n, k)-RS code, any block B' is the linear combination of any other n different blocks B_0, B_1,..., B_{n-1} in the same stripe. We can get $B' = \sum_{i=0}^{k-1} c_i B_i$, where the c_i $(0 \leq i \leq n \leq 1)$ are the decoding coefficients specified by the given RS code. Note that additions and multiplications here are performed in a finite field.

The RS(n, k) coding method enables the system to tolerate a certain number of node failures and rack failures. However, the original data blocks placement of HDFS inevitably causes the recovery of any faulty data block to retrieve available data blocks from other racks, which occupies a large amount of scarce cross-rack bandwidth resources and inevitably delays the recovery process. In order to reduce the cross-rack bandwidth occupation of the erasure code storage system during data recovery, thereby reducing the recovery delay, we have made improvements on the basis of RS(n, k) coding and proposed a new coding method, namely $H^{RS(n,k)}$ - $V^{RS(n',k')}$ abbreviated as H-V. The basic idea of H-V is to add the parity calculation inside the node (called vertical coding in this paper) on the basis of the traditional RS erasure code storage (called horizontal coding), also using RS coding, and store the parity calculation result in the current node. Among them, $H^{RS(n,k)}$ means that the horizontal direction adopts RS(n, k) coding, and $V^{RS(n',k')}$ means that the vertical direction adopts RS(n', k') coding, which will cause data redundancy, in the event of a node failure, the parity block can be decoded from the node itself, thereby reducing the cross-rack traffic generated during recovery. Next, a specific example is given to introduce H-V.

Consider a distributed storage system, such as HDFS [2]. The system is composed of a collection of nodes, and each node stores many blocks. The nodes are divided into data nodes and parity nodes. The data node stores the data block and a little local parity blocks generated by encoding $V^{RS(n',k')}$ within the node, and the parity node stores the parity blocks generated by the cross-node RS encoding, as shown in Fig. 1. The n and k values of horizontal RS can be changed according to the number of nodes. Take H-V as an example, that is, horizontal RS(3,2) and vertical RS(3,1), a total of 5 is used Nodes. DN_0, DN_1, and DN_2 are data nodes, DN_3 and DN_4 are parity check nodes. The 3 data block codes in each row generate 2 parity check blocks and are stored in the parity check node. If any 2 blocks in a row are missing, they can be reconstructed by $H^{RS(3,2)}$. If these 5 nodes are on different racks, the linear characteristics of RS coding can be used to first calculate the linear combination of data blocks in the rack, and then transmit the encoded result across the racks. Inside the node, RS coding is also used. Unlike horizontal, RS(3,1) coding is used for vertical fixation, namely $V^{RS(3,1)}$, every 3 data blocks are coded to generate 1 parity block and stored on the node. Except for the first longitudinal encoding, when encoding the remaining data blocks, the parity block generated in the previous encoding needs to be regarded as a new data block for the next set of encoding. In this way, when a small amount of data is lost in a node, $V^{RS(n',k')}$ can be used to decode and restore without using a parity node, which reduces the number of cross-node and cross-rack data transmissions, thereby reducing the bandwidth occupation between racks.

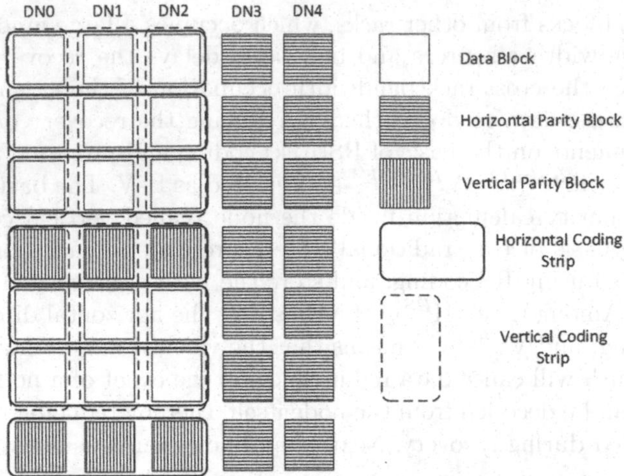

Fig. 1. $H^{RS(3,2)}$ - $V^{RS(3,1)}$ storage architecture.

4 Effect Analysis of $H^{RS(n,k)}$ - $V^{RS(n',k')}$

In this section, we will analyze the effect of H-V from two aspects: data transmission performance and storage redundancy. The data transmission performance mainly analyzes the bandwidth occupied by recovering lost data under different situations and storage redundancy mainly analyzes the disk space occupied by the layout of H-V, and compares and analyzes it with commonly used erasure coding strategies.

4.1 Data Transmission Performance Analysis

Next, using the above example of $H^{RS(3,2)}$ - $V^{RS(3,1)}$ storage architecture as an example, we analyze the difference in bandwidth usage of $H^{RS(3,2)}$ - $V^{RS(3,1)}$ compared to D^3[14] and RS(3,2) in recovering lost data under six different situations. In the fourth situation, which is more complex, it focuses on the amount of data transmitted across-nodes and across-racks when the number of data blocks is different.

Discuss the Bandwidth Usage when Recovering Lost Data by Scenarios.

Case 1: Each node is on the different rack, and a single data block of a node fails. Recovering lost data using D^3 or RS(3,2) requires at least 3 blocks to be accessed and transferred across the rack to the node to be recovered for another calculation, while $H^{RS(3,2)}$ - $V^{RS(3,1)}$ can be recovered using vertical RS(3,1) decoding on the same node, with only one computation and no transmission required.

Therefore, for Case 1, compared with D^3 and RS(3,2), $H^{RS(3,2)}$ - $V^{RS(3,1)}$ reduces the transmission volume from 3 data blocks to 0, while the calculation volume remains unchanged.

Case 2: Each node is on the different rack, and a single node fails. The recovery of D^3 and RS(3,2) is similar to the failure of single data block in Case 1 (suppose there are 4 data blocks on each node), which requires 3 times of cross-rack transmission and 4 times of calculation, while $H^{RS(3,2)}$ - $V^{RS(3,1)}$ requires parity strip transmission of 3 blocks to recover the parity block of the faulty node. Data strip transmission of 3 blocks to recover the data block of any 2 faulty nodes. The remaining unrecovered data blocks are decoded and restored using the parity check block of the faulty node. The encoding calculation is performed for 3 times and the decoding calculation is performed for 1 time.

For Case 2, compared with D^3 and RS(3, 2), $H^{RS(3,2)}$ - $V^{RS(3,1)}$ reduces the transmission volume from 12 data blocks to 9, which is a 25% reduction, and the amount of calculation is not change.

Case 3: Each rack has multiple nodes (e.g., two nodes), and a single data block of a node fails. Using RS(3,2) to recover lost data requires at least three blocks to be accessed. One block is transferred across-node but inner-rack, and two blocks are transferred across the rack to the node to be recovered for calculation. With D^3 recovery, you can transfer data in the same rack to other nodes in the same rack, calculate a linear combination of data, and then transfer the linear combination result across-rack to the new node. That is, to recover a single failed block only requires cross-rack access to one block. In this process, the inter-node transmission and the calculation are twice, the inter-rack transmission is once. And $H^{RS(3,2)}$ - $V^{RS(3,1)}$ can be recovered by using the vertical RS(3,1) decoding on the same node, only one calculation is required and no transmission is required.

Therefore, for Case 3, RS(3,2) transmits 1 block across-node and 2 blocks across-rack, which is calculated once; D^3 transmits 2 blocks across-node and 1 block across-rack, and calculates 2 times; $H^{RS(3,2)}$ - $V^{RS(3,1)}$ reduces the transmission volume to 0, avoiding a large amount of cross-rack bandwidth occupation, and the calculation volume is the same as RS(3,2).

Case 4: Each rack has multiple nodes (e.g., two nodes) and a single node fails. The recovery of D^3 and RS(3,2) is similar to the failure of single data block in Case 3 (suppose there are 4 data blocks on each node). RS(3,2) transfers 4 blocks across-node and 8 blocks across-rack, and calculates 4 times. D^3 transfers 8 blocks across-node and 4 linear combination results across-rack, and calculates 8 times; However, $H^{RS(3,2)}$ - $V^{RS(3,1)}$ requires parity strip to transfer 2 blocks across-node and 1 linear combination result across-rack, and any two rows of data strip to transfer 4 blocks across-node and 2 linear combination result across-rack, and calculate 7 times.

For Case 4, compared with RS(3,2), $H^{RS(3,2)}$ - $V^{RS(3,1)}$ has 25% more cross-node data transmission and 62.5% less cross-rack data transmission, but 1.75

times more computation; compared with D^3, $H^{RS(3,2)}$ - $V^{RS(3,1)}$ has 25% less cross-node and cross-rack data transfer and 12.5% less computation.

Case 5: All nodes are on the same rack, and a single data block of a node fails. Using RS(3,2) to recover lost data requires at least 3 blocks to be accessed and transferred to the node to be recovered for a second calculation, whereas $H^{RS(3,2)}$ - $V^{RS(3,1)}$ can be recovered using vertical RS(3,1) decoding on the same node with a single calculation and no transfer required.

Therefore, for Case 5, compared with D^3 and RS(3,2), $H^{RS(3,2)}$ - $V^{RS(3,1)}$ reduces the transmission volume from 3 data blocks to 0, while the calculation volume remains unchanged.

Case 6: All nodes are on the same rack, and a single node fails. The recovery of D^3 and RS(3,2) is similar to that of a single data block failure in Case 5 (assuming that there are 4 data blocks on each node), and 12 blocks need to be transmitted and calculated 4 times; while $H^{RS(3,2)}$ - $V^{RS(3,1)}$ requires parity strips to transmit 3 blocks across-node, and any two rows of data strips to transmit a total of 6 blocks across-node, which are calculated 4 times.

For Case 6, compared with D^3 and RS(3,2), $H^{RS(3,2)}$ - $V^{RS(3,1)}$ reduces the transmission volume from 12 data blocks to 9, which is a 25% reduction, and the amount of calculation remains unchanged. In summary, in the cases mentioned above, when each node stores 4 data blocks, compared with RS and D^3, $H^{RS(3,2)}$ - $V^{RS(3,1)}$ will reduce the amount of data transmission across nodes and racks, thereby effectively reducing network bandwidth usage. Similarly, in other commonly used RS coding strategies, such as RS(6,3) and RS(10,4), $H^{RS(3,2)}$ - $V^{RS(3,1)}$ also has a similar effect. Table 1 shows the proportion of data transfer results between several different erasure coding strategies and $H^{RS(3,2)}$ - $V^{RS(3,1)}$ when each node stores 10 data blocks. It can be seen from Table 1 that our method can reduce the cross-rack or cross-node bandwidth to 0 to the greatest extent in Cases 1, 3, and 5; in Case 2, each node is on a different rack,$H^{RS(3,2)}$ - $V^{RS(3,1)}$ can use vertical RS(3,1) decoding on the same node to recover lost data, reducing cross-rack data transmission by 30%; Case 6 is similar to Case 2, reducing cross-node data transmission by 30%.

Table 1. Proportion reduction in data transfer volume

	RS(3,2)		RS(6,3)		RS(10,4)		D^3(3,2)	
	Cross-node	Cross-rack	Cross-node	Cross-rack	Cross-node	Cross-rack	Cross-node	Cross-rack
Case1	-	1	-	1	-	1	-	1
Case2	-	0.3	-	0.3	-	0.3	-	0.3
Case3	1	1	1	1	1	1	1	1
Case4	−0.4	0.65	−1.1	0.58	−2.5	0.61	0.3	0.3
Case5	1	-	1	-	1	-	1	-
Case6	0.3	-	0.3	-	0.3	-	0.3	-

Note: "-" means that there is no cross-node and cross-rack data transmission in this case

Analysis of the Impact of the Difference in the Number of Data Blocks on the Amount of Data Transmitted Across Nodes and Racks. The scenario of Case 4 is more complicated. Because $H^{RS(n,k)}$ - $V^{RS(3,1)}$) adds the vertical coding inside the node and uses the linear characteristics of RS(n, k) erasure codes, the same rack is different Node data needs to be transmitted to the same node, so it can reduce cross-rack data transmission, but it will increase a part of cross-node data transmission. In a distributed system, cross-rack traffic is scarcer, but the increase in cross-node data volume is affordable. To this end, we further analyzed the amount of data transmitted across-node and across-rack in Case 4 when the number of data blocks is different, as shown in Fig. 2 and Fig. 3. Figure 2 shows the failure of a single node when the nodes are distributed in different racks, and each node stores 4, 10, and 16 data blocks respectively. $H^{RS(n,k)}$ - $V^{RS(3,1)}$) will transmit more data across-nodes than RS(3,2) and D^3(3,2). In Fig. 3, we can see that $H^{RS(n,k)}$ - $V^{RS(3,1)}$ reduces a lot of cross-rack data transmission compared with other erasure coding strategies. And as the amount of data increases, the advantages of $H^{RS(n,k)}$ - $V^{RS(3,1)}$ become more obvious, therefore, sacrificing part of the cross-node traffic can greatly reduce the cross-rack traffic. This can improve the recovery efficiency of a single node as a whole.

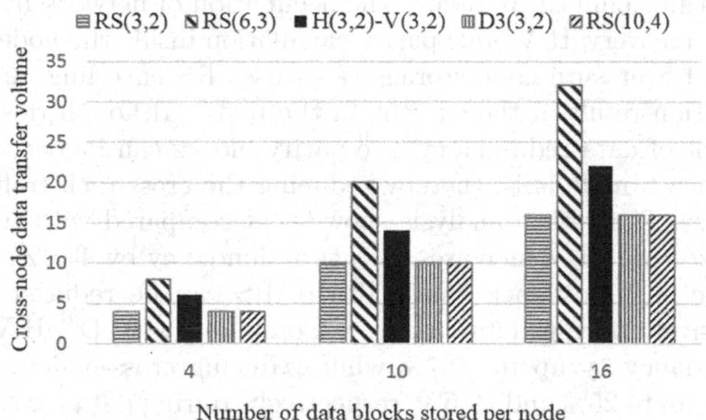

Fig. 2. Case 4 data transmission volume across-nodes.

4.2 Storage Redundancy Analysis

Assuming that 9 strips are stored, in RS(3,2), 3 data nodes, 2 parity nodes, a total of 27 data blocks, 18 parity blocks, and 45 valid blocks can be stored. In D^3, due to intra-rack and inter-rack load balancing, storing the same number of strips will occupy 4 more nodes, and each node has empty space. If 9 strips are stored, it will occupy 81 more blocks, which is about 1.8 times of ordinary RS(3,2). While $H^{RS(3,2)}$ - $V^{RS(3,1)}$ has a parity block calculated vertically on each data node. If 9 valid data blocks are stored, it will occupy 65 blocks of memory, which is about 1.44 times that of RS, which is 19.7% less than D^3.

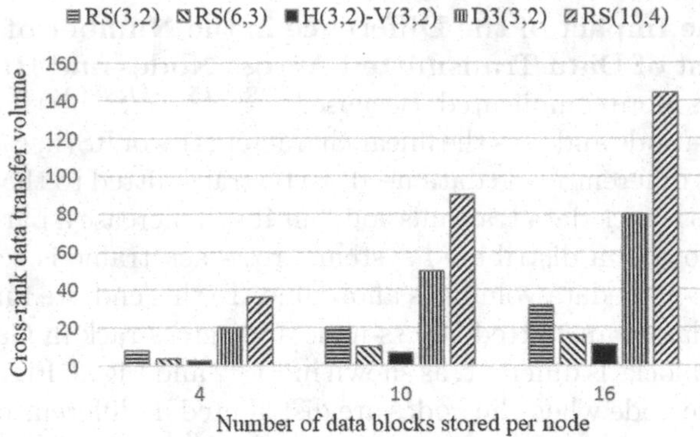

Fig. 3. Case 4 data transmission volume across-racks.

5 Conclusion

Aiming at the reliability problem of erasure correction code system, a new erasure code layout scheme $\mathrm{H}^{RS(n,k)}$ - $\mathrm{V}^{RS(n',k')}$ abbreviated as H-V is proposed in this paper, which can significantly reduce the occupation of network bandwidth and speed up data recovery. H-V adds parity calculation inside the node on the basis of traditional RS erasure code storage, also uses RS encoding, and stores the parity calculation result in the current In the node. Although this will cause a certain amount of data redundancy, the parity blocks can be decoded from the node itself when a node fails, thereby reducing the cross-rack traffic generated during recovery. Theoretical analysis shows that compared with traditional RS erasure code storage, H-V increases storage redundancy by 44.4%, but at most, the cross-node and cross-rack bandwidth of RS can be reduced by 25% and 62.5% respectively during data recovery; Compared with D^3, H-V can reduce storage redundancy by up to 19.7% while reducing cross-node and cross-rack bandwidth by up to 25% and 12.5%, respectively, during data recovery. H-V can minimize the cross-rack or cross-node bandwidth to 0 when a single data block is lost; when the nodes are on different racks, H-V reduces a large amount of cross-rack data transmission compared with other erasure coding strategies, and the advantages of H-V become more obvious as the amount of data in-creases. Although a part of the cross-node traffic is sacrificed, the amount of calculation and disk space occupation is increased, the cross-rack traffic is greatly saved, and the recovery efficiency of a single node is improved as a whole.

In order to further reduce the cost of data recovery, we can improve our method in the following two aspects in the future: 1) By adjusting the execution order of calculations, unnecessary repeated calculations are avoided, the amount of calculations is reduced, and the calculation overhead is reduced. 2) By reasonably selecting the strips used for recovery, the same block of data can be used for the recovery of multiple blocks of data.

Acknowledgements. The research was supported in part by the National Natural Science Foundation of China (Grant No. 61872043), Qin Xin Talents Cultivation Program, Beijing Information Science & Technology University (No. QXTCP B201904), State Key Laboratory of Computer Architecture (ICT, CAS) under Grant No. CARCHA202103, the key scientific and technological projects of Henan Province (Grant No. 202102210174), and the Key Scientific Research Projects of Henan Higher School (Grant No. 19A520043).

References

1. Ghemawat, S., Gobioff, H., Leung, S.T.: The google file system. ACM SIGOPS Oper. Syst. Rev. **37**(5), 29–43 (2003)
2. Shvachko, K., Kuang, H., Radia, S., Chansler, R.: The Hadoop distributed file system. In: Proceedings of the IEEE 26th Symposium on Mass Storage Systems and Technologies (2010)
3. Facebook: HDFS-RAID (2011). https://wiki.apache.org/hadoop/HDFS-RAID
4. Reed, I.S., Solomon, G.: Polynomial codes over certain finite fields. J. Soc. Ind. Appl. Math. **8**(2), 300–304 (1960)
5. Benson, T., Akella, A., Maltz, D.A.: Network traffic characteristics of data centers in the wild. In: Proceedings of 10th ACM SIGCOMM Conference on Internet Measurement, pp. 267–280 (2010)
6. Li, R., Hu, Y., Lee, P.P.: Enabling efficient and reliable transition from replication to erasure coding for clustered file systems. IEEE Trans. Parallel Distrib. Syst. **28**(9), 2500–2513 (2017)
7. Liu, S., Duan, D.: An improved method for HDFS replica recovery: based on SVM algorithm. In: Proceedings of the 2020 4th International Conference on Cloud and Big Data Computing (2020)
8. Zhao, W., Cui, X.: A fast adaptive replica recovery algorithm based on access frequency and environment awareness. In: Proceedings of the 2020 4th International Conference on Cloud and Big Data Computing (2020)
9. Wu, S., Zhu, W., Mao, B., Li, K.C.: PP: popularity-based proactive data recovery for HDFS-RAID systems. Future Gener. Comput. Syst. **86**(SEP), 1146–1153 (2017)
10. Tai, Z., et al.: STORE: data recovery with approximate minimum network bandwidth and disk I/O in distributed storage systems. In: IEEE International Conference on Big Data. IEEE (2015)
11. Li, R., Jian, L., Lee, P.: CORE: augmenting regenerating-coding-based recovery for single and concurrent failures in distributed storage systems. IEEE (2013)
12. Xie, X., et al.: AZ-code: an efficient availability zone level erasure code to provide high fault tolerance in cloud storage systems. In: 2019 35th Symposium on Mass Storage Systems and Technologies (MSST) (2019)
13. Xia, M., et al.: A tale of two erasure codes in HDFS. In: Usenix Conference on File & Storage Technologies USENIX Association (2015)
14. Xu, L., et al.: Deterministic data distribution for efficient recovery in erasure-coded storage systems. IEEE Trans. Parallel Distrib. Syst. **31**(10), 2248–2262 (2020)
15. Caneleo, P., et al.: On improving recovery performance in erasure code based geo-diverse storage clusters. In: International Conference on the Design of Reliable Communication Networks. IEEE (2016)

Astral: An Autoencoder-Based Model for Pedestrian Trajectory Prediction of Variable-Length

Yupeng Diao[1], Yiteng Su[1,2], Ximu Zeng[1], Xu Chen[1], Shuncheng Liu[1], and Han Su[1,2(✉)]

[1] School of Computer Science and Engineering, University of Electronic Science and Technology of China, Chengdu, China
{yupengdiao,yitengsu,ximuzeng,xuchen,liushuncheng}@std.uestc.edu.cn, hansu@uestc.edu.cn

[2] Yangtze Delta Region Institute (Quzhou), University of Electronic Science and Technology of China, Chengdu, China

Abstract. Predicting pedestrian's future behavior in a crowd plays an important role in many fields. Such as autonomous driving, machine navigation, video surveillance, and intelligent security systems. This is very challenging because pedestrian motion can be easily influenced by surrounding pedestrians' interactions. In previous works, researchers use these interactions to make prediction more effective. However, the previous work set fixed-length input in their models. In this way, they ignore shorter pedestrian trajectories. This approach leads to insufficient feature information and inaccurately prediction in some scenarios. In this paper, we propose an Autoencoder-based model for pedestrian trajectory prediction of variable length (**ASTRAL**). At first, we use the autoencoder to process pedestrian data with variable-length trajectories. And then, we use the optimized multi-head attention mechanism to extract the interactions between neighbors. Finally, we use LSTM to decode vectors and make predictions. In particular, we fine-tune the model to make its performance better. We test our model and the state-of-the-art methods on the public benchmark datasets. Compared with others, our model improves ADE (average displacement error) and FDE (final displacement error) by 9% and 33% respectively. Therefore, our model is better than previous works, and we can predict the future trajectory of pedestrians more effectively.

Keywords: Variable length trajectory · Autoencoder · Multi-head attention

1 Introduction

Human motion behavior is becoming more and more important in the fields of autonomous driving and robot intelligent navigation. There exist complex interactive behaviors in predicting pedestrian motions. To promote the development of autonomous driving and navigation, these behaviors should be taken

U. K. Rage et al. (Eds.): DASFAA 2022 Workshops, LNCS 13248, pp. 214–228, 2022.
https://doi.org/10.1007/978-3-031-11217-1_16

into account. For example, autonomous driving uses a series of sensors and algorithms. The algorithm can reasonably predict the future pedestrians' trajectories and effectively avoid traffic accidents [14]. This influences our further exploration in the field of autonomous driving and navigation. In addition, pedestrian trajectory prediction has also been widely studied in the field of algorithms. The application of the algorithm develops the field of trajectory prediction.

At first, in the early research work, the main method is to capture the interactive information between people through hand-made energy functions [8,9]. These studies often need to spend a lot of time on feature engineering, and it is impossible to capture the interaction between people in crowded environments. Second, with the rise of deep learning, pedestrian trajectory research made a breakthrough. Recurrent Neural Networks (RNNs) is widely used in trajectory prediction and have shown excellent results. Social LSTM [1] uses a sequence prediction model based on RNNs and shows the state of space between pedestrians. In [1], the Social-pooling mechanism is used to merge the potential interaction states. However, this grid-based measurement needs to consume a lot of calculations. In addition, this method tends to cause similar impact factors of surrounding pedestrians, which measures the interaction between pedestrians inaccurately. Third, like Sophie [15], which uses the attention mechanism has a wider research environment. [15] weights the interaction between the surrounding pedestrians and the predicted pedestrians. However, this attention mechanism does not fully excavate many potential state information. Based on the attention mechanism, SAPTP [7] uses a spatio-temporal map-based method to associate time and space information. Meanwhile, SCSG [4] adds a multi-head attention mechanism to better extract the interactive information from pedestrians around. However, in feature engineering, SAPTP and SCSG's input requires fixed-length trajectories. Therefore, a lot of pedestrian trajectory data is wasted, and certain trajectory information is also lost.

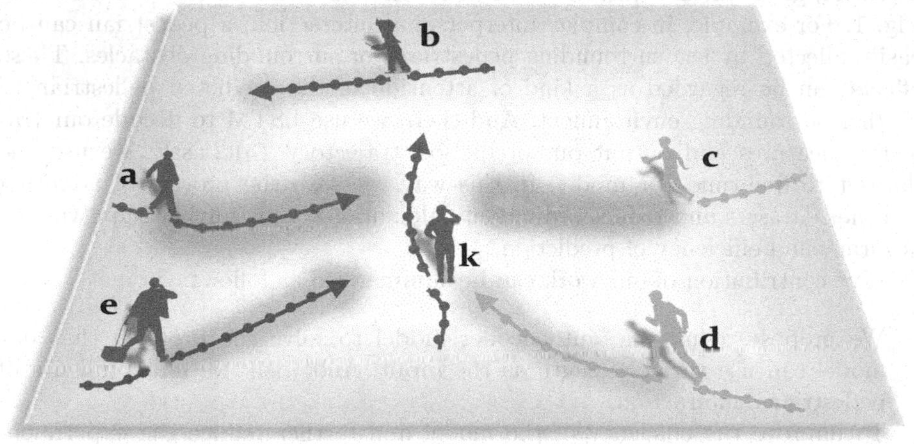

Fig. 1. Surrounding pedestrians with different trajectory length

Based on the problems in the previous works, Astral encounters the following challenges before prediction: (1) how to choose more influential neighboring pedestrians for predicted pedestrians. As shown in Fig. 1, we want to predict the pedestrian k's future trajectory. The previous works select surrounding pedestrians who have longer historical trajectories. That is convenient to set the input trajectory to fixed-length. The previous works choose pedestrian b and ignores pedestrian a. However, it will cause a problem. Obviously, pedestrian a has a certain interaction with pedestrian k. Without considering the interaction of pedestrian a, it is easy to make the prediction results inaccurate. Therefore, such surrounding pedestrians like pedestrian a should be taken into consideration. (2) how to correctly evaluate the interaction between pedestrians under the intricate trajectory influence. Such as pedestrians in Fig. 1, we can see the pedestrians' interaction (pink mark) are different from each other. In other words, we need to effectively identify the interactions that have more influence on the predicted pedestrian. Some pedestrians are far away but fast, while some others are close but slow. We need to find the most influential ones among many factors to make our prediction more accurate. (3) how to improve the training efficiency of the model. Most previous works usually retrain the prediction model when they meet a new dataset, which will affect the efficiency of the model. The model should adjust its adaptability so that it can train data more effectively. The above problems need to be solved.

To address these challenges, we propose an Autoencoder-based model for pedestrian trajectory prediction of variable length (ASTRAL). Our model includes an encoder-decoder and fine-tuning model. We use an autoencoder to extract features from the pedestrian trajectory. (1) In the autoencoder, we calculate the features and expand the feature dimension of each pedestrian. We take into account the surrounding pedestrians with a short-length trajectory. In the feature extraction process, we also use the LSTM method to encode the trajectory information in a sequence. (2) In addition, we also used the optimized attention mechanism to effectively extract the pedestrian interaction shown in Fig. 1. For example, in complex interpersonal interaction, a pedestrian can be easily affected by the surrounding pedestrians or surrounding obstacles. These effects can be regarded as a kind of attention of the predicted pedestrian to his/her surrounding environment. And then, we use LSTM to decode our trajectory features and output our prediction trajectory. (3) Lastly, we use one dataset to fine-tune our model. In this way, we save the process of adapting to other datasets and reduce redundant calculations. This further improves the accuracy and efficiency of prediction.

The contribution of our work can be summarized as follows:

- We propose to use the autoencoder model to solve the problem that our model can use variable-length as the input. And then, we make full use of pedestrians' interaction.
- To improve the efficiency of the model and further reduce the experimental error, we propose a transfer learning idea. It is a way to fine-tune the prediction model of a certain dataset and apply it to other datasets.

- Our model has higher accuracy and better convergence efficiency on the benchmark datasets. Compared with the experiment results of baseline, our model is improved by 9% and 33% on ADE (average displacement error) and FDE (final displacement error) respectively.

The remaining Sections in this paper are as follows: Sect. 2 introduces our problem definition and related description of the model; Sect. 3 introduces the overall framework of our model and the implementation of each algorithm; Sect. 4 shows our experimental results and introduces a case study to analyze the prediction ability of the model; Sect. 5 introduces related work and Sect. 6 shows the conclusion.

2 Problem Definitions

In this section, we briefly introduce the preliminary description of our problem. And then, we summarize the relevant concepts involved in our model.

Trajectory Point. At the certain time, the position information of pedestrians can be expressed by plane geometric coordinates. We use fixed frequency to capture pedestrian trajectory in different scene. Meanwhile, we use each frame based on the video as the time of trajectory prediction. Therefore, in frame t , each captured pedestrian has a position information $P = \{(x_i^t, y_i^t)|t = 1, 2, 3, \dots, n \quad i = 1, 2, 3, \dots, N\}$, where x_i^t represents the transverse position information and y_i^t represents the longitudinal position information, and N is the number of pedestrians in the frame (where n is the last frame). The trajectory point $p_i^t = (x_i^t, y_i^t)$ represents the specific position of the pedestrian i at frame t.

Trajectory. A sequence of trajectory points is a historical trajectory. For example, $T_i^h = [p_i^1, p_i^2, \dots, p_i^n]$, it is the historical trajectory of pedestrian i. We use the historical trajectory T^h of pedestrians to predict the future trajectory of $\hat{T}^f$. In previous work, n is fixed-length, and our model can make n variable.

Trajectory Length. In this paper, we specify that pedestrians have a trajectory point P at each timestep t. In the same scene, the number of pedestrian's P determines the length l of the trajectory. In other words, the number of timesteps n determines the l of their historical trajectory. Therefore, in this paper, the trajectory's l is equal to the number of n in the trajectory.

Problem Definition. In this part, we introduce our problem definition in detail: at any frame t, we assume that we preprocess each scene to obtain everyone's location information at first. The previous works describe the way as obtaining coordinates. Pedestrian trajectories are connected by coordinate points in multiple time frames. At frame t, target pedestrian k's historical trajectory $T_k^h = [p_k^{t-\lambda+1}, \dots, p_k^t]$ and his/her neighbor (Pedestrians around) $T_i^h = [p_i^{t-\lambda+1}, \dots, p_i^t]$ where $i \neq k$. We can predict the future trajectories $\hat{T_k}^f = [p_k^{t+1}, \dots, p_k^{t+\delta}]$ of target pedestrian by the historical trajectories at frame t. Where λ represents the relative length of the historical trajectories, and δ

represents the relative length of the future trajectories. Obstacles in the scene can be regarded as static pedestrians.

Our goal is to make the predicted trajectory $\hat{T_k}^f$ of the target pedestrian k as close to the real trajectory T_k^f as possible.

3 Methodology

3.1 Model Framework

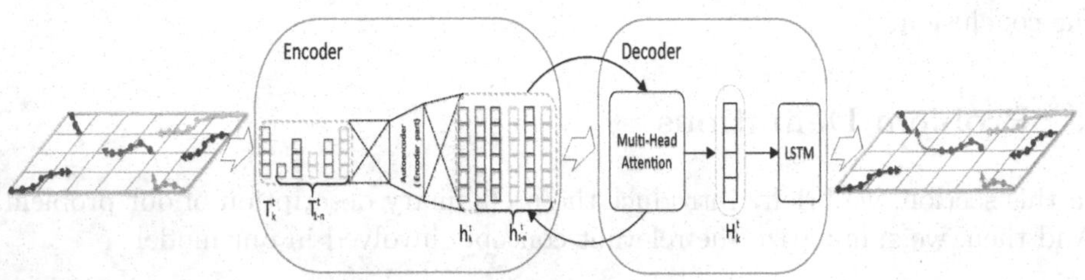

Fig. 2. The Framework of Astral

In this section, we will introduce our model framework. As the Fig. 2 given, our model can be divided into several parts: (1) Offline model (Encoder-Decoder of Astral) (2) Online model (Fine-tuning of Astral). To carry out the work of sequence prediction, we set the target pedestrian k's historical trajectory and his/her neighboring pedestrian i's historical trajectories as trajectories vector T_k^t and T_{1-n}^t, which n is the number of neighboring pedestrians and t refers to the frame t.

At first, the model used pedestrian trajectories as input for the autoencoder, which calculates trajectory data as feature vectors. As the Encoder's output, the feature vectors will be denoted as h_k^t and h_{1-n}^t. And then, the multi-head attention mechanism use feature vectors to make weighted sum H_k^t. After that, we use LSTM to predict the future target pedestrian's trajectory $\hat{T_k}^f$. The above is the training process of the Offline model. And lastly, we fine-tune the model and adjust the offline model process to the online training process. This is the whole process of the Astral model.

Next, we introduce the autoencoder and multi-head attention in Sects. 3.2 and 3.3 respectively. And then, we introduce the Online model in Sect. 3.4.

3.2 Autoencoder

This section will give a brief introduction to the autoencoder applied in this paper. As the Fig. 3 given, in order to generate the variable-length input trajectory of the target pedestrian T_k^t and neighboring pedestrians T_{1-n}^t into a fixed-length vector as h_k^t and h_{1-n}^t, this paper proposes an autoencoder to extend the feature dimension of feature vectors. To learn the latest features of pedestrian trajectories, the proposed autoencoder is composed of three layers of LSTM in

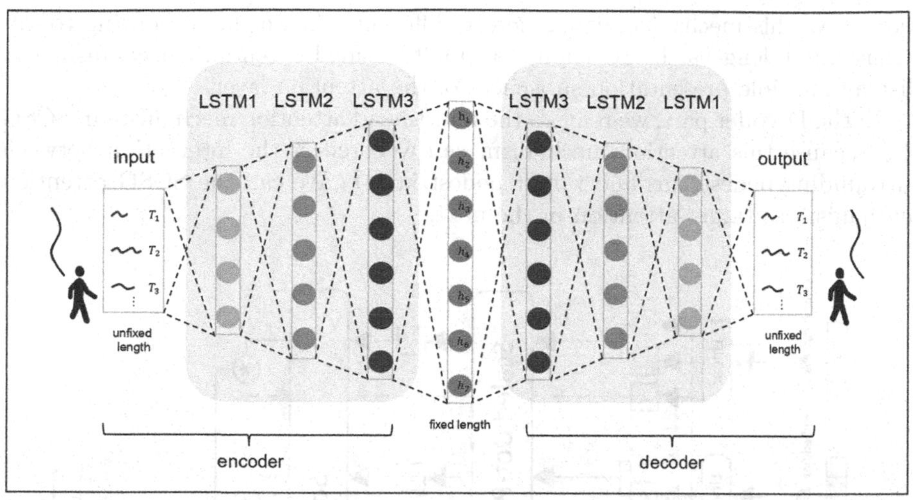

Fig. 3. Autoencoder structure

both the encoder and decoder parts. The encoder part is used to increase the dimension of the trajectory data, while the decoder part is designed to restore the processed feature vectors to the pedestrian trajectory data. And then, the output of the encoder part will be served as input for the multi-head attention mechanism. In this way, the autoencoder can learn the effectively processed feature vectors of a set of data.

There are two reasons for the design of LSTM layers in the autoencoder. For one thing, the deep autoencoder, rather than the two-layer neural network autoencoder, can get good data representation. For another, the LSTM layers have proven to be able to learn the latent feature, especially in sequence prediction which is related to predicting pedestrian trajectories in this paper. On the contrary, using recurrent neural network layers to learn latent features cannot achieve the promising performance of LSTM layers.

LSTM is a variant of recurrent neural network, with Gating Mechanism to control the accumulation speed of information. In this paper, the input trajectory $T_k^h = [p_k^{t-\lambda+1}, ..., p_k^t]$ is put into LSTM layers, which outputs a hidden state as follows:

$$h_k^t = LSTM(e_k^t, h_k^{t-1}; W) \tag{1}$$

where e_k^t is the temporal embedding vector, h^{t-1} denotes the hidden state at $t-1$ frame, and W is the temporal parameters of LSTM layers. The final hidden state h_k^t is calculated by recursively executing LSTM.

3.3 Multi-head Attention

Vaswani et al. [20] proposed a multi-head attention mechanism, which ensembles several self-attention models. When dealing with variable-length vector sequences, this mechanism can generate different join weights according to different input lengths. In addition, the multi-headed attention mechanism can display multiple presentation subspaces of the attention layer.

In the Decoder part, we refer to the multi-head attention mechanism in SCSG [4]. Because this attention mechanism can well reflect the interaction between surrounding pedestrians and target pedestrians k. We call the SCSG attention mechanism as target-attention in the next.

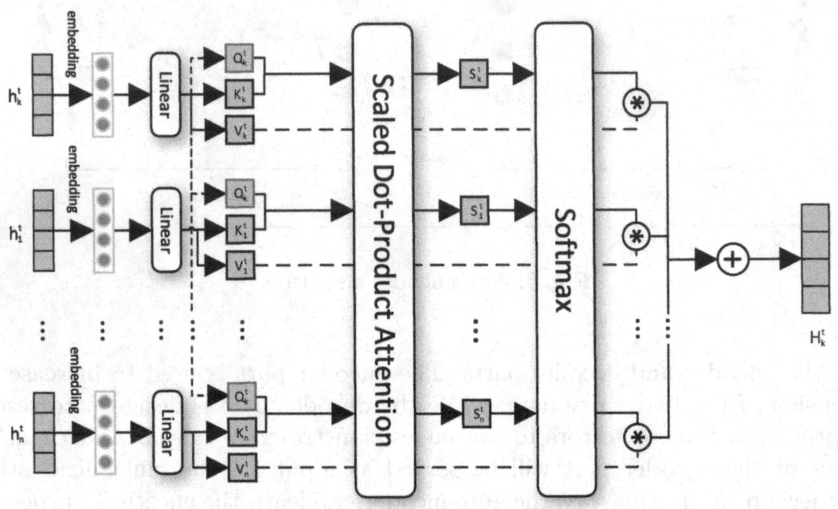

Fig. 4. Attention mechanism

The overall architecture of the target-attention is shown in Fig. 4. Based on the multi-head attention mechanism, the target-attention removes the mutual weight calculation of surrounding pedestrians. We take the encoded pedestrian trajectory feature vector h_i^t as the input of the target-attention mechanism. The i is denoted any pedestrian and all embeddings at the frame t. In order to calculate the attention mechanism for each h_i^t separately, we need to calculate their query vector Q_i^t, key vector K_i^t and value vector V_i^t separately. The definition of vectors is as follows:

$$Q_i^t = \phi_t(h_i^t; W_{qt}) \tag{2}$$

$$K_i^t = \phi_k(h_i^t; W_{kt}) \tag{3}$$

$$V_i^t = \phi_t(h_i^t; W_{vt}) \tag{4}$$

where W_{qt}, W_{kt}, W_{vt} are the parameters of definitions separately.

For the Q_i^t, the target-attention use the target pedestrian vector h_k^t to generate the query matrix W_{qt}^k. the all pedestrians' Q^t are calculation by W_{qt}^k. Target-attention focuses on calculating the interaction between surrounding pedestrians and target pedestrians. On the other hand, we use all pedestrian's historical trajectories embedding into every K^t and V^t. For i-th pedestrian's Q_i^t, we have:

$$Q_i^t = \phi_t(h_i^t; W_{qt}^k) \tag{5}$$

where W_{qt}^k is target pedestrian query matrix.

Next, we need to calculate the interaction between pedestrians and the attention scores. The interaction of neighboring pedestrian i on the target pedestrian k can be expressed as follows:

$$F^{i \to k} = Q_k^t \times K_i^t \tag{6}$$

The score of the target pedestrian k to the surrounding pedestrian i can be defined as follows:

$$S_i^t = Softmax(\frac{[F^{i \to k}]_{i=1:n}}{\sqrt{d_k}}) \times [V_i^t]_{i=1:n} \text{ },i = k, 1:n \tag{7}$$

where d_k represents the dimension of each query and the number of i is equal to the head number h of the attention mechanism.

Lastly, we concatenate weight matrix z_1 to Z_h. The results of each attention mechanism were combined. Therefore, we can use the following calculation function to derive the result:

$$MultiHead(Q_k^t, K_k^t, V_k^t) = W^O \times Concat([S_j^t])_{j=1:h} \tag{8}$$

where W_0 is the matrix representing a linear transformation.

In the end, we get the weighted score $MultiHead(Q_k^t, K_k^t, V_k^t)$ of the interaction of neighboring pedestrians on the target pedestrian.

Outputs. We use the weighted score as the LSTM input, and then through a linear layer output next timestep point of target pedestrian. After 12 timesteps, we get the pedestrian trajectory for the next 12 timesteps as our model output.

3.4 Online Model

There will be a lot of redundant calculations during model training. Because the input information between datasets is similar. To make the model more efficient and further improve the prediction results, we will use the general knowledge N learned during training with a dataset, such as timesteps, data dimensions, and other parameter information. The N is applied to the fine-tuning model of the new dataset for the next training. When training the second dataset, we can use the previous N to adapt the model. When each dataset is put into model training, we need to learn general knowledge first. In the process of learning, the model often consumes a lot of time. So the reuse of N can reduce unnecessary redundant computing. We fine-tune our model and use the trajectory parameters PA trained by the first dataset. Other datasets will also be trained with PA. The Online model uses the basic framework of the Offline model. We fine-tune the model during pretreatment and training. And then, we use the trajectory sequence which is decoded by LSTM as the output for the Online model.

4 Experiments

In this section, we discuss the results of our model and baseline on the benchmark pedestrian datasets. And we use the uniform standards evaluation for the experiment.

4.1 Experimental Settings

Datasets: There are two publicly available datasets which we used to evaluate our method: ETH [13] and UCY [11]. In total, there are 5 datasets which are ETH-ETH, ETH-HOTEL, UCY-ZARA01, UCY-ZARA02, and UCY-UNIV. These datasets are collected from different scenes in the real world. The pedestrian position data information in the datasets covers challenging motion patterns, such as walking together or in a different direction.

Evaluation Metrics: After comparing with our baselines, we decide to use two error metrics: ADE and FDE. ADE and FDE are also the previous works' metrics. We use these metrics to evaluate the prediction results of each model. The experimental results show that our model achieves better results.

1. Average displacement error (ADE): Average distance between the ground truth and predicted trajectories.
2. Final displacement error (FDE): The distance between the ground truth destination and the predicted destination at the last prediction timestep.

Baselines: We compare our model with the previous methods.

1. LSTM (vanilla LSTM model): a simple linear Long Short-Term Memory Networks model.
2. S-LSTM [1]: Social LSTM uses LSTM for each trajectory and proposes a Social pooling layer to share information.
3. S-GAN [6]: Social GAN proposes a GAN-based encoder-decoder framework to predict the trajectory and use a novel pooling mechanism that can learn the social interaction of all people in the same scene.
4. SAPTP [7]: a spatio-temporal graph-based model use complete graph to make prediction.
5. SoPhie [15]: SoPhie uses social attention mechanism with physical attention to extract the most significant of the path and aggregate the information interacted by different agents. This model also uses GAN to generate more realistic samples.
6. SCSG [4]: a self-centered star graph with attention for pedestrian trajectory prediction.

Implementation Details: Our model is based on SCSG attention's input data strategy. We use a target pedestrian and five surrounding pedestrians as data during the training. And then, we take the historical trajectory of pedestrians as the input data. The LSTM in autoencoder is set to 3 layers. The LSTM is set to

2 layers in the Decoder part. The number of heads of the multi-head attention mechanism is 16. We use AdamW to be our optimizer and batch size 64 for 100 epochs. Our Online model's batch size is 64 for 10 epochs. Our model predicts the target sequence in the next 12 timesteps.

We use our model to compare with the state-of-the-art methods as mentioned in baselines. And we reproduced the experimental effect of the vanilla LSTM model. In addition, we test some supplementary experiments based on the results of our model. As shown in Table 1, we will analyze experimental results in the next.

4.2 Evaluation of Trajectory Prediction

In this subsection, we will compare the results of baselines with our model, and then compare the superiority of our model.

Table 1. Quantitative results of baselines and Astral on all datasets

Performance (ADE/FDE)						
Datasets	ETH	HOTEL	UNIV	ZARA1	ZARA2	Average
LSTM	1.31/2.95	0.55/1.35	1.50/2.92	0.42/1.01	0.36/0.93	0.83/1.83
S-LSTM	1.09/2.35	0.79/1.76	0.67/1.40	0.47/1.00	0.56/1.17	0.72/1.54
S-GAN	0.87/1.62	0.67/1.37	0.76/1.52	0.35/0.68	0.42/0.84	0.61/1.21
SAPTP	1.24/2.35	0.48/0.80	0.69/1.45	0.51/1.15	0.56/1.13	0.70/1.38
SoPhie	0.70/1.43	0.76/1.67	0.54/1.24	0.30/0.63	0.38/0.78	0.54/1.15
SCSG	0.58/1.47	0.20/0.65	0.53/1.53	0.20/0.62	**0.19**/0.60	0.34/0.97
Astral-Offline	0.61/1.18	0.34/0.56	0.65/1.12	**0.19**/0.58	0.62/0.95	0.45/0.77
Astral-Online	**0.57/1.08**	**0.19/0.30**	**0.40/0.97**	**0.19/0.42**	0.22/**0.48**	**0.31/0.65**

Comparison of Our Model and Baselines: We use ADE/FDE to evaluate the effect of all model trajectory predictions. All models use the LSTM to predict trajectory sequence. As shown in Table 1, models which use learning methods are better than LSTM-only. The S-GAN shows better than S-LSTM, LSTM, and SAPTP, because it improves S-LSTM's pooling layer and extracts more features. The SoPhie further improves S-GAN and uses both social and physical attention mechanisms to achieve interpretable predictions. Without sacrificing the accuracy of the model, SCSG can produce accurate results in a shorter time [4]. Such as the ZARA02 which has a large amount of data, and SCSG uses spatial modeling to make the performance of the model better. But SCSG is not very well in dense population datasets. And then, Astral-Online uses the idea of transfer learning and reduces the time of repeated calculations Especially the time of parameter learning is reduced relatively, the Astral-Online (Online model) is better than the Astral-Offline (Offline model). Therefore, our Astral-Online comprehensive test is better, and the application scenarios are more changeable. So our Astral-Online improves the performance of the model. We will further evaluate the Online model of Astral next.

4.3 Ablation Studies

We conducted an extensive ablation study on all benchmark datasets to understand the impact of the components in our model.

Effectiveness of Autoencoder: Under each dataset, we measured the performance comparison of models with fixed-length (Fixed-Length) input and variable-length (Variable Length) input. We tested on five datasets. The experimental results show that the model with autoencoder (Variable Length) performs better. Next, we analyze the reasons for the experimental results.

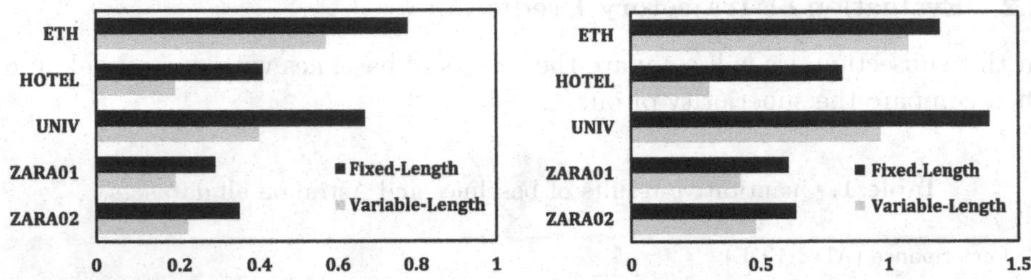

Fig. 5. Comparison of ADE and FDE among different length of Input

We can see the effectiveness of the autoencoder in Fig. 5. In the fixed-length trajectory method (Fixed-Length), the accuracy is relatively low. We analyze that the main reasons. Firstly, it ignores the better pedestrians around the target pedestrian. Although the trajectory information is less, these pedestrians have an impact on the future trajectory of the target pedestrian. And then, lots of "no-effect" trajectory information is concluded. Some pedestrians around that did not influence the target pedestrians. Therefore, our model has performed better under a comprehensive comparison. The ADE and FDE have been improved by 13% and 24% respectively.

Effectiveness of Multi-head Attention: We test the weight value of the target-attention mechanism from the two dimensions of speed and distance. We use the average weight distribution of all datasets to be the experimental results. In the following table, we can see that our model that is more effective to obtain the interactive information of neighboring pedestrians.

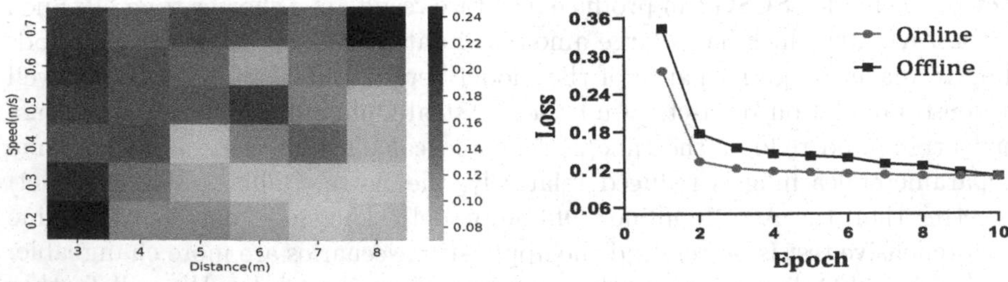

Fig. 6. Attention weight heat map

Fig. 7. Loss of online and offline

As shown in Fig. 6, neighboring pedestrians have different attention weights to target pedestrian. This reason is mainly caused by their walking speed and their distance from the target pedestrian. If he/she give more interaction on target, the map color will be darker. The pedestrians have a strong following tendency, which is more likely to walk close with the target. Such distance equals 3 m, which is more likely to impact on target's future trajectory. However, in the 8 m, the pedestrian who is far away but his/her speed is very fast also has an impact on the target. Because he/she has a certain mutual influence on the target pedestrian's future trajectory. Our model can learn such complex interactions. In other words, this is where our model is better than others. The target-attention can distinguish the strength of the interaction smoothly.

Effectiveness of Online Model: We tested the Online model mentioned in this paper. As shown in Fig. 7, the Online model (Online) compares with the Offline model (Offline) without fine-tuning on stochastic gradient descent. We used the average loss of five benchmark datasets as the experimental loss results. During ten epochs, the Online model shows a better descent effect. The Offline also drops to the same state, but its velocity is slower than the Online model.

The reason for the faster decline rate is that the Offline model needs to retrain the data when a new dataset is to be input. But in the Online model, the first dataset has trained the general knowledge. After training, the Online model only needs to fine-tune the parameters. In other words, the descent velocity is equivalent to the size of the calculation. The Online model reduces repetitive calculation. This way can make the performance of the model is better. Meanwhile, the error is smaller in Fig. 5.

4.4 Case Study

Based on experiment results, we visualize Astral and compare it with the Ground Truth, SCSG, and LSTM. In addition, the trajectory behind the target pedestrian is his/her historical trajectory.

We use the dataset UNIV to visualize pedestrian trajectories. UNIV has more interactions between pedestrians, especially the pedestrians around (pink points) who have a short trajectory and interact with the target pedestrian. As the Fig. 8 given, figure(a) shows the interactions of the surrounding pedestrians who suddenly accelerate to overtake the target pedestrian. Figure(b) shows the interactions of pedestrians walking towards. Figure(c) shows the target pedestrian changes direction. And the pedestrians around approach the target pedestrian at different speeds. Figure(d) shows pedestrians walk in parallel and interact with the target at the same time. In all scenes, there are neighbors with short trajectories and influence on the future trajectory of the target pedestrian.

Although the errors of other prediction methods are relatively small, these methods do not make full use of the trajectory information with a shorter length. This trajectory information is valuable. It has a certain impact on the experimental results. Therefore, our model considers more interactions, so that the prediction of our model is better after a comprehensive comparison.

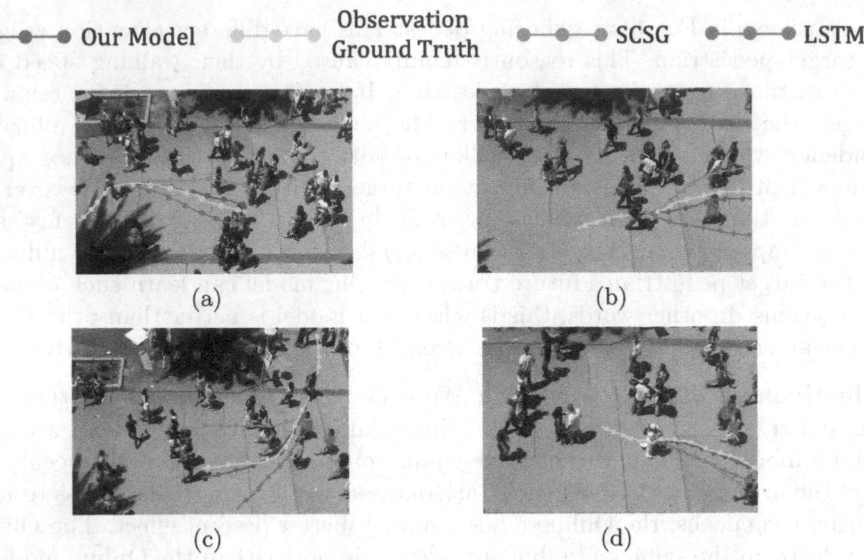

Fig. 8. Visualization of trajectory

5 Related Work

In this paper, we use the concept of crowd interaction and LSTM models for sequence prediction. Next, we introduce the related work of two studies.

5.1 Crowd Interaction

In the previous work, Helbing et al. [9] obtain pedestrian interaction by manually adjusting the energy function. They describe the motion of pedestrians as a social force. This force is pedestrians' internal motivation under certain prejudgmental actions. This model has achieved good results in various fields and has been later revisited and extended to several kinds of research [2,5,19] which also uses hand-crafted rules to model human interactions. For instance, Treuille et al. [19] propose continuum dynamics crowd model in real-time. This model uses a dynamic potential field to integrate global navigation and other pedestrians. In addition, Antonini et al. [2] use a Discrete Choice framework, regarding short-term individuals' behaviors as a response to other pedestrians. However, one limitation of these models is that they particularly rely on hand-crafted features, which have difficulties in generalizing or making the accurate prediction in complex trajectory prediction. In addition, models with merely attractive and repulsive forces can not perform well with complex human interactions [1].

5.2 LSTM for Sequence Prediction

In the application of deep learning, there exists a lot of data in the form of sequences, such as sound, language, and video. Recurrent Neural Networks

(RNNs), such as LSTM [10] and Gated Recurrent Units (GRU), have achieved great success in diverse domains with sequence generation, like machine translation [3,17], speech recognition [12,21], dialogue system [16] to name a few. The experiments of [18] have proved the success of RNNs-based models with a seq2seq structure for making predictions. However, basic RNNs lack spatio-temporal structure and do not model crowd interactions. To address this problem, Alahi et al. [1] propose social pooling layers to share the hidden states between LSTMs of nearby pedestrians; Gupta et al. [6] introduce a pooling module that can jointly aggregate information across pedestrians.

6 Conclusion

In this paper, we propose a variable-length input trajectory prediction model, named Astral. The Astral consists of autoencoder, multi-head attention, and LSTM. And it uses variable-length trajectory information. The Astral model combines with the interaction of the surrounding pedestrians and completes the prediction of the future trajectory of the target pedestrian. Lastly, we fine-tune the Astral to improve the performance of the model. The experimental results show that our model is effective. In future work, we can further improve the attention mechanism algorithm. And then, we can predict multiple pedestrians at the same time.

References

1. Alahi, A., Goel, K., Ramanathan, V., Robicquet, A., Fei-Fei, L., Savarese, S.: Social LSTM: human trajectory prediction in crowded spaces. In: Proceedings of the IEEE Conference on Computer Vision and Pattern Recognition, pp. 961–971 (2016)
2. Antonini, G., Bierlaire, M., Weber, M.: Discrete Choice Models of Pedestrian Walking Behavior. Transp. Res. Part B Methodol. **40**, 667–687 (2006)
3. Bahdanau, D., Cho, K., Bengio, Y.: Neural machine translation by jointly learning to align and translate. arXiv preprint arXiv:1409.0473 (2014)
4. Chen, X., et al.: SCSG attention: a self-centered star graph with attention for pedestrian trajectory prediction. In: Jensen, C.S., et al. (eds.) DASFAA 2021. LNCS, vol. 12681, pp. 422–438. Springer, Cham (2021). https://doi.org/10.1007/978-3-030-73194-6_29
5. Ferrer, G., Garrell, A., Sanfeliu, A.: Robot companion: a social-force based approach with human awareness-navigation in crowded environments. In: 2013 IEEE/RSJ International Conference on Intelligent Robots and Systems, pp. 1688–1694 (2013). https://doi.org/10.1109/IROS.2013.6696576
6. Gupta, A., Johnson, J., Fei-Fei, L., Savarese, S., Alahi, A.: Social GAN: socially acceptable trajectories with generative adversarial networks. In: Proceedings of the IEEE Conference on Computer Vision and Pattern Recognition, pp. 2255–2264 (2018)
7. Haddad, S., Wu, M., Wei, H., Lam, S.K.: Situation-aware pedestrian trajectory prediction with spatio-temporal attention model. arXiv preprint arXiv:1902.05437 (2019)

8. Helbing, D., Buzna, L., Johansson, A., Werner, T.: Self-organized pedestrian crowd dynamics: experiments, simulations, and design solutions. Transpo. Sci. **39**(1), 1–24 (2005)
9. Helbing, D., Molnar, P.: Social force model for pedestrian dynamics. Phys. Rev. E **51**(5), 4282 (1995)
10. Hochreiter, S., Schmidhuber, J.: Long short-term memory. Neural computation **9**(8), 1735–1780 (1997)
11. Leal-Taixé, L., Fenzi, M., Kuznetsova, A., Rosenhahn, B., Savarese, S.: Learning an image-based motion context for multiple people tracking. In: Proceedings of the IEEE Conference on Computer Vision and Pattern Recognition, pp. 3542–3549 (2014)
12. Miao, Y., Gowayyed, M., Metze, F.: EESEN: End-to-end speech recognition using deep RNN models and WFST-based decoding. In: 2015 IEEE Workshop on Automatic Speech Recognition and Understanding (ASRU), pp. 167–174 (2015). https://doi.org/10.1109/ASRU.2015.7404790
13. Pellegrini, S., Ess, A., Van Gool, L.: Improving data association by joint modeling of pedestrian trajectories and groupings. In: Daniilidis, K., Maragos, P., Paragios, N. (eds.) ECCV 2010. LNCS, vol. 6311, pp. 452–465. Springer, Heidelberg (2010). https://doi.org/10.1007/978-3-642-15549-9_33
14. Raksincharoensak, P., Hasegawa, T., Nagai, M.: Motion planning and control of autonomous driving intelligence system based on risk potential optimization framework. Int. J. Autom. Eng. **7**(AVEC14), 53–60 (2016)
15. Sadeghian, A., Kosaraju, V., Sadeghian, A., Hirose, N., Rezatofighi, H., Savarese, S.: Sophie: an attentive GAN for predicting paths compliant to social and physical constraints. In: Proceedings of the IEEE Conference on Computer Vision and Pattern Recognition, pp. 1349–1358 (2019)
16. Shang, L., Lu, Z., Li, H.: Neural responding machine for short-text conversation. In: Proceedings of the 53rd Annual Meeting of the Association for Computational Linguistics and the 7th International Joint Conference on Natural Language Processing (Volume 1: Long Papers), pp. 1577–1586. Association for Computational Linguistics, Beijing, China, July 2015. https://doi.org/10.3115/v1/P15-1152, https://aclanthology.org/P15-1152
17. Stahlberg, F.: Neural machine translation: a review. J. Artif. Intell. Res. **69**, 343–418 (2020)
18. Sutskever, I., Vinyals, O., Le, Q.V.: Sequence to sequence learning with neural networks. In: Advances in neural information processing systems, pp. 3104–3112 (2014)
19. Treuille, A., Cooper, S., Popović, Z.: Continuum crowds. ACM Trans. Graph. **25**(3), 1160–1168 (2006)
20. Vaswani, A., et al.: Attention is all you need. In: Advances in Neural Information Processing Systems, pp. 5998–6008 (2017)
21. Xiong, W., Wu, L., Alleva, F., Droppo, J., Huang, X., Stolcke, A.: The microsoft 2017 conversational speech recognition system. In: 2018 IEEE International Conference on Acoustics, Speech and Signal Processing (ICASSP), pp. 5934–5938 (2018). https://doi.org/10.1109/ICASSP.2018.8461870

A Survey on Spatiotemporal Data Processing Techniques in Smart Urban Rail

Li Jian[1], Huanran Zheng[1], Bofeng Chen[1(✉)], Tingliang Zhou[2], Hui Chen[3], and Yanjun Li[4]

[1] East China Normal University, Shanghai, China
51194501030@stu.ecnu.edu.cn, BofengChen@edu.cn
[2] CASCO Signal Ltd., Shanghai, China
[3] Panzhihua University, Panzhihua, Sichuan, China
[4] Shanghai Normal University, Shanghai, China
1000484331@smail.shnu.edu.cn

Abstract. The smart urban rail has been developed rapidly and widely in recent years. The spatiotemporal data plays an important role in the field of smart urban rail, and is widely used in various application scenarios such as traffic flow prediction. However, there is a lack of systematic reviews of related technologies about spatiotemporal data. Therefore, this article has reviewed the spatiotemporal data and applications in smart urban rail. Firstly, the technologies of spatiotemporal in urban rail data are comprehensively studied. Secondly, the application of AI in smart urban rail is investigated. And the existing intelligent urban rail-related technologies about spatiotemporal data is summarized from four typical applications: intelligent scheduling, intelligent operation platform, intelligent perception, and intelligent train control. Finally, some interesting topics in smart urban rail applications have been listed. And we make a summary for the smart urban rail.

Keywords: Smart urban rail · Spatiotemporal data · Neural networks

1 Introduction

The development of energy-saving and efficient urban rail is an effective way to solve the common problems of big cities and helps to build a green and smart city. At present, digitization, informatization and intelligence have become the general trend of intelligent development of urban rail transit [5]. The application of artificial intelligence to urban rail to endow urban rail with the ability to operate intelligently is a subject with far-reaching research significance.

Urban rail data is obtained from a variety of ways (e.g., station's sensors, surveillance cameras, and operation system). The diversity data source determines

U. K. Rage et al. (Eds.): DASFAA 2022 Workshops, LNCS 13248, pp. 229–243, 2022.
https://doi.org/10.1007/978-3-031-11217-1_17

that the data in the urban rail is highly spatiotemporal correlated, which means that the data is time-series and highly spatially correlated. Urban rail data not only shows strong transitivity in time but also in space. Adjacent stations also influence each other with time, spreading to a more extensive spatial range. Spatiotemporal data analysis has played an essential role in various fields [77], and technologies about spatiotemporal data based on deep learning have been widely adopted in various urban rail applications. Although there are already very comprehensive surveys on artificial intelligence technology in the field of smart urban rail [66,70], there is still a lack of extensive and systematic surveys of application scenarios and a lack of attention to the spatiotemporal characteristics of urban rail data.

In this article, we discuss the concept of spatial and temporal correlation of urban rail data. Then, we summarize the existing technologies related to smart urban rail from intelligent scheduling, intelligent operation platform, intelligent perception, and intelligent train control. And we prospected the artificial intelligence technology in smart urban rail. Our main contributions are summaried as follows:

* The spatial and temporal correlation of urban rail has been investigated comprehensively. Then, some typical spatial and temporal data analysis techniques used in urban rail applications are introduced.
* The application of artificial intelligence technology in smart city rail has been studied carefully. After summarizing and introducing relevant technologies, we classified them in a table.
* Some interesting topics for smart urban rail have been put forward after the analysis of artificial intelligence technology.

2 Spatiotemporal Data Processing Technology

Urban rail data has complex spatial and dynamic temporal dependencies. In this chapter, we introduce the concept and life cycle of spatiotemporal in urban rail data, and some technologies in urban rail data processing, especially in spatiotemporal data.

Spatiotemporal data is essentially unstructured data, including not only time series data, but also map data, such as urban networks, rail networks, etc. Traditional data management parallel computing platform includes MapReduce [20] and Hadoop [6], but the organizational form and data processing methods are not suitable for dealing with spatiotemporal data. In order to analyze and process spatiotemporal data, more reliable, more effective and more practical data management and processing technologies are urgently needed.

Spatiotemporal data has two characteristics: 1. Complex spatial correlation. Different locations have different effects on the predicted location, and the same location has a different impact on the predicted location over time, that means the spatial correlation between locations is highly dynamic. 2. Dynamic time dependence. Observations at different time in precise locations exhibit nonlinear changes [71]. How to mine nonlinear and complex spatiotemporal patterns and select the most relevant historical observation data for accurate traffic data forecasting remains a challenging problem.

Spatiotemporal data processing also has data life cycle, including data collection, data processing, and data application. Next, we will give a detailed introduction to the related technologies of the data life cycle, especially in spatiotemporal data.

Data Collection. For the problem of how to collect data, it is generally necessary to use a variety of sensors to build a sensor network in urban rail system. For example, if an effective sensor network for urban rail trains is to be constructed, sensors for sensing traffic flow information, sensing infrastructure, and sensing train status are often required. Specifically, the traffic flow perception sensor is mainly used to obtain the traffic flow state, including induction coil and road camera, etc. Infrastructure sensors mainly monitor infrastructure conditions such as tunnels and bridges, including temperature sensors and anemometers, etc. The primary function of the train status sensor is to collect the status of various parts of the train in real-time, including vibration sensors and current and voltage sensors, etc. Various countries have used various sensor networks, including the LonWorks network [69] in the United States and the WorldFIP network [68] in France.

There are also some public spatiotemporal data that can be used by researchers. For example, Large-scale taxi pick-up/drop-off data are publicly available for several major cities across the world [10]. These data contain information about each trip made by customers of the taxi service, including the time and location of pick-up and drop-off, and GPS locations for each second during the taxi ride.

Data Processing. The data processing module usually includes three parts: data processing, data quality assessment, and data analysis.

For the processing of spatiotemporal data, it is usually necessary to combine time series data from different sources and unify the spatial data in different coordinate systems. The processing of spatiotemporal data have four main methods. 1. Spatiotemporal statistic: Spatiotemporal statistics [17] usually combine spatial statistics with temporal statistics to process spatiotemporal data. 2. Time series analysis in spatial: Spatial statistics for point reference data have been generalized for spatiotemporal data [41], and methods include spatiotemporal stationarity, spatiotemporal covariance, spatiotemporal variograms, and spatiotemporal Kriging. 3. Spatiotemporal point process: Spatiotemporal point process used in process spatiotemporal data usually generalizes the spatial point process by incorporating the factor of time [26] 4. Time series analysis in lattice (areal) data: Methods used in time series of lattice (areal) data analysis include spatial and temporal autocorrelation, SpatioTemporal Autoregressive Regression (STAR) [17], and Bayesian hierarchical [3].

The data quality assessment part aims to assess the quality level of the collected data, and it is responsible for providing Quality Control (QC) and

Quality Assurance (QA). Specifically, it checks the quality level of the data according to the strategy provided by QC and discards or repairs low-quality data or abnormal data. For example, outlier detection can filter out low-quality data or abnormal data. The intuition behind spatiotemporal outlier detection is that they reflect "discontinuity" on non-spatiotemporal attributes within a spatiotemporal neighborhood. Basic spatial outlier detection methods include visualization based approaches [12] and neighborhood based approaches [62].

The data analysis part performs data analysis algorithms to extract knowledge and discover new insights. Specifically, data analysis methods for spatiotemporal data typically include clustering [14], frequent pattern mining [51], anomaly detection [49], etc.

Data Application. After data collection and data processing, data is used to in various real life applications. The traditional spatiotemporal data model has been improved and expanded to meet the needs of various fields, such as calibration system trajectory, navigation, etc. With the development of spatiotemporal data research, there are many methods to improve trajectory similarity metrics, such as based on spatial geometry, spatiotemporal geometry, spatial models, spatiotemporal models [28], and clustering effectiveness index [64]. Besides, the knowledge representation of time and space have also been applied in the fields of robot vision [1]. And land data has typical temporal characteristics, the related data in land survey and cadastral management are represented as spatiotemporal data, and the study of land management issues through spatiotemporal modeling has also been widely used [31].

3 Applications of Smart Urban Rail

Due to the natural spatial and temporal correlation of urban rail data, in this part, we will discuss the typical applications of artificial intelligence in urban rail about spatiotemporal data from the four dimensions, including intelligent scheduling, intelligent operation platform, intelligent perception, and intelligent train control, and we made a chart to summarize related technologies as Fig. 1. Next, we will discuss this technologies in details.

3.1 Intelligent Scheduling

The traditional scheduling system constantly faces data pressure. Therefore, artificial intelligence technology in scheduling system is becoming more critical [43]. Intelligent scheduling obtains real-time driving information by collecting passenger flow information or rail traffic data, and corrects the deviation of trajectory position through operation map and vehicle information. The purpose of intelligent scheduling is to optimize the allocation of passenger resources and improve efficiency to ensure driving safety [80]. Following, we will introduce related artificial intelligence technologies from two aspects: delay prediction and delay propagation.

Train Delay Prediction. Train delay prediction and analysis directly serve the scheduling and command of high-speed railways. Accurate prediction of train arrival delay time can provide controller with more accurate information and improve scheduling optimization [79]. Due to the occurrence of disturbance events, trains are often delayed, and how to predict the delay information is an important topic.

The Smart Urban Rail	Intelligent Scheduling	Train Delay Forecast	Traditional models: Bayesian algorithm[6,35] Stochastic analysis[68]
			Machine learning: RNN,LSTM:[34,47] LSTM+3DCNN:[27] GRU+LSTM+GCN:. [69]
		Delay propagation	Machine learning: Support Vector Regression (SVR)[74] Mixed Integer Linear Program (MILP)[50] Random Forest[5] GCN[19]
	Intelligent Operation Platform	Traffic Flow Forecast	Traditional models: Linear stochastic algorithm[47,51] Genetic algorithm[22] Particle swarm optimization[15]
			Machine learning: PVAR+NN[78] LSTM[38] RNN+GNN[74,11] GNN+dynamic adjacency matrices[61,62,3]
		Traffic Trajectory Prediction	Traditional models: Sequential pattern mining[4,5] Simple Markov Modeling[17,45]
			Machine learning: RNN, LSTM[12,14, 40,64] GNN+RNN[54,40] Reinforcement Learning[53]
	Intelligent Perception		Traditional models: Support Vector Machine (SVM)[30] Hidden Markov Model (HMM)[58]
			Machine learning: Deep Belief Network(DBN)[8] CNN[47] RNN、LSTM[31] Graph Deviation Network (GDN)[13]
	Intelligent Train Control		Traditional models: Rule based[9,56] Imitation learning[7,23,67]
			Machine learning: CNN[24] Reinforcement learning[23,72] Reinforcement learning+Rule based[68,77] Reinforcement learning+LSTM[26]

Fig. 1. The summary of urban rail technology

Traditional models are based on domain knowledge and the judgments of experts and local authorities. This method mainly uses Stochastic Analysis models [43] and computational theories such as Fuzzy Network and Bayesian Network [11] to predict the propagation of train delays.

In terms of machine learning [42,82], the prediction error can be greatly improved since the spatiotemporal features of each rail station are considered in the prediction model. Traditional machine learning models mainly rely on feature engineering extraction, which is not obvious in feature extraction and spatiotemporal connections, and requires a lot of expert experience. In deep learning methods, CNN and RNN have strong ability in periodic modeling, can capture the periodicity, trend, and proximity of time series. [56] designed a model to predict the delay of the next train by taking into account the different states or characteristics of the k trains in front of the current station and feeding them into a network of RNNs as a sequence.

The temporal dependencies between sequences can be identified by inputting informative features of multiple historical trains, but this approach does not take into account the temporal and spatial dimensions of the interrelated data. There are often two structures for modeling traffic status from a spatial perspective: grid structure and graph structure. The grid structure is often modeled by a CNN-based method, and another is often modeled by a GNN-based method. [36] designed a network named CLF to model the factors related to non-time-series, time-series, and spatiotemporal features with FCNN, LSTM, and 3D CNN for delay prediction. With the development of graph neural network (GNN) on graph-structured data, the co-evolution of the graph is used to model spatial dependencies, and LSTM is used to model time-series dependencies. For example, [80] designed a spatiotemporal graph convolutional network (TSTGCN) model. By calculating the distance between stations, the author designed a weighted adjacency matrix as the spatial structure information, and the train data of different periods as the time series, input the both information into the GCN network with attention to predict the total delay times of the next station.

Delay Propagation. The cascading delays of the current rail network spread from primary sources are imperative for the rail system. The complex nonlinear interactions between various spatiotemporal variables determine the propagation of trains delays quickly spreading throughout the railway network, which will cause trains further severe disruptions [29]. They are inherently difficult to predict, due to the interaction of complex nonlinearities and spatiotemporal variables that govern delay propagation. Besides, GNNs only aggregate the information of the current site's leading sites, and it is difficult to comprehensively consider the effect on the site network by the GNNs.

The methods of traditional algorithms on delay propagation problems mainly use stochastic modeling and discrete distribution methods [4], such as the Monte Carlo sampling approach [78]. For example, [27] uses the Max-plus algebra algorithm to predict and analyze delay propagation problems.

There are many artificial intelligence works in train delay propagation prediction. For example, [8] uses random forest and feedforward neural network to predict arrival time in railway systems. [29] formulates a railway network as a line graph and uses a spatiotemporal graph convolutional network (STGCN) to predict the cascading delay of the entire railway network, where nodes represent railway connections and edges represent connecting stations.

Investigating the effectiveness of replenishment time is a key method for dispatchers to understand the delayed recovery capabilities of railway sections and stations, and it helps dispatcher improve real-time dispatch efficiency. [86] uses Support Vector Regression (SVR) algorithm to study the effectiveness of railway section and station replenishment time Properties. [60] employs a retiming and reordering strategy, utilizing a Mixed Integer Linear Program (MILP) model to minimize the weighted sum of arrival delay times for trains, and trains at all stations Delay times.

3.2 Intelligent Operation Platform

Operation platform refer to providing regular operation services for urban rail lines, which is the basis for ensuring the safe operation of urban rail. The intelligent subway vehicle system is an operation platform based on massive data, which faces the problems of huge data and complex data management. By combining the relevant algorithms of deep learning, it can solve the problems of intelligent data analysis [55], processing and application. Traffic flow prediction and trajectory prediction can help operation system improve efficiency, save train's energy, and make station become safer and more safety.

Traffic Flow Forecast. Traffic forecasting aims to predict future traffic states in a traffic network given a series of historical traffic states and a physical road network. The traffic volume prediction problem can be expressed as:

$$\hat{y}_{t+T'} = f\left([X_{t-T+1}, X_{t-T}, \dots, X_t]\right) \tag{1}$$

The goal is to find the model parameters that minimizing the error between the predicted traffic volume and the observed traffic volume. Accurate and timely traffic flow information can provide early warning information for the operation management system, improve traffic operation efficiency, and ensure stable and efficient traffic operation [37]. The current technologies mainly include experience-based expert algorithms, traditional artificial intelligence algorithms and deep learning algorithms.

The linear prediction methods of traditional traffic flow mainly use the time series model [55], linear programming [59], regression model and moving average method [87]. Nonlinear prediction methods mainly include neural networks [58], genetic algorithm [32], particle swarm optimization (PSO) [23], RBF neural networks [37], etc.

Traffic flow forecasting is a multivariate time series forecasting problem. [88] was the first to consider the advantages of linear and nonlinear models, and proposed a hybrid linear + nonlinear time series analysis model, using panel vector autoregression (PVAR) and neural network (NN) hybrid PVAR-NN forecasting method to predict the passenger flow of the railway system. [45] proposes a short-term passenger flow prediction model based on LSTM, which predicts short-term passenger flow changes by considering the fusion of inbound and passenger flow characteristics on the day.

In traffic flow prediction, graph geural getworks (GNN) is applied to traffic flow prediction by considering the spatiotemporal correlation in rail transit data [18,84]. Since GNN has a reliable ability to handle well-defined graph structures, we can combine GNNs with time series models such as CNNs or RNNs to capture some potential shared patterns of spatiotemporal data. This method have specific effects for modeling spatiotemporal data in some work. However, the predefined graph structure cannot contain complete information about spatial dependencies and it required expert knowledge for artificial definition, which may lead to severe biases. Therefore, self-learning algorithms that dynamically learn to obtain adjacency matrices from sequence data have become the focus parts of current research [73]. [2,72] and others work have designed different graph neural network frameworks taht can automatically extract the one-way relationship between variables through a graph learning module. These frameworks can integrate external knowledge (such as variable properties), process data dynamically without using explicit graph structure, capture graph structure knowledge, and obtain adaptive matrices for graph evolution.

Traffic Trajectory Prediction. The trajectory prediction task aims to predict the future driving trajectory of targets object based on the current or historical trajectory and environmental information state (e.g. pedestrians, vehicles, and other traffic participants). In spatiotemporal data mining, trajectory traffic prediction in various tasks plays an important role, therefore, a variety of methods have been proposed, ranging from sequential pattern mining [7,9] to the latest representative deep learning, from Simple Markov Modeling [25,53] to virtual location-based social networks.

Machine learning (ML) and deep learning (DL) based methods automatically extract relevant features from raw data. Some method [19,24,48,74] analyze the time series based on RNN, LSTM, etc. to predict distribution trajectory patterns. The application of Graph Neural Network (GNN) in trajectory prediction [47,63], mainly builds a spatiotemporal graph centered on the prediction target. It uses a sparse directed temporal graph to simulate the motion trend to estimate the predicted trajectory-the bidirectional Gaussian distribution. [61] uses reinforcement learning to simulate different types of agents of vehicles and pedestrians in the forward trajectory, and the purpose is to predict a large variety of agent trajectories with dynamic and synthetic data (e.g. semantic maps).

3.3 Intelligent Perception

The perception technology based on train can monitor the running status of urban rail trains in real-time and effectively predict potential safety hazards of trains through sensor network technology, which is an essential guarantee for the safe operation of urban rail trains. At present, most researchers use data-driven methods to analyze time-series data collected by sensor networks and predict train failures. And data-driven approaches are divided into statistical methods and deep learning methods.

The statistical methods can use Support Vector Machine (SVM), Hidden Markov Model (HMM), etc. For example, [39] has proposed to use the SVM model to predict faulty nodes in wireless sensor networks. And it is expensive to collect vibration signals from bearings since it requires external sensors. So, [67] proposed a bearing fault diagnosis method based on the motor current signal and introduced a genetic algorithm to select more critical features. Most of the features in statistical methods are manually designed and fixed. These features are not necessarily suitable for all tasks. Therefore, researchers have proposed deep learning methods to let the model learn how to extract effective features from the original data.

Deep learning methods can use models such as Deep Belief Network (DBN), Convolutional Neural Network (CNN), Recurrent Neural Network (RNN), etc. For example, [15] proposed a new multi-sensor data fusion technique. This method inputs features into multiple two-layer Sparse Autoencoder (SAE) neural networks for feature fusion. And [54] proposed a new deep one-dimensional convolutional neural network model (Der-1DCNN). This model draws on the idea of residual learning can effectively alleviate the training difficulties and performance degradation problems of deeper networks. Since RNN has natural advantages for processing time-series data such as vibration signals and the Long Short-Term Memory (LSTM) network improves the long-term dependence problem in RNN. Therefore, [40] introduced the LSTM network into the tasks of failure prediction and remaining service life estimation. This model is capable of diagnosing failures 24 h before the actual failure.

Most of the above methods only use the signal collected by one type of sensor. However, there are usually many kinds of sensors in a sensor network. Therefore, capturing the complex relationship between sensors is one of the critical issues. In response to this problem, [83] proposed a new model based on the graph attention network. This model treats the time series collected by each sensor as an independent feature and uses two graph attention layers to capture the causal and temporal dimension correlations among multiple features, respectively. And [21] proposes a Graph Deviation Network (GDN) by combining structural learning and Graph Neural Networks (GNN). This method captures the unique characteristics of each sensor by introducing a sensor embedding mechanism, and learns the interdependence between sensors through the graph structure learning module.

3.4 Intelligent Train Control

A key technology of intelligent train control is the study of artificial intelligence technology in train energy saving. In order to reduce the traction energy consumption of trains, the current optimization methods for strategies mainly include expert systems, supervision methods, and reinforcement methods.

Expert systems are built by extracting knowledge from human experts [65] and inferring rules from experts to make decisions. For example, [16] establishes an expert knowledge system based on the driving data and control rules of excellent drivers to achieve multi-objective control of trains. Imitation learning is also widely studied in intelligent train control [13,33,34,75]. For example, [75] proposed a new algorithm for intelligent train driving through online optimization by gradient descent. [33] leverages multiple connected agents to handle data availability issues for block chain-based federation, protecting the confidentiality of these data. Reinforcement learning [33,35,50,76,81,85] is also widely used in train control. [76] used reinforcement learning combined with expert system to ensure the security and convergence of the algorithm. This combination between reinforcement learning and expert systems is also explored in [85], where the expert system also ensures the security of the reinforcement learning algorithm.

[35] applies the BLSTM to output the best possible driving decision. By considering the intelligent operation of train control from multiple perspectives, an policy reinforcement based intelligent train control algorithm is proposed, which can reduce the energy consumption of trains, ensure punctuality and passenger comfort [81].

4 Future Work

At present, there is little research on the field of smart urban rail about trustworthy AI. Therefore, in this section, we give some interesting topics for future work of combining trustworthy AI with smart urban rail.

Safety and Robustness. Safety and Robustness means that a artificial intelligence system should achieve stable and high precision in different environments. A growing of work undermines system performance by attacking models [38,57,89]. Therefore, how to ensure the model's robustness has a significant impact on the safety of the smart urban rail.

Interpretability. The performance of smart urban rail systems can be achieved by increasing the complexity of the model [22]. However, deep learning models are black boxes because most models are too complex to be explained [46], which prevents them from being used in medical [52] and other systems [44]. Therefore, it is an interesting topic to establish interpretable models in the field of smart urban rail.

Privacy. Researchers have found that artificial intelligence algorithms can store and expose users' personal information [30]. The research of urban rail often utilizes a large number of users' information and space trajectories. Therefore, how to avoid revealing private information is a interesting topic.

Accountability. Accountability means who we should blame if any part of the AI performs badly. Real-world AI-based systems involve many vital components, including input data, algorithm theory, implementation details, real-time human control. These complex factors further complicate the determination of AI responsibility. It is necessary to define the responsibilities of various parties in the smart urban rail, allowing humans to enjoy the benefits and conveniences of AI with the assurance of accountability.

5 Conclusion

In this paper, we presented a comprehensive analysis of smart urban rail. The contributions of the presented research are diverse. We first presented the relevant background of the spatiotemporal data basis of smart urban rail. Then, we introduced the typical data analysis techniques in urban rail applications. Another key factor of our paper is that we discussed the various applications of artificial intelligence in urban rail from the four dimensions and made a chart to summarize related technologies. Finally, we also listed the interesting topics for the future work of smart urban rail.

References

1. Andrienko, N., Andrienko, G.: A visual analytics framework for spatio-temporal analysis and modelling. Data Min. Knowl. Discov. **27**(1), 55–83 (2013)
2. Bai, L., Yao, L., Li, C., Wang, X., Wang, C.: Adaptive graph convolutional recurrent network for traffic forecasting. arXiv preprint arXiv:2007.02842 (2020)
3. Banerjee, S., Carlin, B.P., Gelfand, A.E.: Hierarchical Modeling and Analysis for Spatial Data. CRC Press, Boca Raton (2014)
4. Berger, A., Gebhardt, A., Müller-Hannemann, M., Ostrowski, M.: Stochastic delay prediction in large train networks. In: 11th Workshop on Algorithmic Approaches for Transportation Modelling, Optimization, and Systems (2011)
5. Bo, W.: Smart urban rail: digital transformation and high-quality development china urban rail transit owners leaders summit 2021 In: Shenzhen Annual Meeting Held. China Metros, vol. 11, p. 4 (2021)
6. Borthakur, D.: The Hadoop distributed file system: architecture and design. Hadoop Pro. Website **11**(2007), 21 (2007)
7. Burbey, I., Martin, T.L.: A survey on predicting personal mobility. Int. J. Perv. Comput. Commun. **8** (2012)
8. Barbour, W., et al.: Prediction of arrival times of freight traffic on us railroads using support vector regression. Transp. Res. Part C Emerg. Technol. **93**, 211–227 (2018)
9. Calabrese, F., Di Lorenzo, G., Ratti, C.: Human mobility prediction based on individual and collective geographical preferences. In: 13th International IEEE Conference on Intelligent Transportation Systems, pp. 312–317 (2010)
10. Castro, P.S., Zhang, D., Chen, C., Li, S., Pan, G.: From taxi GPS traces to social and community dynamics. ACM Comput. Surv. **46**, 1–34 (2013)
11. Chao, W., Zhongcan, L., Ping, H., Rui, T., Weiwei, M., Li, L.: Progress and perspective of data driven train delay propagation. China Saf. Sci. J. **29**(S2), 1 (2019)
12. Chen, D., Lu, C.T., Kou, Y., Chen, F.: On detecting spatial outliers. Geoinformatica **12**(4), 455–475 (2008)
13. Chen, D., Gao, C.: Soft computing methods applied to train station parking in urban rail transit. Appl. Soft Comput. **12**(2), 759–767 (2012)
14. Chen, X.C., Faghmous, J.H., Khandelwal, A., Kumar, V.: Clustering dynamic spatio-temporal patterns in the presence of noise and missing data. In: IJCAI (2015)
15. Chen, Z., Li, W.: Multisensor feature fusion for bearing fault diagnosis using sparse autoencoder and deep belief network. IEEE Trans. Instrum. Measur. **66**(7), 1693–1702 (2017)
16. Cheng, R., Chen, D., Cheng, B., Zheng, S.: Intelligent driving methods based on expert knowledge and online optimization for high-speed trains. Exp. Syst. Appl. **87**, 228–239 (2017)
17. Cressie, N.: Statistics for spatial data. John Wiley & Sons, New York (2015)
18. Cui, Z., Henrickson, K., Ke, R., Wang, Y.: Traffic graph convolutional recurrent neural network: a deep learning framework for network-scale traffic learning and forecasting. IEEE Trans. Intell. Transp. Syst. **21**(11), 4883–4894 (2019)
19. De Brébisson, A., Simon, É., Auvolat, A., Vincent, P., Bengio, Y.: Artificial neural networks applied to taxi destination prediction. arXiv preprint arXiv:1508.00021 (2015)
20. Dean, J., Ghemawat, S.: Mapreduce: simplified data processing on large clusters. Commun. ACM **51**(1), 107–113 (2008)

21. Deng, A., Hooi, B.: Graph neural network-based anomaly detection in multivariate time series. Proc. AAAI Conf. Artif. Intell. **35**(5), 4027–4035 (2021)
22. Doshi-Velez, F., Kim, B.: Towards a rigorous science of interpretable machine learning. arXiv (2017)
23. Dunwei, G., Yong, Z., Jianhua, Z., Yong, Z.: Novel particle swarm optimization algorithm. Control Theory Appl. **25**(1), 5 (2008)
24. Feng, J., et al.: DeepMove: predicting human mobility with attentional recurrent networks. In: Proceedings of the 2018 World Wide Web Conference, pp. 1459–1468 (2018)
25. Gambs, S., Killijian, M.O., del Prado Cortez, M.N.: Show me how you move and I will tell you who you are. In: Proceedings of the 3rd ACM SIGSPATIAL International Workshop on Security and Privacy in GIS and LBS, pp. 34–41 (2010)
26. Gelfand, A.E., Diggle, P., Guttorp, P., Fuentes, M.: Handbook of Spatial Statistics. CRC Press, Boca Raton (2010)
27. Goverde, R.M.: A delay propagation algorithm for large-scale railway traffic networks. Transpo. Res. Part C Emerg. Technol. **18**(3), 269–287 (2010)
28. Han, Z., et al.: Calibrating trajectory data for spatio-temporal similarity analysis. VLDB J. Int. J. Very Large Data Bases **24**(1), 93–116 (2015)
29. Heglund, J.S., Taleongpong, P., Hu, S., Tran, H.T.: Railway delay prediction with spatial-temporal graph convolutional networks. In: 2020 IEEE 23rd International Conference on Intelligent Transportation Systems (ITSC), pp. 1–6 (2020)
30. Henderson, P., et al.: Ethical challenges in data-driven dialogue systems. In: Proceedings of the 2018 AAAI/ACM Conference on AI, Ethics, and Society, pp. 123–129 (2018)
31. Heo, J.: Development and implementation of a spatio-temporal data model for parcel-based land information systems. Ph.D. thesis, The University of Wisconsin - Madison (2001)
32. Hongjiang, C., Kui, F.: Research on clustering search method in collaborative filtering recommendation system. Comput. Eng. Appl. **50**(5), 16–20 (2014)
33. Hua, G., Zhu, L., Wu, J., Shen, C., Zhou, L., Lin, Q.: Blockchain-based federated learning for intelligent control in heavy haul railway. IEEE Access **8**, 176830–176839 (2020)
34. Huang, J., Liu, Y., Xia, Y., Zhong, Z., Sun, J.: Train driving data learning with s-CNN model for gear prediction and optimal driving. In: 2019 Chinese Automation Congress (CAC), pp. 2227–2232 (2019)
35. Huang, J., Zhang, E., Zhang, J., Huang, S., Zhong, Z.: Deep reinforcement learning based train driving optimization. In: 2019 Chinese Automation Congress (CAC), pp. 2375–2381 (2019)
36. Huang, P., Chao, W., Fu, L., Peng, Q., Tang, Y.: A deep learning approach for multi-attribute data: A study of train delay prediction in railway systems. Inf. Sci. **516**, 234–253 (2019)
37. Jie, F., Hong, H.: Prediction of railway passenger traffic volume based on verhulst-RBF. Railway Comput. Appl. **28**(11), 5 (2019)
38. Jin, W., Ma, Y., Liu, X., Tang, X., Wang, S., Tang, J.: Graph structure learning for robust graph neural networks. In: Proceedings of the 26th ACM SIGKDD International Conference on Knowledge Discovery & Data Mining, pp. 66–74 (2020)
39. Kharade, S.S., Khiani, S.: Fault prediction and relay node placement in wireless sensor network-a survey. Int. J. Sci. Res **3**(10), 702–704 (2014)
40. Kizito, R., Scruggs, P., Li, X., Devinney, M., Jansen, J., Kress, R.: Long short-term memory networks for facility infrastructure failure and remaining useful life prediction. IEEE Access **9**, 67585–67594 (2021)

41. Kyriakidis, P.C., Journel, A.G.: Geostatistical space-time models: a review. Math. Geol. **31**(6), 651–684 (1999)
42. Lee, W.H., Yen, L.H., Chou, C.M.: A delay root cause discovery and timetable adjustment model for enhancing the punctuality of railway services. Transp. Res. Part C Emerg. Technol. **73**(Dec.), 49–64 (2016)
43. Lessan, J., Fu, L., Wen, C.: A hybrid Bayesian network model for predicting delays in train operations. Comput. Ind. Eng. **127**, 1214–1222 (2019)
44. Levinson, J., et al.:Towards fully autonomous driving: Systems and algorithms. In: IEEE (2011)
45. Li, Y., Yin, M., Zhu, K.: Short term passenger flow forecast of metro based on inbound passenger plow and deep learning. In: 2021 International Conference on Communications, Information System and Computer Engineering (CISCE), pp. 777–780 (2021). https://doi.org/10.1109/CISCE52179.2021.9446016
46. Linardatos, P., Papastefanopoulos, V., Kotsiantis, S.: Explainable AI: a review of machine learning interpretability methods. Entropy (Basel, Switzerland) **23**(1), 18 (2020)
47. Liu, B., Adeli: Spatiotemporal relationship reasoning for pedestrian intent prediction. In :IEEE Robotics and Automation Letters, pp. 3485–3492 (2020)
48. Liu, Q., Wu, S., Wang, L., Tan, T.: Predicting the next location: a recurrent model with spatial and temporal contexts. In: Thirtieth AAAI Conference on Artificial Intelligence (2016)
49. Liu, W., Zheng, Y., Chawla, S., Yuan, J., Xie, X.: Discovering spatio-temporal causal interactions in traffic data streams. In: KDD (2011)
50. Liu, W., Tang, T., Su, S., Cao, Y., Bao, F., Gao, J.: An intelligent train control approach based on the Monte Carlo reinforcement learning algorithm. In: 2018 21st International Conference on Intelligent Transportation Systems (ITSC), pp. 1944–1949 (2018)
51. Lynch, H.J., Moorcroft, P.R.: A spatiotemporal Ripley's k-function to analyze interactions between spruce budworm and fire in British Columbia, Canada. Cana. J. Forest Res. **38**, 3112–3119 (2008)
52. Miotto, R., Fei, W., Shuang, W., Jiang, X., Dudley, J.T.: Deep learning for healthcare: review, opportunities and challenges. Brief. Bioinform. **19**(6) (2017)
53. Monreale, A., Pinelli, F., Trasarti, R., Giannotti, F.: WhereNext: a location predictor on trajectory pattern mining. In: Proceedings of the 15th ACM SIGKDD International Conference on Knowledge Discovery and Data Mining, pp. 637–646 (2009)
54. Peng, D., Liu, Z., Wang, H., Qin, Y., Jia, L.: A novel deeper one-dimensional CNN with residual learning for fault diagnosis of wheelset bearings in high-speed trains. IEEE Access **7**, 10278–10293 (2018)
55. Peng Hui, Z.Y., Zhanghao, H.: Railway passenger volume forecast based on multiple linear regression model. J. Chong. Insti. Technol. **32**(09), 190–193 (2018)
56. Ping, H., Chao, W., Zhongcan, L., Yuxiang, Y., Qiyuan, P.: A neural network model for real-time prediction of high-speed railway delays. China Saf. Sci. J. **29**(S1), 20 (2019)
57. Qi, L., Wu, L., Chen, P.Y., Dimakis, A.G., Witbrock, M.: Discrete attacks and submodular optimization with applications to text classification. arXiv preprint arXiv:1812.00151 (2018)
58. Qiusheng, T., Peng, C., Na, L.: Short time forecasting of passenger flow in urban railway using GSO-BPNN method. Technol. Econ. Areas Commun. **19**(1), 5 (2017)

59. Qiyuan, P., Jia, N., Gongyuan, L.: Model and algorithm for train platform scheme rescheduling at large high-speed railway station. J. China Railway Soc. **41**(1), 10 (2019)
60. Rößler, D., Reisch, J., Hauck, F., Kliewer, N.: Discerning primary and secondary delays in railway networks using explainable AI. Transpo. Res. Procedia **52**, 171–178 (2021)
61. Salzmann, T., Ivanovic, B., Chakravarty, P., Pavone, M.: Trajectron++: multi-agent generative trajectory forecasting with heterogeneous data for control (2020)
62. Shekhar, S., Lu, C., Zhang, P.: Graph-based outlier detection: algorithms and applications (a summary of results). In: Proceedings of the Seventh ACM SIGKDD International Conference on Knowledge Discovery and Data Mining (2001)
63. Shi, L., et al.: SGCN: sparse graph convolution network for pedestrian trajectory prediction. In: Proceedings of the IEEE/CVF Conference on Computer Vision and Pattern Recognition, pp. 8994–9003 (2021)
64. Takenaka, H., Fujii, Y.: A compact representation of spatio-temporal slip distribution on a rupturing fault. J. Seismol. **12**(2), 281–293 (2008)
65. Tan, C.F., Wahidin, L., Khalil, S., Tamaldin, N., Hu, J., Rauterberg, G.: The application of expert system: A review of research and applications. ARPN J. Eng. Appl. Sci. **11**(4), 2448–2453 (2016)
66. Tedjopurnomo, D.A., Bao, Z., Zheng, B., Choudhury, F., Qin, A.: A survey on modern deep neural network for traffic prediction: Trends, methods and challenges. IEEE Trans. Knowl. Data Eng. **34** (2020)
67. Toma, R.N., Prosvirin, A.E., Kim, J.M.: Bearing fault diagnosis of induction motors using a genetic algorithm and machine learning classifiers. Sensors **20**(7), 1884 (2020)
68. Tovar, E., Vasques, F.: Using worldFIP networks to support periodic and sporadic real-time traffic. In: IECON 1999. Conference Proceedings. 25th Annual Conference of the IEEE Industrial Electronics Society (Cat. No. 99CH37029), vol. 3, pp. 1216–1221 (1999)
69. Tzeng, C.B., Wey, T.S., Ma, S.H.: Building a flexible energy management system with LonWorks control network. In: 2008 Eighth International Conference on Intelligent Systems Design and Applications, vol. 3, pp. 587–593 (2008)
70. Vlahogianni, E.I., Karlaftis, M.G., Golias, J.C.: Short-term traffic forecasting: where we are and where we're going. Transpo. Res. Part C Emerg. Technol. **43**, 3–19 (2014)
71. Wang, X., Zhou, X., Lu, S.: Spatiotemporal data modelling and management: a survey. In: Proceedings 36th International Conference on Technology of Object-Oriented Languages and Systems. TOOLS-Asia 2000, pp. 202–211 (2000)
72. Wu, Z., Pan, S., Long, G., Jiang, J., Chang, X., Zhang, C.: Connecting the dots: multivariate time series forecasting with graph neural networks. In: KDD 2020: The 26th ACM SIGKDD Conference on Knowledge Discovery and Data Mining (2020)
73. Wu, Z., Pan, S., Long, G., Jiang, J., Zhang, C.: Graph waveNet for deep spatial-temporal graph modeling. arXiv preprint arXiv:1906.00121 (2019)
74. Yao, D., Zhang, C., Huang, J., Bi, J.: SERM: a recurrent model for next location prediction in semantic trajectories. In: Proceedings of the 2017 ACM on Conference on Information and Knowledge Management, pp. 2411–2414 (2017)
75. Yin, J., Chen, D.: An intelligent train operation algorithm via gradient descent method and driver's experience. In: 2013 IEEE International Conference on Intelligent Rail Transportation Proceedings, pp. 54–59 (2013)

76. Yin, J., Chen, D., Li, L.: Intelligent train operation algorithms for subway by expert system and reinforcement learning. IEEE Trans. Intell. Transpo. Syst. **15**(6), 2561–2571 (2014)
77. Yin, X., Wu, G., Wei, J., Shen, Y., Qi, H., Yin, B.: Deep learning on traffic prediction: methods, analysis and future directions. IEEE Trans. Intell. Transp. Syst. (2021)
78. Yuan, J., Goverde, R., Hansen, I.: Propagation of train delays in stations. WIT Trans. Built Environ. **61** (2002)
79. Yuan, J., Hansen, I.A.: Optimizing capacity utilization of stations by estimating knock-on train delays. Transpo. Res. Part B Methodol. **41**(2), 202–217 (2007)
80. Zhang, D., Peng, Y., Zhang, Y., Wu, D., Wang, H., Zhang, H.: Train time delay prediction for high-speed train dispatching based on spatio-temporal graph convolutional network. IEEE Trans. Intell. Transp. Syst. **23**, 2434–2444 (2021)
81. Zhang, M., Zhang, Q., Boyuan, Z.: A policy-based reinforcement learning algorithm for inteligent train control. J. China Railway Soc. (2020)
82. Zhang Qi, Chen Feng, Z.T.Y.Z.M.: ntelligent prediction and characteristic recognition for joint delay of high speed railway trains. Acta Automatica Sinica, **45**(12) (2019)
83. Zhao, H., et al.: Multivariate time-series anomaly detection via graph attention network. In: 2020 IEEE International Conference on Data Mining (ICDM), pp. 841–850 (2020)
84. Zhao, L., Song, Y., Zhang, C., Liu, Y., Wang, P., Lin, T., Deng, M., Li, H.: T-GCN: a temporal graph convolutional network for traffic prediction. IEEE Trans. Intell. Transpo. Syst. **21**(9), 3848–3858 (2019)
85. Zhou, K., Song, S., Xue, A., You, K., Wu, H.: Smart train operation algorithms based on expert knowledge and reinforcement learning. In: IEEE Transactions on Systems, Man, and Cybernetics: Systems (2020)
86. Zhou, M., Dong, H., Liu, X., Zhang, H., Wang, F.Y.: Integrated timetable rescheduling for multidispatching sections of high-speed railways during large-scale disruptions. In: IEEE Transactions on Computational Social Systems (2021)
87. Zhu, H.Y.: N days average volume based ARIMA forecasting model for shanghai metro passenger flow. In: 2010 International Conference on Artificial Intelligence and Education (ICAIE) (2010)
88. Zhu, R., Zhou, H.: Railway passenger flow forecast based on hybridPVAR-NN model. In: 2020 IEEE 5th International Conference on Intelligent Transportation Engineering (ICITE) (2020)
89. Zügner, D., Günnemann, S.: Adversarial attacks on graph neural networks via meta learning. arXiv preprint arXiv:1902.08412 (2019)

Fast Vehicle Track Counting in Traffic Video

Ruoyan Qi[1(✉)], Ying Liu[1], Zhongshuai Zhang[1], Xiaochun Yang[2], Guoren Wang[1], and Yingshuo Jiang[3]

[1] Beijing Institute of Technology, Beijing, China
3220190857@bit.edu.cn
[2] Northeastern University, Liaoning, China
yangxc@mail.neu.edu.cn
[3] Shenyang Jianzhu University, Liaoning, China
jiangq16@126.com

Abstract. In order to reduce road congestion, measures such as setting smart signal time schedules are needed. Therefore, it is a key technology to effectively and accurately count the traffic flow of vehicles at various intersections in the surveillance video. The method we propose uses Scaled-YOLOv4 as the vehicle detector, and then implements the vehicle tracking based on the DEEP SORT algorithm. To improve the accuracy and efficiency of the system, we propose a strategy of dynamic frame skipping based on density. We also propose to set key areas, combined with the driving direction, angle, etc., to judge and count the behavior of the vehicle. Experiments show that our method improved system efficiency while remaining high accuracy.

Keywords: Vehicle count · Track count · Traffic monitoring

1 Introduction

With the increase of traffic monitoring and vehicles, vehicle track identification, query and counting become the key concerns to alleviate traffic problems. It is very necessary to recognize and count vehicle tracks, which plays an important role in reducing traffic congestion and improving the use efficiency of signal lights [14]. The objective of vehicle counting by movements of interest (MOI) is the number of vehicles that correspond to the MOI over a period of time on monitored vehicle tracks. The MOI can be pre-defined in combination with possible vehicle states and different motion states. One vehicle query scenario is shown in Fig. 1, which shows 12 tracks (red lines), each of which is an MOI, and we want to find the number of vehicles with similar tracks for all these MOIs. It is still a great challenge to quickly answer the number of tracks corresponding to the movements in the video, due to road conditions, weather conditions, shielding between vehicles and other factors, as well as the large number of monitors.

The traditional vehicle counting algorithms follow two strategies. One strategy is frame-based vehicle counting [3], which only considers the number of vehicles in a single frame, without considering the vehicle states and how to count vehicles in the case of video. Two solutions to this problem can be applied, namely, density sensing strategy and deep learning based target detection strategy. The density sensing algorithm is

U. K. Rage et al. (Eds.): DASFAA 2022 Workshops, LNCS 13248, pp. 244–256, 2022.
https://doi.org/10.1007/978-3-031-11217-1_18

Fig. 1. Visual presentation of MOI. The red line is the target track of the query. (Color figure online)

mainly used to carry out regression on the number of vehicles [1]. In addition, with the popularity of deep learning [9, 10], more and more researches turn to target detection strategies based on deep learning, which detect vehicles first and then count detected vehicles. Another strategy is to instantiate the vehicle count, which is to count vehicles in consecutive frames. Therefore, the knowledge of target re-recognition and vehicle tracking can be used to accurately count the blocked vehicles and unrecognized vehicles. The above counting methods mainly follow the detection-track-counting (DTC) framework [2,7]. Specifically, the targets are detected first, then multi-target tracking is carried out according to the detection results, and finally the vehicles are counted according to the tracking results.

Different from the traditional vehicle counting problem based on query predicates, this paper mainly solves the track counting problem on all specific MOIs in the traffic video. Specifically, tracking and counting tracks require not only the total number of vehicles per MOI, but also the time stamp of vehicles leaving the monitoring area. It takes a huge amount of calculation to detect, track, and count number of vehicles in the video. Through our experiments, we find that vehicle detection module will take approximately over 70% time of the whole system execution time. Also directly adopting the track tracking algorithm could result in unnecessary calculation for tracking some invalid vehicles, which are not needed by our method. The main idea to improve the counting time is to avoid detecting every frame as well as avoid unnecessary areas in the video. On the one hand, the vehicle behavior of adjacent frames is similar, which can be detected by frame skipping. On the other hand, for the track in the MOI that we want to query, we can exclude the area that such track will not go through in advance and do not track the vehicles in this area.

The focus of this paper is to propose a suitable method to improve the efficiency of vehicle track counting tasks. The method proposed in this paper is based on DTC framework, selects Scaled-YOLOv4 [12] as the target detector, and bases on the DEEP SORT [13] for vehicle tracking. On the basis of traditional DTC, this paper firstly determines the key areas of concern, that is, the areas that only the tracks that meet the MOI will pass through. Moreover, since the adjacent frames in the video are similar, we intend to avoid the detection of such similar frames by frame skipping. However, how

(a) Dense vehicles (b) Sparse vehicles

Fig. 2. Changes in traffic flow one minute apart.

to skip frames is a difficult choice. We found that there were moments when there were as many as 50 vehicles in the video, but after a minute it dropped to a dozen. We show it in Fig. 2. If a small frame skipping amplitude is used, there is still room for increasing the amplitude when there are few vehicles. However, if it is increased to be suitable for sparse vehicles, the accuracy of vehicle matching will drop for dense scenes. Therefore, the degree of traffic congestion will affect the size of skip frames. When there are fewer road vehicles, the number of skipped frames can be larger, while when there are more road vehicles, the number of skipped frames should be smaller. At the same time, we use Kalman filter to track the vehicle, and the driving direction is determined by the target's position and speed information. In addition, for the track counting problem, our system finds all the tracks of the target in the video as well as the time stamp when the vehicle leaves the monitoring area. If the corresponding track cannot be determined by key areas and driving direction, we then use the nearest neighbor algorithm to compare it with the typical track of each MOI, judge the behavior of the track, and mark the corresponding track. Our method was evaluated on the AICity 2021 Track-1 dataset. Experimental results show that the proposed method improves the efficiency and accuracy of vehicle tracking counting.

The target of our work is to improve the efficiency of vehicle counting while accurately identifying vehicle tracks with such a large volume of video. We have tried some approaches to this problem, and the main challenges and contributions of this paper are as follows:

- Ideally, we hope to carefully choose partial frames to answer the counting queries. However, it is hard to choose an appropriate fixed skip step on the video. Therefore, we propose a dynamic frame skipping approach that we could adaptively choose frames to answer the query effectively and efficiently.
- Due to the use of dynamic frame skipping or the existence of occlusion, there may be the situations where the newly appeared vehicle is quite different from the previous one, so the simple appearance features (direction and shape) of the vehicles are added to help the matching.
- An efficient and accurate three-step cascade structure is proposed to judge the MOI to which the track belongs. The key areas are set to quickly judge or narrow the range based on whether the vehicle passes by, and then the direction of the vehicle as well as the nearest neighbor algorithm are used for further judgement.

2 Related Work

Recently, there have been several systems proposed to handle queries over massive video data.

Kang et al. [5] proposed a system named NoScope with a cascaded structure. NoScope could quickly identify whether a vehicle was contained in a video frame. NoScope first applies weak classifiers including image difference detector and cheap CNNs on each frame, which has the advantage of increasing object detection speed and reducing the high cost of deep CNNs. Due to their limited judgment ability, sometimes the weak classifiers do not guarantee enough confidence on the accuracy of the results. Noscope then applies deep CNNs on the frames to ensure accuracy. The disadvantage of NoScope is that it can only query simple binary classification tasks for limited object categories due to weak classifiers. Focus [4] fills the gap and divides the entire processing into two phases: ingesting time and query time. When ingesting the live video stream, Focus uses cheap CNNs to build construct approximate indexes (e.g., object categoryies, colors) of all possible objects for later querying. When querying, Focus uses these approximate indexes to answer queries, accelerating the response speed similar to a traditional database. However, the disadvantage of Focus is that the indexes it builds only serve a specific query, which means Focus needs to build new indexes when answering new queries. Kang et al. [6] found an opportunity that many queries only need the same schema to be answered. For example, with a schema including vehicle types and instances in per frame of video, the system can answer both an aggregation query counting the number of cars and a selection query selecting cars in frames. To leverage this opportunity, Kang et al. constructed task-agnostic indexes in TASTI system which can answer many downstream queries efficiently without constructing a new index.

However, although these systems could conduct above query missions well, they could not handle tasks that require the system to view consecutive frames simultaneously. Specifically, the above systems can't handle vehicle track counting tasks. These systems focus on queries that can be conducted within a frame, but vehicle track counting requires matching vehicles across multiple consecutive frames.

In order to count vehicle tracks, in [11], a new approach called RR is proposed. It first detects and tracks multiple vehicle in video frames, and then conducts track counting depending on their movements. To improve system response efficiency, RR adopts frame skipping strategy to speed up. However, as mentioned before, as the density of the vehicles changing frequently, a fixed frame skipping amplitude in RR is either too large or too small. We notice that there is still room for improvement towards the trade-off between system efficiency and accuracy.

3 Preliminary and Problem Statement

To query and count vehicles of specific MOIs, especially for large amounts of surveillance video data, is itself a complex and time-consuming task. Vehicle detection and tracking could take a large fraction of time if all the targets in the video are tracked without selection, since the counting system carries out too much unnecessary calculation. Thus, in this paper, we focus not only the query accuracy, but also the efficiency of the every part of the system.

Formally, the input of the problem can be expressed as: MOIs $M = \{t_1, ..., t_p\}$, where t_j $(1 \leq j \leq p)$ is the j-th pre-marked track representing the j-th MOI and p is the total number of MOIs, as well as a video $v = \{f_1, \ldots, f_n\}$, where f_i $(1 \leq i \leq n)$ is the i-th video frame and n is the total number of frames in the video. The objective of the problem is to count the number of vehicles that have similar tracks with MOI-specific tracks.

And the goal of this paper is to improve the efficiency of reaching the objective while maintaining high accuracy.

4 An Efficient Vehicle Query Counting Method

Intuitively, a basic approach to count the number of vehicles that conforms to specific MOIs is: firstly, using the vehicle detection algorithm to capture the vehicle targets [8], where a frame skipping method will generally be adopted to raise the speed, and then using the vehicle tracking algorithm to match the tracks of vehicles, and finally judging and counting based on the shape of the tracks.

Due to the constantly changing vehicle density, it is difficult to choose an appropriate frame skip amplitude in the detection module. To solve this problem, we propose a density-based dynamic frame skipping method. Besides, in the vehicle tracking module, the CNN appearance features and the position of a vehicle are traditionally used for matching tracks, but the extraction and matching of the CNN appearance features is time-consuming. Therefore, we choose to compare the positions of the vehicle first, after which the remaining unsuccessful matches are corrected based on the simple appearance (direction and shape) of the vehicle. As for vehicle counting, generally speaking, after the vehicle leaves the camera, the complete track of the vehicle could be obtained and compared with the pre-marked track. However, we notice that the vehicle can be determined which MOI it belongs to as early as it travels to some certain area (i.e., the key areas), or at least the possible MOIs it may belong to. The driving direction of the vehicle can further help to narrow the range. After such selection, the unmatched tracks, if any, could be determined in a quite small range of MOIs.

Our system is in tracking-by-detection mode, and the system is divided into three parts: vehicle detection, vehicle tracking, and track query counting. The vehicle detection algorithm first detects the targets in the frame, and then vehicle tracking algorithm matches the tracks of the previous frame. After getting the mark that a track leaves the video, the movement category of the track is judged and counted.

4.1 Adaptively Choosing Frames for Vehicle Detection

In the vehicle detection part, we choose Scaled-YOLOv4 [12] to detect vehicles. We use a pre-trained Scaled-YOLOv4 model that was trained on the COCO dataset, and then prune it based on the manually annotated vehicle dataset. Based on this method, we adopt dynamic frame skipping to improve the speed of vehicle detection.

Dynamic Frame Skipping. Under traditional settings, the vehicle tracking and counting system reads the video frame by frame, and then performs vehicle detection and tracking also for each frame. However, according to Sect. 1, vehicle detection will occupy more than half of the execution time of the entire system, therefore, reducing the number of times of vehicle detection will greatly improve the query speed.

Intuitively, we could skip some of the frames at a certain rate since the video contains a consecutive sequence of frames. However, it is in a dilemma as to choose a good skipping amplitude. If the frame skipping amplitude is too large, the increased distance between the positions of the vehicle in different frames will enlarge the uncertainty in matching tracks, resulting in a drop in accuracy. Meanwhile a small skipping amplitude will otherwise not achieve a proper acceleration effect. Therefore, it is necessary to adjust an appropriate frame skipping frequency through experiments. However, even for the same video, due to traffic lights, time, weather, etc., the traffic flow is constantly changing, so it is more appropriate to skip frames in a dynamic frame-skipping manner. Our work adaptively adjusts the frame skipping amplitude based on the current frame vehicle density. Vehicle density can be divided into four types: empty, scarce, normal and dense. By default, frame skipping is performed based on the normal vehicle density. When the traffic flow is going up or down, the frame skipping amplitude will immediately decrease or increase accordingly. Compared with skipping frames at the same frame skipping amplitude, dynamic frame skipping based on vehicle density can ensure the accuracy and improve the speed.

4.2 Vehicle Tracking Based on Location and Simple Appearance Features

The vehicle tracking part is based on the DEEP SORT [13] algorithm and modified on this basis.

Motion Prediction. Same as DEEP SORT [13], we use Kalman filter to predict the track's center point position (u, v), width to height ratio s, height r, and their respective velocity.

Data Association. After Kalman filter predicts the position, the tracks need to be matched with the vehicles detected by the vehicle detector. Using the IoU metric, the tracks and the targets can be efficiently matched by calculating the coincidence of the predicted position and the detected position. However, due to occlusion, dynamic frame skipping, etc., the vehicle displacement is often too large to be matched. For these unmatched situations, the feature similarity calculation of direction and shape is performed. The features of direction and shape are obtained by Kalman filter.

For a target that has still not been matched, we initialize a track for it. For tracks that are not matched, we mark the disappearance time. Tracks exceeding the disappearance time limit will be deleted.

4.3 Cascade Track Judgment and Counting

Firstly, judging which MOI the track belongs to based on whether the track has appeared in the corresponding key area. If no judgment can be made on the track in the key area part or multiple possibilities are reserved, further judgment is made in combination

(a) Case 1 (b) Case 2

Fig. 3. Two cases of key areas.

with the direction of the track. Finally, nearest neighbor analysis is performed on the pre-marked classic tracks of each MOI with the remaining tracks.

Key Area. As for the selection of key areas, there are two cases that need to be considered: (1) By visualizing the center point of each track, we could find out the unique area of each MOI in the video, and use this area as the key area of the MOI. Once a track passes through this area, mark it belong to this MOI. (2) For complex road conditions such as intersections, where an unique area cannot be found, select some key areas and put the MOIs that may pass through these areas into the marker array. If a track passes through these reduplicative key regions, the MOIs corresponding to different regions are considered. The key areas in these two cases are shown in Fig. 3.

Direction. The possible driving directions of tracks passing through the same area could be completely opposite. Therefore, when judging which MOI the track belongs to, adding the direction and angle of the track can further help the judgment. We can obtain the driving direction and angle by the Kalman filter algorithm.

Nearest Neighbor. When leaving the video, if the MOI of the track is still not determined, the nearest neighbor algorithm is then used to judge. Intuitively, the tracks with the same MOI are more similar. We pre-mark classic tracks of each MOI, and calculate the distance between the classical tracks of the possible MOIs remaining after the previous judgment and the historical center points of the track.

5 Experiments

5.1 Settings

The experiments are carried out on a PC with 2.80 GHz Intel Core i5-8400 CPU, GeForce GTX 1060 GPU, Ubuntu 16.04 LTS platform, 7.70 GB RAM, and 6G GPU memory. The code is implemented using the PyTorch 1.10 framework with CUDA 10.2 and Python 3.7.

We use AI CITY CHALLENGE 2021 track 1 dataset [8]. Videos are captured from 20 different vantage points (including intersection single approaches, full intersections, highway segments and city streets) covering various lighting and weather conditions (including dawn, rain and snow). Videos are 960p or better and most have been captured at 10 frames per second. Since the competition is over, we could only use the data set A (5 h in total) in it for evaluation.

5.2 Metrics

We use F1-score as a metric to evaluate the accuracy of the system. And use the total execution time of the system as the metric of system operation efficiency. When comparing the running time of different methods under different cameras, the running time per unit of video time is selected as the metric because the duration of the video is different.

$$Time\ Rate = \frac{time}{video\_total\_time}$$

To measure the tracking accuracy, we compare tracks produced by an vehicle tracking counting method with hand-labeled tracks in terms of precision and recall. Define TP as the number of correct traces (true positive), FP as the traces that did not blong to but were counted (false positive), and FN as the number of traces that should be counted but were missed (false negative). F1-score is the weighted harmonic mean of precision and recall.

5.3 Experimental Results

We first evaluate the methods in terms of the efficiency and accuracy that they provide for specific MOIs query over the 3 h video dataset where we have hand-labeled query outputs. We evaluate the effciency by measuring the total running time of the methods. The compared approaches are listed as follows.

- RR [11]. It is the winning method of AI CITY CHALLENGE track1, whose source code is available at https://github.com/ThangLehcmute/AIC21-track1-team19.
- TC. It is our proposed approach that sets dynamically skipping frames based on vehicle density, where the skip is set to 2 for normal density scenes, and 7, 3, 1 for empty, scarce and dense scenes. The approach also uses the cascade judgment of key areas, direction and the nearest neighbor method.

Their corresponding precision, recall, F1-score, and the total counting time are showed in Table 1. And we use the bold font to highlight the best item in each metric. TC is better than RR [11] in both accuracy and efficiency. The RR method maintains high accuracy, but the skip amplitude is fixed as 1 or 2, which is still less efficient than dynamic frame skipping based on initial skip equaling to 2.

After the overall efficiency comparison, in order to further analyze the efficiency performance in different modules of each method, we measured the execution time of each key part: vehicle detection, vehicle tracking and track judging. Table 2 shows the execution time of the three modules of the system. Compared with RR, TC has

Table 1. Comparison of different metrics.

Method	Precision	Recall	F1-score	Total counting time
RR	0.9875	0.9751	0.9811	2, 371 s
TC	**0.9894**	**0.9823**	**0.9858**	**2,053 s**

Table 2. Comparison of the time spent in different modules.

Method	Detection time	Tracking tme	Judging time	Total counting time
RR	2, 166 s	169 s	36 s	2, 371 s
TC	**1,873 s**	**150 s**	**30 s**	**2,053 s**

improved the efficiency of each module, among which the efficiency of vehicle detection has improved the most.

In order to compare the impact of dynamic frame skipping on accuracy and efficiency, we measured the accuracy and efficiency of fixed frame skipping and dynamic frame skipping in the videos shot by 20 cameras for comparison. Figures 4 and 5 show the precision and recall scores of different cameras under different frame skipping amplitudes, respectively. From these two figures we can see that among different cameras, the overall best performance of precision and recall is obtained by dynamic skip frame and fixed skip = 1. Figure 6 shows the time rate metric of different cameras under different amplitudes. It can be seen that dynamic frame skipping is much faster than fixed skip = 1. Therefore, we can conclude that dynamic frame skipping has better overall efficiency and accuracy than fixed frame skipping.

5.4 Ablation Experiments

In order to better reflect the impact of dynamic frame skipping and three-level cascade judgment on accuracy and efficiency, we performed ablation experiments. Several cases of ablation experiments are listed below:

- TC_{Basic}. It is the basic version of TC, where neither adaptive frame skipping nor key areas is considered. The fixed skip is set to 2 and only the nearest neighbor method is adopted to judge the MOI to which the track belongs.
- $\text{TC}_{Dynamic}$. It is another version of TC where dynamically skip frames based on vehicle density is considered with the initial skip equals to 2. It only uses the nearest neighbor method to judge the MOI to which the track belongs.
- $\text{TC}_{Cascade}$. It uses the cascade judgment of key areas, direction, and the nearest neighbor method in TC. Only the fix skip (=2) is considered.

Their corresponding precision, recall, F1-score, and the total counting time are showed in Table 3. We can see that TC provides the highest accuracy and efficiency. Compared with TC, although $\text{TC}_{Cascade}$ is also efficient enough, but its accuracy is lower. By comparing the groups of $\text{TC}_{Dynamic}$ and TC, TC_{Basic} and $\text{TC}_{Cascade}$, we

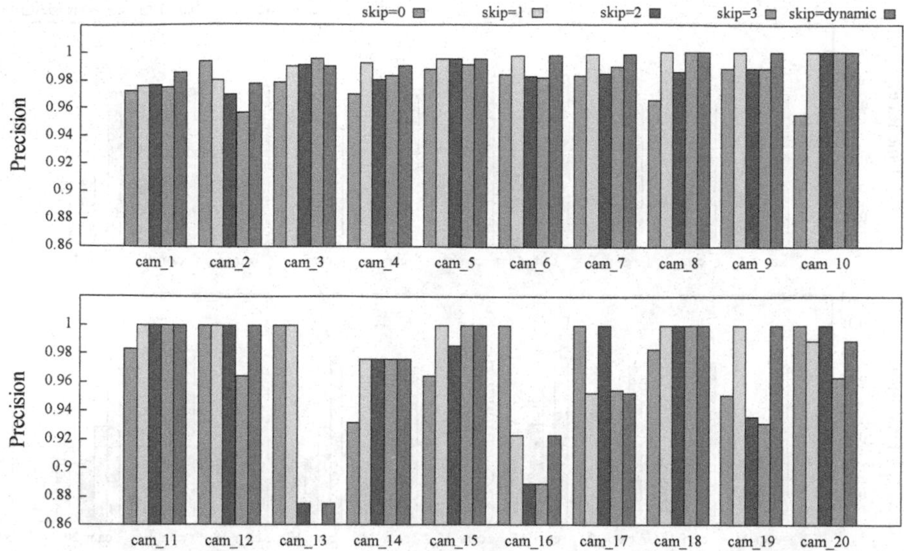

Fig. 4. Precision of different cameras under different frame skipping amplitudes.

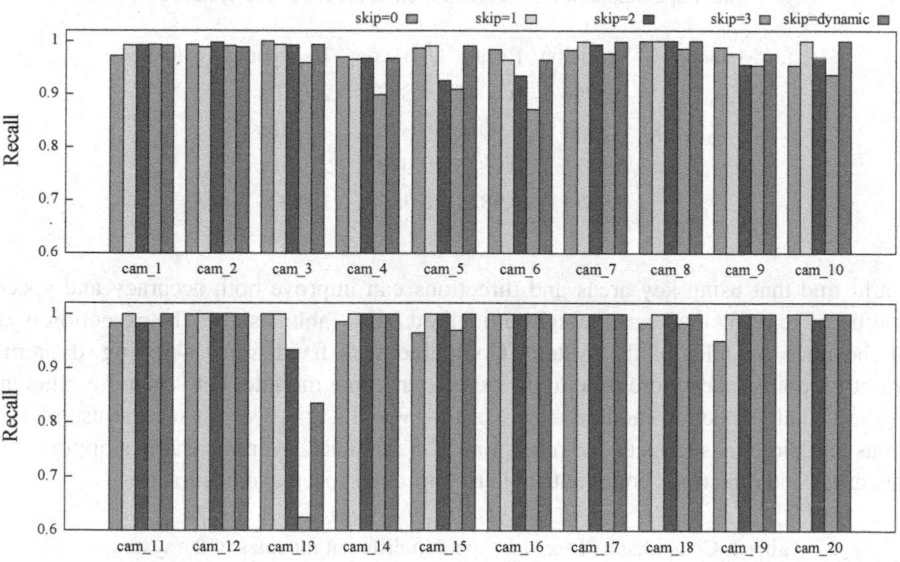

Fig. 5. Recall of different cameras under different frame skipping amplitudes.

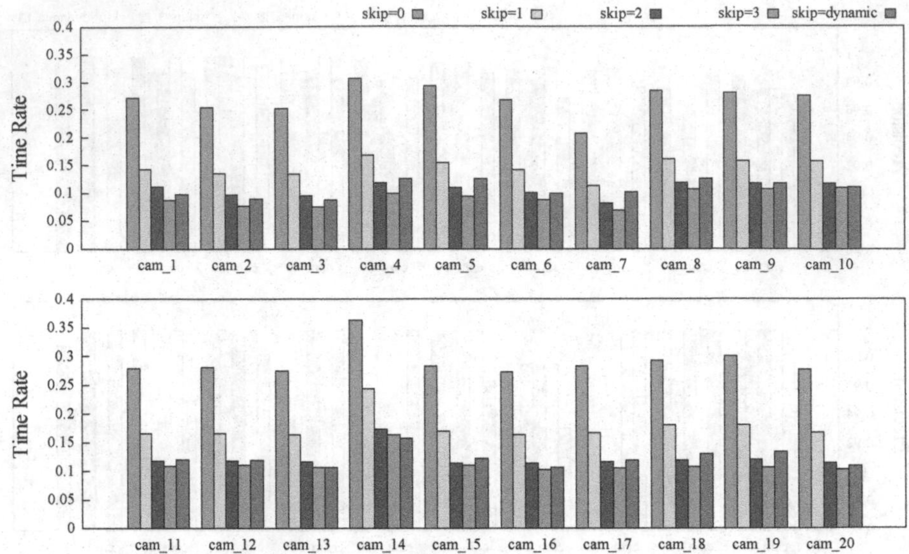

Fig. 6. Time rate of different cameras under different frame skipping amplitudes.

Table 3. Comparison of scores for each metric of the systems.

Method	Precision	Recall	F1-score	Total counting time
TC_{Basic}	0.8266	0.9045	0.8638	2,087 s
$TC_{Dynamic}$	0.8447	0.9068	0.8746	2,073 s
$TC_{Cascade}$	0.9756	0.9815	0.9785	2,064 s
TC	**0.9894**	**0.9823**	**0.9858**	**2,053 s**

could find that using key areas and directions can improve both accuracy and speed compared to using the nearest neighbor method only. Table 4 shows the execution time of the three modules of the system. Compared with fixed frame skipping, dynamic frame skipping takes more time in the vehicle tracking module, but it is more efficient in the vehicle detection module and is faster overall. Also, we can see the use of key areas and directions reduces judgment time. In summary, dynamic frame skipping and key areas and directions both contribute to improved speed and accuracy.

Table 4. Comparison of the time spent in different modules of the system.

Method	Detection time	Tracking time	Judging time	Total counting time
TC_{Basic}	1,900 s	142 s	45 s	2,087 s
$TC_{Dynamic}$	1,875 s	151 s	47 s	2,073 s
$TC_{Cascade}$	1,896 s	**139 s**	**29 s**	2,064 s
TC	**1,873 s**	150 s	30 s	**2,053 s**

6 Conclusion

Vehicle track identification, query and counting are key concerns in alleviating congestion and improving traffic management. In this paper, we proposed a new method to deal with the challenge of vehicle track counting problem. In order to improve the system efficiency while maintaining high accuracy, we used dynamic frame skipping based on variant vehicle density in different frames, in order to reduce the time of vehicle detection in an appropriate rate. We also used a modified DEEP SORT algorithm and set key areas for each video location, so as to quickly track and count the target. Experiments showed that our methods reached a high efficiency without sacrificing too much accuracy comparing with the no skip method and outperforms the RR method on both efficiency and effectiveness.

Acknowledgements. The work is partially supported by the National Key Research and Development Program of China (2020YFB1707900), and Liaoning Distinguished Professor (No. XLYC1902057).

References

1. Aich, S., Stavness, I.: Leaf counting with deep convolutional and deconvolutional networks. arXiv preprint arXiv:1708.07570 (2017)
2. Dai, Z., Song, H., Wang, X., Fang, Y., Li, H.: Video-based vehicle counting framework. IEEE Access **99**, 1–1 (2019)
3. Embleton, K.V., Gibson, C.E., Heaney, S.I.: Automated counting of phytoplankton by pattern recognition: a comparison with a manual counting method. J. Plankton Res. **25**(6), 669–681 (2003)
4. Hsieh, K., et al.: Focus: querying large video datasets with low latency and low cost. In: 13th {USENIX} Symposium on Operating Systems Design and Implementation ({OSDI} 18), pp. 269–286 (2018)
5. Kang, D., Emmons, J., Abuzaid, F., Bailis, P., Zaharia, M.: NoScope: optimizing deep CNN-based queries over video streams at scale. Proc. VLDB Endow. **10**(11), 1586–1597 (2017)
6. Kang, D., Guibas, J., Bailis, P., Hashimoto, T., Zaharia, M.: Task-agnostic indexes for deep learning-based queries over unstructured data. arXiv preprint arXiv:2009.04540 (2020)
7. Liang, H., Song, H., Li, H., Dai, Z.: Vehicle counting system using deep learning and multi-object tracking methods. Transpo. Res. Record. **2674**(4), 114–128 (2020)
8. Naphade, M., et al.: The 5th AI city challenge. In: The IEEE Conference on Computer Vision and Pattern Recognition (CVPR) Workshops, June 2021
9. Pizzo, L.D., Foggia, P., Greco, A., Percannella, G., Vento, M.: Counting people by RGB or depth overhead cameras. Patt. Recogn. Lett. **81**(C), 41–50 (2016)
10. Shih, F.Y., Zhong, X.: Automated counting and tracking of vehicles. Int. J. Pattern Recognit. Artif. Intell . **31**(12), 1750038.1–1750038.12 (2017)
11. Tran, V.H., et al.: Real-time and robust system for counting movement-specific vehicle at crowded intersections. In: Proceedings of the IEEE/CVF Conference on Computer Vision and Pattern Recognition, pp. 4228–4235 (2021)
12. Wang, C.Y., Bochkovskiy, A., Liao, H.Y.M.: Scaled-yolov4: Scaling cross stage partial network. In: Proceedings of the IEEE/CVF Conference on Computer Vision and Pattern Recognition, pp. 13029–13038 (2021)

13. Wojke, N., Bewley, A., Paulus, D.: Simple online and realtime tracking with a deep association metric. In: 2017 IEEE international conference on image processing (ICIP), pp. 3645–3649. IEEE (2017)
14. Yuan, H., Li, G.: A survey of traffic prediction: from spatio-temporal data to intelligent transportation. Data Sci. Eng. **6**(1), 63–85 (2021)

TSummary: A Traffic Summarization System Using Semantic Words

Xu Chen[1,2], Ximu Zeng[2], Shuncheng Liu[2], Zhi Xu[2], Yuyang Xia[2], Ruyi Lai[3], and Han Su[1,2(✉)]

[1] Yangtze Delta Region Institute (Quzhou), University of Electronic Science and Technology of China, Chengdu, China
hansu@uestc.edu.cn

[2] School of Computer Science and Engineering, University of Electronic Science and Technology of China, Chengdu, China
{xuchen,ximuzeng,liushuncheng,zhixu023,yuyangxia}@std.uestc.edu.cn

[3] Southwest Petroleum University, Chengdu, China

Abstract. Due to the popularity of GPS devices, road information and trajectory data are being generated in large quantities and accumulated rapidly. However, raw trajectory data is usually stored in the form of timestamps and locations. As a result, the raw data does not make much sense to humans without semantic representation, because humans cannot get an intuitive view of a group of trajectories that move on the same road, namely traffic of the road. In this work, we propose a partition-and-summarization framework which can automatically generate short text summaries for traffic description, aiming to advance human understanding of traffic information. In the partition phase, we first define a group of features that can be described in summaries, and then partition the roads. The purpose of the partitioning is to make the features of road segments in each partition as homogeneous as possible. In the summarization phase, the most interesting features for each partition are selected, and short text description are generated. For empirical study, experiments are conducted with a real trajectory dataset, and the experiment results have proven that the short text generated by the proposed framework can effectively reflect the important information of traffic.

Keywords: Traffic summarization · Automatic short text description generation · Partition-and-summarization framework

1 Introduction

Driven by advances in sensor technology, a large amount of data related to the motion history of moving objects are recorded by GPS-enabled mobile devices and wireless communications. This phenomena inspires tremendous research on trajectory data ranging from effective data storage and query processings to data mining and visualization. In this paper, we propose a novel information pyramid model to systematically classify the research on traffic data in Fig. 1. In general,

U. K. Rage et al. (Eds.): DASFAA 2022 Workshops, LNCS 13248, pp. 257–271, 2022.
https://doi.org/10.1007/978-3-031-11217-1_19

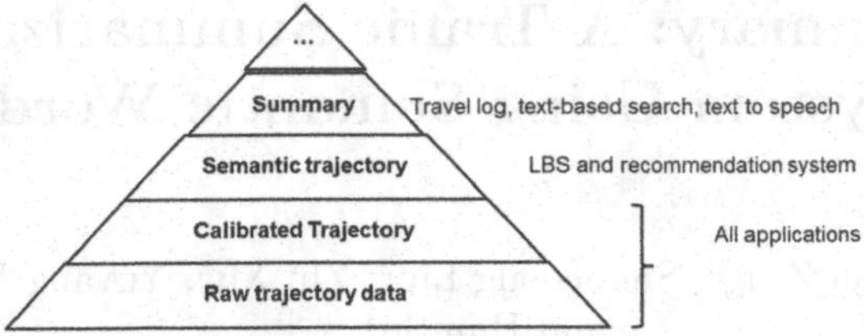

Fig. 1. Information pyramid of trajectory data

the information pyramid model constitutes of a hierarchy of levels from data to intelligence (wisdom) manifesting the human cognitive process. The bottom level of the information pyramid corresponds to raw trajectory data. In theory, a trajectory can be modelled by a continuous function mapping time to space. In practice, a trajectory can only be represented by a discrete sequence of locations as exampled in Table 1. The second level of the information pyramid is cleaned trajectories, which improve upon the quality of raw trajectories by removing redundant sample points and inserting important missing sample points. The above two levels can serve as the foundation for most existing trajectory-based queries and analytical tasks [1–3, 6, 9, 11, 16, 17]. In order to facilitate better interpretation of raw trajectories, researchers have proposed several models by linking GPS locations to semantic entities such as POIs, roads, regions, resulting in semantic trajectories or annotated trajectories [13, 18], which constitute the third level in our information pyramid. Figure 2(a) demonstrates the semantic level expression of the raw trajectory in Table 1.

Table 1. Trajectory in database

Latitude	Longitude	Time-stamp
39.9383	116.339	20131102 09:17:56
39.9382	116.337	20131102 09:18:02
...	...	...
...	...	...
39.9259	116.310	20131102 09:33:26
39.9253	116.310	20131102 09:34:31

Nevertheless, all these three levels contain too much redundant information, which is hard for human to read and interpret. In this work, we build a new and higher level – "*summary*" – to the information pyramid to provide a user friendly expression of traffic data. The summary should use human readable language to express key information of traffic data. Figure 2 exemplifies the expected traffic summarization. We find there are several benefits by translating a raw trajectory into text. Comparing to the semantic level expression, the summary level expression can highlight the important parts of the trajectories such as significant landmarks, e.g., Beijing Shangri-la Hotel, important roads, e.g., W 3rd Ring Road, and travel behaviors, e.g., low speed, while the summary level expression is more concise. The summary level expression can be used in many applications, such as violation reminder, travel generator, and TTS (Text To Speech) techniques.

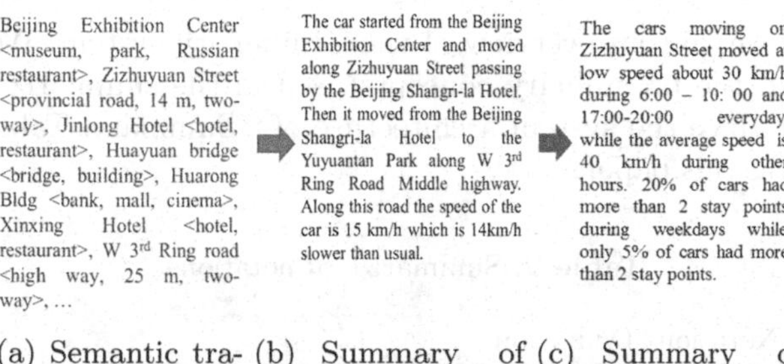

(a) Semantic trajectory (b) Summary of the trajectory (c) Summary of the traffic

Fig. 2. Semantic trajectory level and summary level

In order to realize the newly added summary level, we propose a prototype system – TSummary [14]. Given several trajectories moving on a same road, TSummary can generate a summary of the traffic condition of the road, namely *traffic summarization*. Nevertheless, in order to generate informative and concise summary, this summarization task is faced with several challenges. First, the TSummary needs to extract features, e.g., grade of road, road width, etc., since the trajectory data lacks the semantic information. Second, the trajectory and the road should be partitioned into several parts and described separately, since a road can stretch to 10 km and the traffic condition can be quite different in various segments. Third, even within a partition, properly choosing summary contents is non-trivial. In traffic summarization, how the features change with time may interest users the most. To tackle the aforementioned challenges, we propose a partition-and-summarization framework: (1) In the partition phase, by minimizing the variation of predefined features for road segments within the same partition, it finds an optimal partition with concise representation that can be applied in summarizing each partition. (2) In the summarization phase, to monitor how features change over time, it exploits the common patterns learned from historical trajectories.

In sum, we make the following major contributions in this paper.

- We define the information pyramid of trajectory data, and introduce a new level 'summary' to the information pyramid.
- We propose a partition-and-summarize framework, TSummary, to address several challenges of our proposal and to generate traffic summarization.
- Experiments are conducted on real dataset and demonstrate that the textual descriptions generated by TSummary are easier for humans to understand.

The remainder of the paper is structured as follows. The preliminary concepts and the traffic summarization framework are introduced in Sect. 2. Sections 3 demonstrates the details of the traffic summarization. The experimental results are shown in Sect. 4. A review of related work and conclusion are presented in Sect. 5 and Sect. 6 respectively.

2 System Overview

In this section, we give an overview of the TSummary system. We will explain the features and the preliminary concepts used in the summarization process, and briefly introduce the system architecture of TSummary. Table 2 shows the notations used in this paper.

Table 2. Summarize of notations

Notation	Definition
T	A raw trajectory
l	A landmark in the space
$l.s$	The significance of a landmark l
$\overline{T}$	A symbolic trajectory
f	A feature of trajectory
$\mathbb{F}_{\overline{T}}$	The concerning features of a trajectory $\overline{T}$
$\overline{R}$	A symbolic road
$\overline{RS}_i$	The road segment connecting two consecutive landmarks l_i and l_{i+1} of $\overline{R}$
$\mathbb{F}$	A frequency of a feature
$\mathbb{F}_f(\overline{RS})$	The frequencies of feature f of $\overline{RS}$

2.1 Feature Extraction

The features considered in TSummary can be mainly divided into two types: routing features, which are spatial-only features and describe where the moving object travels, and moving features, which are spatial-temporal features and describe how the moving object travels. In the TSummary system, we identify and use 3 kinds of road information (*'grade of road'*, *'road width'* and *'direction'*) as the routing features. We also propose 3 types of moving features (*'speed'*, *'number of stay points'*, and *'number of U-turns'*) to describe the motion behaviour of a moving object.

2.2 Preliminary Concepts

Definition 1 (Raw Trajectory). *A trajectory T is a discrete sequence of locations, which are sampled from the original continuous movement of a moving object with associated time-stamps, i.e.,* $T = [(p_1, t_1), (p_2, t_2), \cdots, (p_n, t_n)]$.

Definition 2 (Landmark). *A landmark l is a geographical point and is stable and independent of trajectories, such as a Point Of Interest (POI) and a turning point of the road network.*

Definition 3 (Road). *A road R is a finite sequence of road intersections on the road, i.e., $R = [l_1, l_2, \cdots, l_n]$.*

Since road intersections can be treated as landmarks in the network, we use l to denote road intersections. A road R is always static in a road network.

Definition 4 (Road Segment). *A road segment $\overline{RS}_i$ is a sub-road which connects two consecutive road intersections l_i and l_{i+1}.*

Definition 5 (Traffic Summarization). *Given a set of trajectories $\mathbb{T}$ on road R and a trajectory feature set $\mathbb{F}$, traffic summarization is to use a text to summarize the feature values of $\mathbb{F}$ on R.*

2.3 Structure of TSummary

TSummary is a traffic summarization system, whose functionalities can be divided into two parts: (1) generating a summary of the moving behavior of a single trajectory, and (2) generating a summary of the traffic of a specific road given a set of trajectories on the road. TSummary comprises five components: trajectory calibration, feature extraction, data partitioning, content selection and summary generation. (1) The trajectory calibration module preprocesses the input trajectories by rewriting the raw trajectories into symbolic ones. (2) The feature extraction module extracts features from commercial maps, POIs, check-in data and offline trajectory data. (3) Then TSummary conducts a partitioning to split the road into several non-overlapping parts to ensure the features within each partition is similar. (4) After that, given the fact that there are too many features to describe or so many values for each feature to describe, content selection part will choose the most significant features within each partition. (5) At last, the summary generation component will plug the selected features or feature values into the predefined templates to form the summary for each data partition. Then a summary of the traffic of a road will be given to the user.

3 Traffic Summarization

The traffic can be summarized from two dimensions: (1) per trajectory, and (2) per road. Specifically, to summarize the traffic of a road, we summarize the statistical moving patterns of all the trajectories via that road. This process can be divided into two phases: partition and summarization.

Given a road $R = [l_1, l_2, \cdots, l_n]$ and a trajectory dataset $\mathbb{T}$, a trajectory $\overline{T} = [l'_1, l'_2, \cdots, l'_k] \in \mathbb{T}$ overlaps a road segment $RS = [l_i, l_{i+1}]$, if $\exists 1 \leq a < b \leq k$ such that $l'_a = l_i$ and $l'_b = l_{i+1}$. We use $\mathbb{T}_{RS}$ to denote all trajectories in $\mathbb{T}$ that overlap RS. In order to summarize the traffic on RS, a straightforward approach is to generate the summary of all trajectories in $\mathbb{T}_{RS}$, and then use some text clustering method to extract common keywords shared in these summaries (with frequencies higher than a threshold). The selected keywords can roughly describe the traffic condition of RS. However, this solution has mainly two disadvantages in terms of expressiveness of temporal information and efficiency. Firstly, the traffic condition of a road always changes with time, but the naive approach cannot express how traffic changes with time. For example, traffic during the rush hour may have more stay points and lower speed than usual. Secondly, the time complexity of the naive approach is proportional to the size of $\mathbb{T}_{RS}$. For a road that many trajectories overlap, the summary can only be generated after summarizing all the trajectories in $\mathbb{T}_{RS}$.

3.1 Periodic Feature Detection

To address the aforementioned drawbacks and emphasize the temporal characteristics of the traffic, we propose to study the traffic behaviors of a road at different time periods, and focus on the potential periodic patterns that reflect time-related events. For instance, an ideal summary of a road's speed feature could be "cars on Changan Road move with an average speed of 25 km/h on weekdays, while move with an average speed of 40 km/h on weekends". In TSummary, we allows users to summarize in a monthly, weekly or daily fashion, focusing on periodic patterns of different time granularities.

We formulate the problem of detecting periodic patterns in features of a road segment $\overline{RS}$ as a frequency analysis problem on time series data. Formally speaking, a feature f of $\overline{RS}$ is a function of time, and thus can be viewed as a time series. We denote this by $f(\overline{RS}) = x(t)$. For each time point t, we compute $f(\overline{RS})$ by averaging feature f of all the trajectories that overlap with $\overline{RS}$ at time t. In practice, we use discretized time intervals (e.g. per hour) to compute a sampled approximation to the true continuous $f(\overline{RS})$. To perform frequency analysis on $x(t)$, we apply *Fast Fourier Transform* (FFT) to carry out efficient Fourier Transform calculation on the feature record sequence of $\overline{RS}$. To meet the criteria of online analysis, a sliding window can be employed to constrain our analysis on the most recent historical records, so that we can perform the DFT analysis on the window data and update the results continuously. Once we obtain the DFT values, only those significant frequency components $\mathbb{F}_1, \cdots, \mathbb{F}_n$ with amplitude above a specific threshold σ will be kept for further analysis in road summarization. We use $\mathbb{F}_f(\overline{RS}) = \{\mathbb{F}_1, \cdots, \mathbb{F}_n\}$ to denote the significant frequencies of the moving feature f of the road segment $\overline{RS}$. For a road $R = [l_1, l_2, \cdots, l_n]$ consisting of $n-1$ segments, we repeat this frequency analysis on each road segments $[l_i, l_{i+1}]$ for $i \in [1, n-1]$.

Below we exemplify this frequency analysis using an example. In this example, we mimic a periodic behavior on the feature '# stay point' of a road during a day, by duplicating an 8-h worth of feature data 3 times with randomly added noises. The generated dataset is depicted in Fig. 3(a). We plot its DFT analysis result in Fig. 3(b). Figure 3(b) shows that the frequency $\frac{1}{8}$ Hz is significantly higher than the other frequencies[1], which indicates that the traffic moving on Zhizhuyuan Road exhibits a period of about 8 h, agreeing with our simulation. Combining with the curve in Fig. 3(a), the traffic peak arrives at time 8, 16 and 24 of a day.

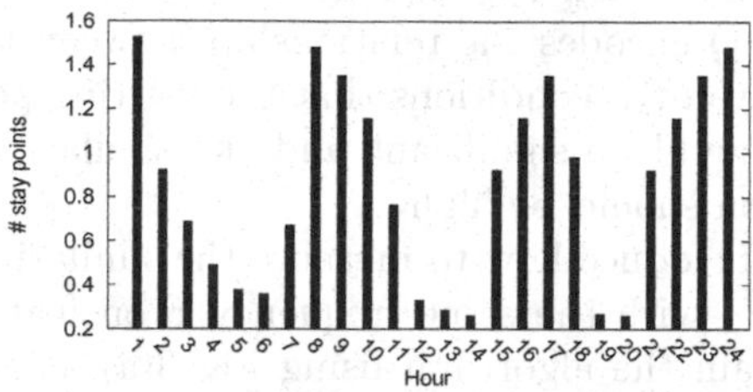

(a) Moving Feature Records of Zizhuyuan Road.

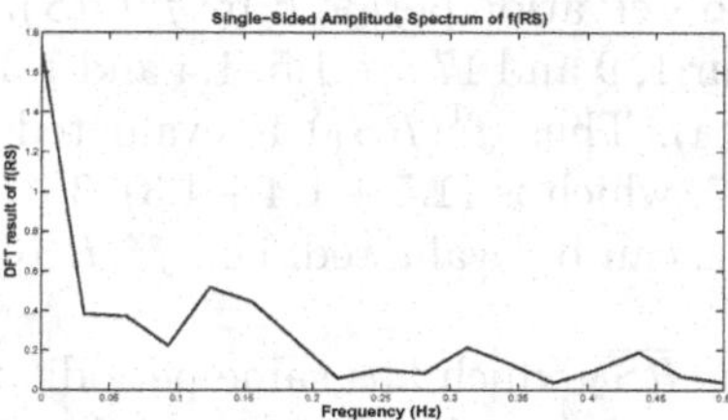

(b) Frequency Analysis of Zizhuyuan Road.

Fig. 3. An example of DFT frequency analysis on the feature of # stay points on Zizhuyuan Road within a day.

Last but not least, we have a few remarks on some customization issues in our system. First, the frequency analysis can be performed under different granularities, e.g., we can calculate the value of $f(\overline{RS})$ per hour, per day, per month to detect daily, weekly or monthly feature patterns. Second, we can adjust the window size to tune our system for short term or long term analysis, e.g., we can retrieve the traffic status of the recent one week or one month.

3.2 Road Partitioning

A road $R = [l_1, l_2, \cdots, l_n]$ is formed by multiple road segments. The traffic condition on different segments of the same road can vary a lot. Thus, in order to generate a short and concrete summary of the traffic condition of a road, we divide the road into several partitions by grouping road segments together based on their moving features and periodic features. Notably, a road partition is a group of road segments, while a trajectory partition is a group of trajectory

[1] We ignore 0 Hz because it stands for the direct current offset.

segments. In the rest of the paper, we use partition to refer to either trajectory partition or road partition when the context is clear.

We conduct the road partitioning by assigning each road segment a tag and aggregating the segments with the same tag into a single road partition. Formally, a random variable $\mathbb{Y}_i$ is assigned to road segment $\overline{RS}_i$, that $\mathbb{Y}_i$ is the tag of $\overline{RS}_i$, and $\mathbb{Y}$ denotes the whole tag sequence of the road. The potential function $\Phi(\mathbb{Y}_i, \mathbb{Y}_{i+1}, \overline{RS}_i, \overline{RS}_{i+1})$ encodes the relationship between tags of two consecutive road segments, given the conditions that the starting point and destination of a road partition should be significant and the similarity between segments within a same partition should be high.

In the sequel, we introduce how to measure the similarity between two road segment $\overline{RS}_1$ and $\overline{RS}_2$ with the same frequency $\mathbb{F}$ on feature f. Without loss of generality, we explain the algorithm using Fig. 3(a) as an example. Assume Fig. 3(a) show the # stay points feature of $\overline{RS}_1$, which is repeated every 8 h. Let us denote the average feature value of feature f on road segment $\overline{RS}$ during the k-th time interval of the observation period $\mathbb{F}$ by $f^k(\overline{RS})$. Specifically, the # stay points of $\overline{RS}_1$ during hour 1, 9 and 17 are 1.5, 1.4 and 1.3 respectively, according to the records in Fig. 3(a). Thus $f^1(\overline{RS}_1)$ is evaluated as the average # stay points of hour 1,9 and 17, which is $(1.5+1.4+1.3)/3 = 1.4$. Similarly, $f^k(\overline{RS}_1)$ of other k's time intervals can be evaluated, i.e., $f^2(\overline{RS}_1) = 0.9$, $f^3(\overline{RS}_1) = 0.6$, $\cdots$, $f^8(\overline{RS}_1) = 1.3$.

Given a road segment $\overline{RS}_2$ which has same periodic frequencies $\mathbb{F}$, $\overline{RS}_1$ and $\overline{RS}_2$ are regarded as similar if on each time interval they have similar feature values. Therefore we can use histogram distance to measure the feature f's distance between $\overline{RS}_1$ and $\overline{RS}_2$ on frequency $\mathbb{F}$. The algorithm is shown as follows.

$$d^{\mathbb{F}}(\overline{RS}_1, \overline{RS}_2) = \frac{\sum_{k=1}^{\mathbb{F}} |f^k(\overline{RS}_1) - f^k(\overline{RS}_2)|}{C \cdot \mathbb{F}} \tag{1}$$

where C is a normalization factor defined as $C = \max f^k(\overline{RS}_i)$.

For a feature f, there could be multiple significant frequencies of f. In order to measure the similarity between two segments $\overline{RS}_1$ and $\overline{RS}_2$ on a feature f, we need to compare $\overline{RS}_1$ and $\overline{RS}_2$ for each significant frequencies respectively.

$$S_f(\overline{RS}_1, \overline{RS}_2) = \sum_{\mathbb{F} \in \mathbb{F}_f(\overline{RS}_1) \cap \mathbb{F}_f(\overline{RS}_2)} 1 - d^{\mathbb{F}}(\overline{RS}_1, \overline{RS}_2) \tag{2}$$

In summary, the similarity between two road segments $\overline{RS}_1$ and $\overline{RS}_2$ can be evaluated by the mean of their similarities of all features as below:

$$S(\overline{RS}_1, \overline{RS}_2) = \frac{\sum_{f \in \mathbb{F}} S_f(\overline{RS}_1, \overline{RS}_2)}{|\mathbb{F}|} \tag{3}$$

where $\mathbb{F}$ is moving feature set.

3.3 Temporal Merge

It could be very verbose when describing a feature within a single feature period. For instance, Fig. 3(a) shows that the # of stay points feature has a repeating period of 8 h, we need to describe the # stay points feature for each hour within a single 8 h period. However, if the feature values of several consecutive hours are similar, we can aggregate these feature values by merging smaller time intervals together into longer ones. As an example, the first hour of a period in Fig. 3(a) (i.e., hour 1, 9 and 17) has similar feature values as the second hour (i.e., hour 2, 10 and 18) and the third hour (i.e., hour 3, 11 and 19). Merging the first separate three hours into one can significantly reduce redundancy of summary.

Next we will introduce how to merge short time intervals with similar features to longer time intervals. Again we merge small time intervals based on tagging. We assign a label for each small time interval and aggregate the road segments with the same label into a longer time interval. For the road segment in Fig. 3(a) 8 labels need to be assigned. Mathematically, a random variable $\mathbb{Z}_k$ is assigned to the k-th time interval of the feature period and $\mathbb{Z}$ denotes the whole label sequence of a period. The potential function $\Phi(\mathbb{Z}_k, \mathbb{Z}_{k+1}, t_k, t_{k+1})$ encodes the relationship between tags of two consecutive time intervals. Ideally, two time interval should have similar feature values if merged together, and have quite different feature values otherwise. Thus the potential function can be defined as below:

$$\Phi(\mathbb{Z}_k, \mathbb{Z}_{k+1}, t_k, t_{k+1}) = \begin{cases} -|f^k(\overline{RS}) - f^{k+1}(\overline{RS})|, \\ \qquad\qquad\qquad if\ \mathbb{Z}_k = \mathbb{Z}_{k+1} \\ |f^k(\overline{RS}) - f^{k+1}(\overline{RS})|, \\ \qquad\qquad\qquad if\ \mathbb{Z}_k \neq \mathbb{Z}_{k+1} \end{cases} \tag{4}$$

4 Experiment

In this section, to validate the effectiveness of our partition-and-summarization framework, we conduct extensive experiments based on real traffic dataset of a large city—Beijing—provided by a third-part company. The experiments are run on a computer equipped with Intel Core i7-2600 CPU (3.40 GHz) and 8 GB memory.

4.1 Experiment Setup

Landmark Dataset: The landmark dataset includes two parts: the turning point dataset and the POI dataset, which are both provided by a collaborating company, We extract approximately 32,000 turning points and 510,000 raw POI points. We use DBSCAN [8] to cluster the raw POI dataset into about 17,000 clusters, which are presented with the geometric centers of each clusters as the landmarks.

Trajectory Dataset: A real-world trajectory dataset is used in the experiments. In detail, the trajectory dataset is generated by 33,000+ taxis in Beijing over three months, containing more than 100,000 trajectories (a full version of Geolife dataset [20]). The raw trajectory dataset is randomly split into two parts: a training dataset and a testing dataset. More specifically, the training dataset consists of 50,000 trajectories, which is used to exploit popular routes between the landmarks and to build the historical feature map; and the testing dataset includes the rest trajectories, which is used to test the effectiveness of the proposed framework.

4.2 Evaluation Approach

In this section, we present the evaluation approach of studying both the effectiveness and efficiency of our partition-and-summarization framework. In all experiments, the weight of the landmark significance in the potential function is set as 0.5, the feature weight as 1 and the irregular rate threshold for a selected feature as 0.2.

For the evaluation of traffic summarization, we use 3 features in the experiments, which are Spe, Stay and U-turn. To study the effectiveness, we study the following 2 aspects of the summarization: (1) whether the summarization reveals the real feature frequency. (2) whether users get intuitive view of the traffic condition of a road. We compare the time costs between generating a traffic summary from summaries of trajectories and generating a traffic summary utilizing the methods introduced in Sect. 3.

4.3 Performance Evaluation

Case Study. Before conducting the performance evaluation, we first present a case study of the proposed summarization system in Fig. 4.

We have two ways to generate traffic summarization. The first way is that as mentioned in the beginning of Sect. 3, by conducting text clustering methods on summaries of trajectories to extract keywords about the hottest landmarks and common behaviors. Figure 4(a) and Fig. 4(b) demonstrates the keywords of three main roads in Beijing—the No. 2, 3, and 4 Ring Road—at 5:00 am and 5:00 pm, respectively. We can see that the common behaviors change a lot at different time. E.g., 'moving faster' is the most popular traveling behaviors of moving objects at 5:00 am, whereas 'moving slower' appears very frequently at 5:00 pm. The second way is generating a traffic summary utilizing the methods introduced in Sect. 3. Examples of the summary are *'Cars on Section 1, 2, Third Ring Road moved with an average speed of 28* km/h *on weekdays. Cars moved with an average speed of 40* km/h *on weekends. 10% of the Cars took U-Turn on weekdays. From 06:00–12:00 h, 8% of cars had stay points. From 12:00–18:00 h, 1% of cars had stay points. From 18:00–00:00 h, 3% of cars had stay points.' (with the granularities of day)* Obviously, the text reveals more information, including the periodic information and the quantitative information while the keywords provide less information. And the keywords only provides the traffic information of a certain time, otherwise these keywords may counterattack each other, e.g., containing 'faster' and 'slower' at the same time.

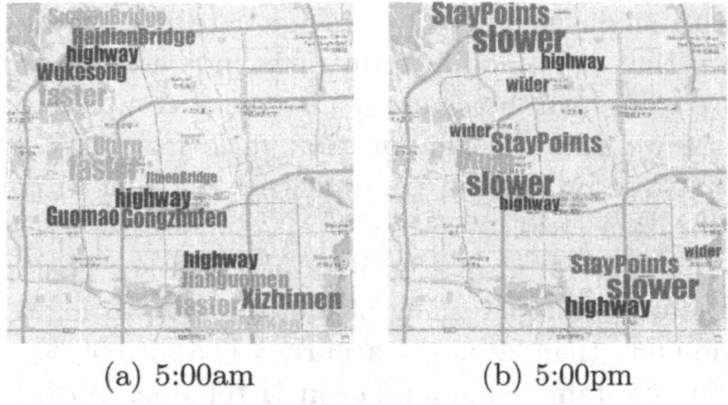

(a) 5:00am (b) 5:00pm

Fig. 4. Keywords of traffic summarization

Summaries in Different Weather. Weather is an important factor which greatly affects the patterns of traffic. We roughly categorize four weather types: sunny, cloudy, rainy and snowy. Among the three months during which the trajectory dataset was collected, there are 25, 35, 19 and 7 days respectively with each weather type. We analyze how FF of various features change under the four weather types. Figure 5(a) shows the results of all the eight features considered in the experiments.

As shown, FF of each feature in sunny days is similar to that in cloudy days, all under 20%. On the contrary, for trajectories in rainy and snowy days, FF of all features increase dramatically compared to sunny/cloudy days. For instance, FF of the speed feature in snowy days even reaches 70%. This observation agrees to our common sense that the traffic in snowy and rainy days is very different from that in sunny and cloudy days. It verifies our expectation that our summarization framework well reflects the change in weather.

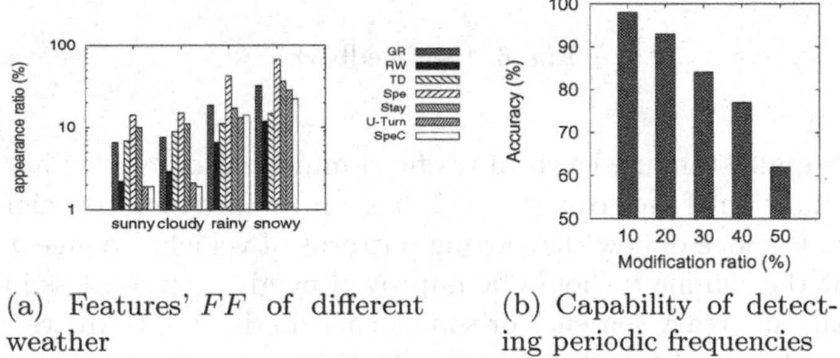

(a) Features' FF of different weather (b) Capability of detecting periodic frequencies

Fig. 5. Effectiveness of traffic summarization

Effectiveness of Traffic Summarization. Whether the traffic summarization is able to detect the accurate frequency of features is a key issue to evaluate the effectiveness of the traffic summarization method. Thus we manipulate the variation of features of a road during tens hours by duplicating the first t hours' feature values k times with slight change. The change rate is controlled by parameter *modification ratio*, that 10% modification ratio indicates that during duplicating 10% of feature values will be changed instead of completely copy. So in this set of experiments, we tune the modification ratio from 10% to 50% with the step of 10%. Then we record the rate of detecting frequency t correctly with the changing of modification ratio.

The result is demonstrated in Fig. 5(b). We can see that the accuracy of detecting t reduces with the increasing of modification ratio. When the modification is slight, no more than 20%, the accuracy is above 90%, which means the frequency detecting of traffic summarization is reliable. With the modification ratio increasing to 50%, the accuracy drops to 60%. This is caused by that 50% modification ratio is a big change to the feature input. Thus their frequencies have high possibility to change.

Impressions of Users. The primary purpose of traffic summarization is to give users an intuitive understanding of the traffic condition of roads. Therefore, in this experiment, we test whether users have an intuitive understanding after reading the summaries. In order to assess how well a user understands the traffic and to measure a summary's amount of information and readability, four understanding levels are used in experiments.

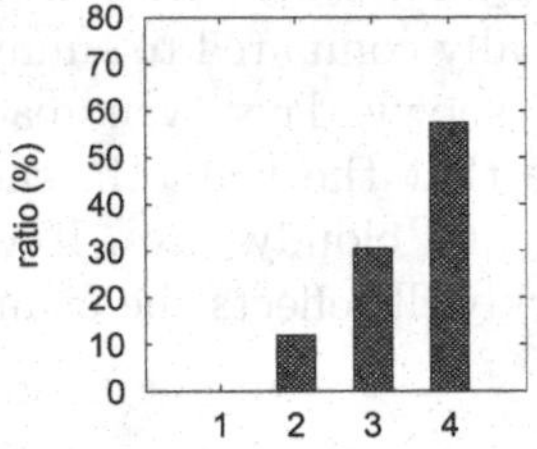

Fig. 6. User feedback

The four understanding levels of traffic summarization are (1) has no idea of the traffic condition of the given road; (2) has a little idea of how vehicles move on the road; (3) has idea of how the moving patterns of vehicles change with time of the road but the summary should be improved by giving more/less information, improving the summary sentence or some other methods; (4) knows clearly how the moving patterns of vehicles change with time of the road, and the summary is well presented. We randomly select 150 traffic summaries and ask thirty volunteer users to read nine traffic summaries each. Then each user is asked to classify their understanding of the traffic into one of the four levels. Figure 6 compares the users feedback of four understanding levels. It can be seen that nearly 90% (grade 3 and 4) summaries can give users an intuitive understanding of the traffic condition, and the understanding level marked at grade 4 accounts for roughly 60% of randomly selected 150 summaries. This result illustrates that the proposed traffic summarization algorithms can give users an intuitive view of the traffic condition of roads and can thus achieve the primary purpose.

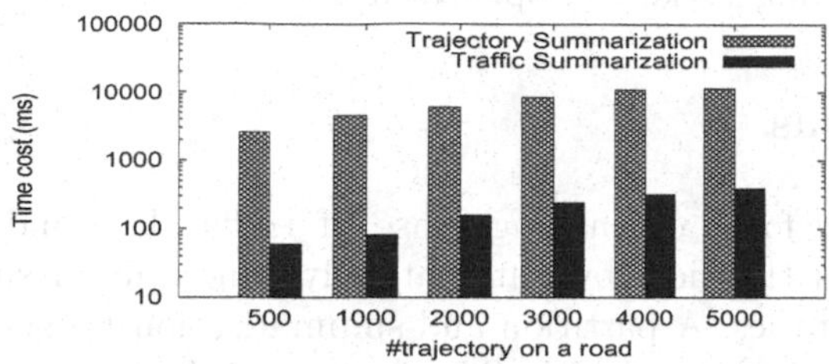

Fig. 7. Average time cost for summarizing the traffic of a road

Summarization Time Cost. The time cost of the proposed traffic summarization algorithm is also evaluated, since time cost is a key performance evaluation for online summarization systems.

Traffic summarization can be generated in two ways, keywords extracted from summaries of trajectories and TSummary. So we record the time costs of traffic summarization from these two ways. We tune the trajectory size of the road from 500 to 5000. The result is shown in Fig. 7, from which we observe that TSummary generates traffic summary summarized within 200 milliseconds, while the keywords method takes more than 10 times of time costs of TSummary.

5 Related Work

To the best of our knowledge, there is no existing work that uses text to summarize traffic conditions. Therefore, we review several works on text and multimedia summarization in this section, since these works share similar inspirations to our work.

Text Summarization. There are many works related to text summarization. [5] utilized a hidden Markov model and made use of the relations between sentences to extract summaries of documents. [19] used modified corpus-based approach and LSA-based T.R.M. approach to find a summary sentence in a document, but it failed to leverage the relation between the sentences. [12] regarded the text summarization task as a sequence labeling task, and used a Conditional Random Field framework to partition a document and to identify the summarization sentences.

Multimedia Summarization. Summarization of audio/video data have been increasingly important, because the large amount of various multimedia data are available and the requirements for multimedia applications increase nowadays. [15] extracted a sequence of stationary images and moving images from a video as the video summarization. [4] conducted research in user interface designs for video browsing and summarization. [10] presented a different algorithm for exploiting both audio and video information. [7] proposed a system to automatically generate natural language to summarize video information, which focused on extracting visual concept features from video.

6 Conclusions

In this paper, we focus on making sense of traffic data and take an important step towards this field by automatically generating a summary text for a group of trajectories. A partition-and-summarization framework is proposed to generate traffic summarization. The framework first splits a road into several partitions which are similar in travel behaviour in each partition. Then it chooses interesting contents for each partition to generate summarization. The experiment results show the effectiveness and efficiency of the proposed summarization framework. Our framework can reflect representative features of traffic with generated traffic summarization.

Acknowledgements. This work is supported by NSFC (No. 61802054), scientific research projects of Quzhou Science and Technology Bureau, Zhejiang Province (No. 2020D010, 2021D022).

References

1. Cai, Y., Ng, R.: Indexing spatio-temporal trajectories with chebyshev polynomials. In: SIGMOD, pp. 599–610 (2004)
2. Chakka, V., Everspaugh, A., Patel, J.: Indexing large trajectory data sets with SETI. In: CIDR (2003)
3. Chen, L., Özsu, M., Oria, V.: Robust and fast similarity search for moving object trajectories. In: SIGMOD, pp. 491–502 (2005)
4. Christel, M.G.: Automated metadata in multimedia information systems: creation, refinement, use in surrogates, and evaluation. In: Synthesis Lectures on Information Concepts, Retrieval, and Services, vol. 1, no. 1, pp. 1–74 (2009)
5. Conroy, J.M., O'leary, D.P.: Text summarization via hidden Markov models. In: Proceedings of the 24th Annual International ACM SIGIR Conference on Research and Development in Information Retrieval, pp. 406–407. ACM (2001)
6. Cudre-Mauroux, P., Wu, E., Madden, S.: Trajstore: an adaptive storage system for very large trajectory data sets. In: ICDE, pp. 109–120 (2010)
7. Ding, D., et al.: Beyond audio and video retrieval: towards multimedia summarization. In: Proceedings of the 2nd ACM International Conference on Multimedia Retrieval, p. 2. ACM (2012)
8. Ester, M., Kriegel, H., Sander, J., Xu, X.: A density-based algorithm for discovering clusters in large spatial databases with noise. In: KDD, pp. 226–231 (1996)
9. Frentzos, E., Gratsias, K., Pelekis, N., Theodoridis, Y.: Nearest neighbor search on moving object trajectories. In: Bauzer Medeiros, C., Egenhofer, M.J., Bertino, E. (eds.) SSTD 2005. LNCS, vol. 3633, pp. 328–345. Springer, Heidelberg (2005). https://doi.org/10.1007/11535331_19

10. Li, Y., Merialdo, B.: Multi-video summarization based on AV-MMR. In: 2010 International Workshop on Content-Based Multimedia Indexing (CBMI), pp. 1–6. IEEE (2010)
11. Ni, J., Ravishankar, C.: Indexing spatio-temporal trajectories with efficient polynomial approximations. TKDE **19**(5), 663–678 (2007)
12. Shen, D., Sun, J.T., Li, H., Yang, Q., Chen, Z.: Document summarization using conditional random fields. In: IJCAI, vol. 7, pp. 2862–2867 (2007)
13. Spaccapietra, S., Parent, C., Damiani, M.L., de Macedo, J.A., Porto, F., Vangenot, C.: A conceptual view on trajectories. Data Knowl. Eng. **65**(1), 126–146 (2008)
14. Su, H., Zheng, K., Zeng, K., Huang, J., Zhou, X.: STMaker-a system to make sense of trajectory data. Proc. VLDB Endow. **7**(13), 1701–1704 (2014)
15. Truong, B.T., Venkatesh, S.: Video abstraction: a systematic review and classification. ACM Trans. Multimed. Comput. Commun. Appl. (TOMCCAP) **3**(1), 3 (2007)
16. Vlachos, M., Kollios, G., Gunopulos, D.: Discovering similar multidimensional trajectories. In: ICDE, pp. 673–684. IEEE (2002)
17. Wang, H., Zheng, K., Xu, J., Zheng, B., Zhou, X., Sadiq, S.: SharkDB: an in-memory column-oriented trajectory storage. In: Proceedings of the 23rd ACM International Conference on Information and Knowledge Management, pp. 1409–1418. ACM (2014)
18. Yan, Z., Spaccapietra, S., et al.: Towards semantic trajectory data analysis: a conceptual and computational approach. In: VLDB PhD Workshop (2009)
19. Yeh, J.Y., Ke, H.R., Yang, W.P., Meng, I., et al.: Text summarization using a trainable summarizer and latent semantic analysis. Inf. Process. Manag. **41**(1), 75–95 (2005)
20. Zheng, Y., Xie, X., Ma, W.Y.: GeoLife: a collaborative social networking service among user, location and trajectory. IEEE Data Eng. Bull. **33**(2), 32–39 (2010)

Attention-Cooperated Reinforcement Learning for Multi-agent Path Planning

Jinchao Ma and Defu Lian(✉)

School of Computer Science and Technology,
University of Science and Technology of China, Hefei, China
majinc@mail.ustc.edu.cn, liandefu@ustc.edu.cn

Abstract. Multi-agent path finding (MAPF), in multi-agent systems, is a challenging and meaningful problem, in which all agents are required to effectively reach their goals concurrently with not colliding with each other and avoiding the obstacles. Effective extraction from the agent's observation, effective utilization of historical information, and efficient communication with neighbor agents are the challenges to completing the cooperative task. To tackle these issues, in this paper, we propose a well-designed model, which utilizes the local states of nearby agents and obstacles and outputs an optimal action for each agent to execute. Our approach has three major components: 1) observation encoder which uses CNN to extract local partial observation and GRU to make full use of historical information, 2) communication block which uses attention mechanism to combine the agent's partial observation with its neighbors, and 3) decision block with the purpose to output the final action policy. Based on the three major components, all agents formulate their own decentralized policies to apply. Finally, we use success rate and extra time rate to measure our approach and other well-known algorithms. The results show that our method outperforms the baselines, demonstrating the efficiency and effectiveness of our approach, especially in the case of large scale in the world.

Keywords: Multi-agent path finding (MAPF) · Reinforcement learning · Decentralized planning · Deep learning

1 Introduction

Multi-agent path finding (MAPF) is a significant problem in multi-agent planning problems, whose goal is to effectively plan paths for multiple agents with the constraint that agent need to avoid colliding with obstacles and other agents [1]. MAPF is a common problem in many practical scenarios. For example, MAPF is deployed in automated warehouse [2,3], airplane taxiing [4,5], automated guided vehicles (AGVs) [6,7] and so on. Indeed, MAPF is full of challenges because solving for optimal method is NP-hard [8], and when the size of environments, the number of agents, and the density of obstacles increase sharply, a large number of potential path conflicts may occur.

U. K. Rage et al. (Eds.): DASFAA 2022 Workshops, LNCS 13248, pp. 272–290, 2022.
https://doi.org/10.1007/978-3-031-11217-1_20

Generally, there are two major categories in the multi-agent system: centralized methods and decentralized methods. Centralized methods apply a single learner to discover joint solutions (team learning), while decentralized methods use multiple simultaneous learners, usually one for each agent (concurrent learning) [9]. The same classification also exists on MAPF issues [1]. In the centralized MAPF, all agents' partial observations are known and collected to generate collision-free paths for them, while in decentralized MAPF it is no necessary to know all the states of all agents to determine each agent, and each agent makes decisions independently according to its own local observation. With the growth of the number of agents, world scale and obstacle density, computational complexity will be an important concern, especially for centralized methods. So, in this paper, we keep focusing on decentralized methods, where effective and efficient communications between agents are crucial. What's more, the information for communication also needs to be carefully coded, and it needs to consider not only the effective coding of local observations, but also the effective combination of historical state information. So, we use Convolutional Neural Networks (CNN) with residual structure to encode local observations and Gated Recurrent Unit (GRU) cell to integrate historical information.

In the past few decades, decentralized solutions to this problem have attracted extensive interest. In traditional methods, the reaction method is intuitive and widely adopted benchmarks. Take local repair A*(LRA*) [10] for example, in which each agent searches for its own path to its own goal using the A* algorithm, ignoring all other agents except for its current neighbors. Obviously, there will be many conflicts, so dealing with conflicts effectively is the key work of reaction. Whenever a collision is about to occur, the agent needs to avoid this situation by recalculating the remainder of its route and considering the occupied grids. Besides reaction-based methods, there are also conflict-based methods (CBS [11], MA-CBS [12], ECBS [13], ICBS [14]). In general, CBS [11] is a two-level search algorithm. While CBS calculates the path of each agent independently at the low level, at the high level it detects conflicts between agent pairs and solves the conflict by splitting the current solution into two related subproblems, each of which involves re planning a single agent. By dividing the subproblem into two subproblems to solve the conflict recursively, the search tree is implicitly defined. Advanced search searches this tree with best-first search and terminates when the conflict-free leaf is extended.

In recent years, there has been some seminal work using deep architectures to automatically find optimal or sub-optimal solutions for planning problems. These approaches vary from supervise learning to deep reinforcement learning, while their structures contain CNN [15,16], LSTM [17], GNN [18,19]. In the previous approaches, some work (for example, PRIMAL [15], PRIMAL_2 [16]) do not think the communication among agents is necessary and do not make any effort to design communication structure, while some work (for example, LSTM [17], GNN [18], GAT [19]) think the communication is very important and they take all the effort to build the communication structure for agents to share their state information. Inspired by MAAC which uses an attention mechanism in multi-agent reinforcement learning to select relevant information for each agent at every time step, we use an attention mechanism to share neighbors' information for each agent. The reinforcement learning methods also face the problem of solving long-horizon tasks with sparse reward [20], especially when the environment is very large, the training process will be inefficient. To speed up the training process in reinforcement learning, some of the previous works also use imitation learning for the cold start dilemma in Reinforcement Learning [15,16], while the curriculum learning for environment setting also makes sense where the easy task will take some inspiration for the hard task and human also explore complex task from easy step by step [16].

The major contributions of this paper are:

1. We refine the observation encoder by using Residual CNN to extract local observations and Gated Recurrent Unit (GRU) to make full use of historical information.
2. We build the interaction layer by using attention mechanism to combine the agent's partial observation with its neighbors.
3. Experiment shows that our model outperforms most of the traditional methods and previous learning models in terms of the success rate and the extra time rate.

The structure of his paper is as follows. Section 2 introduces related works about multi-agent path finding. Section 3 provides problem formulation. The detail of our approach is introduced in Sect. 4, followed by Sect. 5 which presents the experiment results in various environment settings. Finally, we give a brief conclusion in Sect. 6.

2 Related Work

2.1 Classical Path Planning Methods

As mentioned above, there are two major categories in MAPF approaches: coupled (centralized) and decoupled (decentralized). There will be a catastrophe for centralized approaches when the size of the world and the number of agents grows sharply, and the performance of centralized approaches will be in the struggle because of tremendous search space. Almost all traditional methods tend to use centralized or coupled methods, which use the complete information of all agents and world environment to plan the path globally. And among the traditional methods, the most famous and well-used decoupled methods will be search-based (for example, LRA* and CBS) methods. A*-based (for example, LRA* and WHCA*) algorithms rely on complete observations and use A* to calculate full paths for each agent, which could work in both centralized (CBS, WHCA, ORCA) and decentralized manner [6]. In A*-based methods, the reaction trick is intuitive and widely adopted, so there are also some works that take the reaction-based methods as a branch of approaches. Take LRA* (Local Repair A*) [10] for example, in which each agent searches for its own path to its own goal using the A* algorithm, ignoring all other agents except for its current neighbors. Obviously, there will be many conflicts, so dealing with conflicts effectively is the key work of reaction. Whenever a collision is about to occur, the agent needs to avoid this situation by recalculating the remainder of its route and considering the occupied grid [10]. Besides search-based methods, there are also Conflict-Based Search and its variants (CBS [11], MA-CBS [12], ECBS [13], ICBS [14]), which make a plan for each single agent and construct a set of constraints to find the optimal or near optimal solution without exploring the high-dimensional spaces. In Conflict-Based approaches, CBS [11] is the most original and influential method, which is a two-level search algorithm. While CBS calculates the path of each agent independently at the low level, at the high level it detects conflicts between agent pairs and solves the conflict by splitting the current solution into two related subproblems, each of which involves re planning a single agent. By dividing the subproblem into two subproblems to solve the conflict recursively, the search tree is implicitly defined. Advanced search searches this tree with best-first search and terminates when the conflict-free leaf is extended.

Besides the search-based methods and the conflict-based methods, there are also thousands of trick to solve the problem, such as Direction Map [21], Subgraph Structure [22], Flow Annotation Replanning [23], Increasing cost tress search [24–26], Satisfiability [27–30], Integer linear programming [31–33], Answer set programming [34] and so on. In Direction Map, there is a direction map (DM) which stores information about the direction that agents have traveled in each portion of a map. And in the planning process, agents then use this information, in which movement that runs counter to the pattern incurs additional penalties, thus encouraging delegates to move more evenly throughout the environment. In Subgraph Structure method, they propose a new abstract form to plan more effectively, in which the key of this approach is to divide the mapping into subgraphs with known structure. And these known subgraphs have entry and exit constraints, and can be compactly represented. Then, planning becomes a search in a smaller subgraph configuration space. Once an abstract plan is found, it can be quickly decomposed into correct (but possibly suboptimal) concrete plans without further search. Increasing cost tress search (ICTS) interweaves two search processes. The first is called advanced search, which aims to find the size of the agent's single agent plan in the optimal solution of a given MAPF problem. The second is called low-level search, which accepts a plan size vector and verifies whether there is an effective solution for a given MAPF problem. Finally, there are some methods that transform the MAPF problem into a different problem for which good solvers exist, such as Satisfiability [27–30], Integer linear programming [31–33], Answer set programming [34].

In fact, solving the optimality is NP hard. Although significant progress has been made in reducing the amount of computation, the scalability of these methods in an environments with a large number of potential path conflicts is still very poor. In order to reduce the number of hand-tuning parameters and deal with sensing uncertainty, some researchers proposed learning-based methods to solve the planning problem [20].

2.2 Learning Based Methods

Thanks to the rapid development of deep learning technology in recent years, learning based method is considered to be a promising and promising direction to solve the task of path planning. Reinforcement learning has always been a powerful tool to solve planning problems, and reinforcement learning successfully solves the path planning problem, in which the agent completes the task through repeated trial-and-error. ORCA [35] adjusts the speed, size and direction of agents online to avoid collision. On the separately planned single agent path, the recent work focuses on the obstacle avoidance method of reinforcement learning Sartoretti et al. [15,16] have proposed a hybrid learning-based method called PRIMAL for multi-agent path-finding that integrated imitation learning and multi-agent reinforcement learning. In addition, there is a method of reinforcement learning combined with evolutionary thought. The approaches of Sartoretti et al. did not consider inter-robot communication, and thus, did not exploit the full potentials of the decentralized system [20]. In other words, communication is important, especially for decentralized approaches, and it is difficult to fully estimate the intention of adjacent decision agents without communication. Qingbiao et al. [18] use CNN to extract adequate features from local observations, and GNN to communicate these features among agents. Then train the model by imitating an expert algorithm, and use the model online in decentralized planning involving only local communication and local observations. In [19], they think that vanilla GNN relies on simplistic message aggregation mechanisms that prevent agents from prioritizing important information, so they extend the previous work that use vanilla Graph Neural Networks (GNNs) to Graph Attention Network (GAT). Their Message-Aware Graph Attention neTwork (MAGAT) is based on a key-query-like mechanism that determines the relative importance of features in the messages received from various neighboring robots. Everettz et al. [17] propose a decentralized multi-agent collision avoidance algorithm based on deep reinforcement learning, in which introduces a strategy using LSTM that enables the algorithm to use observations of an arbitrary number of other agents, instead of previous methods that have a fixed observation size.

3 Problem Formulation

We model the multi-agent path finding (MAPF) under the Markov decision processes (MDPs) framework, which converts MAPF to a sequential decision-making problem that each agent need to take the instant action option at time t, with two goals: 1) quickly reach the goal contently, 2) make great efforts to avoid collision among agents.

Environment. Consider a 2-dimensional discrete environment $E \subset \Re^2$ with size $W \times W$ and the cell $C \in E$. For each cell C, the cell will have three states: free, busy (occupied by someone agent) and There are a set of N_o obstacles $C_o = \{o_1, ..., o_{N_o}\}$, where $o_i \in E$ represents that there is the ith obstacle in the environment, and a set of N_a agents $A = \{a_1, ..., a_{N_a}\}$, where $a_i \in E$ represents that the ith agent can travel between the free cells And the free cells of the world can be represented by $C = E \backslash O$. For i^{th} agent, there are unique start position c_{s_i} and unique goal position c_{g_i}, with the constrain that $c_{s_i} \in C$ and $c_{g_i} \in C$. Our goal is to find a decentralized planner, which plans the ith agent's motion path by taking the local observation and the neighbor agents' position of the ith agent, besides quickly find the path to dispatch, it is also important to guarantee the safeness of the solution, i.e. try our best to avoid collision among agents.

Local Observation. For each agent in the environment, it will have a perception of its local environment and the information in its perception take an important role to plan the agent's motion path. In detail, at each time step t, the agent a_i will take its local observation around it which contains the information on obstacle location information near the current agent L_c^i, the other agents' position L_o^i, the agent goal position (if the goal in its neighboring) L_g^i and the other agents' goals as input L_{og}^i. Local observations of agent a_i can be formally expressed as $L_{ob}^i = [L_c^i; L_o^i; L_g^i; L_{og}^i]$.

Communications. Because there is no global positioning of the agents, the communication (or sharing information) between the agents are significant. Addressing the problems of what information should be sent to whom and when is crucial to solving the task effectively [19]. We try to design a communication block, which takes the local observation of the agent and its neighbor agents as input and outputs a tensor condensing the information that will redound to do a decision-making. And the local observation of the agent a_i and its neighbor agents can be formally expressed as $I^i = \{L_{ob}^j \mid j \in N_i\}$, while the output tensor can be recorded as S^i. To design more detail about the communication block, we should design a reasonable structure for the input and the output. Take an example, the agent a_i has three neighbor agents at some time. If the agent a_i should take all four agents' observations as input, the four observation information will have different degrees of importance. And the attention mechanisms, which can improve the ability of neural networks to process long sequence information, can select only some key information inputs to process.

4 Approach

To complete the description of an MDP process, we will detail the state's structure, the action spaces, the reward design, and the model architecture. To formally express the whole process, we also introduce the symbols declare in the previous section then we give the whole pseudo-code for the training process.

4.1 State's Structure

We consider that every agent partially observes the environment $(W \times H)$, where agents only observe the grid world in a limited field of view (FOV), the radius of which is defined as R_{FOV}, and 9 is actually used in our experiment. As mentioned in [15], partially observing the world is an important step towards the deployment of robots in the real world. Only in limited and small scenarios where a complete map of the environment is available (e.g., automated warehouses), agents with the full observation of the system can be trained by using a large enough FOV. Additionally, assuming a fixed FOV, the strategy can be extended to any world size, which also helps to reduce the input dimension of the neural network and reduce model complexity [15].

As mentioned above, the local observation of agent a_i can be formally expressed as $L_{ob}^i = [L_c^i; L_o^i; L_g^i; L_{og}^i]$, where static obstacle near the current agent $L_c^i \in W_{FOV} \times H_{FOV}$, the other agents' position $L_o^i \in W_{FOV} \times H_{FOV}$, the agent goal position (if exist) $L_g^i \in W_{FOV} \times H_{FOV}$, and the other agents' goals $L_{og}^i \in W_{FOV} \times H_{FOV}$. The observation structure is shown in Fig. 1. Due to the limitation of FOV, it's difficult to know the agent's destination when the agent's goal is far away from it and beyond FOV. So it's necessary to guide the goal when beyond FOV. Compared to [20,36] where use global guide path to guide the position of goal, this method implies finding the path through other methods. In other words, the guide path will need to be recomputed when the agents are out of the guide path and will make over-compute. In our experiment, using a guide path can not make sense and does not produce better results. So, we apply the setting similar to [15,16] where use unit vector pointing to its goal.

States Definition: Different from previous work [15,16], which only use the agent local observation as states, we think that the agents around the agent are very important to influence its action decision. Therefore, the information about the agents around the agent is also part of the current agent status, which take an important role in action decision. As mentioned above, local observations of the agent a_i and its neighbor agents can be formally expressed as $I^i = \{L_{ob}^j \mid j \in N_i\}$, and the local guide directions can be formally expressed as $V^i = \{v^j \mid j \in N_i\}$. Local observations I^i and local vectors V^i will be the input of our neural network, in which the optimal action will be estimated step by step.

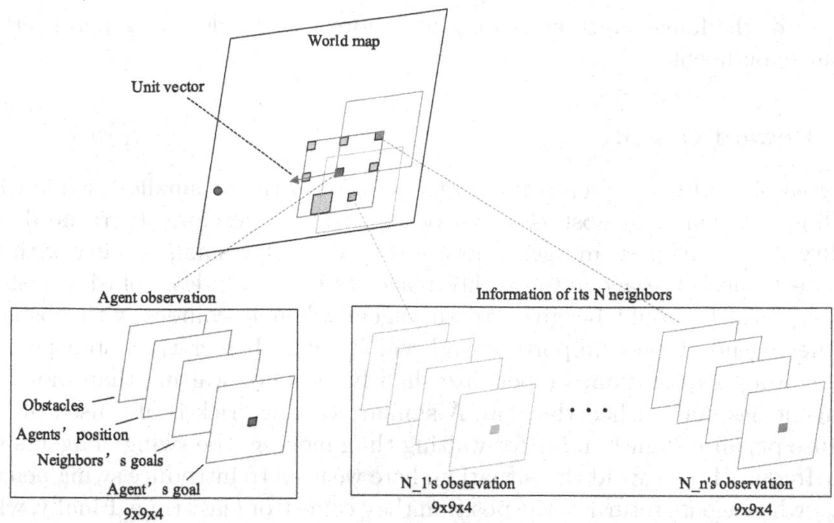

Fig. 1. State's structure of each agent (here, for the red agent). Agents represented as a red square are positioned in the squares, while their goals are represented as a solid circle of the same color. For the current agent (red agent), its local observation contains four channel information: positions of obstacles, positions of nearby agents, goal positions of these nearby agents, and position of its own goal (if exist in the FOV). In addition to the current (red) agent status, other nearby agents' states are also required (here, take the yellow agent and blue agent for example). (Color figure online)

4.2 Action Space

We describe the Multi-agent Path Finding (MAPF) problem as a sequence classification problem, which selects an optimal action from action space in each time step. And in our experiment, we consider a 4-connected grid environment, which means the agent can take 5 discrete actions in the grid world: moving a cell in one of the four basic directions or staying stationary. However, if the target mesh is already occupied by other agents or obstacles, the agent will not be able to move and will stay in their current position.

In some previous works, their experiment settings consider the grid environment is 8-connected, but there is only a little difference between the two settings because the movement in the 8-connected setting can be split into two steps movement in the 4-connected setting. For example, when the North and East grids of the agent are occupied and it want to go to the northeast grid, it can't go directly to the northeast grid in the 4-connected setting but can do it in the 8-connected. Something like a continuous slash in the real-world is discretized into a grid world. More specifically, a slender obstacle lays on the ground obliquely. When the agent needs to pass from one side to the other, the agent will encounter the above situation. At this time, it should not go directly through the slender obstacle and the 4-connected is more suitable for this situation.

In a word, the four-connected setting is a more reasonable and general setting for our experiment.

4.3 Reward Design

The goal of MAPF is to reach the target position with the smallest stride while avoiding collision with obstacles and other agents. Therefore, there need step penalty r_{step} that after an agent move a step will get a small penalty with the purpose to push the agent to quickly reach its goal. Besides, collision penalty $r_{collision}$, which should be given to the agent when it collides with obstacles or other agents, is also important and will be little bigger than step penalty. To encourage exploration, we penalize slightly more for waiting than moving if the agent has not reached the goal. A similar training trick is also used in [20]. Because penalize slightly more for waiting than moving, the swing of agents will occur frequently. To avoid this situation, here we need to introduce swing penalty r_{swing} when agents return to the position they come from last time. Finally, when the agent reach its goal, the goal-reaching reward r_{goal} will be given to the agent. The detailed values of these reward components in our experiment can be found in Table 1.

Table 1. Reward design

Action	Reward
Step penalty (move)	−0.3
Step penalty (wait)	−0.5
Collision penalty	−2.0
Swing penalty	−1.0
Goal-reach reward	+40.0

4.4 Model Architecture

In this section, we will explain how the agent outputs an action based on the inputs. The whole model architecture is shown in Fig. 2. First, the agents locally observe their environment to get the surrounding information, and through an observation encoder to get the expression of its local observation. Then, the agent will communicate with its neighbor agents to cooperate, in which an attention-based communication block is used to exchange the local observation, so the communication block can also be called the interaction layer. Finally, the tensor output from the communication block will be inputted to a decision block which is considered as the learned strategy and estimates the optimal action at that time.

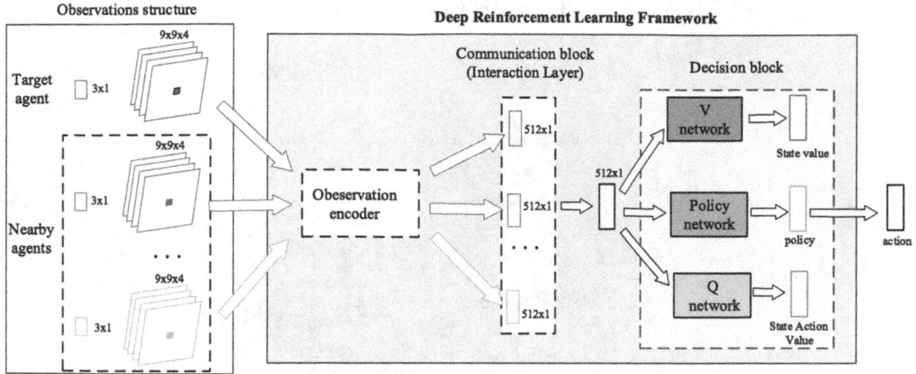

Fig. 2. Model architecture. For the input observations, there are local observations of the target agent (red agent) that need to be planned and other agents nearby the target agent (blue agent, yellow agent, etc.). For our deep reinforcement learning framework, there are three components: observation encoder, communication block, and decision block. Finally, the estimated optimal action from the policy network is taken as the output of the model. (Color figure online)

Observation Encoder: Because the motivation of this module is to extract and encode the local observations, it also can be called Feature Layers. In detail, at each time step t, the agent a^i will take its local observation around it, which be formally expressed as

$$L_t^i = [L_C^i; L_{OP}^i; L_G^i; L_{OG}^i] \tag{1}$$

where L_C^i is the obstacle location information near the current agent a^i, L_{OP}^i is the other agents' position, L_G^i is the agent goal position (if the goal in its neighboring) and L_{OG}^i is the other agents' goals. Besides observation information, it is significant to introduce the local guide direction formally expressed as v_t^i with the purpose to point the goal position especially when the grid world is too big and the goal is beyond the FOV. By using the two local state as input, the observation encoder will output an intermediate expression h_t^i:

$$h_t^i = ObservationEncoder(L_t^i, v_t^i) \tag{2}$$

The whole model structure of the Observation Encoder can be found in Fig. 3. The observation encoder is refined from previous works. For [15,16,20,36], they also use very deep architecture for Convolutional Neural Networks but a degradation problem has been exposed: with the network depth increasing, accuracy gets saturated and then degrades rapidly [37]. So we refine the structure with Residual CNN [37]. Except for this modification of observation encoder, it is worth mentioning that many previous works [15–17,36] put history states or history trajectory in an important position. Here, we store a tensor for history state and use the GRU cell to combine current state h_t^i and history state h_{t-1}^i. Compared to LSTM cell used in [15,16], GRU cell has few parameters, which reduces the risk of over fitting.

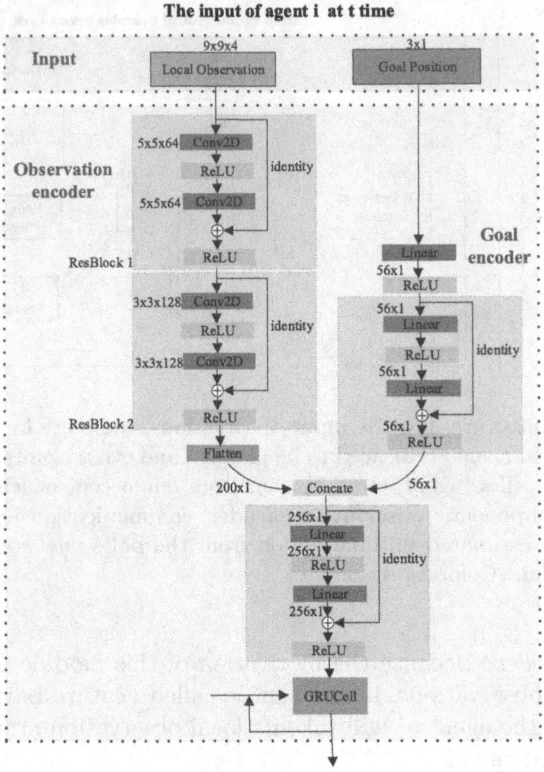

Fig. 3. The structure of observation encoder

Communication Block: With the For a certain agent a^i, it will communicate with its neighbors N^i and share their intermediate expressions h^j $(j \in N^i)$. Then communication block will output the comprehensive expression s_t^i:

$$s_t^i = CommunicationBlock(h_t^j), h_t^j \in N_i \tag{3}$$

Communication block is the well-known attention mechanism:

$$\alpha(h_t^i, h_t^j) = \frac{exp(score(h_t^i, h_t^j))}{\sum_{j\prime \in N_i} exp(score(h_t^i, h_t^{j\prime})))} \tag{4}$$

where h_t^i (h_t^j) is the output of observation encoder which takes the local observation and the guiding vector of agent a^i (a^j) as input at t time, *score* is the similarity of its two input tensor (in our experiment, the result of the dot product is the score), N_i is the nearby agents of the agent a^i, $\alpha(h_t^i, h_t^j)$ is the attention weight between h_t^i and h_t^j. The comprehensive expression s_t^i will be the weighted sum of $h_t^j(j \in N_i)$, and it contains local states and nearby agents' states so it will help the target agent to make action decision.

$$s_t^i = \sum_{j \in N_i} \alpha(h_t^i, h_t^j) \cdot h_t^j \tag{5}$$

Decision Block: Based the comprehensive expression s^i of communication block, decision block will need output the action policy for agent a^i at t time:

$$a_t^i = DecisionBlock(s_t^i) \tag{6}$$

We choose SAC to train our model because of its stability and exploratory, and what's more, we compare with other algorithms (e.g. VPG, TRPO, PPO, DDPG, TD3) but do not achieve the same good effect as SAC. It is very hard to prove and explain which one of the popular DRL algorithms, such as VPG, TRPO, PPO, DDPG, TD3, and SAC, is best. And our focus of work does not tend to fully compare the algorithms, so we did not insist that SAC was the best one, while the SAC algorithm may be more suitable for our environment setting. SAC has a parameterized state value function $V_\psi(s_t)$, soft Q function $Q_\theta(s_t, a_t)$, and the strategy function π_ϕ, where ψ, θ, ϕ are their parameters. And SAC is still based on actor-critic architecture, while the value function and the Q function are related, which can be regarded as critic network. The soft value function is trained to minimize the squared residual error:

$$J_V(\psi) = E_{s_t \sim D}\ [\frac{1}{2}(V_\psi(s_t) - E_{a_t \sim \pi_\phi}\ [Q_\theta(s_t, a_t) - log\pi_\phi(a_t|s_t)])^2] \tag{7}$$

where a_t is obtained by current strategy based on the current state s_t, D is experience replay buffer, $log\pi_\phi(a_t|s_t)$ is the entropy of strategy. The soft Q-function parameters can be trained to minimize the MSE loss between Q value and target $\hat{Q}$ value:

$$J_Q(\theta) = E_{(s_t, a_t) \sim D}\ [\frac{1}{2}(Q_\theta(s_t, a_t) - \hat{Q}(s_t, a_t))^2] \tag{8}$$

where (s_t, a_t) is sampled from experience replay buffer D, target $\hat{Q}$ value and state value V is related:

$$\hat{Q}(s_t, a_t) = R(a_t, s_t) + \gamma E_{s_{t+1}}\ [V_{\bar{\psi}}(s_t)] \tag{9}$$

$V_{\bar{\psi}}$ is the target network which can reduce sample correlation. Finally, the policy parameters θ can be learned by directly minimizing the expected KL-divergence

$$J_\pi(\phi) = E_{s_t \sim D}[D_{KL}(\pi_\phi(\cdot|| \frac{exp(Q_\theta(s_t, \cdot))}{Z_\theta(s_t)})] \tag{10}$$

In SAC, the updating of actor network π_ϕ is realized by minimizing KL divergence. The original paper uses the reparameterization trick $a_t = f_\psi(\varepsilon_t, s_t)$. The final loss function can be obtained as follows:

$$J_\pi(\phi) = E_{s_t \sim D, \varepsilon_t \sim \Delta}\ [log\pi_\phi(f_\phi(\varepsilon_t, s_t)|s_t) - Q_\theta(s_t, f_\phi(\varepsilon_t, s_t))] \tag{11}$$

where ε_t is noise sampled from Gaussian distribution, π_θ is defined implicitly in terms of f_θ [38].

Algorithm 1. Off-policy Training of Proposed Framework.

Initialize the capacity of replay memory D;
Initialize the weights of observation encoder and decision block;
for each iteration **do**
  for each environment step t **do**
    for each agent i **do**
      get local observation L_t^i, guiding vector v_t^i and its nearby agents N_i
      $h_t^i = ObservationEncoder(L_t^i, v_t^i)$
      $s_t^i = CommunicationBlock(h_t^j), j \in N_i$
      $a_t^i = DecisionBlock(s_t^i)$
      $s_{t+1} \sim p(s_{t+1}|s_t, a_t)$
      $D = D \cup (s_t, a_t, r(s_t, a_t), s_{t+1})$
    end for
  end for
  for each gradient step **do**
    Sample a minibatch data from D
    Update framework by loss Eq. (7), Eq. (9), Eq. (11)
  end for
end for

5 Experiments

5.1 Experiment Setting

We evaluate our approach in a grid world simulation environment, which is similar to [15,16]. The size of the square environment is randomly selected at the beginning of each episode to be either 20, 50, and 100. And the obstacle density is randomly selected from 0%, 10%, or 30%. The placement of obstacles, agents, and goals is uniformly at random across the environment, with the caveat that each agent had to be able to reach its goal. Here we draw lessons from [15] and each agent is initially placed in the same connected region as its goal. The actions of the agents are executed sequentially in random order at each timestep to ensure that they have equal priority.

Inspired by curriculum learning, our training procedure is divided into a few stages and starts from easier tasks to more difficult tasks. In the easy scene, we begin by initializing a small population of agents and dynamic obstacles, and sample goals within a certain distance to let agents learn a short-range navigation policy. Then we increase the agents and dynamic obstacles number, and sample goals in the whole map.

5.2 Training Details

We use a discount factor (γ) of 0.95. We use different length episodes in different size world, e.g. in 20-size world episode length is 128, in 50-size world episode is 256, in 100-size world episode length is 512. And the batch size is 128 so that integer multiple times of gradient updating can be performed each episode per agent. And we use Pytorch to realize the model and use RAdam [6] with a learning rate beginning at $3 \cdot 10^{-4}$.

5.3 Metrics

1. *Success Rate* $= n_{success}/n$, where $n_{success}$ is the number of agents that completed its travel and reached the goal and n is the total number of the agents that need to be planned, is the ratio of the number of agents reaching their goals within a certain time limit over the total number of robots.
2. *Extra Time rate* $= (T - T^*)/T^*$ is the difference between the averaged travel time on all robots and the lower bound of the travel time, where T is the averaged travel time on all robots and T^* is the lower bound of the travel time. The lower bound is the needed time for the agent's tour ignoring other agents.

5.4 Baselines

We introduce two traditional algorithms and three learning-based models:

1. LRA* [9]: Local Repair A* (LRA*) is a simple re-plan method, in which each agent will search for its own path to its goal ignoring other agents, but when an agent will encounter collision the algorithm will recalculate the remainder of its route. Local Repair A* may be an adequate solution for simple environments with few obstacles and few agents. But with more complex environments, LRA* will fall into the dilemma of too many recalculations.
2. CBS [11]: Conflict Based Search (CBS) is a two-level algorithm. At the high level, a search is performed on the Conflict Tree (CT), which is a tree-based on conflicts between individual agents, whose each node represents a set of constraints for the agents' motion. And at the low level, fast single agent searches are performed to meet the constraints imposed by the high level CT node.
3. PRIMAL [15,16]: PRIMAL is a hybrid learning-based method for MAPF that uses both imitation learning (based on an expert algorithm) and multi-agent reinforcement learning. But in PRIMAL it does not take inter-agent communication into consideration. As mentioned in [18], the key to solving the MAPF problem is learning what, how, and when to communicate.
4. IL_GNN [18,19]: They use a convolutional neural network to extract features from local observations, and a graph neural network to communicate these features among agents. But in their work, they use imitation learning to train models, so there will be a lack of exploration of the environment and it's hard to converge.
5. G2RL [36]: This algorithm combines the global guidance path and trains the model through the framework of reinforcement learning. In G2RL, it does not consider inter-agent communication and add the trick of guide path into every agent's state.

We also conduct ablation experiments that remove part of components of our method, which are removing the moving history state in observation (w/o his.) and removing the communication block (w/o comm.). We evaluate the performance of each method in terms of the success rate and the extra time rate in different experiment settings.

5.5 Results

In this section, all the results are shown and there will be some comparisons among our model, baseline models, and our ablation model.

Table 2. Results for success rate. Values are listed as "mean" across 100 instances. The highest (best) values are highlighted.

Environment setting			Success rate							
Map size	Agents number	Obstacles density	LRA*	CBS	PRIMAL	IL_GNN	G2RL	Ours	Ours w/o his.	Ours w/o comm.
20 × 20	4	0%	**1.00**	**1.00**	0.965	0.960	0.990	0.990	0.990	0.945
		10%	**0.985**	**0.985**	0.875	0.915	0.960	**0.985**	0.980	0.925
		30%	0.975	**0.985**	0.920	0.910	0.955	**0.985**	0.975	0.915
50 × 50	10	0%	0.986	**0.992**	0.746	0.848	0.852	0.988	0.940	0.914
		10%	0.972	**0.990**	0.774	0.926	0.826	**0.990**	0.924	0.848
		30%	0.986	0.984	0.816	0.924	0.830	**0.988**	0.912	0.862
100 × 100	20	0%	0.961	0.974	0.678	0.896	0.783	**0.981**	0.890	0.901
		10%	0.962	0.973	0.681	0.894	0.783	**0.975**	0.879	0.892
		30%	0.950	0.966	0.662	0.868	0.764	**0.970**	0.863	0.871

The results of the success rate can be found in Table 2. In our experiment, we took a search-based method (LRA*) and a conflict-based method (CBS) as traditional models to be part of baseline. Compared to traditional models (e.g. LRA*, CBS), the success rate of our model is very similar to that of the two baselines when the grid world is small. With the growth of the number of agents, world scale and obstacle density, we can find that the success rate of baseline has decreased obviously, but the success rate of our model decreases more slowly than the two baselines. However, we found that the traditional methods can get best success rate in small scale world, which is caused by the fact that if the world is small and the number of agent is small, it may be no need to do any communication in this situation. And what's more, if the world is small and the targets are very close to the agents, the information of FOV already contains most of the information of the environment, so the decentralized partial observation problem can be regarded as centralized full observation problem. Then our model may lose its advantages and will not achieve the high success rate of the traditional methods. Fortunately, the application scenarios in reality will not be particularly small. Besides search-based and conflict-based methods, we also take some learning-based model into consideration (PRIMAL, IL_GNN, G2RL).

Compared with the previous learning based models, our model has the best success rate in any case. There is also an interesting discovery that as the problem becomes more complex, the models without the global guide path (PRIMAL) will fast decline, while the models with the global guide path (IL_GNN, G2RL) will slowly decline. In ablation experiments, we found that the history state's effect is small when the size of the problem is simple while the influence of the history state increases with the expansion of the environment size. Because when the world is small scale and simple, there is no need to do long-horizon decision, so the history state do not have a great influence on path planning.

Table 3. Results for extra time rate. Values are listed as "mean/standard deviation" across 100 instances. The lowest (best) values are highlighted.

Environment setting			Extra time rate			
Map size	Agents number	Obstacles density	PRIMAL	IL_GNN	G2RL	Ours
20 × 20	4	0%	3.825/2.007	2.314/2.897	1.512/1.340	**0.320/0.547**
		10%	4.200/3.628	4.097/1.892	1.790/2.470	**0.430/0.201**
		30%	4.485/4.912	4.139/2.321	2.915/1.983	**0.601/0.381**
50 × 50	10	0%	6.146/7.132	5.614/1.563	5.614/1.563	**0.572/0.243**
		10%	6.850/8.324	6.810/1.522	4.502/2.801	**0.729/0.257**
		30%	6.200/10.373	6.894/2.518	8.294/1.489	**0.821/0.681**
100 × 100	20	0%	7.301/12.613	5.401/5.201	6.401/5.201	**0.901/0.412**
		10%	9.812/8.979	6.942/4.975	7.823/8.231	**1.981/1.034**
		30%	10.272/8.346	8.241/7.515	8.241/7.515	**2.492/1.341**

The results of extra time rate can be found in Table 3. We think that the success rate can not completely measure the quality of the model, because it only focuses on whether the agent has achieved its goal Within a given time ignoring how good the path it finds. Therefore, we use the extra time rate to indicate how much extra time the method needs, in which the low bound of a path length is referred to a global path computed in a static environment ignoring other agents. If the path length found through the model is closer to the global path, we think the solution will be better and the Extra Time Rate metric will be smaller. Obviously, we found that our model can reach the low extra time rate than other learning-based methods. In addition, compared with other learning based models, the standard deviation of the Extra Time Rate in our model is the lowest, which proves the stability of the solution of our model.

6 Conclusion

This paper proposes a new model for multi-agent path finding in the partially observable environment, which is formalized into DEC-POMDP process and trained in the form of deep reinforcement learning through repeated trial-and-error. And our model contains an observation encoder which can extract features from local observations and integrate its history information, a communication block which can complete communication in nearby agents, and a decision block which can output the estimated optimal action. The experiment result shows that our model outperforms traditional methods in most cases, and outperforms the previous learning-based models in all cases in terms of the success rate and the extra time rate among various experiment settings. What's more, the ablation experiment shows that the components exactly work and the performance of our model will deteriorate without them.

However, based on the fact that our model do not much better than traditional models when the world size is small, we think that if the world is small and the number of agent is small, it may be no need to do any communication in this situation. Fortunately, the application scenarios in reality will not be particularly small. And there is another fact that removing the history state does not affect result too much in small scale world, so we think we can reorganize the history state design to improve the result of small scale world in future work.

References

1. Stern, R., et al.: Multi-agent pathfinding: definitions, variants, and benchmarks. arXiv preprint arXiv:1906.08291 (2019)
2. Hönig, W., Kiesel, S., Tinka, A., Durham, J.W., Ayanian, N.: Persistent and robust execution of MAPF schedules in warehouses. IEEE Robot. Autom. Lett. **4**(2), 1125–1131 (2019)
3. Wurman, P.R., D'Andrea, R., Mountz, M.: Coordinating hundreds of cooperative, autonomous vehicles in warehouses. AI Mag. **29**(1), 9 (2008)
4. Balakrishnan, H., Jung, Y.: A framework for coordinated surface operations planning at Dallas-Fort Worth International Airport. In: AIAA Guidance, Navigation and Control Conference and Exhibit, p. 6553 (2007)
5. Baxter, J.L., Burke, E., Garibaldi, J.M., Norman, M.: Multi-robot search and rescue: a potential field based approach. In: Mukhopadhyay, S.C., Gupta, G.S. (eds.) Autonomous Robots and Agents. SCI, vol. 76, pp. 9–16. Springer, Heidelberg (2007). https://doi.org/10.1007/978-3-540-73424-6_2
6. Zhang, Y., Qian, Y., Yao, Y., Hu, H., Xu, Y.: Learning to cooperate: application of deep reinforcement learning for online AGV path finding. In: Proceedings of the 19th International Conference on Autonomous Agents and MultiAgent Systems, pp. 2077–2079 (2020)
7. Arques Corrales, P., Aznar Gregori, F.: Swarm AGV optimization using deep reinforcement learning. In: 2020 The 3rd International Conference on Machine Learning and Machine Intelligence, pp. 65–69 (2020)
8. Yu, J., LaValle, S.: Structure and intractability of optimal multi-robot path planning on graphs. In: Proceedings of the AAAI Conference on Artificial Intelligence, vol. 27 (2013)
9. Panait, L., Luke, S.: Cooperative multi-agent learning: the state of the art. Auton. Agent. Multi-Agent Syst. **11**(3), 387–434 (2005). https://doi.org/10.1007/s10458-005-2631-2
10. Silver, D.: Cooperative pathfinding. In: AIIDE, vol. 1, pp. 117–122 (2005)

11. Sharon, G., Stern, R., Felner, A., Sturtevant, N.R.: Conflict-based search for optimal multi-agent pathfinding. Artif. Intell. **219**, 40–66 (2015)
12. Sharon, G., Stern, R., Felner, A., Sturtevant, N.R.: Meta-agent conflict-based search for optimal multi-agent path finding. In: SoCS, vol. 1, pp. 39–40 (2012)
13. Barer, M., Sharon, G., Stern, R., Felner, A.: Suboptimal variants of the conflict-based search algorithm for the multi-agent pathfinding problem. In: Seventh Annual Symposium on Combinatorial Search. Citeseer (2014)
14. Boyarski, E., et al.: ICBS: improved conflict-based search algorithm for multi-agent pathfinding. In: Twenty-Fourth International Joint Conference on Artificial Intelligence (2015)
15. Sartoretti, G., et al.: PRIMAL: pathfinding via reinforcement and imitation multi-agent learning. IEEE Robot. Autom. Lett. **4**(3), 2378–2385 (2019)
16. Damani, M., Luo, Z., Wenzel, E., Sartoretti, G.: PRIMAL_2: pathfinding via reinforcement and imitation multi-agent learning-lifelong. IEEE Robot. Autom. Lett. **6**(2), 2666–2673 (2021)
17. Everett, M., Chen, Y.F., How, J.P.: Motion planning among dynamic, decision-making agents with deep reinforcement learning. In: 2018 IEEE/RSJ International Conference on Intelligent Robots and Systems (IROS), pp. 3052–3059. IEEE (2018)
18. Li, Q., Gama, F., Ribeiro, A., Prorok, A.: Graph neural networks for decentralized multi-robot path planning. arXiv preprint arXiv:1912.06095 (2019)
19. Li, Q., Lin, W., Liu, Z., Prorok, A.: Message-aware graph attention networks for large-scale multi-robot path planning. IEEE Robot. Autom. Lett. **6**(3), 5533–5540 (2021)
20. Liu, Z., Chen, B., Zhou, H., Koushik, G., Hebert, M., Zhao, D.: MAPPER: multi-agent path planning with evolutionary reinforcement learning in mixed dynamic environments. In: 2020 IEEE/RSJ International Conference on Intelligent Robots and Systems (IROS), pp. 11748–11754. IEEE (2020)
21. Jansen, R., Sturtevant, N.: A new approach to cooperative pathfinding. In: Proceedings of the 7th International Joint Conference on Autonomous Agents and Multiagent Systems, vol. 3, pp. 1401–1404. Citeseer (2008)
22. Ryan, M.R.K.: Exploiting subgraph structure in multi-robot path planning. J. Artif. Intell. Res. **31**, 497–542 (2008)
23. Wang, K.-H.C., Botea, A., et al.: Fast and memory-efficient multi-agent pathfinding. In: ICAPS, pp. 380–387 (2008)
24. Aljalaud, F., Sturtevant, N.R.: Finding bounded suboptimal multi-agent path planning solutions using increasing cost tree search. In: SOCS (2013)
25. Sharon, G., Stern, R., Goldenberg, M., Felner, A.: The increasing cost tree search for optimal multi-agent pathfinding. Artif. Intell. **195**, 470–495 (2013)
26. Walker, T.T., Sturtevant, N.R., Felner, A.: Extended increasing cost tree search for non-unit cost domains. In: IJCAI, pp. 534–540 (2018)
27. Surynck, P.: On propositional encodings of cooperative path-finding. In: 2012 IEEE 24th International Conference on Tools with Artificial Intelligence, vol. 1, pp. 524–531. IEEE (2012)
28. Surynek, P.: Compact representations of cooperative path-finding as sat based on matchings in bipartite graphs. In: 2014 IEEE 26th International Conference on Tools with Artificial Intelligence, pp. 875–882. IEEE (2014)
29. Surynek, P.: Reduced time-expansion graphs and goal decomposition for solving cooperative path finding sub-optimally. In: IJCAI, pp. 1916–1922 (2015)
30. Surynek, P., Felner, A., Stern, R., Boyarski, E.: Efficient SAT approach to multi-agent path finding under the sum of costs objective. In: Proceedings of the Twenty-Second European Conference on Artificial Intelligence, pp. 810–818 (2016)

31. Yu, J., LaValle, S.M.: Multi-agent path planning and network flow. In: Frazzoli, E., Lozano-Perez, T., Roy, N., Rus, D. (eds.) Algorithmic Foundations of Robotics X. STAR, vol. 86, pp. 157–173. Springer, Heidelberg (2013). https://doi.org/10.1007/978-3-642-36279-8_10
32. Yu, J., LaValle, S.M.: Optimal multirobot path planning on graphs: complete algorithms and effective heuristics. IEEE Trans. Robot. **32**(5), 1163–1177 (2016)
33. Lam, E., Le Bodic, P., Harabor, D.D., Stuckey, P.J.: Branch-and-cut-and-price for multi-agent pathfinding. In: IJCAI, pp. 1289–1296 (2019)
34. Erdem, E., Kisa, D., Oztok, U., Schüller, P.: A general formal framework for pathfinding problems with multiple agents. In: Proceedings of the AAAI Conference on Artificial Intelligence, vol. 27 (2013)
35. Chen, Y.F., Liu, M., Everett, M., How, J.P.: Decentralized non-communicating multiagent collision avoidance with deep reinforcement learning. In: 2017 IEEE International Conference on Robotics and Automation (ICRA), pp. 285–292. IEEE (2017)
36. Wang, B., Liu, Z., Li, Q., Prorok, A.: Mobile robot path planning in dynamic environments through globally guided reinforcement learning. IEEE Robot. Autom. Lett. **5**(4), 6932–6939 (2020)
37. He, K., Zhang, X., Ren, S., Sun, J.: Deep residual learning for image recognition. In: Proceedings of the IEEE Conference on Computer Vision and Pattern Recognition, pp. 770–778 (2016)
38. Haarnoja, T., Zhou, A., Abbeel, P., Levine, S.: Soft Actor-Critic: off-policy maximum entropy deep reinforcement learning with a stochastic actor. In: International Conference on Machine Learning, pp. 1861–1870. PMLR (2018)

Big Data-Driven Stable Task Allocation in Ride-Hailing Services

Jingwei Lv, Nan Zhou(✉), and Shuzhen Yao

School of Computer Science and Engineering, Beihang University, Beijing 100191, China
{lvjingwei,nzhou,yaoshuzhen}@buaa.edu.cn

Abstract. Ride-hailing is important for urban transportation and greatly improves the efficiency of public transportation. In the ride-hailing platform, it is a core work to match the passengers and drivers scientifically and quickly. The interests of passengers, drivers and the platform should all be considered in the task assignment process. We first proposed a baseline algorithm which can satisfy stable matching, but the global pickup distance (the total distance of all taxis to pick up passengers) is relatively large, so we invented a chain algorithm which can optimize the global pickup distance, and the most of the results satisfy stable matching at the same time. Moreover, we verified the practical effect of this algorithm through experiments.

Keywords: Ride-hailing · Stable matching · Global optimal distance

1 Introduction

In recent years, ride-hailing platforms are becoming more and more important for public transportation. The core task of ride-hailing platforms is task assignment, which is to assign passengers to drivers. Most of the existing researches model this problem as bipartite graph matching problem. In a graph $G = (U; V; E)$, the set of nodes U and V can represent the workers (drivers) and tasks (passengers) respectively. The set of edges E can represent the utility or cost between different tasks and workers.

Thus, we need to find a matching on G to achieve different goals. The first goal is to maximize revenue, or minimize expense. In this respect, Greedy methods [1,2] are used to reduce the computation cost in an actual scene (a real-world scene). However, all of them are from the perspective of global optimization, and don't reflect the fairness of the interests between passengers and drivers. In other words, global optimization often fails to meet the maximum preferences of both passengers and drivers. Therefore, some researchers proposed to apply the theory of stable matching to improve the fairness and meet the preferences of each passenger and driver to the maximum extent. By defining the preferences of passengers and drivers, Zhao et al. [3] addressed the task allocation in ride-hailing platforms which considers both dynamic scenario and the stability of the

U. K. Rage et al. (Eds.): DASFAA 2022 Workshops, LNCS 13248, pp. 291–300, 2022.
https://doi.org/10.1007/978-3-031-11217-1_21

matching. Fenoaltea et al. [4] analyzed the difference between global optimization and stable matching from the perspective of energy optimization, and it is clearly pointed out that both cannot be achieved at the same time from the perspective of physicists. Therefore, this paper proposes how to balance stable matching and global distance optimization in ride-hailing task allocation. We set up a model of ride-hailing task allocation, attempted equilibrium stable matching and global distance optimization, and proposed a chain algorithm to achieve the goal.

Our main contributions can be summarized as follows:

1. We invent an equilibrium optimization problem of stable matching and global distance optimization in ride-hailing task allocation.
2. A baseline algorithm satisfying stable matching is proposed, and then a chain algorithm satisfying stable matching with distance constraint is proposed.
3. We model the problem of equilibrium optimization between stable matching and global distance optimization as a function, and find the optimal value of constraint conditions through experiments. We get a good equalization effect with the proportion of stable matching and the global optimal distance. The higher the stable matching ratios, the greater the global distances, and vice versa.

In the rest of this paper, we discuss some related work in Sect. 2, show concept definition and propose the baseline algorithm in Sect. 3. We propose the Chain algorithm with distance constraint, solve and the Equilibrium problem of stable matching and global distance by establishing benefit function in Sect. 4. We verify the above method through data and compare it with baseline algorithm in Sect. 5 and conclude in Sect. 6.

2 Related Work

Global Distance Optimization. It is usually optimized from the perspective of reducing the total distance traveled. Ahuja et al. [5] transferred the problem to the minimum-cost maximum-flow problem. Long et al. [6] explored a swap chain algorithm to find the optimal answer of matching with maximum cardinality. Tong et al. [7,8] studied the problem to maximize the total utility and achieved the minimum total expenditure.

Stable Marriage. The stable marriage problem was firstly introduced in [9], in which some classic concepts like Stable were proposed. Since then, this matching model has been applied to lots of situations [10,11]. Similar to our problem, the online stable matching problem was researched in [12]. Khuller et al. [13] brings out an online weighted bipartite matching problem as an online stable marriage model. In addition, some effort was made on house-roommates stable matching to maximize the social welfare [14].

3 Problem Definition

We will introduce the problem definition and propose a baseline method for stable matching.

3.1 Preliminaries and Definition

Definition 1 (Task). *A task, denoted by*

$$t = <A_t, s_t, p_t> \tag{1}$$

appears at time s_t and at location A_t, needs to be served with a delivery worth p_t.

Definition 2 (Worker). *A worker, denoted by*

$$t = <A_w, s_w, w - threshold> \tag{2}$$

appears at time s_w and at location A_w, it will only accept the task to which the distance from the worker is no more than w-threshold. In other words, if there is a task r and the distance between r and w is more than w-threshold, r cannot be assigned to w in any instance.

Normally, a task request is often put forward with a destination rather than the travel cost, which is easy to calculate with both locations (the starting point and the destination) in the 2D space. We assume that workers only consider the travel cost related to the tasks, and then the travel costs serve as the same role as the travel distance to workers, simplifying the following discussion.

Definition 3 (Distance). *The distance, denoted by d(t, w), can be defined as Euclidean Distance*

$$d(t, w) = l(|t - w|) \tag{3}$$

Definition 4 (Matching). *If T and W are a set of tasks and a set of workers respectively, a matching, denoted by $M \in T \times W$, insists of binary pairs $<t, w>$. Certain t or w don't appear twice in different pairs.*

Definition 5 (Blocking Pair). *A blocking pair, denoted by*

$$<t, w> \in T \times W \tag{4}$$

satisfies following condition: There is another pair $<t^*, w^*>$, and $p_t^* < p_t$ or w is unmatched, $d(t, w^*) < d(t, w)$ or t is unmatched.

As presented in the definition, when a task selects a worker, it only considers the pickup distance, it prioritizes the workers closest to it. When a worker selects a task, it only considers the price, and the higher the price is, the higher the priority is.

Definition 6 (Stability). *A matching M is stable if there are no blocking pairs in M.*

Definition 7 (Total pickup distance). *Given a matching M, the Total pickup distance of M is denoted by*

$$L = \sum <t, w> \in Md(t, w) \tag{5}$$

3.2 A Baseline Approach

As for Algorithm 1 Baseline, given parameters T and W, in line 1, the algorithm firstly initializes the M as an empty set. In lines 2–9, as long as T is not empty, select t from T who offers the highest price, then set W_t containing all workers in capable to t and w is the one who is closest to t in W_t. In lines 6–8, add pair (t, w) to M, then delete t from T and w from W, then turn to line 2 for check. When the cycle ends, return M as answer in line 10.

Algorithm 1 satisfies stable matching, but the total pickup distance is relatively large. As described in [4], global distance optimization and stable matching cannot be achieved at the same time. We can explain the above problem by the following example.

Table 1. The algorithm description of baseline.

```
Algorithm 1: Baseline (T, W)
   Input: Tasks: T, Worker: W
   Output: Matching: M
1  M ← ∅;
2  while T != ∅ do;
3      t ← argmax t ∈ T p_t;
4      W_t ← {w∈W —d(t,w) ≤ t_{w−threshhold}};
5      w ← argmin w∈ W_t d(t,w);
6      Insert (t,w) into M;
7      Remove t from T;
8      Remove w from W;
9  end while;
10 return M;
```

In Fig. 1, there are two tasks, t_1 and t_2, whose prices are 5 and 3 respectively, and two workers w_1 and w_2. The distance between any two objects is shown in the figure. At this point, there are two schemes. According to Algorithm 1, the matching result should be based on the red dotted line with a total distance of $9 + 4 = 13$, while according to the global optimal distance, it should be based on the blue dotted line with a total distance of $2 + 10 = 12$.

Therefore, we consider about the following question, if stable matching is broken according to the global distance optimal, then can we equilibrium stable matching and global distance optimal? We will try to propose an algorithm for it.

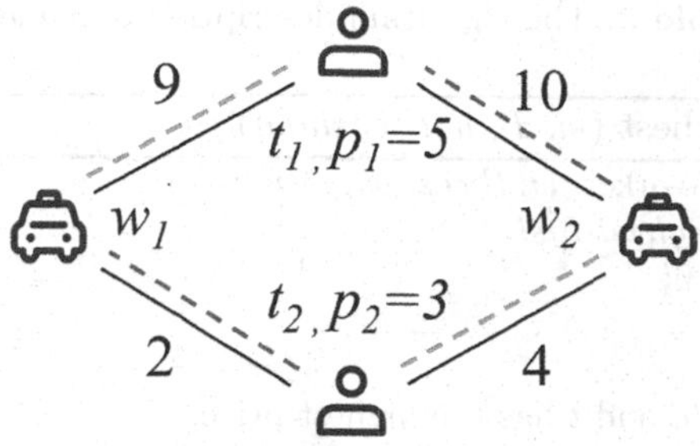

Fig. 1. Stable matching and global distance

4 Equilibrium Stable Matching and Global Distance Optimization

4.1 Chain Algorithm

In order to explore a method that can balance stable matching and global optimal distance, we need to establish a new matching process. Inspired by [15], we establish a chain algorithm, as Algorithm 2.

Table 2. The algorithm description of chain.

Algorithm 2: Chain (T, W)

```
Input: Tasks: T, Worker: W
Output: Matching: M
M ← ∅;
while T!= ∅ do;
  select a random object t ∈ T;
  C ← {t} // C is a chain;
  while C!= ∅ do;
  x ← the last element of C;
  if x ∈ T then ;
    y ← the Nearest neighbor of x in W;
    if y = the previous element of x in C then;
      remove y and x from C;
      insert (y, x) into M ;
      remove y and x from W and T, respectively;
    else;
      insert y into C;
  else;
    y ← Highest (x, T, x-threshold) ;
    if y = the previous element of x in C then;
      remove y and x from C;
      insert (x, y) into M ;
      remove x and y from W and T, respectively ;
    else;
      insert y into C;
  end while;
end while;
return M;
```

Table 3. The algorithm description of highest.

Algorithm 3: Highest (w, T, *w-threshold*)

```
Input: Tasks: T, Worker: w, threshold of w
Output: task: Highest-task
Highest − task = ∅;
for t in T;
   d1 = d(t,w);
   if d1 ≤ threshold and t has the highest price;
      Highest-task=t;
   if Highest-task=∅;
      Highest-task ← Nearest neighbor (w, T);
end for ;
return Highest-task;
```

M is initially an empty set in line 1. As long as T is a non-empty set, line 2 enters the loop, takes a random task t from T and stores it in linked list C. In line 5, as long as list C is a non-empty set, take the last element of the list and assign it to x. If x is a task, assign x's nearest neighbor in W to y. If the previous element of x in list C happens to be y, store them in M, and remove them from the list as well as T and W respectively. Otherwise, y continues to be stored in linked list C. If x is a worker, at line 16, we call a **Highest** subroutine (Algorithm 3), find the link from the worker to the next element, and instead of the nearest neighbor as mentioned earlier, the worker gets the most expensive task within a certain distance, and the rest of the process is the same. It can be seen that this linked list has a feature that the nodes stored in it are from the two sets T and W and are interlaced, which means the adjacent nodes must come from the two different sets above. This implements staggered mutual selection between sets of two elements.

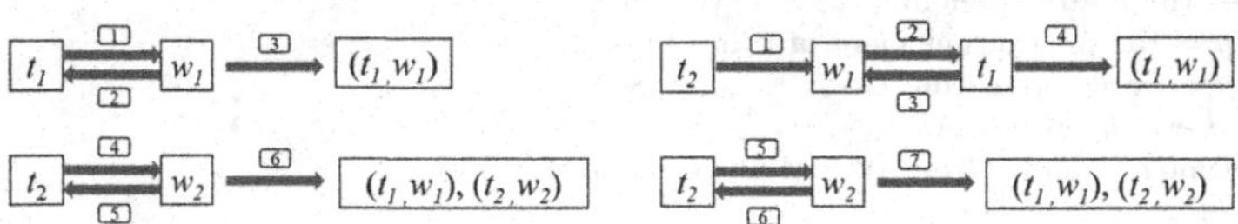

Fig. 2. The process of randomly picking t_1 and t_2 respectively for the first time without setting a threshold using chain

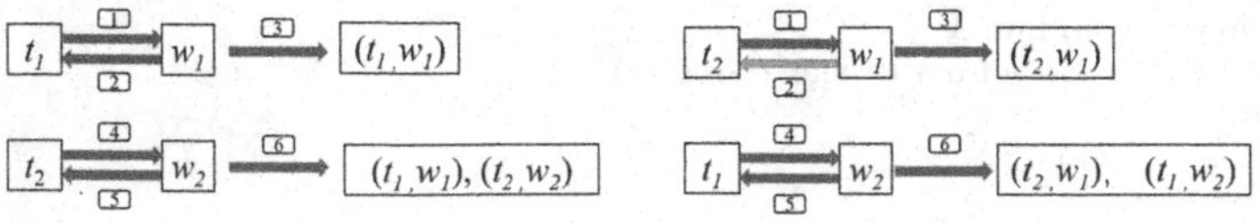

Fig. 3. The process of randomly picking t_1 and t_2 respectively for the first time setting a threshold of 8 using chain

4.2 The Benefit Function for Equilibrium

The parameter *w-threshold* in Algorithm 3 means when a worker chooses a task, it will not directly select the task which has the highest price, but chooses the task with the highest price in a certain scope for a distance threshold. In this condition, the matching result is not to satisfy the strict stability, but this constraint is more practical, which is different from Algorithm 1. It also gives more opportunities to shorten the global distance. Next, we take the example in Fig. 1 to detail the operation process of the chain algorithm. We can explain why the Chain algorithm can shorten the global distance with this example. **In the first case**, let's take a look at Fig. 2, if the distance threshold is not set when the worker selects the task, then the whole process is as follows. If we randomly pick task t_1, the execution process is the same as Algorithm 1, t_1 matches w_1, and t_2 matches w_2. If the first one randomly picks t_2, then w_1 continues in the list, w_1 picks t_1, t_1 picks w_1, t_1 matches w_1, and t_2 matches w_2. It can be seen that the results of Algorithm 2 and Algorithm 1 are consistent without setting the threshold. **In the second case**, let's take a look at Fig. 3, if we set **a threshold of 8** and randomly pick task t_1, t_1 matches w_1, and then t_2 matches w_2, we get the same result as in the first case. If t_2 is randomly selected for the first time, w_1 will continue to be stored in the linked list, but now w_1 will not select t_1, because t_1 does not meet the threshold requirements. At this point, w_1 will select the nearest t_2, t_2 matches w_1, and finally t_1 matches w_2. The total distance of this result is 12, which is less than the strictly stable matching distance with the total distance 13.

In the above example, we set the threshold at random, so how do we achieve the equilibrium of stable matching and global distance optimization when performing the process using chain algorithm? We define a function to represent the relationship among stable matching, global distance, and computation time. The goal is that the higher the proportion of stable matching in the matching result is, the shorter the global distance is and the shorter the calculation time is. When the function is at its maximum, the best threshold value can be obtained.

Definition 8 (The interests of the function)

$$IF = S/L - log(time) \tag{6}$$

In formula 6, IF represents the value of the benefit function, S represents the number of tasks and workers matched which satisfied stable matching in the matching result, L represents the global pickup distance, and *time* represents the time of calculating matching. In the overall goal, stable matching and global distance are the main factors, and the calculation time is also one of our goals, but not as important as the above two factors, so the first two are based on multiplication and division, then the logarithm of the calculation time is subtracted. In practical applications, we can obtain the maximum value of the benefit function through experiments, inversely deduce the distance threshold in **Highest** at this time, and the matching result at this time is the expected optimal matching result.

5 Experimental Study

The algorithm was implemented using VScode and Python, on a machine with CPU 3.00 GHz, 16 GB main memory. We used random distribution to construct the locations of tasks and the workers, the number from 1000, 2000 to 10000, Table 4 shows different running effect of Algorithm 1 and Algorithm 2, which have the stable matching percentage, and the global pick up distance, and calculation time. The calculation time of Algorithm 2 is about 3–4 times of Algorithm 1, but the total calculation time is acceptable in practice. In the case that the stable matching ratio decreases by less than 20%, the global pickup distance can be reduced by about 10%, which is of great significance for the majority of vehicles in the city to reduce empty driving and save fuel consumption although some matching results do not satisfy the strict stable matching. Figure 4 shows the comparison of global pickup distance and running time between two algorithms.

Table 4. Performance comparison between Algorithm 1 and Algorithm 2 under different data volume scales

Spend time for calculation (S)										
Number of tasks (number of workers)	1000	2000	3000	4000	5000	6000	7000	8000	9000	10000
Algorithm 1-baseline	0.33	1.27	2.89	5.13	8.04	11.68	16.26	20.50	27.23	33.15
Algorithm 2-chain	0.98	5.49	12.41	22.33	29.26	51.47	69.35	91.34	121.56	141.21
Algorithm 2/Algorithm 1	2.97	4.33	4.30	4.35	3.64	4.41	4.26	4.46	4.46	4.26
Total pickup distance										
Number of tasks (number of workers)	1000	2000	3000	4000	5000	6000	7000	8000	9000	10000
Algorithm 1-baseline	137643.42	5490912.35	5462875.39	7688472.67	1426516.37	9561045.23	9876729.32	9229110.37	10810949.90	6620067.14
Algorithm 2-chain	132424.19	5284084.09	4960434.83	6929788.67	1225164.26	8224093.29	8868032.40	8109400.50	9388456.43	5773017.83
Range optimization ratio	3.79%	3.77%	9.20%	9.87%	14.11%	13.98%	10.21%	12.13%	13.16%	12.80%

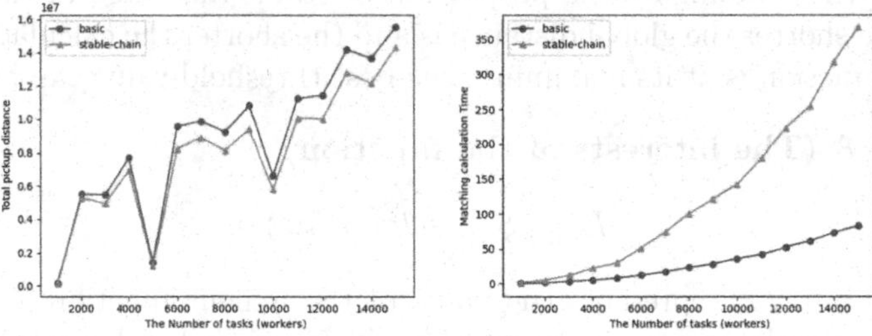

Fig. 4. Comparison of global pickup distance and running time between Algorithm 1 and Algorithm 2

Figure 5 shows if there are 1000 tasks (workers), and the benefit function value change with the threshold values ranging from 1 to 200. It can be found that when the threshold values is about 110 or so, we can get the best value of the interest function which is the matching results. In practical engineering applications, urban maps are often divided into equal size cells, and the matching tasks in each cell are carried out in parallel. Therefore, about 5000 tasks (workers) in a cell are enough, and we do not need to worry about the time-consuming of Algorithm 2 in matching calculation. Although this paper only studies the problem under the static matching case, which is not described under the online scenario, and we can transform it from the static scenario to the dynamic scenario by dividing a period of time into equal intervals, by setting the length of time window 2 min which is commonly used. We will accumulate all the tasks in each time window and matching workers, tasks or workers that fail to match in this time window go directly to the next time pane. This allows us to move our approach to an online dynamic task assignment scenario.

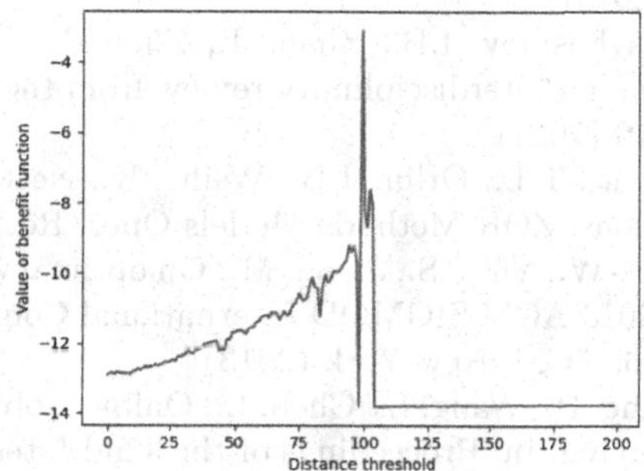

Fig. 5. Threshold value and benefit function value

6 Conclusion

For the ride-hailing task allocation, it is proposed for equilibrium stable matching and the global pick up distance, we proposed a baseline algorithm which satisfy stable matching, and then invented a chain algorithm with a distance threshold. The Chain algorithm will cost much more time than the baseline algorithm, the matching results proportion of stable matching has decreased, and has little influence on the whole calculation. The global pickup distance is significantly optimized, which is of great practical importance for saving resources. The effectiveness of this method is verified by experiments. By constructing the benefit function, we find the optimal equilibrium point of stable matching, global pickup distance and calculation time for a real data set. Our method also can be translated into dynamic scenario.

References

1. To, H., Fan, L., Tran, L., Shahabi, C.: Real-time task assignment in hyperlocal spatial crowdsourcing under budget constraints. In: 2016 IEEE International Conference on Pervasive Computing and Communications, Piscataway, NJ, pp. 1–8. IEEE (2016)
2. Tran, L., To, H., Fan, L., Shahabi, C.: A real-time framework for task assignment in hyperlocal spatial crowdsourcing. ACM Trans. Intell. Syst. Technol. **9**(3), 37 (2018)
3. Zhao, B., Xu, P., Shi, Y., Tong, Y., Zhou, Z., Zeng, Y.: Preference-aware task assignment in on-demand taxi dispatching: an online stable matching approach. In: Proceedings of the AAAI Conference on Artificial Intelligence, Menlo Park, pp. 2245–2252. AAAI (2019)
4. Fenoaltea, E.M., Baybusinov, I.B., Zhao, J., Zhou, L., Zhang, Y.-C.: The stable marriage problem: an interdisciplinary review from the physicist's perspective. Phys. Rep. **917**, 1–79 (2021)
5. Ahuja, R.K., Magnanti, T.L., Orlin, J.B., Weihe, K.: Network flows: theory, algorithms and applications. ZOR-Methods Models Oper. Res. **41**(3), 252–254 (1995)
6. Long, C., Wong, R.C.-W., Yu, P.S., Jiang, M.: On optimal worst-case matching. In: Proceedings of the 2013 ACM SIGMOD International Conference on Management of Data, pp. 845–856. ACM, New York (2013)
7. Tong, Y., She, J., Ding, B., Wang, L., Chen, L.: Online mobile micro-task allocation in spatial crowd-sourcing. In: Proceedings of the 32nd International Conference on Data Engineering, Piscataway, pp. 49–60. IEEE (2016)
8. Tong, Y., She, J., Ding, B., Chen, L., Wo, T., Ke, X.: Online minimum matching in real-time spatial data: experiments and analysis. Proc. VLDB Endow. **9**(12), 1053–1064 (2016)
9. Gale, D., Shapley, L.S.: College admissions and the stability of marriage. Am. Math. Monthly **69**(1), 9–15 (1962)
10. Gusfield, D., Irving, R.W.: The Stable Marriage Problem-Structure and Algorithms. Foundations of Computing Series, MIT Press, Cambridge (1989)
11. David, M.: Algorithmics of Matching Under Preferences. World Scientific, Singapore (2013)
12. Lee, H.: Online stable matching as a means of allocating distributed resources. J. Syst. Archit. **45**(15), 1345–1355 (1999)
13. Khuller, S., Mitchell, S.G., Vazirani, V.V.: On-line algorithms for weighted bipartite matching and stable marriages. Theor. Comput. Sci. **127**(2), 255–267 (1994)
14. Huzhang, G., Huang, X., Zhang, S., Bei, X.: Online roommate allocation problem. In: Proceedings of the 26th International Joint Conference on Artificial Intelligence, Menlo Park, pp. 235–241. AAAI (2017)
15. Wong, R.C.-W., Tao, Y., Fu, A.W.-C., Xiao, X.: On efficient spatial matching. In: Proceedings of the 33rd International Conference on Very Large Data Bases. VLDB Endowment, Trondheim, Norway, pp. 579–590 (2007)

Weighted Mean-Field Multi-Agent Reinforcement Learning via Reward Attribution Decomposition

Tingyu Wu[1], Wenhao Li[1], Bo Jin[1], Wei Zhang[2], and Xiangfeng Wang[1(✉)]

[1] School of Computer Science and Engineering, East China Normal University, Shanghai, China
xfwang@cs.ecnu.edu.cn

[2] School of Information and Communication Engineering, University of Electronic Science and Technology of China, Sichuan, China

Abstract. Existing MARL algorithms have low efficiency in many-agent scenarios due to the complex dynamic interaction when agents growing exponentially. Mean-field theory has been introduced to improve the scalability where complex interactions are approximated by those between a single agent and the mean effect from neighbors. However, only considering the averaged actions of neighborhood at last step and ignoring the dynamic influence of neighbors leads to unstable training procedures and sub-optimal solutions. In this paper, the Weighted **M**ean-**F**ield Multi-Agent Reinforcement Learning via **R**eward **A**ttribution **D**ecomposition (MFRAD) framework is proposed by differentiating heterogeneous and hysteresis neighbor effect with weighted mean-field approximation and reward attribution decomposition. The multi-head attention is employed to calculate the weights which formulate the weighted mean-field Q-function. To further eliminate the impact of hysteresis information, reward attribution decomposition is integrated to decompose weighted mean-field Q-value, improving the interpretability of MFRAD and achieving fully decentralized execution without information exchanging. Two novel regularization terms are also introduced to guarantee the consistency of temporal relationship among agents and unambiguity of local Q-value with no agents. Numerical experiments on many-agent scenarios demonstrate the superior performance against existing baselines.

Keywords: Multi-agent reinforcement learning · Weighted mean-field approximation · Reward attribution decomposition

1 Introduction

In recent years, multi-agent reinforcement learning (MARL) has registered great potential in multi-agent systems, showing extraordinary performance in various scenarios, such as multi-player games [27], resource allocation [19,32], and network routing [13,14]. When the number of agents increases, classic MARL algorithms are far less effective than expectation due to the non-stationary issue [15]

U. K. Rage et al. (Eds.): DASFAA 2022 Workshops, LNCS 13248, pp. 301–316, 2022.
https://doi.org/10.1007/978-3-031-11217-1_22

that agents not only interact with the environment but also with other agents. Besides, the increasing action space resulting in the curse of dimensionality brings new challenges. Therefore, some efficient approaches [6,16] are proposed recently to solve above problems, especially the investigation of centralized training and decentralized execution (CTDE) framework [10–12,20].

However, many real-world scenarios [1,31] contain hundreds of agents cooperating and struggling with each other, bringing about the more difficult learning procedure due to the enormous action space and exponential dynamic interaction. The CTDE framework can not directly adapt to such many-agent scenarios due to the existence of centralized critic. [26] takes advantage of mean-field theory to realize scalability, transferring the many-agent interaction into the interactions between every ego agent and the approximated mean-field effect of the overall population. It should be noted that the mean-field approximation strongly relies on the introduction of the mean action of neighbors, which is directly averaged by all the neighbor actions, ignoring the different influence may caused by different locations, types or any attribute of neighbors. Unfortunately, most of derivative works [3,5,9] based on mean-field theory either focus on determining the accurate accessible neighbors or simply extend the same type agents to multi-types. Though weighted information [2,18,25] has also been considered, they still formulate the mean-field function with hysteresis information that same as original mean-field MARL, i.e., generating the mean action from the last step. The theoretical idea or the empirical results [4,18] indicate that, the miscalculation of the mean-field effect caused by equal treatment of neighbors and use of hysteresis information may leads to the wrong direction of optimization, which matches with the massive oscillations during training.

In order to eliminate the unexpected effect of mean action in mean-field approximation, we first introduce attention mechanism, similar with existing methods [2,25], to differentially process neighbors' information. Then we decompose the weighted local Q-value via reward attribution decomposition which is inspired by [29], formulating the weighted mean-field approximation as a joint optimization over an implicit reward assignment among the ego agent and its neighbors. After decomposition, not only in the training phase, the ego agent can better distinguish the impact of the neighbors' hysteresis information, but also the execution phase is now fully decentralized without any information exchanging, especially dropping the effect from hysteresis information utilization. Moreover, it distinguishes from the value decomposition methods [17,22] owing to the latter is decomposed from team perspective which only adapted in cooperative settings, while the former from individual point of view under the guidance of different reward assignment, improving the interpretability to some extent.

To this end, we proposed the Weighted **M**ean-**F**ield MARL via **R**eward **A**ttribution **D**ecomposition (MFRAD) framework by differentiating heterogeneous and hysteresis neighbor effect with weighted mean-field approximation and reward attribution decomposition. Specifically, we first achieve weighted mean-field approximation by calculating the weighted state-action embeddings of neighbors nearby the ego agent. Then, in reward attribution decomposition, considering the effect of the ego agent's interaction with its neighbors caused by its own actions and neighbors' actions, the pairwise local Q-function is decomposed as two terms: the SELF-term that only relies on the agent's own state, and the neighbor term that is related to the weighted mean-field effect of other agents which combined with multi-head attention mechanism. Intuitively, the relationship between agents maintains temporarily stable even if the characteristics of the neighbor agents change at a certain moment. Thus we propose a novel regularization term named temporal relationship regularization to maintain the temporary difference of attention weights between timesteps. Moreover, decomposing pairwise local Q-function with a simple addition, the solution of two terms might not be unique. Drawing the inspiration from previous work [29], we introduce an extra regularization term to guarantee the unambiguity of ego agent's local Q-value with no neighbors. Main contributions are listed as follows: 1) We propose the weighted mean-field approximation that captures the fine-grained neighbor information and employ multi-head mechanism to calculate the dynamic mean-field effect; 2) The idea of reward attribution decomposition is introduced to reduce the negative effect of antique signal from calculating the delayed mean action of neighbors, transforming the Q-function of each agent into summation of local Q-function and weighted mean-field Q-function that related to neighbors from individual perspective; 3) Multiple many-agent experiments on MAgent and CityFlow are conducted to verify the proposed MFRAD algorithm can achieve higher return and stable performance in both cooperative and competitive tasks, and has certain scalability in real-world scenarios where cooperation and struggle coexist.

2 Related Work

Mean-Field Games. Introducing mean-field theory into MARL has gained wide attention recently, which approximates the complex interactions between agents into the interaction between ego agent and the neighboring agent distribution [7], eliminating the dimensional disaster. It also effectively alleviates exploration noise caused by multiple agents so that each agent can efficiently make beneficial local decisions. [26] firstly proposes a model-free scheme for learning the optimal action based on mean-field theory. [4] relaxes the assumption on the neighbor range in [26], establishing the mean-field effect of accessible agents which in a predefined observation range or visible distribution. When it comes with more complex game settings, [3] approximates the joint action of N agents to N mean actions while [25] approximately estimates the inter-type and intra-type interactions between agents without exact number. Also, weighted information [25] and graph neural network with attention mechanism [8] has been introduced to model the neighbor relationship, while existing works mainly focus on the weighted action distribution to formulate the pairwise mean-field Q-function directly.

Value Function Decomposition. The most straightforward way to train a MARL task is to learn each agent's Q-function independently [23], while it ignores the dynamic influence of other agents that leads to the non-stationary environment especially in many-agent scenarios. Value function decomposition(VFD) methods, e.g., VDN [22], QMIX [17], QTRAN [21], adopt CTDE paradigm to rewrite the joint Q-function as $Q^{\pi}(s, \mathbf{a}) = \phi\left(s, Q^1\left(o^1, a^1\right), \ldots, Q^N\left(o^N, a^N\right)\right)$ where the formulation of ϕ differs in each method. Although these VFD methods successfully solve the non-stationary issue, none of them is well adapted to many-agent scenarios where large-scale numbers of agent exist.

3 Preliminaries

3.1 Markov Decision Process and Markov Game

The Markov Game with N agents which generalizes from Markov Decision Process is formalized by the tuple $\Gamma \triangleq \left(\mathscr{S}, \mathscr{A}^1, \ldots, \mathscr{A}^N, r^1, \ldots, r^N, p, \gamma\right)$, where $\mathscr{S}$ represents the state space and $\mathscr{A}^j$ denotes the actions of the agent $j \in \{1, \ldots, N\}$. The reward function is $r^j : \mathscr{S} \times \mathscr{A}^1 \times \cdots \times \mathscr{A}^N \rightarrow \mathbb{R}$. All agents maximize their discounted sum of rewards with the discount factor $\gamma \in [0, 1)$. p is the transition probability $\mathscr{S} \times \mathscr{A}^1 \times \cdots \times \mathscr{A}^N \rightarrow \Omega(\mathscr{S})$ where $\Omega(\mathscr{S})$ is the collection of state space probability distributions. For agent j, the corresponding policy is defined as $\pi^j : \mathscr{S} \rightarrow \Omega\left(\mathscr{A}^j\right)$ and each agent is trying to maximize its return over the consideration of others' behaviors, where $\Omega\left(\mathscr{A}^j\right)$ represents the set of probability distributions on the agent's j action space $\mathscr{A}^j$. The joint policy of all agents can be denoted as $\boldsymbol{\pi} \triangleq \left[\pi^1, \ldots, \pi^N\right]$. Considering the initial state s, the value function of agent j under the joint policy $\boldsymbol{\pi}$ is formulated as the expected future cumulative discount reward: $v_{\boldsymbol{\pi}}^j(s) = \sum_{t=0}^{\infty} \gamma^t \mathbb{E}_{\boldsymbol{\pi}, p}\left[r_t^j \mid s_0 = s, \boldsymbol{\pi}\right]$. The Q-function of agent j under the joint policy $\boldsymbol{\pi}$ can be formalized as: $Q_{\boldsymbol{\pi}}^j(s, \boldsymbol{a}) = r^j(s, \boldsymbol{a}) + \gamma \mathbb{E}_{s' \sim p}\left[v_{\boldsymbol{\pi}}^j\left(s'\right)\right]$, where s' represents the next state.

3.2 Mean-Field Reinforcement Learning

Mean-field MARL approximates the complicated interactions in many-agent scenarios into the bilateral estimation of two agents where the second agent corresponds to the mean effect of the overall population. The Q-function $Q^j(\boldsymbol{s}, \boldsymbol{a})$ in mean-field MARL will be decomposed by using only local bilateral interactions:

$$Q^j(\boldsymbol{s}, \boldsymbol{a}) = \frac{1}{N^j} \sum_{k \in \mathcal{N}(j)} Q^j\left(\boldsymbol{s}, a^j, a^k\right), \tag{1}$$

where $\mathcal{N}(j)$ represents the sequence number set of agent j's neighbors with size $N^j = |\mathcal{N}(j)|$. After decomposing the Q-function through the bilateral estimation of the agent and its neighbors, it dramatically reduces the interaction complexity in the large-scale scenarios. So this decomposition converts the joint Q-function into the mean field formulation $Q_{\mathrm{MF}}^j\left(\boldsymbol{s}, a^j, \bar{a}^j\right)$ where the mean action $\bar{a}^j$ is calculated according to the neighboring agent set $\mathcal{N}(j)$. Considering the small disturbance, a^k is denoted as: $a^k = \bar{a}^j + \delta a^{j,k}$, where $\bar{a}^j = \frac{1}{N^j} \sum_{k \neq j} a^k$. The Q-function is updated in a recurrent manner:

$$Q_{t+1}^j\left(\boldsymbol{s}, a^j, \bar{a}^j\right) = (1 - \alpha) Q_t^j\left(\boldsymbol{s}, a^j, \bar{a}^j\right) + \alpha\left[r^j + \gamma v_t^j\left(\boldsymbol{s}'\right)\right], \tag{2}$$

where α is the learning rate and r^j is the obtained reward. $\boldsymbol{s}$ and $\boldsymbol{s}'$ represents the old state and resulting state respectively. The value function $v_t^j(\boldsymbol{s}')$ for agent j at time t is formulated as:

$$v_t^j(\boldsymbol{s}') = \sum_{a^j} \pi_t^j\left(a^j \mid \boldsymbol{s}', \bar{a}^j\right) \mathbb{E}_{\bar{a}^j(\boldsymbol{a}^{-j}) \sim \pi_t^{-j}}\left[Q_t^j\left(\boldsymbol{s}', a^j, \bar{a}^j\right)\right], \tag{3}$$

with $\bar{a}_t^j = \frac{1}{N^j}\left(\sum_{k \neq j} a_t^k\right)$, $a_t^k \sim \pi^k\left(\cdot \mid \boldsymbol{s}_t, \bar{a}_{t-1}^k\right)$, and

$$\pi_t^j\left(a_t^j \mid \boldsymbol{s}_t, \bar{a}_{t-1}^j\right) = \frac{\exp\left(-\beta Q^j\left(\boldsymbol{s}_t, a_t^j, \bar{a}_{t-1}^j\right)\right)}{\sum_{a_t^{j'} \in A^j} \exp\left(-\beta Q^j\left(\boldsymbol{s}_t, a_t^{j'}, \bar{a}_{t-1}^j\right)\right)}, \tag{4}$$

where β is the Boltzmann parameter and π denotes the Boltzmann policy.

4 Algorithm

In this section, we introduce the proposed MFRAD algorithm illustrated in Fig. 1. Considering the limitations of mean-field MARL, we firstly extend the existing mean-field approximation to the form with weight information and give the detailed mathematical derivation. Secondly, inspired by [29], we transform the joint Q-function into the integration of ego agent's individual Q-function and weighted mean-field Q-function of its neighbors, which called *reward attribution decomposition*, utilizing the multi-head attention to calculate the weights.

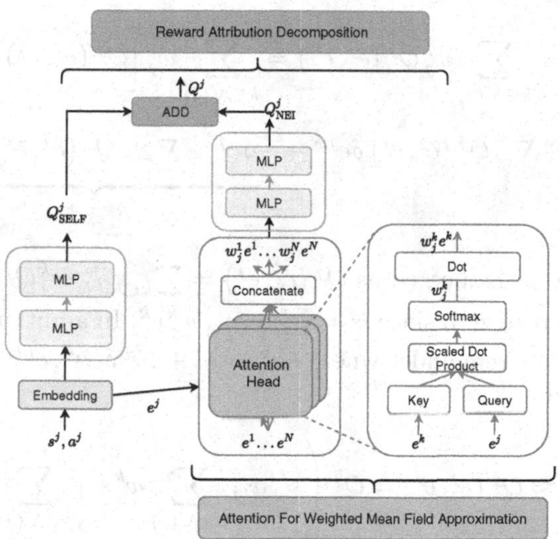

Fig. 1. Architecture of MFRAD. Each agent calculate its Q_{SELF}^j based on state-action embeddings which consists of local observation and action. Meanwhile, multi-head attention module receives the state-action embeddings of ego agent's neighbors as input, calculating attention weights as mean-field weights to construct the weighted mean-field effect Q_{NEI}^j. Finally, these two items constitute the decentralized Q^j.

4.1 Weighted Mean-Field Approximation

Drawing inspiration from existing works [2,18], we rewrite the original mean-field approximation formula (1) into a form with weight information

$$Q^j(\boldsymbol{s},\boldsymbol{a}) = \sum_{k\in\mathcal{N}(j)} w_j^k Q^j\left(s^j, a^j, s^k, a^k\right), \tag{5}$$

where w_j^k represents the weight of each neighbor's effect on ego agent j, and $0 \leq w_j^k \leq 1$, $\sum_{k\in\mathcal{N}(j)} w_j^k = 1$. $\mathcal{N}(j)$ represents the sequence number set of agent j's neighbors with size $N^j = |\mathcal{N}(j)|$. As for clarity, we denote $e^j \triangleq e^j\left(s^j, a^j\right)$ which performed with embedding operation, thus (5) can be reformulate as

$$Q^j(\boldsymbol{s},\boldsymbol{a}) = \sum_{k\in\mathcal{N}(j)} w_j^k Q^j\left(e^j, e^k\right). \tag{6}$$

Similar as the deviation in mean-field approximation, the weighted mean-field approximation is still based on the weighted average effect of the state-action pair $\bar{e}^j$ from the adjacent agent set $\mathcal{N}(j)$. We represent the local state-action information of each neighbor as the sum of weighted average effect $\bar{e}^j$ and a small disturbance $\delta e^{j,k}$, that is, $e^k = \bar{e}^j + \delta e^{j,k}$, where $\bar{e}^j = \sum_{k\in\mathcal{N}(j)} w_j^k e^k$ can be interpreted as neighborhood state-action distribution. Then according to Taylor's theorem, if the bilateral weighted Q-function of agent k is twice-differentiable, then (6) can be expanded as

$$\begin{aligned} Q^j(\boldsymbol{s},\boldsymbol{a}) &= \sum_{k\in\mathcal{N}(j)} w_j^k Q^j\left(e^j, e^k\right) = \sum_{k\in\mathcal{N}(j)} w_j^k \Big[Q^j\left(e^j, \bar{e}^j\right) \\ &+ \nabla_{\bar{e}^j} Q^j\left(e^j, \bar{e}^j\right)\delta e^{j,k} + \underbrace{\frac{1}{2}\delta e^{j,k} \cdot \nabla^2_{\tilde{e}^{j,k}} Q^j\left(e^j, \tilde{e}^{j,k}\right)\delta e^{j,k}}_{R^j_{e^j}\left(e^k\right)} \Big], \end{aligned} \tag{7}$$

where the first term is merged as $Q^j\left(e^j, \bar{e}^j\right) = \sum_{k\in\mathcal{N}(j)} w_j^k Q^j\left(e^j, \bar{e}^j\right)$, and the second term equals to zero since $\bar{e}^j = \sum_{k\in\mathcal{N}(j)} w_j^k e^k$. In addition, $R^j_{e^j}\left(e^k\right)$ is the Taylor polynomial's remainder where $\tilde{e}^{j,k} = \bar{e}^j + \epsilon^{j,k}\delta e^{j,k}$, $\epsilon^{j,k} \in [0,1]$, so (7) is finally reduced to

$$Q^j(\boldsymbol{s},\boldsymbol{a}) \approx Q^j\left(e^j, \bar{e}^j\right) = Q^j\left(s^j, a^j, \sum_{k\in\mathcal{N}(j)} w_j^k s^k, \sum_{k\in\mathcal{N}(j)} w_j^k a^k\right). \tag{8}$$

Therefore, based on the weighted mean effect, the bilateral interaction between agent j and its neighbor agent k is simplified as the local pairwise interaction between the ego agent and the mean-field agent, and the latter is abstracted from the weighted mean effect of neighborhood state-action information.

4.2 Reward Attribution Decomposition

Though weighted information is introduced, the another drawback of mean-field MARL that the mean-field effect of neighbor is generated from obsolete information, which is unreasonable to choose actions according to the generated policy. Drawing inspiration from [29], we decompose the Q-value of ego agent into its own part and that of neighbor agent via reward assignment mechanism. Therefore, we realize the decentralized execution without historical information sharing among agents, alternatively, the weighted mean-field effect of neighbors is dexterously converted into the centralized training process. The intuition of this decomposition is that the effect of the ego agent's interaction with its neighbors is caused by two factors, that is, the action taken by the ego agent based on its local observation and the actions taken by neighbors based on their local observations, and all these actions are chosen under the guidance of reward assigning. We will explain in more detail later why the proposed MFRAD is able to achieve fully decentralization in execution. As a result, the weighted mean-field Q-value for each agent can be effectively decomposed, which decoupled the description of global information under the partially observed assumption. It also realizes high scalability in many-agent scenarios from individual perspective, making up for the limitation of mean-field MARL in calculating mean action of neighborhood during the decentralized execution phase. Specifically, according to the weighted mean-field approximation (8), we have

$$\begin{aligned} Q^j(\boldsymbol{s}, \boldsymbol{a}) &= Q^j\left(s^j, a^j, \sum_{k\in\mathcal{N}(j)} w_j^k s^k, \sum_{k\in\mathcal{N}(j)} w_j^k a^k\right) \\ &= \mathbb{E}\left[\sum_{t=0}^{\infty}\gamma^t r^j\left(s_t^j, a_t^j, \sum_{k\in\mathcal{N}(j)} w_j^k s_t^k, \sum_{k\in\mathcal{N}(j)} w_j^k a_t^k\right)\middle| s_0 = \mathbf{s}, a_0 = \mathbf{a}\right], \end{aligned} \tag{9}$$

followed with the principle of reward attribution decomposition that explained in [29], agent acts according to a state not only because the reward to itself, but also because it is more rewarding than other agents. Therefore, we split the reward of an agent from the ego agent's point of view, that is, the reward is not only derived from itself, but also influenced by that of neighbors.

$$r^j\left(s_t^j, a_t^j, \sum_{k\in\mathcal{N}(j)} w_j^k s_t^k, \sum_{k\in\mathcal{N}(j)} w_j^k a_t^k\right) = r^j\left(s_t^j, a_t^j\right) + r^j\left(\sum_{k\in\mathcal{N}(j)} w_j^k s_t^k, \sum_{k\in\mathcal{N}(j)} w_j^k a_t^k\right), \tag{10}$$

then $Q^j(\boldsymbol{s}, \boldsymbol{a})$ can be further decomposed

$$\begin{aligned} Q^j(\boldsymbol{s}, \boldsymbol{a}) &= \mathbb{E}\left[\sum_{t=0}^{\infty}\gamma^t r^j\left(s_t^j, a_t^j\right)\middle| s_0 = \mathbf{s}, a_0 = \mathbf{a}\right] \\ &\quad + \mathbb{E}\left[\sum_{t=0}^{\infty}\gamma^t r^j\left(\sum_{k\in\mathcal{N}(j)} w_j^k s_t^k, \sum_{k\in\mathcal{N}(j)} w_j^k a_t^k\right)\middle| s_0 = \mathbf{s}, a_0 = \mathbf{a}\right] \\ &\approx Q_{\mathrm{SELF}}^j(s^j, a^j) + Q_{\mathrm{NEI}}^j\left(\{w_j^k, s^k, a^k\}_{k\in\mathcal{N}(j)}\right). \end{aligned} \tag{11}$$

Finally, the weighted average Q-function is transformed into the summation of the local Q-function Q_{SELF} of the ego agent and the neighbor agent Q-function Q_{NEI} with weighted information.

4.3 Network Architecture

The overall architecture of the proposed MFRAD is illustrated in Fig. 1. As discussed above, the local Q-function of each agent j consists of two parts: SELF-Q network and NEI-Q network. For detail, on the one hand, SELF-Q network parameterized by θ^j_{self} is separated for calculating the Q^j_{SELF} of each agent j based on its local observation s^j and action a^j. Noted that for each agent j, its local observation and action pair (s^j, a^j) is encoded as e^j via a state-action encoder[1] before fed into SELF-Q network to reduce the noise redundancy of original coarse information. On the other hand, the NEI-Q network parameterized by θ^j_{nei} employs the multi-head attention module (will be explained soon) to model the weighted neighbour state-action distribution, which is further processed for calculating the weighted effect Q^j_{NEI} of neighbors.

Concretely, the multi-head attention module introduces the attention mechanism to calculate the weights in the weighted mean-field approximation. The embedding vector e^j is regarded as the *query* vector, while e^k which contains neighbour state-action information is regarded as the *key* vector. The mean-field weights are then calculated by comparing the key vectors and query vector in terms of their dot similarity, which is evaluated through a softmax function

$$w^k_j \propto \exp\left(e^j\left(s^j, a^j\right)^T W^T_{\text{key}} W_{\text{query}} e^k\left(s^k, a^k\right)\right), \tag{12}$$

and we denote the parameters of attention module $(W_{\text{key}}, W_{\text{query}})$ of each agent j as θ^j_{att}.

Rather than the single-head attention, here we use multi-head attention mechanism to comprehensively utilize all aspects of the information of the agent from multiple angles and extract more abundant feature representation. Each attention head corresponds to a separate set of weight parameters $(W^m_{\text{key}}, W^m_{\text{query}})$ where $m \in [M]$ and M is the number of heads. For clarity, $(W^m_{\text{key}}, W^m_{\text{query}})$ is denoted as $\theta^{j,m}_{\text{att}}$ and $\Theta^j_{\text{att}} := \{\theta^{j,m}_{\text{att}}\}^M_{m=1}$ refers to parameters of all attention heads. The final weights are then obtained by averaging the attentions of multiple attention heads. That is, the observation-action distribution of neighbors e^{-j} is calculated as follows

$$e^{-j} = \frac{1}{M}\sum_m^M \sum_k w^{k,m}_j e^k. \tag{13}$$

The obtained e^{-j} is then fed into the NEI-Q network, and the Q-value of neighbors after the fusion of attention mechanism is calculated, which describes the comprehensive effect of neighbors on the ego agent with better explanation and representation. Finally, we perform the summation operation on Q^j_{SELF} and Q^j_{NEI} to formulate the $Q^j(\boldsymbol{s}, \boldsymbol{a})$.

[1] Without causing confusion, we incorporate the parameters of this encoder into θ^j_{self}.

4.4 Overall Optimization Objective

Intuitively, the interaction between agents is not a transient process, the relationship between agents maintains temporarily stable even if the characteristics of the neighbor agent change at a certain moment. Therefore, the attention weight distribution should also remain stable in a short period of time. We use KL divergence to measure the difference of attention weights between timesteps as in [8], in order to keep the consistency of the temporal relationship. Therefore, the regularization term about temporal relationship, called temporal relationship regularization (TRR), is added to the loss function

$$\Omega(\theta^j_{\text{self}}, \Theta^j_{\text{att}}; s^l, a^l, s^{l,'}, a^{l,'}) = \frac{1}{\mathrm{M}} \sum_{m=1}^{\mathrm{M}} D_{\mathrm{KL}} \left[w_j^{k,m}(s^l, a^l; \theta^j_{\text{self}}, \theta^{j,m}_{\text{att}}) \| w_j^{k,m}(s^{l,'}, a^{l,'}; \theta^j_{\text{self}}, \theta^{j,m}_{\text{att}}) \right], \tag{14}$$

where $l := \mathcal{N}(j) \cup j$, (s^l, a^l) and $(s^{l,'}, a^{l,'})$ are state-action pairs at two consecutive timesteps, $D_{\mathrm{KL}}[\cdot\|\cdot]$ denotes the KL-divergence operator.

Moreover, revisiting (11), with a simple addition, the solution of Q^j_{SELF} and Q^j_{NEI} might not be unique. Indeed, we might add any constant to Q^j_{SELF} and subtract that constant from Q^j_{NEI} to yield the same local Q-value Q^j. Drawing the inspiration from previous work [29], we introduce an extra regularization term, Q^j_{NEI}. Intuitively, we hope that during the training process, Q^j_{NEI} can gradually converge to 0, so that there is a unique optimal solution for two terms in (11). From another perspective, the introduction of this regularization term is also similar to the teacher-student framework in transfer learning [28]. As learning progresses, Q^j_{NEI} gradually distills knowledge into Q^j_{SELF}. This enables the ego agent to adaptively process hysteresis information from neighbors. Further, this also enables MFRAD to only make decisions based on Q^j_{SELF} during the execution phase, without the need to communicate with the neighbors to calculate the average actions of them, enabling fully decentralized execution. Compared with the existing work based on mean-field theory, MFRAD has better scalability. We also observe that with NEI objective, training is much stabilized in following numerical experiments. By the way, since the occurence that no agents exists in the neighborhood may create ambiguity, this regularization term also making the guarantee that $\arg\max_{a^j} Q^j = \arg\max_{a^j} Q^i_{\text{SELF}}$.

In order to make MFRAD have faster convergence speed and better scalability, the parameters of all agents are shared. Therefore we denote the parameters of SELF-Q network, NEI-Q network, and multi-head attention module of any agent as θ_{self}, θ_{nei} and Θ_{att} respectively. Finally, the overall optimization objective for each agent j that integrates the regularization terms is shown as follows:

$$\begin{aligned} \mathrm{L}(\theta_{\text{self}}, \theta_{\text{nei}}, \Theta_{\text{att}}) = \frac{1}{N} \sum_{j=1}^{N} \mathbb{E}_{s,a,s',a'} \Big[& \underbrace{\left(y^j - \left(Q^{\text{SELF}}_{\theta_{\text{self}}}(s^j, a^j) + Q^{\text{NEI}}_{\theta_{\text{nei}}}(s^k, a^k)\right)\right)^2}_{\text{DQN Objective}} \\ & + \underbrace{\lambda_1 \left(Q^{\text{NEI}}_{\theta_{\text{nei}}}(s^j, a^j)\right)^2}_{\text{NEI Objective}} + \underbrace{\lambda_2 \Omega(\theta_{\text{self}}, \Theta_{\text{att}}; s^l, a^l, s^{l,'}, a^{l,'})}_{\text{TRR Objective}} \Big], \end{aligned} \tag{15}$$

where $k \in \mathcal{N}(j)$, $l := \mathcal{N}(j) \cup j$ and λ_1 and λ_2 represent the relative importance of two regularization terms against the optimization direction of original Q-function respectively.

5 Experiments

We consider many-agent scenarios in this section and evaluate the performance of the proposed MFRAD framework. Firstly, three different tasks will be discussed based on MAgent [30] platform, including a competitive task (i.e., *Gather Game*), a cooperative task (i.e., *Predator-prey Game*) and a mixed cooperative-competitive task (i.e., *Battle Game*). Additionally, more detailed analysis on the *Battle Game* will be conducted to verify the effects of attention mechanism and two regularization terms. Moreover, we choose a real-world task on traffic flow to demonstrate the scalability of MFRAD which outperforms both existing rule-based and value function decomposition methods.

5.1 Results and Analysis

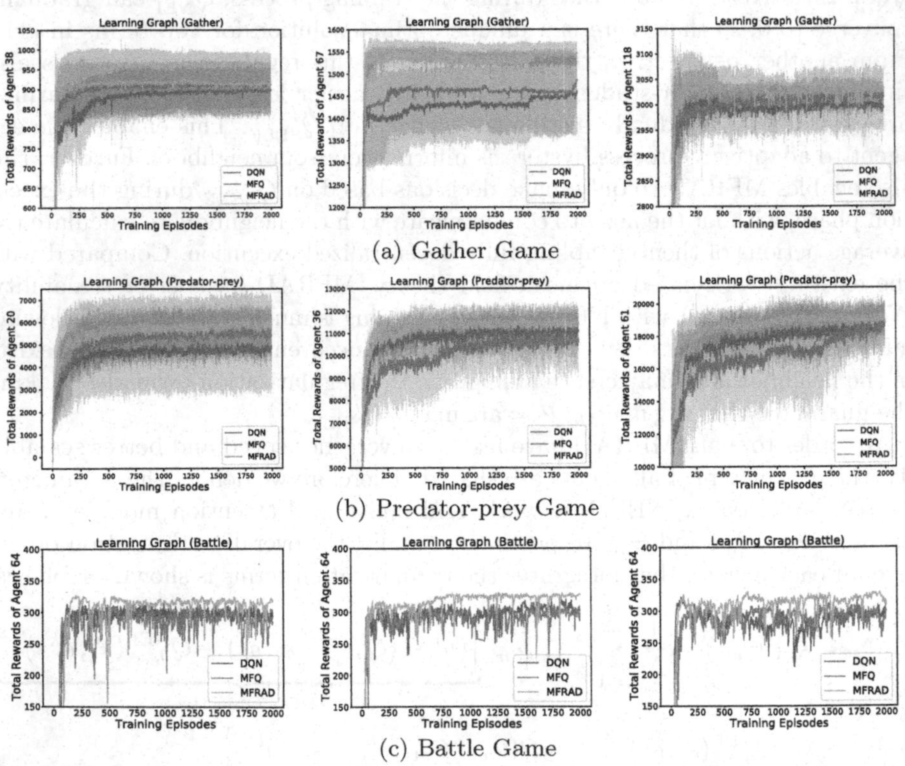

(a) Gather Game

(b) Predator-prey Game

(c) Battle Game

Fig. 2. In all the learning graphs of three MAgent games with different agent numbers, MFRAD shows the best performance. (Color figure online)

Experiments on Gather Game. In this scenario, agents compete for limited food resources as much as possible and can kill other agents to maximize their own survival time. The average total rewards with a standard deviation of 20 experiments for each algorithm is proposed in Fig. 2(a) where the agent number scales. MFRAD consistently outperforms with higher total reward. It is similar to other algorithms when in small scale, while gradually has faster convergence speed than MFQ and DQN nearly 100 episodes though starts with relatively slow growth rate.

Experiments on Predator-Prey Game. Predator-prey Game is a fully cooperative task where the predators cooperate to capture as many preys as possible. As shown in Fig. 2(b), MFRAD estimates the influence of other agents accurately through the reward attribution mechanism when number of agent increasing. However, MFQ is inferior to MFRAD and DQN in terms of convergence speed and total reward, related to approximating the mean-field effect among indiscriminate neighbors. In addition, we visualize the pursuit process in Fig. 3(a) and find that MFRAD predator cooperate to capture alone rather than continue to chase preys that have been observed and chased by others. Then, we record the pursuing results of different methods which fight with each other for 200 episodes in Fig. 3(b). Obviously, MFRAD with weighted neighbor information always defeats other methods.

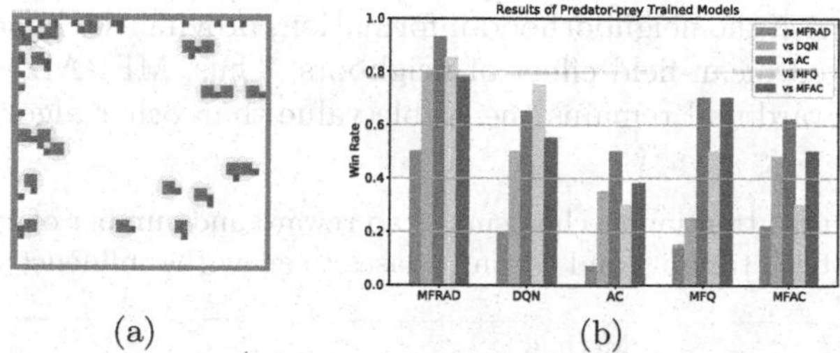

Fig. 3. Results in testing phases. (a) In the Predator-prey Game, each agent cooperate to catch preys as much as possible. (b) Win rate among MFRAD and other baselines.

Experiments on Battle Game. The learning graph of Battle Game in which red army of 64 agents fight with the similar blue army is shown in Fig. 2(c). Since only the total reward corresponding to the red army is recorded when it is higher than that of the blue army, the timestep of each record in each round cannot be aligned synchronously. Therefore, calculating the mean and standard deviation of the total reward lacks physical meaning and cannot reflect the actual performance. We randomly select 3 results from 50 experiments to show the robustness of each algorithm to different random seeds. Although the convergence speed of MFRAD in the initial stage is slightly slower, it outperforms MFQ and DQN with higher cumulative reward and minor variance after convergence, showing the stable performance in multiple trials. This phenomenon proves that the introduction of weighted mean-field effect after decomposition by reward has a significant impact on the stability of the training process.

In addition, the average individual reward in 2000 episodes and number of opponents killed by each algorithm are shown in Table 1. Noted that we add MFAC and AC algorithm based on actor-critic architecture to enrich comparison information, however, MFRAD performs significantly better than AC-family methods. Besides, Multi-head attention mechanism plays an significant role in

Table 1. Mean agent reward and number of opponents killed in different scales.

Method	Agent36 vs 36		Agent 64 vs 64		Agent 196 vs 196		Agent 324 vs 324	
	Reward	Kill	Reward	Kill	Reward	Kill	Reward	Kill
DQN	0.13	35.31	0.18	62.9	0.09	193.43	0.08	**316.99**
AC	0.14	24.19	0.12	43.94	0.11	149.1	0.07	247.27
MFQ	0.14	35.21	0.12	62.92	0.09	193.3	0.08	315.32
MFAC	**0.15**	14.31	0.06	44.21	0.11	153.59	0.08	232.13
MFRAD	0.14	**36.47**	**0.19**	**63.3**	**0.12**	**194.97**	**0.09**	316.91

describing how important effect the neighbors make on the ego agent. From the ablation results in Table 2, the proposed MFRAD benefits from multi-head attention to enrich the neighborhood information, proving the benefit of approximating weighted mean-field effect of neighbors. Thus, MFRAD converges to a higher total reward and remains the stable value than other algorithms.

Table 2. Impact of attention mechanism. Mean reward and number of opponents killed is evaluated on both training and testing phases to show the influence of attention.

Metric		Method		
		w/o attention	Single-head attention	Multi-head attention
Training	Total reward	290.47	298.13	**302.62**
	Individual reward	0.12	0.18	**0.19**
	Kill number	58.4	62.9	**63.3**
Testing	Total reward	216.1	239.74	**243.13**
	Kill number	39.0	45.2	**48.9**

We also study the importance of NEI Loss and TRR loss by removing it from MFRAD in scenarios where both army contains 196 agents, as shown in Table 3. Using these two regularization terms boosts the performance and stabilizes the training process, consistent with the proposed reward attribution decomposition.

Experiments on Traffic Flow Control. In order to investigate the scalability of MFRAD to more complex real-world task, we choose the traffic flow control

Table 3. Impact of regularization term of loss function. Mean total reward and kill-death ratio is evaluated to show that MFRAD with refactoring loss is better at killing opponents and protecting themselves.

Method	NEI objective	TRR objective	Battle scenario metrics	
			Mean reward	Kill-Death ratio
OL			887.92	1.88
OL-NEI	√		925.81	2.13
OL-NEI-TRR	√	√	**928.49**	**2.26**

task and experiment on the large-scale traffic flow platform named CityFlow. Following existing studies, we model each intersection as an RL agent and realize the communication by sharing information among agents. It is a typical mixed cooperative-competitive scenario.

We compare our model with the following two categories of methods: rule-based method (i.e., *Max-Pressure* [24]) and RL methods (i.e., *IQL,* QMIX, *VDN*) based on value function decomposition. The average travel time, which calculates the average travel time of all the vehicles spent from waiting in the queue and leaving the intersection, is chosen to evaluate the performance of different methods. MFRAD achieves consistent performance improvements on both synthetic data and real-world data, it costs less travel time not only compared with rule-based method but also RL (VFD) methods (Table 4).

Table 4. Performance on synthetic data and real-world data w.r.t average travel time

Model	Arterial $_{1\times6}$	Grid $_{6\times6}bi$	Grid $_{6\times6}uni$	NewYork $_{16\times3}$	Hangzhou $_{4\times4}$
Max-pressure	122.98	204.72	186.06	405.69	431.53
IQL	143.95	269.18	244.39	254.93	472.38
QMIX	110.27	542.63	678.27	226.15	562.39
VDN	131.53	468.94	630.82	198.24	358.73
MFRAD	**72.98**	**171.51**	**168.25**	**183.82**	**310.07**

Ablation study is conducted to further analyze the effect of attention mechanism. Temporal distribution of attention in Grid $_{6\times6}$ roadnet learned by MFRAD is demonstrated in Fig. 4. Similarly as traffic jam, flow of Inter I0 changes greatly that flow from Inter I4 to Inter I0 decreases while increases from Inter I3 to Inter I0. As shown in Fig. 4(b) the score of SELF-attention occupies the largest blue area while that of I4 and I2 decreases and I3 and I1 increases, indicating that the attention scores match with the real traffic condition. Thus, MFRAD is verified to capable of approximating accurate weighted mean-field neighbor effect.

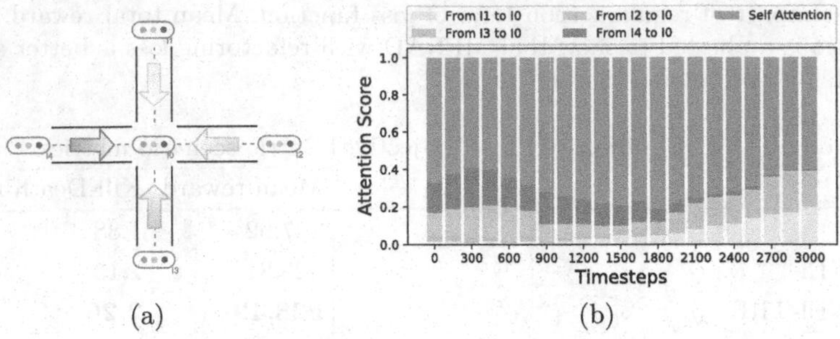

Fig. 4. Temporal distribution of attention score when facing with changeable traffic flow.(a) Roadnet of Inter I0. (b) Temporal distribution of attention score of Inter I0.

6 Conclusion

In this paper, we develop a weighted mean-field multi-agent reinforcement learning algorithm via reward attribution decomposition, which incorporates weighted information of neighbors with attention mechanism to capture the dynamic influence of others. Considering the negative effect of hysteresis information, reward attribution decomposition is integrated to decompose the pairwise mean-field Q-function as SELF-term and neighbor term which represents the local effect and mean-field effect of neighbors respectively, realizing the fully decentralized execution without any information exchanging. Experiments in various many-agent scenarios demonstrate that MFRAD boosts the performance and stabilizes the training process and also has great scalability to real-world tasks.

Acknowledgment. This work was supported in part by the National Key Research and Development Program of China (No. 2020AAA0107400), STCSM (No. 18DZ2271000 and 19ZR141420), NSFC (No. 12071145) and the Fundamental Research Funds for the Central Universities.

References

1. Chen, C., et al.: Toward a thousand lights: decentralized deep reinforcement learning for large-scale traffic signal control. AAAI **34**(04), 3414–3421 (2020)
2. Fang, B., Wu, B., Wang, Z., Wang, H.: Large-scale multi-agent reinforcement learning based on weighted mean field. In: Sun, F., Liu, H., Fang, B. (eds.) ICCSIP 2020. CCIS, vol. 1397, pp. 309–316. Springer, Singapore (2021). https://doi.org/10.1007/978-981-16-2336-3_28
3. Ganapathi Subramanian, S., Poupart, P., Taylor, M.E., Hegde, N.: Multi type mean field reinforcement learning. In: AAMAS (2020)
4. Ganapathi Subramanian, S., Taylor, M.E., Crowley, M., Poupart, P.: Partially observable mean field reinforcement learning. In: AAMAS (2021)
5. Guo, X., Hu, A., Xu, R., Zhang, J.: Learning mean-field games. In: NeurIPS (2019)
6. Gupta, J.K., Egorov, M., Kochenderfer, M.: Cooperative multi-agent control using deep reinforcement learning. In: AAMAS (2017)

7. Jeong, S.H., Kang, A.R., Kim, H.K.: Analysis of game bot's behavioral characteristics in social interaction networks of MMORPG. ACM SIGCOMM Comput. Commun. Rev. **45**(4), 99–100 (2015)
8. Jiang, J., Dun, C., Huang, T., Lu, Z.: Graph convolutional reinforcement learning. In: ICLR (2020)
9. Li, M., et al.: Efficient ridesharing order dispatching with mean field multi-agent reinforcement learning. In: WWW (2019)
10. Li, W., Wang, X., Jin, B., Sheng, J., Hua, Y., Zha, H.: Structured diversification emergence via reinforced organization control and hierarchical consensus learning. In: AAMAS (2021)
11. Li, W., Wang, X., Jin, B., Sheng, J., Zha, H.: Dealing with non-stationarity in MARL via trust region decomposition. In: ICLR (2022)
12. Lowe, R., Wu, Y., Tamar, A., Harb, J., Abbeel, P., Mordatch, I.: Multi-agent actor-critic for mixed cooperative-competitive environments. In: NeurIPS (2017)
13. Mao, H., et al.: Neighborhood cognition consistent multi-agent reinforcement learning. In: AAAI (2020)
14. Mao, H., Zhang, Z., Xiao, Z., Gong, Z., Ni, Y.: Learning multi-agent communication with double attentional deep reinforcement learning. Autonom. Agents Multi-agent Syst. **34**(1), 1–34 (2020). https://doi.org/10.1007/s10458-020-09455-w
15. Matignon, L., Laurent, G.j., Le fort piat, N.: Review: Independent reinforcement learners in cooperative Markov games: a survey regarding coordination problems. Knowl. Eng. Rev. **27**(1), 1–31 (2012)
16. Mnih, V., et al.: Human-level control through deep reinforcement learning. Nature **518**(7540), 529–533 (2015)
17. Rashid, T., Samvelyan, M., Schroeder, C., Farquhar, G., Foerster, J., Whiteson, S.: Qmix: Monotonic value function factorisation for deep multi-agent reinforcement learning. In: ICML (2018)
18. Ren, W.: Represented value function approach for large scale multi agent reinforcement learning. Arxiv (2020)
19. Sheng, J., et al.: Learning to schedule multi-NUMA virtual machines via reinforcement learning. Pattern Recogn. **121**, 108254 (2022)
20. Sheng, J., et al.: Learning structured communication for MARL. ArXiv (2020)
21. Son, K., Kim, D., Kang, W.J., Hostallero, D.E., Yi, Y.: QTRAN: learning to factorize with transformation for cooperative multi-agent reinforcement learning. In: ICML (2019)
22. Sunehag, P., et al.: Value-decomposition networks for cooperative multi-agent learning based on team reward. In: AAMAS (2018)
23. Tan, M.: Multi-agent reinforcement learning: independent vs. cooperative agents. In: ICML (1993)
24. Varaiya, P.: Max pressure control of a network of signalized intersections. Transp. Res. Part C Emerg. Technol. **36**, 177–195 (2013)
25. Yang, F., Vereshchaka, A., Chen, C., Dong, W.: Bayesian multi-type mean field multi-agent imitation learning. In: NeurIPS (2020)
26. Yang, Y., Luo, R., Li, M., Zhou, M., Zhang, W., Wang, J.: Mean field multi-agent reinforcement learning. In: ICML (2018)
27. Ye, D., et al.: Towards playing full MOBA games with deep reinforcement learning. In: NeurIPS (2020)
28. Yim, J., Joo, D., Bae, J., Kim, J.: A gift from knowledge distillation: Fast optimization, network minimization and transfer learning. In: CVPR (2017)
29. Zhang, T., et al.: Multi-agent collaboration via reward attribution decomposition. Arxiv (2020)

30. Zheng, L., Yang, J., Cai, H., Zhou, M., Zhang, W., Wang, J., Yu, Y.: MAgent: a many-agent reinforcement learning platform for artificial collective intelligence. In: AAAI (2018)
31. Zhou, M., et al.: Multi-agent reinforcement learning for order-dispatching via order-vehicle distribution matching. In: CIKM, pp. 2645–2653 (2019)
32. Zimmer, M., Glanois, C., Siddique, U., Weng, P.: Learning fair policies in decentralized cooperative multi-agent reinforcement learning. In: ICML (2021)

BDQM

Evaluating Presto and SparkSQL with TPC-DS

Yinhao Hong[1,2(✉)], Sheng Du[2], and Jianquan Leng[2]

[1] Renmin University of China, Beijing, China
hongyh@ruc.edu.cn
[2] Beijing Kingbase Information Technology Co., Ltd., Beijing, China
{dusheng,jqleng}@kingbase.com.cn

Abstract. From the perspective of the development trend of database technology and the application of big data, the unified management and analysis of relational data and non-relational data is a new trend. New relational computing engines, such as SparkSQL and Presto, provide parallel processing and analysis of distributed relational data and non-relational data, which can effectively improve the performance of data analysis and data quality maintenance scenarios. The purpose of this work is to compare the performance of Presto and SparkSQL using TPC-DS as a benchmark to determine how well Presto and SparkSQL perform in the same scenario. TPC-DS is a benchmark test developed by the Transaction Processing Performance Council (TPC). It contains complex applications such as data statistics, report generation, online query, and data mining, and also has data skew and can effectively reflect system performance in real scenarios. In test results, Presto performed better than SparkSQL in many query scenarios, and in the most significant test results, Presto performed three times better than SparkSQL.

Keywords: Big data · Presto · Spark · TPC-DS

1 Introduction

With the development of Internet applications, such as e-commerce and social networks, the demand for big data applications has exploded [7,8,13,15,24,28,29], and the processing and analysis of non-relational data has been fully developed in big data applications [6,9,11,25–27]. With the deepening of informatization transformation, all data scenarios are interconnected, and the collaboration between relational data and non-relational data has become a new trend.

Traditional relational database technology cannot support OLAP operations in big data scenarios, so a number of new relational computing engines, such as SparkSQL and Presto, emerge. The characteristics of these computing engines

This work is supported by the National Key Research and Development Program of China (No. 2018YFB1004401), National Natural Science Foundation of China (No. 61972402, 61972275, 61732014 and 62072459).

U. K. Rage et al. (Eds.): DASFAA 2022 Workshops, LNCS 13248, pp. 319–329, 2022.
https://doi.org/10.1007/978-3-031-11217-1_23

include providing distributed parallel processing and analysis capabilities to efficiently compute large-scale data sets; In addition, it provides the common computing capability of multiple data models and can effectively deal with the data quality problems of heterogeneous multi-data sources through mapReduce and other computing models.

However, to the best of our knowledge, no current study has analyzed performance differences between these relational computing engines. SparkSQL and Presto were selected based on the popularity and similarity of the computing engines, and a 3-node cluster was created for each engine for comparison testing. Our experiments used TPC-DS [17,18], a benchmark developed by the Transaction Processing Performance Council (TPC). The execution time of each test is the only performance metric that this article focuses on.

The rest of this paper is organized as follows: Sect. 2 introduces the background about the computing engines and databases we involved in the study. The experimental setup is in Sect. 3 followed by experimental results and analysis of Presto and SparkSQL in Sect. 4. Section 5 describes related work. Finally, we conclude and explain our future work.

2 Background

In this section, we describe some background about each query engines and databases we involved in this paper.

Hadoop is a well-known distributed system infrastructure in the field of big data [1]. Users can use a cluster of multiple physical machines to develop and run large-scale data calculation and storage without understanding the details of the underlying implementation of the distributed. In this experiment, we use the Hadoop system to form a distributed cluster to maximize the performance of the computing engine [4].

YARN is a new Hadoop resource manager [23]. As a general resource management system, it can provide unified resource management and scheduling for upper-layer applications. It brings huge benefits to the cluster in terms of utilization rate, unified resource management and data sharing. We used Yarn in this experiment to provide the resource scheduling function for the Spark computing engine.

Hive is a Hadoop-based data warehouse tool used for data extraction, conversion, and loading [22]. This is a softwarethat can store, query, and analyze large-scale data stored in Hadoop [10]. The Hive data warehouse tool can map structured data files into a database table. In this experiment, we use Hive as the data source to provide query data for the computing engine.

Spark is a fast and universal computing engine designed for large-scale data processing, which is an open source Hadoop MapReduce-like general parallel framework [2]. It saves the intermediate output in memory, and no longer needs to read and write HDFS, so Spark has better performance than Hadoop. SparkSQL is a module for Spark to process structured data. In this experiment, we chose SparkSQL as the data query interface, and used the Spark engine to complete the calculation [3].

Presto is an open source distributed SQL query engine, suitable for interactive analysis query. Presto is a system that can operate independently and does not depend on any other external system [20]. At the same time, Presto has a rich plug-in interface, which can perfect dock with external storage systems, or add user define functions. In this experiment, we used Presto as a query engine to compare with SparkSQL.

3 Experimental Setup

In this experiment, we used 3 servers on Alibaba Cloud to serve as the underlying physical server resources for our test. Each cloud server uses four cores of Intel's 2.5 GHz Xeon Platinum 8269 processor and has 16G memory. The disk read/write bandwidth is 148 MB/s and 147 MB/s respectively. We used CentOS 8 as the operating system.

The various software versions used in our evaluation are listed below (Table 1):

Table 1. The software versions

Software	Version
Hadoop	2.7.7
HBase	1.4.13
Hive	2.3.7
Spark	2.3.4
Presto	0.233.1

We used TPC-DS as our evaluating framework. TPC-DS is a decision support benchmark that includes both query and data maintenance tests. The benchmark measures response time in single-user mode, query throughput in multi-user mode, and other metrics [17]. In this evaluation, we focused on response time because of the complexity of the query and the limitations of hardware configuration. The data set generated by the official TPC tool contains 7 fact tables and 17 latitude tables, with an average of 18 columns.

In preparation, data is imported to Hive in Parquet format. Both query engines can use Hive as data source and read data in Parquet format. Parquet is a columnar storage format that supports a nested structure, which is very suitable for OLAP scenarios. Parquet has a high compression ratio, can effectively reduce disk space, and can use predicate pushdown and map pushdown, reduce unnecessary data scanning.

The workload for generating query plans contains 99 SQL queries, covering the core parts of SQL'99 and SQL 2003. However, for compatibility reasons, not all query scenarios are supported by both query engines. We selected 30 of queries to cover the application scenarios of big data sets such as statistics, report generation, online query and data mining as comprehensively as possible.

Memory allocation can significantly affect the performance of a query engine and even cause it to fail. We ran some queries beforehand to adjust the memory allocation configuration of the two query engines. Presto divides the amount of memory it manages into two categories: user memory and system memory. User memory is related to user data. For example, reading user OSS data takes up corresponding memory, and the amount of memory occupied is strongly related to the size of user's underlying data. System memory is the memory allocated by the operator itself in order to execute the query, such as table scan and aggregation, which are not strongly correlated with the user's data. In Presto's configuration file, memory configuration includes three parts: the maximum (user + system) memory allowed for a single worker query, the maximum user memory allowed for a single worker query, and the memory set for the third-party library. After the test, we set the three to 6 GB, 4 GB and 1 GB respectively. For SparkSQL, we used Yarn as the resource manager of the cluster. A total of 40G memory were allocated to Spark.

4 Experimental Result

4.1 TPC-DS on Presto

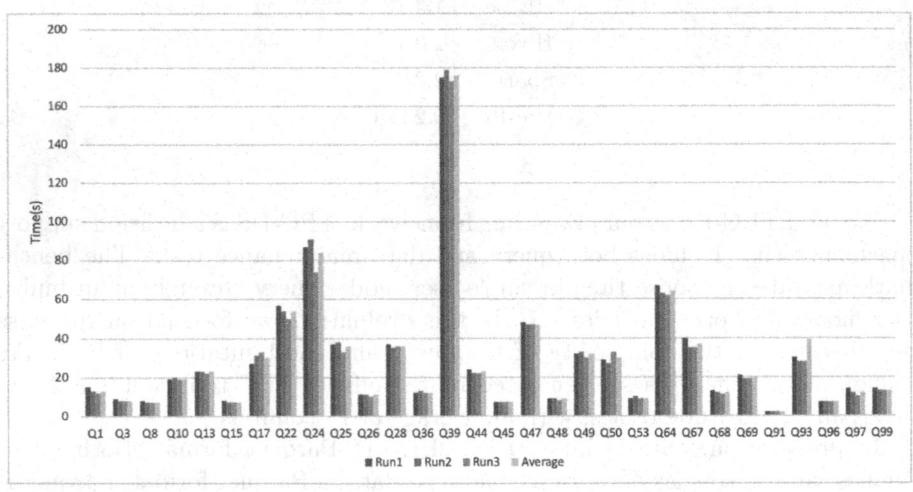

Fig. 1. The execution time of Presto

Figure 1 shows the results of our test on Presto against the TPC-DS benchmark, which tested 30 of the 100 queries generated by the TPC-DS tool. In order to reduce the volatility of query time caused by system resource preemption and other factors, we performed three tests for each query and took the average as the final test result. The impact of query characteristics on Presto query time is our focus, and we select queries based on query time differences to analyze Presto performance characteristics.

We chose Query22 and Query91 for our analysis. The overall structure of the two queries is similar, involving multi-table joins, conditional filtering, sorting, and grouping. However, through the analysis of the query results, we found that the number of rows traversed by the two queries was very different. We give the statistics of each table in Table 2.

Table 2. Number of rows in TPC-DS tables for the tested scale factors

TableName	Rows	TableName	Rows
call_center	24	promotion	500
catalog_page	12000	reason	45
catalog_returns	1439749	ship_mode	20
catalog_sales	14401261	store	102
customer	500000	store_returns	2875432
customer_address	250000	store_sales	28800991
customer_demographics	1920800	time_dim	86400
date_dim	73049	warehouse	10
household_demographics	7200	web_page	200
income_band	20	web_returns	719217
inventory	133110000	web_sales	7197566
item	102000	web_site	42

Therefore, we believe that table size and query operation together affect query efficiency. The reason for this is that Presto is a memory-based query engine designed for massively parallel processing. It executes queries in a pipelined fashion, with all intermediate data stored in memory. Compared with Spark, which stores intermediate results to disk, this design reduces the I/O overhead of queries. However, when memory resources are limited, the query efficiency is reduced.

Based on this idea, we counted the number of result rows returned by these 30 queries in TPC-DS at 10G data size. Table 3 reflects this statistic. Consistent with the statistics, Query22 uses larger tables and repeatedly performs join calculations, while Query91 involves tables that are one order of magnitude smaller. Obviously, larger table joins consume more resources and time.

Table 3. Query rows

Query	Row count	Query	Row count
Q1	6.4 M	Q45	8.22 M
Q3	29 M	Q47	86.9 M
Q8	29.9 M	Q48	31 M
Q10	53.3 M	Q49	55.7 M
Q13	31.1 M	Q50	31.8 M
Q15	15.2 M	Q53	29 M
Q17	46.4 M	Q64	105 M
Q22	133 M	Q67	29 M
Q24	65.1 M	Q68	29.9 M
Q25	46.4 M	Q69	53.3 M
Q26	16.5 M	Q91	4.19 M
Q28	173 M	Q93	31.7 M
Q36	29 M	Q96	28.9 M
Q39	267 M	Q97	43.3 M
Q44	115 M	Q99	14.5 M

4.2 TPC-DS on SparkSQL

As with the scheme tested on Presto, we tested the same query three times and calculated the query average as the final test result to reduce the query time error caused by various factors. As you can see from the test results in Fig. 2, SparkSQL also performed well in the TPC-DS test, with most queries completed in less than 1 min. Careful analysis of the relationship between specific query statements and test results shows that Spark has excellent query performance for specific types of queries, such as multi-table association. However, for the query tasks associated with a single table or a few tables with complex filtering conditions, the query performance is relatively poor. We take Query93 as an example for analysis below.

Query93 statement is a complex statement with multiple tables joins, which involves nested joins of three tables, and has many complicated filtering conditions, grouping, and sorting. Its query process mainly connects the store_sales table, reason table and store_returns table, and groups the result table to the ss_customer_sk table, and finally, sort by some fields.

This query takes an average of 20 s. By monitoring Spark, we can see the various stages of this query. As a query associated with multiple tables. Spark divides tasks into 7 stages. The first stage is mainly in the process of map-partition, that is, mapping partition data. In the second, third, and fourth phases, the mapping processing of the store_sales table, store_returns table, and reason table is performed, respectively. In the fifth stage, the left external join of the store_sales table and the store_returns table is executed, and the results are mapped again.

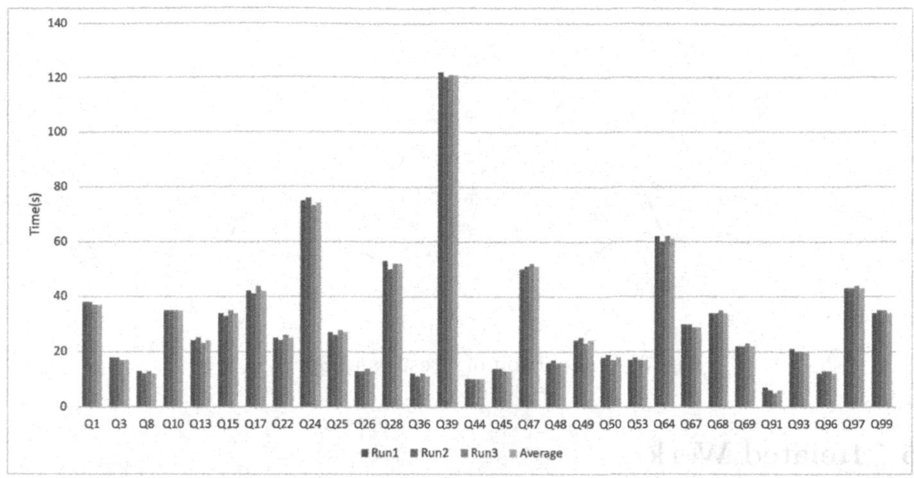

Fig. 2. The execution time of SparkSQL

The sixth stage joins the above result table with the processed reason table and then performs the mapping operation. The seventh stage performs grouping and sorting operations on the final result table and then performs a mapping process to obtain a final result set.

In SparkSQL, each stage are distributed to downstream machines for calculation through shuffle operations [5]. This process is somewhat similar to the idea of MapReduce, but thanks to Spark's RDD design, the results of intermediate operations will be stored in memory, which makes Spark's next iteration calculations much faster. This is why Spark performs well when performing complex queries like the one above for multi-table joins. At the same time, the result set generated in each stage will be persistently stored. This process guarantees the fault tolerance of Spark computing [21]. When the task is wrong, Spark does not need to recompute all tasks, only need to find the failed stage to continue. This design not only increases fault tolerance but also provides high efficiency for task recovery.

4.3 Comparison

As shown in Fig. 3, Presto was superior in overall time and average performance when performing paging queries, but SparkSQL left Presto far behind when it needed full data and statistics.

In Query15 and Query68, Presto is far superior to SparkSQL, while SparkSQL is superior to Query22 and Query39. For this result, we briefly analyzed the four query characteristics. Query15 and Query68 have a large number of conditional queries and use small tables, while Query22 and Query39 frequently use large tables for join operations.

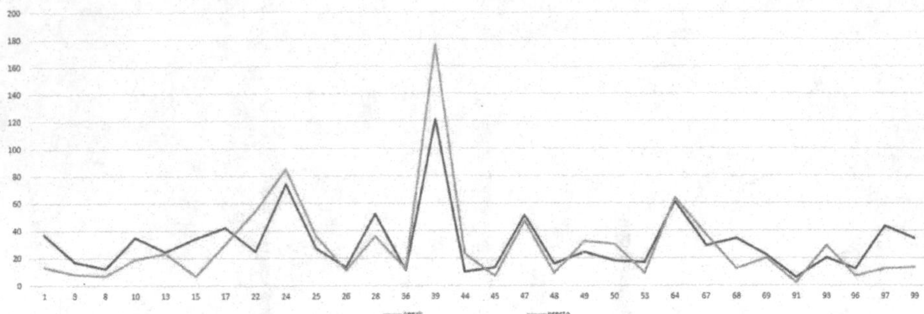

Fig. 3. Comparison of Presto and SparkSQL

5 Related Work

Presto. Some works have performed TPC-DS-based query tests and analyzed how presto should be optimized to achieve better performance [14,19,20]. These efforts explored not only how to optimize Presto for engineering implementation, but also ways to improve Presto performance through query optimization and leveraging modern storage.

SparkSQL. There has been a lot of work comparing the performance of SparkSQL with other systems, and one of the primary systems is Hive [12,16,30]. These works compare the two systems on TPC-H, BigBench and other different data sets, and analyze the impact of different file formats on query performance. Hive is based on MapReduce and differs from SparkSQL in design. Presto and SparkSQL both serve as in-memory computing engines. They handle computing tasks in a similar way, but have different scheduling strategies, resulting in performance differences.

6 Conclusion

By comparing Presto with Spark through 30 queries in TPC-DS, we found that Presto had better query execution performance overall than SparkSQL, especially for conditional queries involving fewer tables. SparkSQL performs better than Presto for complex queries and large table calculations. In the future, we plan to compare the performance differences of query engines based on heterogeneous data sources, so as to better understand the performance characteristics of each query engine and provide reference for various data management scenarios.

References

1. Apache Hadoop. http://hadoop.apache.org/. Accessed 12 Feb 2022
2. Apache spark^{TM} - unified engine for large-scale data analytics. http://spark.apache.com. Accessed 12 Feb 2022
3. Armbrust, M., et al.: Spark SQL: relational data processing in spark. In: Sellis, T.K., Davidson, S.B., Ives, Z.G. (eds.) Proceedings of the 2015 ACM SIGMOD International Conference on Management of Data, Melbourne, Victoria, Australia, May 31–June 4 2015, pp. 1383–1394. ACM (2015). https://doi.org/10.1145/2723372.2742797
4. Borthakur, D.: The Hadoop distributed file system: architecture and design. Hadoop Project Website **11**(2007), 21 (2007)
5. Davidson, A., Or, A.: Optimizing shuffle performance in spark. University of California, Berkeley-Department of Electrical Engineering and Computer Sciences, Technical report (2013)
6. Feng, B., Wang, Y., Chen, G., Zhang, W., Xie, Y., Ding, Y.: EGEMM-TC: accelerating scientific computing on tensor cores with extended precision. In: Lee, J., Petrank, E. (eds.) PPoPP 2021: 26th ACM SIGPLAN Symposium on Principles and Practice of Parallel Programming, Virtual Event, Republic of Korea, 27 February–3 March 2021, pp. 278–291. ACM (2021). https://doi.org/10.1145/3437801.3441599
7. Feng, B., Wang, Y., Ding, Y.: Saga: sparse adversarial attack on EEG-based brain computer interface. In: IEEE International Conference on Acoustics, Speech and Signal Processing, ICASSP 2021, Toronto, ON, Canada, 6–11 June 2021, pp. 975–979. IEEE (2021). https://doi.org/10.1109/ICASSP39728.2021.9413507
8. Feng, B., Wang, Y., Geng, T., Li, A., Ding, Y.: APNN-TC: accelerating arbitrary precision neural networks on ampere GPU tensor cores. In: de Supinski, B.R., Hall, M.W., Gamblin, T. (eds.) SC 2021: The International Conference for High Performance Computing, Networking, Storage and Analysis, St. Louis, Missouri, USA, 14–19 November 2021. pp. 37:1–37:13. ACM (2021). https://doi.org/10.1145/3458817.3476157
9. Feng, B., Wang, Y., Li, G., Xie, Y., Ding, Y.: Palleon: a runtime system for efficient video processing toward dynamic class skew. In: Calciu, I., Kuenning, G. (eds.) 2021 USENIX Annual Technical Conference, USENIX ATC 2021, 14–16 July 2021, pp. 427–441. USENIX Association (2021). https://www.usenix.org/conference/atc21/presentation/feng-boyuan
10. George, L.: HBase - The Definitive Guide: Random Access to Your Planet-Size Data. O'Reilly (2011). http://www.oreilly.de/catalog/9781449396107/index.html
11. Ivanov, T., Korfiatis, N., Zicari, R.V.: On the inequality of the 3v's of big data architectural paradigms: a case for heterogeneity. CoRR abs/1311.0805 (2013). http://arxiv.org/abs/1311.0805
12. Li, X., Zhou, W.: Performance comparison of hive, impala and spark SQL. In: 2015 7th International Conference on Intelligent Human-Machine Systems and Cybernetics, vol. 1, pp. 418–423. IEEE (2015)
13. Manyika, J., et al.: Big Data: The Next Frontier for Innovation, Competition, and Productivity. McKinsey Global Institute (2011)
14. Margoor, A., Bhosale, M.: Improving join reordering for large scale distributed computing. In: Wu, X., et al. (eds.) 2020 IEEE International Conference on Big Data (IEEE BigData 2020), Atlanta, GA, USA, 10–13 December 2020, pp. 2812–2819. IEEE (2020). https://doi.org/10.1109/BigData50022.2020.9378281

15. Pan, Z., et al.: Exploring data analytics without decompression on embedded GPU systems. IEEE Trans. Parallel Distrib. Syst. **33**(7), 1553–1568 (2022). https://doi.org/10.1109/TPDS.2021.3119402
16. Poggi, N., Montero, A., Carrera, D.: Characterizing bigbench queries, hive, and spark in multi-cloud environments. In: Nambiar, R., Poess, M. (eds.) TPCTC 2017. LNCS, vol. 10661, pp. 55–74. Springer, Cham (2018). https://doi.org/10.1007/978-3-319-72401-0_5
17. Pöss, M., Floyd, C.: New TPC benchmarks for decision support and web commerce. SIGMOD Rec. **29**(4), 64–71 (2000). https://doi.org/10.1145/369275.369291
18. Pöss, M., Smith, B., Kollár, L., Larson, P.: TPC-DS, taking decision support benchmarking to the next level. In: Franklin, M.J., Moon, B., Ailamaki, A. (eds.) Proceedings of the 2002 ACM SIGMOD International Conference on Management of Data, Madison, Wisconsin, USA, 3–6 June 2002, pp. 582–587. ACM (2002). https://doi.org/10.1145/564691.564759
19. dos Reis, V.L.M., Li, H.H., Shayesteh, A.: Modeling analytics for computational storage. In: Amaral, J.N., Koziolek, A., Trubiani, C., Iosup, A. (eds.) ICPE 2020: ACM/SPEC International Conference on Performance Engineering, Edmonton, AB, Canada, 20–24 April 2020, pp. 88–99. ACM (2020). https://doi.org/10.1145/3358960.3375794
20. Sethi, R., et al.: Presto: SQL on everything. In: 35th IEEE International Conference on Data Engineering, ICDE 2019, Macao, China, 8–11 April 2019, pp. 1802–1813. IEEE (2019). https://doi.org/10.1109/ICDE.2019.00196
21. Shanahan, J.G., Dai, L.: Large scale distributed data science using apache spark. In: Cao, L., Zhang, C., Joachims, T., Webb, G.I., Margineantu, D.D., Williams, G. (eds.) Proceedings of the 21th ACM SIGKDD International Conference on Knowledge Discovery and Data Mining, Sydney, NSW, Australia, 10–13 August 2015, pp. 2323–2324. ACM (2015). https://doi.org/10.1145/2783258.2789993
22. Thusoo, A., et al.: Hive - a petabyte scale data warehouse using Hadoop. In: Li, F., et al. (eds.) Proceedings of the 26th International Conference on Data Engineering, ICDE 2010, 1–6 March 2010, Long Beach, California, USA, pp. 996–1005. IEEE Computer Society (2010). https://doi.org/10.1109/ICDE.2010.5447738
23. Vavilapalli, V.K., et al.: Apache Hadoop YARN: yet another resource negotiator. In: Lohman, G.M. (ed.) ACM Symposium on Cloud Computing, SOCC 2013, Santa Clara, CA, USA, 1–3 October 2013, pp. 5:1–5:16. ACM (2013). https://doi.org/10.1145/2523616.2523633
24. Wang, Y., Feng, B., Ding, Y.: DSXplore: optimizing convolutional neural networks via sliding-channel convolutions. In: 35th IEEE International Parallel and Distributed Processing Symposium, IPDPS 2021, Portland, OR, USA, 17–21 May 2021, pp. 619–628. IEEE (2021). https://doi.org/10.1109/IPDPS49936.2021.00070
25. Wang, Y., et al.: GNNAdvisor: An adaptive and efficient runtime system for GNN acceleration on GPUs. In: Brown, A.D., Lorch, J.R. (eds.) 15th USENIX Symposium on Operating Systems Design and Implementation, OSDI 2021, 14–16 July 2021, pp. 515–531. USENIX Association (2021). https://www.usenix.org/conference/osdi21/presentation/wang-yuke
26. Zhang, F., Chen, Z., Zhang, C., Zhou, A.C., Zhai, J., Du, X.: An efficient parallel secure machine learning framework on GPUs. IEEE Trans. Parallel Distrib. Syst. **32**(9), 2262–2276 (2021). https://doi.org/10.1109/TPDS.2021.3059108
27. Zhang, F., Zhai, J., He, B., Zhang, S., Chen, W.: Understanding co-running behaviors on integrated CPU/GPU architectures. IEEE Trans. Parallel Distrib. Syst. **28**(3), 905–918 (2017). https://doi.org/10.1109/TPDS.2016.2586074

28. Zhang, F., Zhai, J., Shen, X., Mutlu, O., Du, X.: POCLib: a high-performance framework for enabling near orthogonal processing on compression. IEEE Trans. Parallel Distrib. Syst. **33**(2), 459–475 (2022). https://doi.org/10.1109/TPDS.2021.3093234
29. Zhang, F., et al.: TADOC: text analytics directly on compression. VLDB J. **30**(2), 163–188 (2021). https://doi.org/10.1007/s00778-020-00636-3
30. Zhang, M., Liu, F., Lu, Y., Chen, Z.: Workload driven comparison and optimization of hive and spark SQL. In: 2017 4th International Conference on Information Science and Control Engineering (ICISCE), pp. 777–782. IEEE (2017)

Optimizing the Age of Sensed Information in Cyber-Physical Systems

Yinlong Li[1], Siyao Cheng[1(✉)], Feng Li[1], Jie Liu[2], and Hanling Wu[3]

[1] Harbin Institute of Technology, Harbin 150006, Heilongjiang, China
{liyinlong,csy,feng.li}@hit.edu.cn

[2] Harbin Institute of Technology (Shenzen), Shenzen 518055, Guangdong, China
jieliu@hit.edu.cn

[3] Beijing Institute of Astronautical Systems Engineering, Beijing 100076, China
hlwu8000@sina.com

Abstract. With the wide spread of IoT applications, the timeliness of sensed information becomes more and more important. Recently, the researchers proposed to use the Age of Information (AoI) to evaluate the timeliness of sensed data, and a series of algorithms have been proposed to optimize the AoI in IoT and cyber-physical systems. There algorithms are efficient for the systems with identical sensors. However, they are not very suitable for the systems containing different sensors since the different variations of data are not sufficiently considered by them. In order to evaluate the AoI of different sensors data more fairly, we propose to use the data queue length instead of time to denote it in this paper. Based on such new metric, the problem of minimizing the max AoI is provided. Finally, the optimized scheduling algorithm is given for solving the above problem. Extensive experimental results are carried out and show that the proposed algorithm has higher performance comparing with the baseline and existing works.

Keywords: Cyber-physical systems · Age of information · Sensed data

1 Introduction

With the rapid development of sensing and communication techniques, the applications of cyber-physical system [1,2], such as smart transportation [6], smart homes [3], smart agriculture [7,8,10] *etc.* have been more mature and widely used. In these cyber-physical systems, the sensors that can collect the sensed values from the monitoring environment and objects are primary components to ensure the normal operations of the systems. In order to guarantee the cyber-physical system makes a correct decision, the freshness of sensed values should be kept as much as possible.

In the existing works, the researchers proposed to use the age of information (AoI for short) to evaluate the freshness of sensed values [11,12]. Then, the problem of how to shorten the AoI for each sensor in the network aroused widely attention especially for the scenario that the bandwidth between the sink and sensors is

U. K. Rage et al. (Eds.): DASFAA 2022 Workshops, LNCS 13248, pp. 330–346, 2022.
https://doi.org/10.1007/978-3-031-11217-1_24

quite limited. A series of sensor scheduling algorithms have been proposed in order to minimize the AoI. The early works considering AoI were given by [13,14], the authors discussed how to shorten the age of information under energy replenishment constraints, and proposed the algorithm to optimize the process of generating information to minimize AoI in the case of limited sensor energy. The design of the data transmission model and scheduling strategy is ignored, although the author considers the issue of energy constraints. To improve the transmission fairness for different sensors, the authors in [15] discussed how to minimize the expected sum of AoIs in the whole network while timely-throughput constraints were met. The authors proved that such problem is NP-hard, and approximation algorithm with low-complexity was provided to deal with it. However, minimize the expected sum of AoIs is not reasonable because not all devices need to transmit data urgently. Different queueing systems that be used as a model to minimize the AoI are analyzed in [16–18]. These models focus more on changes in time than changes in data. In [19–21], the authors tried to evaluate and optimize age for multiple sources sharing a queue or simple network. Paper [20] assumed that the channels between the nodes and the sink are unreliable and allow a node transmit more than once to counter channel uncertainty. In [21], the authors only computed a closed form expression for the average age and the average peak age of each stream, but not giving a policy.

Although the above algorithms tried to improve AoI, they are more suitable for the systems with identical sensors since these algorithms only use the data sampling time to evaluate the AoIs of different sensors, and ignore the different variation rate of the sensed data. The sensors in a cyber-physical system always have different information variation rate. Therefore, the variation rate of sensed data sampled by different sensor may be quite different with each other and should be considered.

Example 1: In an application of automatic driving, various sensors are deployed in a smart car in order to monitor the status of the car and control it. Obviously, the sensor monitoring the speed of car has the higher data variation rate comparing with that sampling the temperature inside the car. That is, the data sampled by the temperature sensor will be valid for quite a long time, on the other hand, the data generated by the speed monitoring sensor will be expired soon. For unmanned aerial vehicle (UAV) trajectory planning, the location of UAV and the data collected by UAV are both important. The joint optimization of power consumption and age performance is necessary to improve the efficiency and quality of UAV-assisted data collection [9].

Example 2: As shown in Fig. 1, there exist multiple types of sensors for a smart agriculture application. The variation rates of different sensed data are also indicated in Fig. 1. Based on our observation, the temperature data of air and soil always vary quickly, while the PH data of the soil seems more stable. Therefore, if we only use the data received time to evaluate the AoIs for different sensors, the bandwidth and energy will be wasted for transmitting the unimportant data for a CPS.

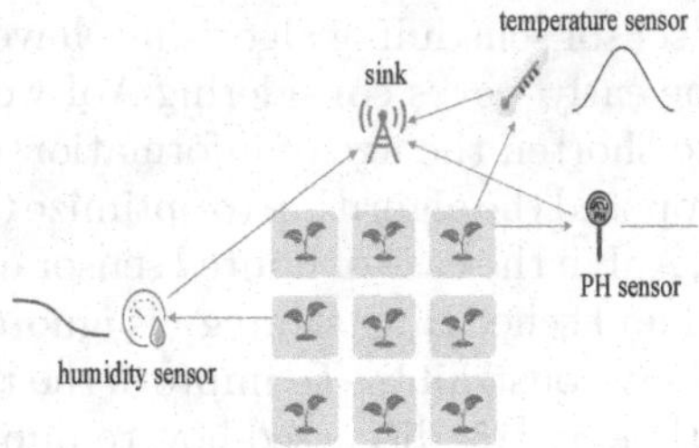

Fig. 1. Smart agricultural data collection system

Based on the above two examples, the diverse variation rates of the sensed data should be fully considered and a new metric that evaluates the AoIs in a CPS is required in order to ensure that the important values are submitted to the sink on time when the bandwidth is limited. Due to such reason, we use the number of data packets in the queue to measure the AoIs of sensed data in this paper. When its sensed data dramatically change and cannot be forecast for a sensor node, a new data packet will be generated and insert into a unsubmitted queue. It means that the number of the data packets that the data packets are different in the unsubmitted queue is the data update times of a sensor. Actually, the length of unsubmitted queue is the number of data packets and can be used to denote the AoIs for a sensor node. The new measure of AoIs should sufficiently take the data variation rate under consideration, so that the limited bandwidth is fully used to transmit the important sensed values that changes dramatically. Based on the newly proposed AoIs, an algorithm which can minimize the max AoI is provided to scheduling sensors for transmission when the bandwidth is limited. Meanwhile, further improvement is also given under the small probability event model. In summary, the contributions of the paper are as follows.

(1) A new metric for evaluating AoIs in a cyber-physical system with different types of sensors is proposed, and the problem of minimizing the max AoI is also given.
(2) We prove that it is a polynomial-time problem when the number of new data packets in the following time slot can be exactly predicted, and provide the optimal algorithm with $O(n^2)$ complexity to solve it. We also discuss the problem of how to forecast the number of newly coming data packets as well.
(3) The extensive experimental results were carried out, where the experimental results show that our proposed algorithm has high performance than the existing one.

The rest of paper is organized as follows. The system model and problem definition are given in Sect. 2. The algorithm for minimizing the max AoI is discussed in Sect. 3. Meanwhile, Sect. 4 shows the experimental results and the performance evaluation. Finally, Sect. 5 concludes the whole paper.

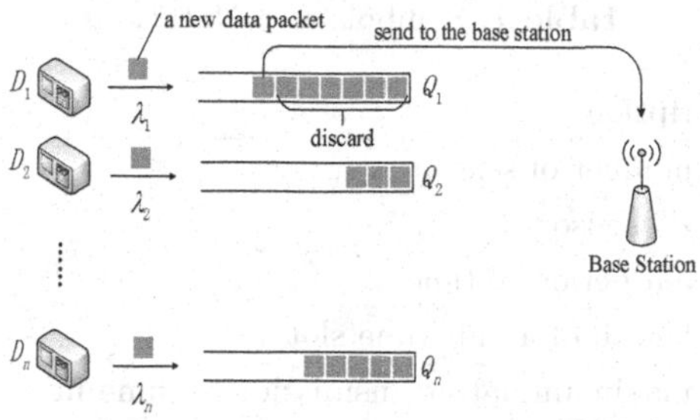

Fig. 2. The data schedule model

2 Problem Definition

2.1 System Model

We assume that there exist n sensors and a base station in the considered system. Similar to [22,23], the sensors in the system are responsible for collecting the sensed data from the monitoring objects, and report their data to the base station, where $s_{i,t}$ is used to denote the the sensed value generated by sensor node i at time t.

Let t_c be the current time, then the whole period of $[0, t_c]$ can be divided into a series of time slots, which are denoted by $T_1, T_2, ..., T_l$, where $\cup_{j=1}^{l} T_l = [0, t_c]$, and $T_p \cap T_q = \emptyset$. Since the transmission bandwidth among sensors and base station is limited according to the above discussion, we use M to denote such bandwidth, i.e. the transmission rate should not exceed M, so that not all the sensed data are able to be submitted to the base station. The system model is as shown in Fig. 2.

Furthermore, as shown in Fig. 2, the packets of sensed values contained in different unsubmitted queues are not the same since sensors are monitoring different objects and their variations are quite diverse. Thus, in this paper, we considered to use the number of unsubmitted packet to define the *Age of Information* (AoI for short). The formal definition is given as follow.

Definition 1 (AoI for a Cyber-Physical System). *Let $t_{i,j}$ be the time when $s_{i,t_{i,j}}$ was submitted to the base station. The definitions of Age of a Sensed Value, Average AoI and Max AoI of a system are determined as follow.*

1. AoI for a Sensed Value: The AoI for a sensed value of sensor node i, denoted by $Age(t_{i,j})$, equals to the number of unsubmitted data packets between $t_{i,j}$ to the next data submitted time of sensor node i (or t_c).
2. Average AoI of a Sensor Node: Given the period, $[0, t_c]$, the average AoI of sensor node i, denoted by $Age_i(t_c)$, determined by the average AoIs of all the submitted values of Sensor i.
3. Max AoI of the Whole System: the Max of the whole system in the period $[0, t_c]$, denoted by $MaxAge(t_c)$, satisfies that

$$MaxAge(t_c) = \max_{1 \le i \le n} Age_i(t_c)$$

Table 1. Symbols and Notations

Symbol	Description
n	The number of sensors
i	The ith sensor
$[0, t_c]$	A given period of time
$\bar{T}$	The length of a unit time slot
M	The maximum data transmission volume in a unit time period $\bar{T}$
L_i	The data packet length of sensor i
λ_i	The poisson distribution arrival frequency of sensor i
$Age_i(t)$	The average AoI of sensor i at time t
$MaxAge(t_c)$	The maximum value of all $Age_i(t)$
AoI_{OPT}	The value obtained by minimizing $MaxAge(t_c)$

2.2 Problem Definition

In a cyber-physical system, the max AoI is significant because they provide the metrics to evaluate the worst currency guaranteed by a system. Therefore, the aim of our algorithm is to minimize the max AoI of the whole system during scheduling different types of sensors. Let L_i $(1 \leq i \leq n)$ denotes the packet length of sensor i, and x_{it} be an indicator variable to show whether sensor node i is permitted to report its sensed value to the base station at time t, where a submitted sensed value is contained in a packet, $1 \leq i \leq n$ $t \in T_p$, and T_p $(1 \leq p \leq l)$ is a time slot belonging to $[0, t_c]$. Then, x_{it} satisfying the following formula

$$x_{it} = \begin{cases} 1, \text{If Sensor } i \text{ is schedulded to transmit at } t \\ 0, \text{Otherwise} \end{cases} \tag{1}$$

Input:

1. A system with n sensor nodes;
2. A period $[0, t_c]$;
3. M, which is the upper bound of the bandwidth limitation;
4. The data packet length of each sensor, *i.e.* $L_1, L_2, ..., L_n$.

Output: $X = \{(i, t)|x_{it} = 1\}$ for all $1 \leq i \leq n$ to make $MaxAge(0, t_c)$ be minimized while the following conditions are satisfied.

$$\sum_{i=1}^{n} x_{it} \leq 1 \text{ For } \forall t \in [0, t_c] \tag{C1}$$

$$\sum_{i=1}^{n} \sum_{t \in [0, t_c]} \frac{L_i}{M} x_{it} \leq t_c \text{ For } \forall 1 \leq p \leq l \tag{C2}$$

where Condition C_1 means there at most exists one sensor to report its sensed value to base station due to communication conflict, and Condition C_2 means that total time for transmission is no more than the length of $[0, t_c]$. In order to make the paper to be easy to understand, the symbols used in it is summarized in Table 1.

3 Max AoI Optimizing Algorithm in Cyber-Physical Systems

3.1 Determining the AoI of Sensed Data

In order to optimize the AoI of the whole system, the age of each submitted sensed values should be determined firstly. Based on the system model given in Sect. 2, we use the *Unsubmitted Queue*, denoted by Q_{it}, to contain the packets of sensed values that has not been transmitted to the sink for each sensor node i. Such queue can be utilized for determining the AoI of the sensed values.

Then, the age of $Age(t_{i,j})$ can be determined by the following formula.

$$Age(t_{i,j}) = \begin{cases} |Q_{it_c}|, & \text{if } j = k_i \\ |Q_{i,t_{i,j+1}}|, & \text{Otherwise} \end{cases} \tag{2}$$

where k_i is the number of total transmission times of sensor i and $t_{i,1}$, $t_{i,2}$, ..., t_{i,k_i} are the time that the sensor node i submits the values to the base station during $[0, t_c]$.

According to the update strategy of *Unsubmitted Queue*, it will be cleared up after a successful submission. Therefore, all the unsubmitted packets of sensed data between $t_{i,j}$ and $t_{i,j+1}$ are contained by the queue $Q_{i,t_{i,j+1}}$ when $1 \leq j < k_i$. Then, $Age(t_{i,j}) = |Q_{i,t_{i,j+1}}|$ equals to the number of unsubmitted data packets between $t_{i,j}$ to $t_{i,j+1}$, which meets the Definition 1. Meanwhile, when $j = k_i$, all the unsubmitted data packets are included in Q_{it_c}, so that the age $Age(t_{i,j})$ is determined by $|Q_{it_c}|$, which also meets the requirement of Definition 1.

Comparing with the traditional AoI definition [13,14,24], the metric given in Definition 1 has better performance to reflect diverse variation of sensed data during evaluating AoI of sensed values. Based on Formula (2) and Definition 1, $Age_i(t_c)$ can be calculated by the following formula.

$$Age_i(0, t_c) = \frac{\sum_{i=1}^{k_i-1} Age(t_{i,j}) + |Q_{it_c}|}{k_i} \tag{3}$$

Based on Formula (3), if sensor node i totally submitted k_i values to the sink, the age of $k_i - 1$ submitted sensed values are stable since the previous submitted time slots, $t_{i,1}$, $t_{i,2}$, ..., t_{i,k_i}, are determined. And the Average AoI of a Sensor Node, $Age_i(t_c)$, is only varying with the length of Q_{it_c}.

3.2 Greedy Strategy for Each Available Time

Let t_a represents a current available time for the sink to receive data pack, t_{a+1} and t_{a-1} be the next and previous available time being adjacent to t_a, and $Age_i(t_a)$, $Age_i(t_{a-1})$ and $Age_i(t_{a+1})$ denote the AoIs of sensor node i at t_a, $t_{a-1}$and t_{a+1} respectively, where $1 \leq i \leq n$.

The aim of each scheduling selection is to make the max AoI at next available time to be minimized, so that $MaxAvg(t_c)$ can be minimized according to induction. Based on such motivation, we can adopt the greedy selection strategy if we can obtain the number of new data packets coming in next available time.

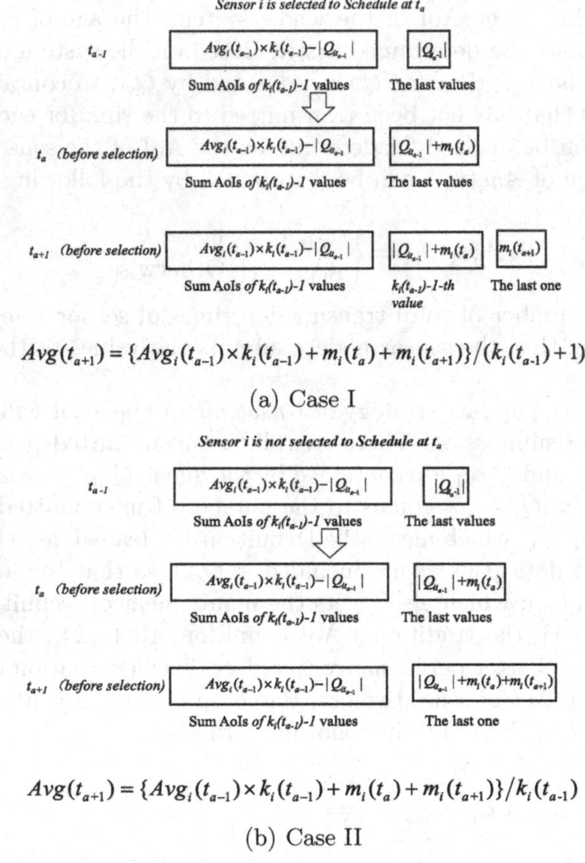

Fig. 3. The relationship between $Age_i(t_a)$ and $Age_i(t_a+1)$

Specifically, let $m_i(t_a)$ and $m_i(t_{a+1})$ denote the number of newly coming data packets in $[t_{a-1}, t_a)$ and $[t_a, t_{a+1})$, respectively. As shown in Fig. 3, $Age_i(t_{a+1})$ can be determined according to the situation whether node i is selected to schedule at t_a, which is also presented in formula (4).

$$Age_i(t_{a+1}) = \begin{cases} \frac{Age_i(t_{a-1})\times k_i(t_{a-1})+m_i(t_a)+m_i(t_{a+1})}{k_i(t_{a-1})+1}, \\ \text{if } i \text{ is selected to transmit at } t_a \\ \frac{Age_i(t_{a-1})\times k_i(t_{a-1})+m_i(t_a)+m_i(t_{a+1})}{k_i(t_{a-1})}, \\ \text{Otherwise} \end{cases} \tag{4}$$

Therefore, we have the following Theorem 1.

Theorem 1. *At each time t_a, if the sensor node i that has the max $(Age_i(t_{a-1}) \times k_i(t_{a-1}) + m_i(t_a) + m_i(t_{a+1}))/k_i(t_{a-1})$ is scheduled to submit its sensed value to the sink, then $MaxAge(t_{a+1})$ is minimized, where $MaxAge(t_{a+1})$ denoted the Max AoI of the system in period $[0, t_{a+1})$.*

Proof. Let $i = \underset{1\le i\le n}{\text{argmax}} \frac{Age_i(t_{a-1})\times k_i(t_{a-1})+m_i(t_a)+m_i(t_{a+1})}{k_i(t_{a-1})}$, then we have

$$\begin{aligned} &\frac{Age_i(t_{a-1}) \times k_i(t_{a-1}) + m_i(t_a) + m_i(t_{a+1})}{k_i(t_{a-1})} \\ \ge &\frac{Age_j(t_{a-1}) \times k_j(t_{a-1}) + m_j(t_a) + m_j(t_{a+1})}{k_j(t_{a-1})} \end{aligned} \tag{5}$$

for any $1 \le j \le n$ and $j \ne i$.

Suppose that in the optimal selection, denoted by S_O, it is sensor node r instead of sensor node i to be selected for transmission at t_a, then we have that

$$\begin{aligned} Avg_r(t_{a+1}) &= \frac{Age_r(t_{a-1}) \times k_r(t_{a-1}) + m_r(t_a) + m_r(t_{a+1})}{k_r(t_{a-1}) + 1} \\ &\le \frac{Age_r(t_{a-1}) \times k_r(t_{a-1}) + m_r(t_a) + m_r(t_{a+1})}{k_r(t_{a-1})} \le Avg_i(t_{a+1}) \end{aligned} \tag{6}$$

Based on formulas (5) and (6), the max AoI generated by S_O in period $[0, t_{a+1})$, denoted by $MavAge^{(S_O)}(t_{a+1})$ satisfying that

$$\begin{aligned} MavAge^{(S_O)}(t_{a+1}) &= \max_{1\le p\le n} Avg_p(t_c) = Avg_i(t_{a+1}) \\ &= \frac{Age_i(t_{a-1}) \times k_i(t_{a-1}) + m_i(t_a) + m_i(t_{a+1})}{k_i(t_{a-1})} \end{aligned} \tag{7}$$

On the other hand, we can construct S_G based on greedy selection. That is, in S_G, we select i at t_a for transmission, then we use $MavAge^{(S_G)}(t_{a+1})$ to denote the max AoI caused by S_G in period $[0, t_{a+1})$, then we have

$$\begin{aligned} &MavAge^{(S_G)}(t_{a+1}) \\ &= \max\{\frac{Age_i(t_{a-1}) \times k_i(t_{a-1}) + m_i(t_a) + m_i(t_{a+1})}{k_i(t_{a-1}) + 1}, \\ &\quad \max_{1\le j\le n, j\ne i}\{\frac{Age_j(t_{a-1}) \times k_j(t_{a-1}) + m_j(t_a) + m_j(t_{a+1})}{k_j(t_{a-1})}\}\} \\ &\le \frac{Age_i(t_{a-1}) \times k_i(t_{a-1}) + m_i(t_a) + m_i(t_{a+1})}{k_i(t_{a-1})} \\ &= MavAge^{(S_O)}(t_a) \end{aligned} \tag{8}$$

based on formulas (5) and (7).

We also have $MavAge^{(S_G)}(t_c) \geq MavAge^{(S_O)}(t_c)$ since S_O is the optimal selection. So it also can obtain the optimal scheduling result at any time t_c by greedy selection for $MavAge^{(S_G)}(t_c) = MavAge^{(S_O)}(t_c)$.

Based on Theorem 1, the greedy algorithm can be adopted for scheduling sensor node if we can forecast the number of data packets coming in next available time exactly, *i.e.* $m_i(t_{a+1})$ for each $1 \leq i \leq n$. However, it is hard to get the exact values of $\{m_i(t_{a+1})|1 \leq i \leq n\}$ since it depends not only on the data variation but also on the selection at t_a when $\{m_i(t_{a+1})|1 \leq i \leq n\}$ are not available.

Let λ_i be the parameter of poisson process for sensor node i, then the number of data packets coming in a period of $[t_a, t_{a+1}]$ can be estimated by a Poisson Distribution with parameter $\lambda_i \times (t_{a+1} - t_a)$. That is,

$$\Pr\{m_i(t_{a+1}) = k\} = e^{-\lambda_i \times (t_{a+1} - t_a)} \frac{(\lambda_i \times (t_{a+1} - t_a))^k}{k!} \tag{9}$$

When we have λ_i, one of the most common way is to use its expectation to estimate $m_i(t_{a+1})$. Suppose that sensor node k is selected to transmit at t_a, then $t_{a+1} = t_a + \frac{L_k}{M}$, where L_k denotes the packet length of sensor node k. Therefore, $m_i(t_{a+1})$ can be estimated by

$$m_i(t_{a+1}) = \lambda_i(t_{a+1}) \times \frac{L_k(t_a)}{M} \tag{10}$$

Since $MaxAge(t_c) = \max_{1 \leq i \leq n} Age_i(t_c)$, then the lower bound of $MaxAge(t_c)$ can be calculated as follow,

$$\sum_{i=1}^{n} \lambda_i t_c L_i \leqslant M t_c MaxAge(t_c) \tag{11}$$

that is $MaxAge(t_c) \geq \frac{\sum_{i=1}^{n} \lambda_i L_i}{M}$.

3.3 Scheduling Algorithm

Based on the analysis in above two sections, the scheduling algorithm to minimize max AoI of the system in $[0, t_c]$ can be summarized as the following steps.

Steps 1. Initialize the parameters and the base station schedules the first data packet of each sensor in turn. Set $t_a = 0$.

Steps 2. For each sensor node i ($1 \leq i \leq n$), assume that it is selected to transmit at t_a. For each $1 \leq j \leq n$, the following steps will be execute:

1. Let $t_{a+1} = t_a + \frac{L_i}{M}$;
2. Obtain $m_j(t_{a+1})$ and $Age_j(t_{a+1})$ according to formulas (10) and (4) in above two sections.

Steps 3. Set $MaxAvg(t_{a+1}) = \max_{1 \le j \le n}(Age_j(t_{a+1}))$. If the selection of node i can make $MaxAvg(t_{a+1})$ to be decreased, then record i and the smaller Max AoI.

Steps 4. Repeat the above two steps until $i = n$. Choose the k that can make $MaxAvg(t_{a+1})$ to be smallest, and schedule sensor node k at t_a. Update t_a with $t_a = t_a + L_k/M$

Steps 5. Repeat step 2 to step 4 until $t_a = t_c$

The complexity of sink to apply the above scheduling algorithm is $O(n^2)$ at each time slot since two loop with size n are involved.

Algorithm 1: Scheduling Algorithm for Minimizing Max AoI

Input: $L_1, L_2, ..., Ln, \lambda_1, \lambda_2, ..., \lambda_n, M, t_c$
Output: X

Initialize $L_1, L_2, ..., L_n, \lambda_1, \lambda_2, ..., \lambda_n, M, t_c$;
$T_{remaining} \leftarrow t_c, t \leftarrow 0, Age(0, t_c) \leftarrow 0, X \leftarrow \varnothing$;
while $T_{remaining} - t > 0$ **do**
  $AM = +\infty$;
  for $i = 1; i \le n$ **do**
    for $j = 1; j \le n$ **do**
      $t_{a+1} = t_a + \frac{L_i}{M}$; Estimate $m_j(t_{a+1})$ according to formula (10);
      Update $Age_j(t_{a+1})$ according to formula (4);
    $MaxAge = \max_{1 \le j \le n}(Age_j(t_{a+1}))$;
    if $MaxAge < AM$ **then**
      $k = i$;
      $AM = MaxAge$;
  Schedule k for transmission, and the latest data packet from the queue Q_k are submitted to the sink;
  $Q_k = \emptyset$;
  $t = t + L_k/M$;
  Update X;
Return X.

---

4 Experiment Results

In this section, we use simulation experiments to evaluate the performance of $algorithm1(SAMMA)$. There are some parameters that need to be analyzed to measure how the experiment results are affected by these parameters. We will compare $SAMMA$ with three baselines which will be shown in the following and $Juventas$ algorithm because the application scenario of this algorithm is very similar to the model in this paper. We will use the arrival frequency λ_i of the sensor as the sensor weight w_i of the $Juventas$ algorithm [23] because λ_i can correspond to the urgency of the sensor to transmit data. In addition, the AoI of $Juventas$ algorithm is different from this paper. When we calculate the AoI of $Juventas$ algorithm, it will be calculated according to the definition of this paper.

4.1 Experiment Settings

The data to be used in the experiment are shown in Table 2 and Table 3. The data in Table 2 is asynchronous data, which means that no two sensors have the same parameter λ_i or L_i. The data in Table 3 is synchronous data, which means that the parameters λ_i and L_i of two or more sensors are the same. $L_i(1)$, $L_i(2)$ and $L_i(3)$ are three types data with different degrees of synchronization and the degree of synchronization is increased in sequence. In the following experimental analysis, we will show the influence of different parameters on the experimental results.

Table 2. Asynchronous parameter value

Sensor	D_1	D_2	D_3	D_4	D_5	D_6	D_7	D_8
L_i	7	15	12	6	8	7	20	10
λ_i	16	5	4	7	18	11	3	20
M	40–100							
t_c	50							
$\bar{T}$	1							

Table 3. Synchronous parameter value

Sensor	D_1	D_2	D_3	D_4	D_5	D_6	D_7	D_8
$L_i(1)$	2	2	4	4	6	6	8	8
$L_i(2)$	4	4	4	4	8	8	8	8
$L_i(3)$	6	6	6	6	6	6	6	6
λ_i	5	5	5	5	5	5	5	5
M	40–100							
t_c	50							
$\bar{T}$	1							

4.2 Baseline

The *SAMMA* algorithm provides a data packet scheduling policy with limited bandwidth. In order to show the effectiveness of our algorithm, we compare the results of the *SAMMA* algorithm with the following three baselines.

Random. The BS schedules the data packets randomly.

Period. The BS schedules the data packets based on a specific time period. As long as the data packets are scheduled repeatedly in the order from Q_1 to Q_n until the end of time.

Lower Bound. The lower bound derived in above section is $\frac{\sum_{i=1}^{n} \lambda_i L_i}{M}$. We this lower bound as a baseline.

Besides, *Juventas* algorithm provides an excellent scheduling policy for the problem of minimizing the weighted average AoI and will be compared with *SAMMA*. The λ_i is set as sensor weight w_i of the *Juventas* algorithm.

4.3 Performance of SAMMA Algorithm

In this section, we will evaluate the performance of our *SAMMA* algorithm by comparing with three baseline, i.e. *Random*, *Period* and *LowerBound* since existing research does not have an algorithm for minimizing the AoI of the system based on data packet changes. In addition, the closeness of the algorithm results to the lower bound will be able to visually demonstrate the superiority of the algorithm.

1) The Effect of Synchronous or Asynchronous Data Packet: When the L_i and λ_i of some sensors are the same, then the data packets of these sensors are synchronous, otherwise they are asynchronous. We need to research the effect of data packets on the experimental results under synchronous and asynchronous conditions. The synchronization parameters are shown in Table 3.

The asynchronous parameter settings are shown in Table 2 and the synchronous parameter settings are shown in Table 3. According to the enhancement of data synchronization, we divide the data into three types of synchronization, of which type (1) has the weakest synchronization and type (3) has the strongest synchronization. We compare the results of these sensors under the algorithm *SAMMA* scheduling. Since the lower bounds of the these sensors are different, we take the lower bound of the weakest synchronization sensor as the baseline.

Figure 4(a) shows the standard deviation(SD) of $Age_i(t_c)$ at different synchronization degree, which can show the consistency of the $Age_i(t_c)$. Figure 4(b) show the AoI_{OPT} generated by algorithm *SAMMA* at different synchronization. It can be found that no matter what the value of M is, the AoI_{OPT} of the strongly synchronized sensors is larger than that of weakly synchronized sensors. Our algorithm is more likely to make partial wrong choices when dealing with strongly synchronized sensors, which leads to large differences in the $Age_i(t_c)$ of each sensor. When dealing with weakly synchronized sensors, the $Age_i(t_c)$ difference of each sensor is less. The scheduling results obtained by our algorithm are very close to the lower bound in both the case of strong synchronization and weak synchronization.

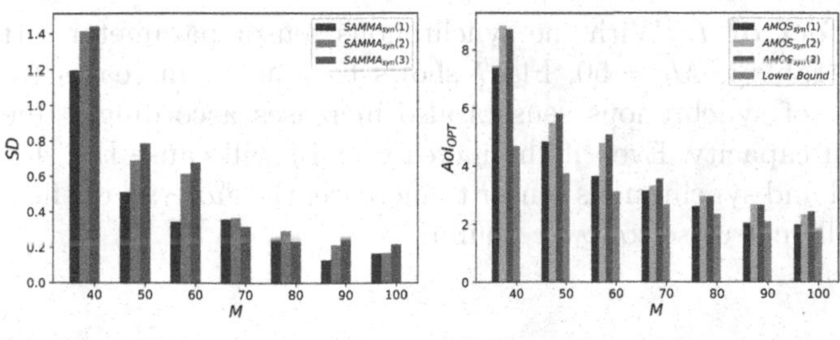

(a) The Effect of Synchronization on SD (b) The Effect of Synchronization on AoI

Fig. 4. The effect of synchronization

2) The Effect of M: Figure 5 investigates the impact of M on the AoI_{OPT} of $SAMMA$. Actually, changing the value of M may have different effects on the experimental results. If M is very large, it may takes very little time to transmit all the data packets in the queue. The $Age(t_c)$ is 1 because no packet is dropped. This is meaningless and not the case we considered. The situation we have to consider is that M is much smaller than the amount of data packets that need to be transmitted in each time slot.

The parameter settings are shown in Table 2. The value of M is increased from 40 to 100. Figure 5 shows that $SAMMA$ can achieve a better performance than *Random*, *Period* and $Juventas$. The result of $SAMMA$ is very close to lower bound. As M increases, the AoI_{OPT} gradually decreases because the increase in the transmission volume allows the data packet of the base station to be updated earlier.

3) The Effect of n: With the asynchronous sensor parameter settings λ_i, L_i, t_c, $\bar{T}$ given in Table 2, $M = 50$. According to the order of sensor in Table 2, we add 4 sensor each time and run the algorithm. Figure 6 shows that as n increases, the AoI_{OPT} also increases accordingly since more sensors come to seize the limited transmission. The AoI_{OPT} of the synchronous sensor is closer to the lower bound. The results of *Random*, *Period* and $Juventas$ are larger than $SAMMA$ and *LowerBound* because the increase of sensors will greatly increase the probability of *Random*, *Period* and $Juventas$ making wrong decisions.

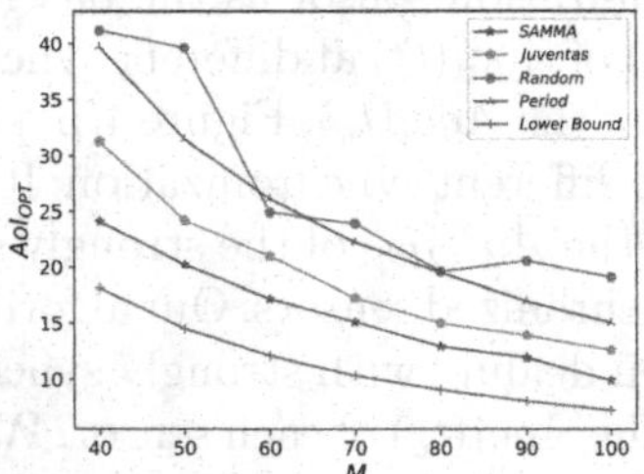

Fig. 5. The effect of M on *AoI*

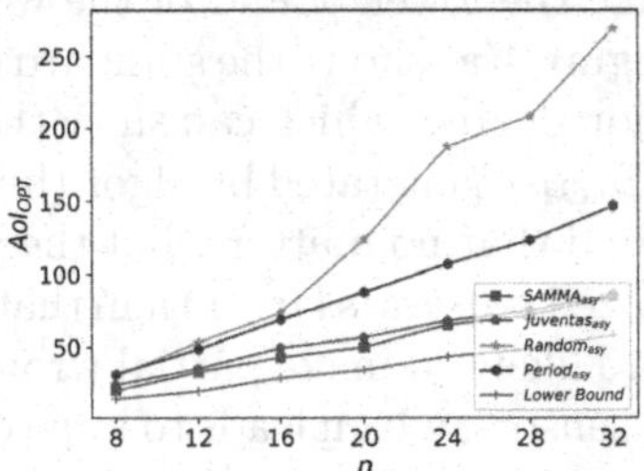

Fig. 6. The effect of n on *AoI*

4) The Effect of L: With the synchronous sensor parameter settings λ_i, t_c, $\bar{T}$ given in Table 3, $M = 50$, Fig. 7 shows that as L_i increases from 6 to 14, the AoI_{OPT} of synchronous sensors also increases accordingly due to limited transmission capacity. Even if the increase of L_i will cause the AoI_{OPT} of the lower bound and synchronous sensor to increase, the AoI_{OPT} of the synchronous sensor is still very close to lower bound.

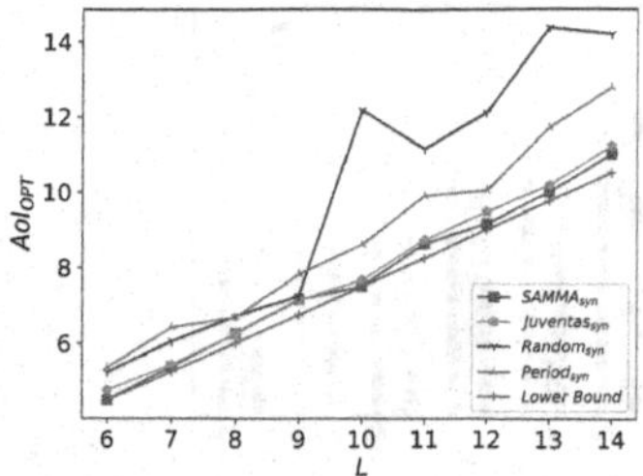

Fig. 7. The effect of L on AoI

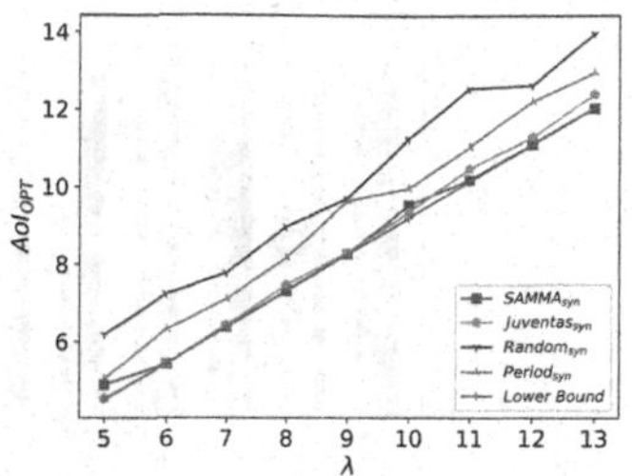

Fig. 8. The effect of λ on AoI

5) The Effect of λ: With the synchronous sensor parameter settings $L_i(3)$, t_c, $\bar{T}$ given in Table 3, $M = 50$, Fig. 8 shows that as λ_i increases from 5 to 13, the AoI_{OPT} of synchronous sensors also increases accordingly. Even if the increase of λ_i will cause the AoI_{OPT} of the lower bound and synchronous sensor to increase, the AoI_{OPT} of the synchronous sensor is still very close to lower bound.

The experiments in Fig. 7 and Fig. 8 is run on synchronized data because we want to make sure that there is only one parameter is changed. The synchronized data makes the wrong decisions of *Random*, *Period* and *Juventas* not highlighted and looks closer to *SAMMA* and *LowerBound*.

From Fig. 6 to Fig. 8, we can find that the proposed algorithm has high stability and will not deviate from lower bound due to the increase of parameters n, L_i, and λ_i. *Random*, *Period* and *Juventas* scheduling policies are highly volatile and may cause $Age(t_c)$ to be much larger than the $Age(t_c)$ output by the *SAMMA* algorithm and lower bound.

6) The Effect of t_c: In order to study the effect of the total transmission time $[0, t_c]$ on AoI_{OPT}, we use the ratio of AoI_{OPT} and lower bound to measure the gap between the AoI_{OPT} output by our algorithm and the lower bound. The sensor parameter settings in this group is show in Table 2. In addition, we use the standard deviation to measure the difference in the average AoI of each sensor under different values of t_c. The ratio is calculated as follow.

$$ratio = \frac{AoI_{OPT}}{C^*} \tag{12}$$

where C^* is the lower bound.

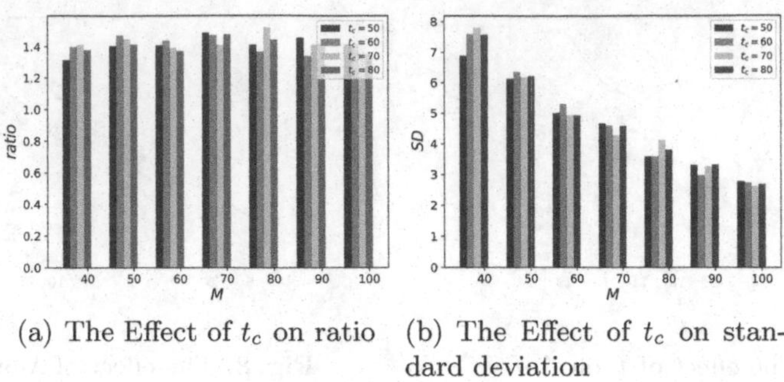

(a) The Effect of t_c on ratio (b) The Effect of t_c on standard deviation

Fig. 9. The effect of t_c

From Fig. 9(a), we can find that when t_c changes from 50 to 80, under different bandwidths, the ratio of AoI_{OPT} and lower bound obtained by our algorithm are very close. Figure 9(b) shows that the standard deviation of $Age_i(t_c)$ is very close when t_c takes different values. This means that the change of t_c will not be a great fluctuation of the AoI_{OPT} and the $Age_i(t_c)$ of each sensor.

The $Age(t_c)$ achieves the minimum value and is very close to lower bound. The AoI_{OPT} decreases when M increases and increases when L_i, λ_i and n increase. This can be understood as an increase in L_i, λ_i and n will cause M to be insufficient, resulting in more dropped data packets. Besides, Strongly synchronized sensors may make the algorithm more likely to make wrong decisions. Finally, we conclude that our algorithm has a good performance in minimizing $Age(t_c)$.

5 Conclusion

In this paper, a new metric to evaluate the age of sensed information is proposed in cyber-physical systems. Based on such metric, the problem of minimizing the max AoI in a bandwidth limited system is studied. We propose an optimal algorithm for it and discuss how to forecast the number of the newly coming data packets as well. Furthermore, the lower bound of max AoI and the improve strategies of the algorithm are also provided. Both the theoretical analysis and the experimental results show that the proposed algorithm has the high performance, and the results returned by it are very close to the lower bound.

Acknowledgements. This work is partly supported by the National Key Research and Development Plan Project of China under Grant No. 2019YFE0125200, the National Key R&D Program of China under Grant No. 2021ZD0110900, the Programs for Science and Technology Development of Heilongjiang Province under Grant No. 2021ZXJ05A03, the National Natural Science Foundation of China under Grant No. 61972114, 62106061, the National Natural Science Foundation of Heilongjiang Province under Grant No. YQ2019F007, and the Key Science Technology Specific Projects of Heilongjiang Province under Grant No. 2019ZX14A01.

References

1. Humayed, A., Lin, J., Li, F., Luo, B.: Cyber-physical systems security-a survey. IEEE Internet Things J. **4**(6), 1802–1831 (2017)
2. Hussain, B., Du, Q., Sun, B., Han, Z.: Deep learning-based DDoS-attack detection for cyber-physical system over 5G network. IEEE Trans. Industr. Inf. **17**(2), 860–870 (2020)
3. Fang, Y., Lim, Y., Ooi, S.E., Zhou, C., Tan, Y.: Study of human thermal comfort for cyber-physical human centric system in smart homes. Sensors **20**(2), 372 (2020)
4. Kockemann, U., et al.: Open-source data collection and data sets for activity recognition in smart homes. Sensors **20**(3), 879 (2020)
5. Limbasiya, T., Das, D.: Searchcom: vehicular cloud-based secure and energy-efficient communication and searching system for smart transportation. In: Proceedings of the 21st International Conference on Distributed Computing and Networking, pp. 1–10 (2020)
6. Zichichi, M., Ferretti, S., D'Angelo, G.: Are distributed ledger technologies ready for smart transportation systems? arXiv preprint arXiv:2001.09018 (2020)
7. Namani, S., Gonen, B.: Smart agriculture based on IoT and cloud computing. In: 2020 3rd International Conference on Information and Computer Technologies (ICICT), pp. 553–556. IEEE (2020)
8. Jin, X.B., Yang, N.X., Wang, X.Y., Bai, Y.T., Su, T.L., Kong, J.L.: Hybrid deep learning predictor for smart agriculture sensing based on empirical mode decomposition and gated recurrent unit group model. Sensors **20**(5), 1334 (2020)
9. Sun, M., Xu, X., Qin, X., Zhang, P.: AoI-energy-aware UAV-assisted data collection for IoT networks: a deep reinforcement learning method. IEEE Internet Things J. **8**(24), 17275–17289 (2021)
10. Garca, L., Parra, L., Jimenez, J.M., Lloret, J., Lorenz, P.: IoT-based smart irrigation systems: an overview on the recent trends on sensors and IoT systems for irrigation in precision agriculture. Sensors **20**(4), 1042 (2020)
11. Kaul, S., Gruteser, M., Rai, V., Kenney, J.: Minimizing age of information in vehicular networks. In: 2011 8th Annual IEEE Communications Society Conference on Sensor, Mesh and Ad Hoc Communications and Networks, pp. 350–358. IEEE (2011)
12. Kaul, S., Yates, R., Gruteser, M.: Real-time status: how often should one update? In: 2012 Proceedings IEEE INFOCOM, pp. 2731–2735. IEEE (2012)
13. Bacinoglu, B.T., Ceran, E.T., Uysal-Biyikoglu, E.: Age of information under energy replenishment constraints. In: 2015 Information Theory and Applications Workshop (ITA), pp. 25–31. IEEE (2015)
14. Bacinoglu, B.T., Uysal-Biyikoglu, E.: Scheduling status updates to minimize age of information with an energy harvesting sensor. In: 2017 IEEE International Symposium on Information Theory (ISIT), pp. 1122–1126. IEEE (2017)
15. Kadota, I., Sinha, A., Modiano, E.: Optimizing age of information in wireless networks with throughput constraints. In: IEEE INFOCOM 2018-IEEE Conference on Computer Communications, pp. 1844–1852. IEEE (2018)
16. Champati, J.P., Al-Zubaidy, H., Gross, J.: On the distribution of AoI for the GI/GI/1/1 and GI/GI/1/2* systems: exact expressions and bounds. In: IEEE INFOCOM 2019-IEEE Conference on Computer Communications, pp. 37–45. IEEE (2019)
17. Moltafet, M., Leinonen, M., Codreanu, M.: On the age of information in multi-source queueing models. IEEE Trans. Commun. **68**(8), 5003–5017 (2020)

18. Kam, C., Kompella, S., Nguyen, G.D., Ephremides, A.: Effect of message transmission path diversity on status age. IEEE Trans. Inf. Theory **62**(3), 1360–1374 (2015)
19. Jiang, Z., Krishnamachari, B., Zheng, X., Zhou, S., Niu, Z.: Decentralized status update for age-of-information optimization in wireless multiaccess channels. In: ISIT, pp. 2276–2280 (2018)
20. Yates, R.D., Kaul, S.K.: Status updates over unreliable multiaccess channels. In: 2017 IEEE International Symposium on Information Theory (ISIT), pp. 331–335. IEEE (2017)
21. Najm, E., Telatar, E.: Status updates in a multi-stream M/G/1/1 preemptive queue. In IEEE INFOCOM 2018-IEEE Conference On Computer Communications Workshops (INFOCOM WKSHPS), pp. 124–129. IEEE (2018)
22. Moltafet, M., Leinonen, M., Codreanu, M.: An approximate expression for the average AoI in a multi-source M/G/1 queueing model. In: 2020 2nd 6G Wireless Summit (6G SUMMIT), pp. 1–5. IEEE (2020)
23. Li, C., Li, S., Hou, Y.T.: A general model for minimizing age of information at network edge. In: IEEE INFOCOM 2019-IEEE Conference on Computer Communications, pp. 118–126. IEEE (2019)
24. Kadota, I., Sinha, A., Modiano, E.: Scheduling algorithms for optimizing age of information in wireless networks with throughput constraints. IEEE/ACM Trans. Netw. **27**(4), 1359–1372 (2019)

Aggregate Query Result Correctness Using Pattern Tables

Nitish Yadav(✉), Ayushi Malhotra, Sakshee Patel, and Minal Bhise

Distributed Databases Group, DAIICT, Gandhinagar, India

201911035@daiict.ac.in

Abstract. The state-of-the-art techniques for aggregate query results correctness works well only when a reference table is available. We are proposing a technique, which will work well even when the reference table is absent. This technique uses pattern tables for checking the correctness of aggregate queries. It is demonstrated on Sofia Air Quality Dataset where complete clusters for aggregate queries are identified. The results show a reduction of 71% in average query execution time for Pattern Table Method PTM over Data Table Method DTM. Further, the scaled data results till 5X show that PTM scales linearly while DTM scales linearly only till 3X. The algorithm execution time is analyzed for scaled data, the number of levels, and the number of NULLs. Our algorithm is well behaved till 5X for scaled data. The correctness of aggregate queries will help in ensuring the correctness of the analytics built on it.

Keywords: Pattern table · Reference table · Pattern tree · Pattern table algorithm · Data table method · Pattern table method

1 Introduction

The data revolution is supported by technical advances, new algorithms and abundant data storage capabilities, allowing the creation of new services, tools and applications, which produce a large amount of data. It is estimated that the Data volume grows by a factor of 75, starting from 2 ZB up to 149 ZB for the period 2010–2024 [18]. More and more businesses around the world have started analyzing big data to make more business informative decisions and introduce new business models to gain an upper hand over their competitors. This process of analyzing big data to draw inferences is called big data analytics. Big Data analytics is applied to domains like Healthcare, Manufacturing, Media & Entertainment, IoT and Decision Support Systems [3]. The task of big data analytics involves the execution of multiple queries which includes aggregate queries. Aggregate queries create a cluster based on the group by attributes upon which an aggregate function is applied. The completeness of data impacts the accuracy of inference drawn. In this paper, we have worked on identifying and representing the completeness of data using pattern tables which when queried using aggregate queries produce accurate results.

U. K. Rage et al. (Eds.): DASFAA 2022 Workshops, LNCS 13248, pp. 347–362, 2022.
https://doi.org/10.1007/978-3-031-11217-1_25

As the data volume increases, the problem of dataset incompleteness arises inadvertently ranging from a few missing values in a tuple (represented by NULL) to the entire tuple set as missing. Data incompleteness emanates in a dataset for reasons like Human Errors, Lack of Information sources and Hardware Failures [11]. Aggregate query when executed on incomplete datasets produces partially correct results. If the cluster upon which aggregation function is applied is complete, the results obtained will be correct, however, if the cluster is incomplete, an incorrect result will be obtained. Therefore, it is necessary to identify which clusters are complete and for this NULLs and missing tuple sets needs to be identified and represented. NULLs can be identified by simply scanning the dataset. However, the missing tuple set can be identified only by comparing the dataset with a complete dataset called as reference Table [5].

Throughout this paper, a dummy dataset D (shown in Table 1) containing sensor observations will be used to explain all the concepts. This dummy dataset contains Temperature T, Pressure P and Humidity H readings which are recorded for spatial attribute Location and Temporal attributes Year, Month (Mth) and Day by the sensors. D is an incomplete dataset because it contains NULLs (represented in green) as well as a missing tuple (represented in blue). NULLs can be identified by scanning the dataset. However, to identify the missing tuple, a comparison with the reference table is required. The reference table used for the comparison is shown in Table 2.

Table 1. Data Table D containing sensor observations

Loc	Year	Mth	Day	T	P	H
A	2019	June	Mon	32.1	95270.27	62.48
A	2019	June	Tue	25.1	98578.24	74.55
A	2019	July	Mon	18.1	95268.45	80.55
A	2019	July	Tue	NULL	35875.77	NULL
A	2020	June	Mon	11.5	NULL	25.25
A	2020	June	Tue	10.5	NULL	34.25
A	2020	July	Mon	NULL	98452.44	65.55
A	2020	July	Tue	X	X	X
B	2019	June	Mon	12.8	89999.25	NULL
B	2019	June	Tue	20.5	90253.35	NULL
B	2019	July	Mon	21.5	96253.25	75.24
B	2019	July	Tue	27.3	NULL	35.15

Incomplete datasets affect aggregate query results when queried upon. To understand the effect on the query result, consider an analyst who wants to executes an aggregate query Q on D, which calculates the average temperature [year, month, and location] wise and is shown in Fig. 1.

Table 2. Reference Table R for sensor observations in D

Loc	Year	Mth	Day
A	2019	June	Mon
A	2019	June	Tue
A	2019	July	Mon
A	2019	July	Tue
A	2020	June	Mon
A	2020	June	Tue
A	2020	July	Mon
A	2020	July	Tue
B	2019	June	Mon
B	2019	June	Tue
B	2019	July	Mon
B	2019	July	Tue

```
select Loc, Year, Mth, avg(T) as AVG
from D
group by Loc, Year, Mth;
```

Fig. 1. Query Q for calculating average temperature

The output of Q is shown in Table 3. The tuples in red have either incorrect results or NULL under the column AVG. Average for the cluster [A, 2019, July] is calculated by just considering one record, i.e. [A, 2019, July, Mon, 18.1] and the other record containing NULL is simply ignored by SQL. This is incorrect as wrong conclusions can be drawn because of incomplete information. Tuples in red should be removed from the output as an analyst can draw erroneous conclusions.

Table 3. Q's output containing incorrect values

Loc	Year	Mth	AVG
A	2019	June	28.6
A	2019	June	18.1
A	2019	July	11.0
A	2019	July	NULL
A	2020	June	16.65
A	2020	June	24.4

To generate the correct result (free from tuples in red) for Q in Table 3, standard Relational Algebra can be used to identify clusters that are complete but will require a lot of operations (join, subtract, union) which will be costly when done on large datasets. This approach is termed as Data Table Method DTM. An alternative approach could be to identify and list all the clusters in advance that are complete so that whenever an aggregate query comes, only the clusters that are complete w.r.t a query are selected to generate the correct result. To represent complete clusters, the concept of a pattern table was developed [6,13]. Pattern table represents complete clusters in a dataset compactly by using a reference table. For D, the pattern table for T is shown in Table 4.

Table 4. Pattern table for temperature

Loc	Year	Mth	Day
B	*	*	*
*	*	June	*
*	2019	*	Mon

A tuple in a Pattern table is called a pattern and '*' is called a wildcard that can take any value from the domain of the attribute in which it is present. [B,*,*,*] represents that all the tuples containing Location as B are complete. For Q, clusters that are complete are [B,*,*,*] and [*,*,June,*]. A join between these patterns and D will produce the values of Temperature that represent complete clusters upon which Q will be applied, thus producing the result shown in Table 3 without tuples in red. This approach is termed Pattern Table Method PTM. Both DTM and PTM methods will produce the same output.

The state-of-the-art technique for generating pattern tables uses FoldData Algorithm [6]. The FoldData algorithm needs a reference table for generating pattern tables. These reference tables are created by the database administrator at the time of dataset creation as part of the dataset specification. Creating a reference table requires a deep knowledge base about the possible values an attribute can take. Situations in which this knowledge base is incomplete lead to an absence of a reference table with the dataset. Examples of datasets that have a missing reference table are Sofia Air Quality Dataset [8], Bank Marketing Dataset [16], and Dataset for river flow and flash flood forecasting [9]. All three datasets contain NULL values. NULLs are present in [16] and [9] due to the Lack of Information sources and in [8] due to Hardware Failures. Missing tuple sets cannot be identified in any of the three datasets because no reference table is available to identify them. FoldData cannot be used to generate pattern tables when the reference table is absent. One can suggest creating a virtual reference table (Universe Table [12]) that contains all possible combinations of domain values an attribute can take. However, this will lead to the generation of fictitious tuples that can never occur in the dataset. Pattern table generated by FoldData using the Universe Table will lead to incorrect pattern sets due to fictitious tuples which affect query results.

To avoid the creation of incorrect patterns, we proposed Pattern Table Algorithm PTA that generates pattern tables for the dataset with an absent reference table. The generated pattern tables are used for executing aggregate query workload using PTM. These techniques are demonstrated using Sofia Air Quality Dataset [8]. The pattern table generated by PTA has less expressive power as compared to the pattern table generated by FoldData because missing tuples are not represented in our case due to the absence of a reference table. However, our method of generating pattern table allows us to get correct results for an incoming aggregate query q on a dataset with missing reference table by identifying the clusters that are complete w.r.t Q.

2 Related Work

The study of the incompleteness of data is as old as the relational model itself. Razniewski [13] has described the reason for data incompleteness in cloud databases, Collaborative databases and network data warehouses. In all these databases, completeness information regarding what parts of the dataset is complete is given either in the form of annotations, partitioning information or check-pointing with time. This completeness information helps an analyst to determine whether to run a query on the dataset or to check for other information sources to estimate missing data or to execute queries.

Dataset completeness is categorized using three settings based on the knowledge base available regarding datasets. These settings are Closed World Assumption CWA, Open World Assumption OWA, and Partially Closed World Assumption PCWA. CWA states that, a tuple is true if it exists in the database else, it does not exist in the real world [15]. OWA states that, tuples that do not exist in the dataset can exist in the real world, and such datasets are incomplete [15]. PCWA states that, some part of data can be classified as complete provided a completeness statement is provided, and some part of the dataset is incomplete [12]. To identify which part of dataset is incomplete, a concept of reference table was introduced by Fan and Greets [4,5].

The concept of m-tables [19] inspired from c-tables [7] is used for determining generalized tuples as probable under CWA and OWA. This concept deals with appending certainty information to generalized tuples as well as propagation of it to query answers. Other works such as Denecker [2], Cortes–Calabuig [1] and Razniewski [14] used database instances for reasoning about query result completeness using highly expressive language for completeness description and/or queries but inadvertently suffered from Co-NP data complexity or π_2^P query complexity.

Under PCWA setting, Motro [12] showed how Completeness information for some query answers could be obtained by studying Completeness information for existing query answers. Levy [10] extended this concept to represent information regarding Completeness for a table using existing Completeness information from known tables. Razniewski [13], inspired from Levy and Motro, laid the framework for defining Pattern Tables using virtual reference table and Pattern Algebra that can be used to represent and obtain completeness information for a data table and query result using existing completeness information from known tables and query results.

Fan and Geerts [4,5] proposed the concept of reference table which acts as a master table or upper bound database instance that can be used to assess dataset completeness just by comparison. Reference table can be categorized as Real-World reference table and Universe Reference table [12]. Hannou [6] extended the concept of pattern tables to represent available and missing tuples in datasets using a real-world reference table. To generate pattern tables Fatma proposed FoldData algorithm. The pattern tables obtained from FoldData are queried upon using a sound pattern algebra that uses folding and unfolding operations for each query. Generating pattern tables using FoldData depends entirely on the availability and correctness of reference tables. If the reference table is missing, FoldData cannot generate pattern tables and if the reference table is a universes reference table that may contain fictitious tuples (don't exist in real world), incorrect pattern tables will be obtained thus affecting aggregate query results.

The Completeness of clusters generated by aggregate queries can be checked by simply checking that all the elements belonging to a cluster exists or not. This acts as the framework of DTM method. Shoshani [17] stated that the correctness of aggregate query is guaranteed when executed on the cluster that is complete. In our case pattern table represents all the clusters that are complete (i.e., no NULL is present). Any aggregate query, when executed on such a pattern table, will give a correct result. Our objectives are 1) Generate a Pattern Table Algorithm (PTA) that creates a minimal Pattern table representing non-NULL tuples for datasets with missing reference table. 2) Execute Aggregate queries using DTM and PTM. 3) Examining which method among PTM and DTM executes queries faster. Our contributions are 1) PTA, which generates pattern tables that represent non-NULL tuples, which can be used to identify clusters that are complete for a query. These pattern tables are obtained for datasets which do not have a reference table. 2) We also propose two methods for obtaining correct aggregate query results: DTM and PTM. We have shown that PTM produces faster results when compared with DTM for datasets containing NULL.

3 Important Concepts

3.1 Reference Table

Identifying available and missing data for a data table is relative as it can be computed only by comparing a data table with a complete dataset. This complete dataset is called Reference Table [4,5]. The attributes in the reference table are called referenced attributes [6]. The reference table is created by the database administrator at the time of dataset creation. This requires a deep knowledge base about the possible values a reference attribute can take. Reference tables are of two types: Universe Reference Table U and Real-World Reference Table W [12].

- U: It is an instance of all possible tuples and is obtained by performing a cross between domains of all the referenced attributes. Creating this leads to the creation of fictitious tuples that do not exist in the real world.
- W: It is an instance of all TRUE tuples that exist in the real world.

3.2 Pattern Table and Pattern Tree

Pattern tables act as a tool to represent completeness information. Fatma [6] has used this to represent available and missing data with the help of a reference table. Attributes that are used for generating patterns are termed referenced attributes. A tuple in a Pattern table is called a partition pattern, and '*' is called a wildcard that can take any value from the domain of the attribute in which it is present. For D, the pattern table for Temperature is shown in Table 4. [B,∗,∗,∗] represents that all the tuples containing Location as B are complete. [∗,∗,June,∗] represents all the tuples containing Month as June are complete. [∗,2019,∗,Mon] represents all the tuples containing Year as 2019 and Day as Mon are complete.

Patterns among themselves follow a specialization and generalization concept [6]. A pattern tree is created with the most generalized pattern on top and subsequent levels containing more and more specialized patterns and it is traversed in level order. The instance of a pattern means how many tuples in the dataset can be satisfied by that pattern. For example, In D, p = [B,∗,∗,∗] satisfies 4 tuples.

3.3 Aggregate Query Execution

Big data analytics involves the execution of multiple queries which includes aggregate queries. An aggregate query is a method of deriving clusters by analyzing individual data tuples present in the dataset. An aggregate SQL query contains two components, aggregate function and group by clause. A group by clause is used to create clusters upon which aggregate function like Average, Minimum, Maximum, Count and Sum is applied to draw inferences. The completeness of data impacts the accuracy of inference drawn using aggregate queries. Therefore, completeness information needs to be represented. Pattern tables act as a tool to represent completeness information of clusters in incomplete datasets. For incomplete datasets, pattern tables which when queried produce accurate results enabling an analyst to draw correct inferences. To obtain correct results two methods have been proposed: Data Table Method DTM and Patter Table Method PTM.

Data Table Method DTM. In this method, we use the Relational Algebra (RA) table manipulation techniques to identify and remove tuples with NULL and its accomplice tuples from D. In query Q, both Mon and Tue records are required to get a correct result. Suppose the Tue record is NULL, then the Mon record, which is non-NULL, becomes an accomplice record that needs to be removed to get correct results. The table obtained after the removal of such tuples is complete, and Q can be directly applied to get the correct result. Steps to get correct the result for Q are shown in Fig. 2:

$$\begin{aligned}
&Step\ 1{:}\ Res2 = \pi_{Loc,Year,Month}(D) - \pi_{Loc,Year,Month}(\sigma_{Temperature=NULL}(D)) \\
&Step\ 2{:}\ Res1 = {}_{Loc,Year,Month}G_{avg(Temperature)\ as\ AVG}(D) \\
&Step\ 3{:}\ \pi_{Res1.Loc,Res1.Year,Res1.Month,AVG}(Res1 \bowtie_c Res2) \\
&\text{where } c = Res1.Loc = Res2.Loc\ and \\
&Res1.Year = Res2.Year\ and \\
&Res1.Month = Res2.Month
\end{aligned}$$

Fig. 2. Steps involved DTM method

Pattern Table Method PTM. In this method, we use the PTA to create a pattern table for D. On this, we apply the modified query that takes into account the semantics of ∗. Accomplice tuples are the tuples that are not NULL in an incomplete cluster. To identify the accomplice tuples, we identify the set of attributes present in the query that are being grouped. For Q, these attributes are Loc, Year, and Month. From Table 4, we obtain the patterns that have ∗ under Day. The instances of these patterns will not have any accomplice tuples in them. Steps to get the correct result for Q are shown in Fig. 3:

$$\begin{aligned}
&Step\ 1{:}\ Res2 = \pi_{*}(\sigma_{Day=*}(T)) \\
&Step\ 2{:}\ Res1 = {}_{Loc,Year,Month}G_{avg(Temperature)\ as\ AVG}(D) \\
&Step\ 3{:}\ \pi_{Res1.Loc,Res1.Year,Res1.Month,AVG}(Res1 \bowtie_c Res2) \\
&\text{where } c = (Res1.Loc = Res2.Loc\ or\ Res2.Loc ='*')\ and \\
&(Res1.Year = Res2.Year\ or\ Res2.Year ='*')\ and \\
&(Res1.Month = Res2.Month\ or\ Res2.Month ='*')
\end{aligned}$$

Fig. 3. Steps involved in PTM method

4 Pattern Table Algorithm

Pattern Table Algorithm PTA generates patterns for non-NULL tuples in the dataset without using a reference table. The algorithm shown in Fig. 4 creates a pattern tree and traverses it in level order. At each level, patterns are generated using print_powerset and patterns functions and are stored in the database in the form of a table. All the generated patterns at that level are checked using patterncheck function. Selected patterns are stored in levelpattern and finalpattern. Former contains the list of patterns selected at a particular level while the latter stores the patterns selected throughout the execution. After this deletion of all the patterns in levelpattern happens using the pruning function. The selected patterns stored in the finalpattern are returned by the algorithm. At any level, if the entire dataset comprises of tuples with NULL under the aggregate attributes, the algorithm stops executing and returns finalpattern. The algorithm uses four functions:

- print_powerset (A, level): This function generates the powerset for pattern attributes (A) at a particular level and returns it in B.
- patterns (B, D, A): This function generates patterns for B (powerset at level i) using A and D. The result is stored in the database.
- patterncheck (D, A): Fig. 5 shows the patterncheck function's algorithm that fetches the generated patterns from the database and stores in lis. It then checks the instance of p belonging to lis in D using A. If any instance of p contains a NULL for the aggregate attribute for which the pattern table is being created, then that pattern will be rejected. Selected patterns are added to a python list variable templis and returned by the function. In D, p = [B,∗,∗,∗] is selected because it satisfies four tuples and none of them have NULL in Temperature. p = [A,∗,∗,∗] is not selected because it satisfies eight tuples which contain three NULLs for Temperature.
- pruning (levelpattern, A): Fig. 6 shows the pruning function's algorithm. All the patterns present in the levelpattern are accessed by variable i. All the instances of these patterns need to be removed from D so that no specializations of selected patterns in a particular level are selected in the subsequent levels. In D, p = [B,∗,∗,∗] is selected. All the instances of p i.e. [B,2019,June,Mon], [B,2019,June,Tue], [B,2019,July, Mon], [B,2019,July,Tue] should be removed from D, so that in the next level no specialization of p are selected. This allows us to avoid checking already selected patterns.

Input: A (Type: List), D (Type: String)

```
finalpattern=[ ] #Contains list of all the patterns selected from all levels
levelpattern=[ ] #Contains list of all patterns selected at each level
algobreakflag=1
for level in range(0,len(A)+1):
    if algobreakflag==0:
        break
    for i in print_powerset(A, level):
        patterns(i, D, A)
        var1=Number of tuples where aggregate attribute is NULL
        var2=Total number of tuples in D
        if var1 != var2:
            templis=patterncheck(D, A)
            If templis is not empty:
                Add content of templis to finalpattern
                Add content of templis to levelpattern
        else:
            algobreakflag=0
            break
    pruning(levelpattern, A)
return finalpattern
```

Fig. 4. Pattern table algorithm

Complexity Analysis: We have identified four factors on which Algorithm Execution Time (AET) depends. These factors are the number of levels of pattern tree accessed (n), time taken (in seconds) to generate the patterns in a level (F_{pg}), time taken (in seconds) to check the generated patterns in order to select them (F_{pc}) and time taken (in seconds) to delete the instances of selected patterns (F_{pp}). The AET is given by:

$$AET(in\ seconds) = \sum\nolimits_{i=0}^{n} (F_{pg} + F_{pc} + F_{pp})$$

Input: D, A

```
lis=[ ] # this stores the list of all patterns generated for this level
templis=[ ] # this stores selected patterns from lis
for i in lis:
    if i ==[*]:
        Add i to templis
    else:
        Check instances(I') of i in D using A for NULL
        if I'==0:
            Add i to templis
return templis
```

Fig. 5. Algorithm for patterncheck

Input: levelpattern, A

```
for i in levelpattern:
    create a delete command using A to delete instances of pattern i in D
    Execute the command on D
```

Fig. 6. Algorithm for pruning

$F_{pg} = f$ (Group (D_i)) where $Group\,(D_i)$ is the number of patterns generated at level i and D_i represents the cardinality of domains of pattern attributes in A.

$Group\,(D_0)\,isGroupof0atlevel0 \rightarrow 1$
$Group\,(D_1)\,isGroupof1atlevel1 \rightarrow A_1 + A_2 + A_3 + A_4 \ldots.$
$Group\,(D_2)\,isGroupof2atlevel2 \rightarrow A_1A_2 + A_2A_3 +$
$A_3A_4 + A_1A_4 + \ldots\ldots$
$Group\,(D_3)\,isGroupof3atlevel3 \rightarrow A_1A_2A_3 + A_2A_3A_4 +$
$A_1A_2A_4 + \ldots\ldots$
soon....

$F_{pc} = f(N)$ where N is the number of NULLs in the dataset.
$F_{pp} = f(P, |D|)$ where P is the number of patterns selected by patterncheck and $|D|$ is the size of the dataset from where instance of P are to be deleted.

5 Experimental Setup

We ran our experiment on an i5, 10th gen processor with 1.6 Ghz base frequency having four cores, 8 GB DDR4 RAM, 1TB standard storage, and 6 MB L3 cache. We have used Python 3.8.3 for implementing the PTA and PostgresSQL Version 10 for storing and managing datasets and pattern tables. To connect the python code with PostgresSQL, we have used the psycopg2 library. The operating system used for the experiment is Windows 10. We have used Sofia Air Quality Dataset [8] which contains temporal as well as spatial attributes without a reference table.

The evaluation parameters have been categorized as Input parameters and Output Parameters. For Input Parameters we have considered Size of the Dataset and Number of NULLs in the dataset. For Output Parameters we have considered Algorithm Execution Time AET and Query Execution Time QET.

6 Results

6.1 Basic Query Execution

In this experiment, we took datasets with and without NULLs to understand which method (DTM and PTM) works better under what scenario. QW was executed on these datasets and the average QET in seconds was calculated by simply averaging all the queries QETs. Figure 7(a) shows a reduction of 71% in average QET for PTM over DTM when the dataset contains NULLs. The reason for this is that accomplice tuples can directly be identified from pattern tables by simply searching for * under non-group by attributes. However, in DTM, attributes need to be scanned for NULL and then removed from the dataset using the RA table manipulation technique which is time-consuming. Figure 7(b) shows a reduction of 82% in average QET for DTM over PTM when the dataset contains no NULLs. The pattern obtained is [*]. In DTM no join is required. A simple aggregate query will give the result. However, in PTM a join between dataset and [*] is required thus increasing QET.

This experiment shows that PTM outperforms DTM when the dataset has NULL values in them. However, for datasets without NULL creating pattern tables and using PTM acts as an overhead as a simple aggregate query can be applied to get the correct result.

6.2 Data Scaling for Query Execution

In this experiment, QW was executed using DTM and PTM on scaled data from 1X to 5X, and average QET in seconds obtained by averaging the QET of queries of QW. Figure 8 shows that the average QET for DTM increases linearly up to 3X and then increases rapidly beyond it. For PTM the average QET increases linearly up to 5X.

6.3 Algorithm Execution

In this section, we experimented on the PTA, to understand the effect of scaling w.r.t., size of the dataset taken, number of levels accessed, and number of NULLs in the dataset.

Scaling w.r.t. Size of the Dataset. In this experiment, we kept the number of NULLs in the dataset as constant and increased the size of the dataset by introducing new non-NULL tuples. Figure 9 shows the AET obtained for scaled data having NULLs as constant. We can observe that as the size of the dataset increases, the pattern set generated at each level increases because of an increase in the domain values of pattern attributes. This also increases the checking and pruning operation time for each level accessed by the algorithm, thus increasing

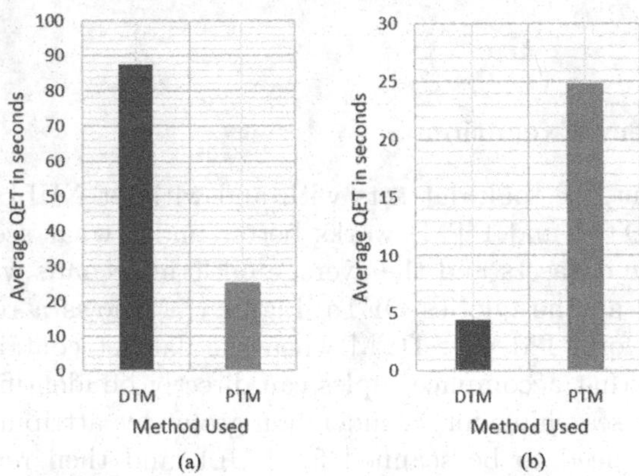

Fig. 7. Average QET for QW when executed on a dataset (a) with NULLs (b) without NULLs

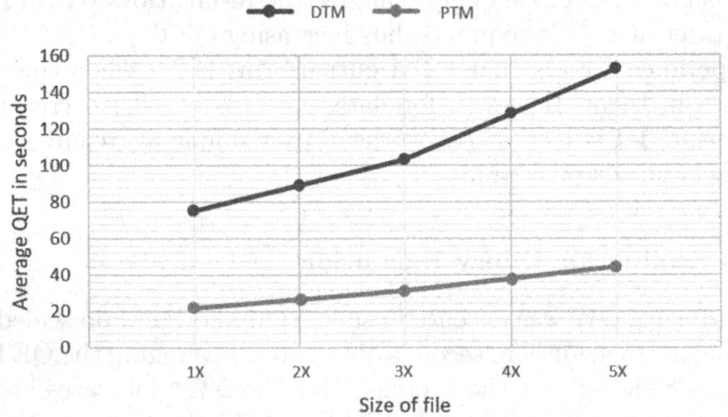

Fig. 8. Average QET vs. scaled data

AET. The AET increases linearly till 3X, it is sublinear till 5X and after 5X it increases exponentially. It indicates that our algorithm is well behaved till 5X.

Scaling w.r.t. Number of Levels. In this experiment, we executed the PTA on a dataset containing NULLs. PTA generated patterns by accessing four levels of the pattern tree. At the end of the fourth level, only tuples with NULL under p1 remained (Stopping condition of the algorithm). Time taken by the algorithm in each level is shown in Fig. 10 and we can observe that algorithm is well behaved up to level 3. Beyond level 3 it increases rapidly. The reason for this is that as the number of levels accessed by the algorithm increases, more patterns will be generated, resulting in checking and pruning operation for that increased level, thus increasing AET.

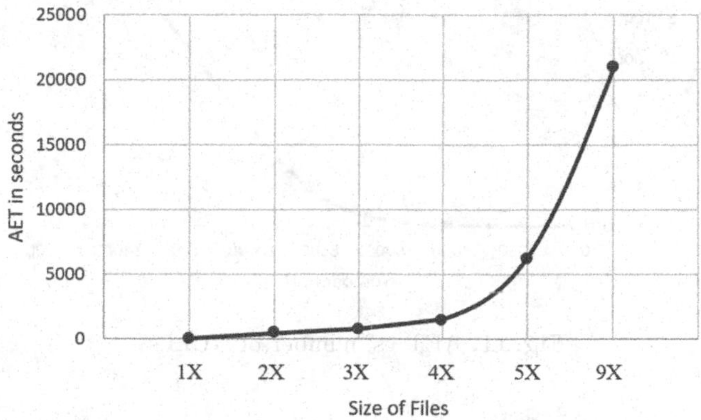

Fig. 9. AET vs. scaled data

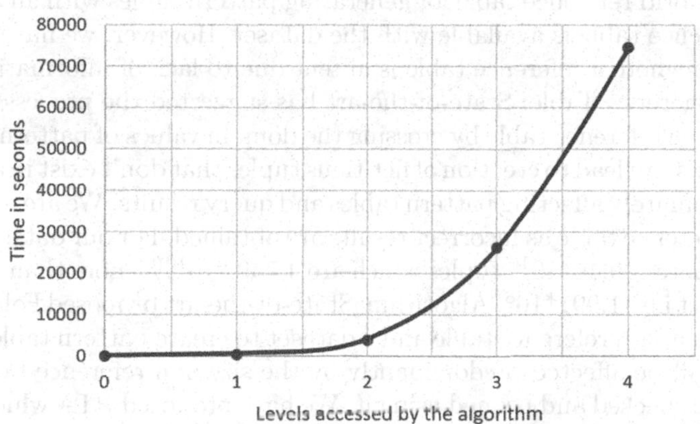

Fig. 10. AET in seconds vs. number of levels accessed by algorithm

Scaling w.r.t. Number of NULLs in the Dataset. In this experiment, we took a dataset without NULLs and then increased the number of NULLs (by retaining previous NULLs). For every increase in NULL, we calculated the AET obtained by executing the PTA. Figure 11 shows the result obtained. We can observe that if we do a successive increase in the number of NULLs (retaining previous NULLs) in the dataset, the AET increases as levels accessed by the algorithm increases to get more specialized patterns. If we increase the NULLs randomly without retaining previous NULLs, then AET may or may not increase. The AET in Fig. 11 increases linearly till 9000. Beyond 9000, AET increases rapidly.

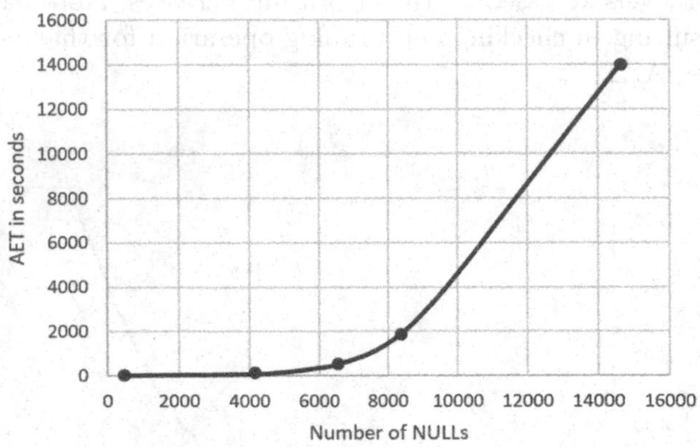

Fig. 11. AET vs. number of NULLs

Comparison with the State-of-the-Art: This section contains comparison between our work and the state-of-the-art [6] Reference Table: State-of-the-art uses a real-world reference table for generating pattern tables with an assumption that a reference table is available with the dataset. However, we have considered the scenario where a reference table is absent due to lack of information sources. Universe Reference Table: State-of-the-art has suggested the process of generating a Universe reference table by crossing the domain values of pattern attributes. However, this can lead to creation of fictitious tuples that don't exist in a real-world scenario ultimately affecting pattern tables and query results. We are not using the universe reference table as incorrect results are obtained. For our dataset, the Universe will contain $9.46*10^{16}$ tuples which are $4.749*10^{10}\%$ more than total tuples in the dataset i.e., $1.991*10^{8}$. Algorithm: State-of-the-art proposed FoldData algorithm which uses a reference table and a dataset to create pattern tables. AET for FoldData will be affected predominantly by the size of a reference table because patterns are checked and pruned using it. We have proposed PTA which uses only the dataset to generate pattern tables. AET for PTA depends on the size of the dataset and not on the size of the reference table. Patterns: State-of-the-art generates Patterns for available and missing data. NULLs in the dataset are considered as missing tuples and are removed from the dataset during preprocessing of data. Also the dataset considered by the state-of-the-art contains only one aggregate attribute. In our experiment we have generated Patterns for non-NULL tuples. Patterns for missing tuples cannot be generated as the reference table is missing. For datasets with more than one aggregate attribute, NULLs cannot be considered as missing tuples because their removal leads to data loss. Query Result Accuracy: For the situation where a reference table is present, state-of-the-art will have a higher query result accuracy as the pattern tables generated represent both available and missing data. However, for the situation where a reference table is absent, no result will be generated by state-of-the-art. In this scenario our approach will be able to generate correct query results for the available data. Storage Space: A lot of storage space will be required to store the reference table and dataset in the case of state-of-the-art. However, in our case, less storage space will be required because only the dataset has to be stored. For Sofia Air Quality dataset, if we generate a universe reference table then size of the reference table will be 72 times more than that of the dataset.

7 Conclusion

In this thesis, we proposed an approach for querying aggregate queries on datasets for which no reference table is available to identify missing tuples. We have considered such datasets to be in the CWA setting. We have used the PTA to generate pattern tables that act as a compact representation of non-NULL tuples in the dataset. These pattern tables are used to identify clusters that are complete for a given aggregate query using group by clause. The identified complete clusters are used for querying aggregate queries using DTM and PTM. The results produced by these queries are complete and correct because they are obtained from complete clusters.

PTM is better for datasets with NULL as it offers a 71% reduction in average QET over DTM. However, for datasets without NULL, DTM is better as it offers an 82% reduction in average QET over PTM. QW when executed on variable-sized data (1X-5X), results in a linear increase for PTM up to 5X, whereas for DTM it increases linearly up to 3X and then increases rapidly up to 5X.

The AET is analyzed for scaled data, the number of levels, and the number of NULLs. Our algorithm is well behaved till 5X for scaled data. The behavior of the algorithm beyond level 3 needs to be investigated further. With an increase in NULLs, the AET increased linearly to 9000 NULLs. Beyond 9000, AET increased rapidly.

References

1. Cortés-Calabuig, Á., Denecker, M., Arieli, O., Bruynooghe, M.: Representation of partial knowledge and query answering in locally complete databases. In: Hermann, M., Voronkov, A. (eds.) LPAR 2006. LNCS (LNAI), vol. 4246, pp. 407–421. Springer, Heidelberg (2006). https://doi.org/10.1007/11916277_28
2. Denecker, M., Cortés-Calabuig, A., Bruynooghes, M., Arieli, O.: Towards a logical reconstruction of a theory for locally closed databases. ACM Trans. Database Syst. **35**(3), 1–60 (2008). https://doi.org/10.1145/1806907.1806914

3. Edureka: Real-Time Big Data Applications. https://www.edureka.co/blog/big-data-applications-revolutionizing-various-domains/#Big_Data_in_IoT. Accessed 20 May 2021
4. Fan, W., Geerts, F.: Capturing missing tuples and missing values. In: Proceedings of the Twenty-Ninth ACM SIGMOD-SIGACT-SIGART Symposium on Principles of Database Systems, pp. 169–178. Association for Computing Machinery, New York (2010). https://doi.org/10.1145/1807085.1807109
5. Fan, W., Geerts, F.: Relative information completeness. ACM Trans. Database Syst. **35**(4), 1–44 (2010). https://doi.org/10.1145/1862919.1862924
6. Hannou, F.-Z., Amann, B., Baazizi, M.-A.: Explaining query answer completeness and correctness with partition patterns. In: Hartmann, S., Küng, J., Chakravarthy, S., Anderst-Kotsis, G., Tjoa, A.M., Khalil, I. (eds.) DEXA 2019. LNCS, vol. 11707, pp. 47–62. Springer, Cham (2019). https://doi.org/10.1007/978-3-030-27618-8_4. https://hal.sorbonne-universite.fr/hal-02310582
7. Imieliński, T., Lipski, W.: Incomplete information in relational databases. J. ACM **31**(4), 761–791 (1984)
8. Kaggle: Sofia Air Quality Dataset. https://www.kaggle.com/hmavrodiev/sofia-air-quality-dataset. Accessed 14 Feb 2021
9. Kaggle: FlowDB Sample for river flow and flash flood forecasting. https://www.kaggle.com/isaacmg/flowdb-sample. Accessed 16 Mar 2021
10. Levy, A.Y.: Obtaining complete answers from incomplete databases. In: Proceedings of the 22th International Conference on Very Large Data Bases, VLDB 1996, pp. 402–412. Morgan Kaufmann Publishers Inc., San Francisco (1996)
11. Li, D., Zhou, Y., Hu, G., Spanos, C.J.: Handling incomplete sensor measurements in fault detection and diagnosis for building HVAC systems. IEEE Trans. Autom. Sci. Eng. **17**(2), 833–846 (2020). https://doi.org/10.1109/TASE.2019.2948101
12. Motro, A.: Integrity = validity + completeness. ACM Trans. Database Syst. **14**(4), 480–502 (1989). https://doi.org/10.1145/76902.76904
13. Razniewski, S., Korn, F., Nutt, W., Srivastava, D.: Identifying the extent of completeness of query answers over partially complete databases. In: Proceedings of the 2015 ACM SIGMOD International Conference on Management of Data, SIGMOD 2015, pp. 561–576. Association for Computing Machinery, New York (2015). https://doi.org/10.1145/2723372.2750544
14. Razniewski, S., Nutt, W.: Completeness of queries over incomplete databases. PVLDB **4**, 749–760 (2011). https://doi.org/10.14778/3402707.3402715
15. Reiter, R.: On Closed World Data Bases. In: Gallaire, H., Minker, J. (eds.) Logic and Data Bases, pp. 55–76. Springer, Boston (1978). https://doi.org/10.1007/978-1-4684-3384-5_3
16. UCI Machine Learning Repository: Bank Marketing Dataset. https://archive.ics.uci.edu/ml/datasets/Bank+Marketing. Accessed 16 Mar 2021
17. Shoshani, A.: Olap and statistical databases: similarities and differences. In: Proceedings of the Sixteenth ACM SIGACT-SIGMOD-SIGART Symposium on Principles of Database Systems, PODS 1997, pp. 185–196. Association for Computing Machinery, New York (1997). https://doi.org/10.1145/263661.263682
18. Statista: Volume of data/information created, captured, copied, and consumed worldwide from 2010 to 2024. https://www.statista.com/statistics/871513/worldwide-data-created/. Accessed 20 May 2021
19. Sundarmurthy, B., Koutris, P., Lang, W., Naughton, J., Tannen, V.: m-tables: representing missing data. In: ICDT (2017)

Time Series Data Quality Enhancing Based on Pattern Alignment

Jianping Huang[1], Hao Chen[1], Hongkai Wang[2], Jun Feng[2(✉)], Liangying Peng[2], Zheng Liang[3], Hongzhi Wang[3], Tianlan Fan[3], and Tianren Yu[3]

[1] State Grid Zhejiang Electric Power Co., Ltd., Hangzhou, China
[2] State Grid Zhejiang Information and Telecommunication Branch, Hangzhou, China
fengjun@zj.sgcc.com.cn
[3] Harbin Institute of Technology, Harbin, China
wangzh@hit.edu.cn

Abstract. We have witnessed the rapid evolution of data intelligence benefiting the decision making of complex multi-equipment systems. Collected by sensors on the equipment temporally, such data indicates the opportunity of real-time analysis and workflow optimization, while bringing data quality challenges to the specialists. The usage of low quality data could lead to misleading analysis results and decisions in face of machine breakdown, sensor record failure or working status malfunction. Faced with the urgency of time series data quality problem, we propose a pattern alignment-based method, combining multiple constraints for speed-dominated, stable, and pattern-dominated time series to describe the target time series. Using a segment similarity graph to represent DTW distance information of data segments divided by SST, we utilize community search to conduct pattern mining and alignment. We also propose a data repairing strategy for addressing the problem of low-quality data in pattern-dominated time series. It is shown in the experimental results that our approach can work effectively on time series from different domain and distributions.

Keywords: Time series · Data quality · Pattern alignment

1 Introduction

The flourish of big data management has enabled data-driven decision making and workload balancing in complex multi-equipment system. Data generated by sensors on equipment and machine reflects the operation status of the equipment and implies potential opportunities of intelligent resource schedule. Domain experts require an efficient data quality enhancing module, which can give rise to the early warning of productivity lost, precise anomaly pattern information, etc. [6]. These modules focus on time series collected by sensors on the multi-equipment pipeline, the data quality of which decides the reliability of such intelligent systems.

The research filed of time series data quality management are mostly constraint-based [1,2]. [1] proposes Speed constraints, where a speed constraint

U. K. Rage et al. (Eds.): DASFAA 2022 Workshops, LNCS 13248, pp. 363–375, 2022.
https://doi.org/10.1007/978-3-031-11217-1_26

interval is used to detect and repair low-quality data by calculating the speed changes of adjacent timestamps. [2] proposes the variance constraint, using the variance within a sliding window and the threshold for detection and the auto regressive smoothing method for repair. [3] calculates the correlation between heterogeneous time series data streams, and identify data points that violate correlation constraints with other attributes as low-quality data points.

It might seem that the time series data quality problem in complex multi-equipment systems can be tackled with existing algorithms. However, in spite of the clear definition of data quality, the ability to represent domain knowledge, and the potential of mining new knowledge [4], constraint-based approaches [1–3] are highly biased, and require further improvement for better robustness [5]. In our work, we present a pattern alignment-based time series data quality enhancing approach. The advantages of our approach are as follows.

- **Adaptability.** We combine multiple constraints for speed-dominated, stable, and our newly-proposed pattern-dominated time series to repair the low-quality data segments the newly-arrived time series. Combining the frequent dependence of constraint violations with mechanism and patterns, domain knowledge is enriched, thus our approach can handle more scenarios.
- **Scalability.** We use multi-scale community search on the segment similarity graph to represent the DTW distance information of subsequences divided with Singular Spectrum Transformation(SST). For data quality problem detected at different scale, we propose a guided matching strategy to guide segments for pattern-dominated time series data quality enhancing. Our approach can easily process statistical properties and pattern on time series of different scale.
- **Interpretability.** We propose the pattern alignment methods to help explain the work status with time series data. With such pattern annotation, one can understand the decision of data quality enhancing with statistical properties or guide segments.

The rest of our paper is organized as follows. In Sect. 2, we give some basic concepts of our approach, including problem definition and approach overview. In Sect. 3, we introduce our DTW similarity graph representing SST-divided subsequences, community search strategy for pattern alignment, and propose the guided matching strategy on the data quality problem to fulfill the task. In Sect. 4, we evaluate our approach with case study and different circumstances. In Sect. 5, we conclude our work by addressing its advantages and drawbacks, and discuss future work.

2 Overview

In this section, we give some basic definition related to the problem we focus on. Then, we give an overview of our approach, and demonstration of each module.

2.1 Preliminaries

Definition 1 (Speed-dominated time series): Given a time series $T = t_1, t_2, ...,$ t_N, $V = t_2 - t_1, t_3 - t_2, ..., t_N - t_{N-1}$ is the successive difference sequence of T. For all subsequences $V_{s,l}$ of length l in V $(1 \leq s \leq N-l)$, the top-k smallest element set of $V_{s,l}$ is G_1 and the top-k smallest element set is G_2. If the confidence of a randomly sampled value pair from the elements in G_1,G_2 less than a threshold δ is c_1, T is identified as a speed-dominated time series with confidence c_1.

Definition 2 (Stable time series): Given a time series $T = t_1, t_2, ..., t_N$, if the following conditions are satisfied with the confidence of c_2

- All subsequences $T_{s,l}$ of length l of $T(1 \leq s \leq N-l)$ have equal mean $= \frac{t_s + t_{s+1} + ... + ts + l - 1}{l}$.
- All subsequences $T_{s,l}$ of length l of $T(1 \leq s \leq N-l)$ have equal variance d$=\frac{(t_s - mean)^2 + ... + (t_{s+l-1} - mean)^2}{l}$.
- The correlation coefficient r between any two of the above subsequences $T_{s_1,l}$ and $T_{s_2,l}$ is only related to the time difference $\mid s_1 - s_2 \mid$ and decays rapidly to 0 as the time difference increases.

we call T a stable time series with confidence c_2.

Definition 3 (Pattern-dominated time series): Given a time series $T = t_1, t_2, ..., t_N$, split T into n segments $\{T_{1,c_1-1}, T_{c_1,c_2-c_1}, ..., T_{c_n,c_n-c_{n-1}}\}$ and label each segment $T_{c_i,c_{(i+1)}-c_i}$ with a label l_i.

$$l_i \begin{cases} = 0, \ T_{c_i,c_{(i+1)}-c_i} \ \textit{belongs to the } l_i\textit{th pattern class} \\ \neq 0, \ s_i \ \textit{is not a mode segment} \end{cases} \tag{1}$$

For the set of subsequences $\{T_{(c_i,c_{(i+1)}-c_i)}\}$ of all $l_i \neq 0$. If it satisfies:

$$\sum_i len(T_{(c_i,c_{(i+1)}-c_i)}) > c * len(T) \tag{2}$$

we identify T as a pattern-dominated time series, and the preset parameter c indicates the minimum proportion of the sum of pattern-dominated subsequence lengths.

Definition 4 (Number of subsequence layers): For a certain time series T and a segmentation strategy, the subsequence $T_{s,s+l}$ falls into the i-th layer when this subsequence is obtained by T after i segmentation operations $\{c_1, c_2, ...c_i\}$. At the same time, we define the number of levels of the partition operation c_i as i as well.

For each layer of partitioning of T forming the set of subsequences $\{T_{s,s+l}\}$, we can then describe the class information of the subsequences contained in the set. A running example described in 2 layers is given below.

This time series cannot be regarded as speed-dominated or stable, so our segmentation strategy splits it into 12 segments and labels the category

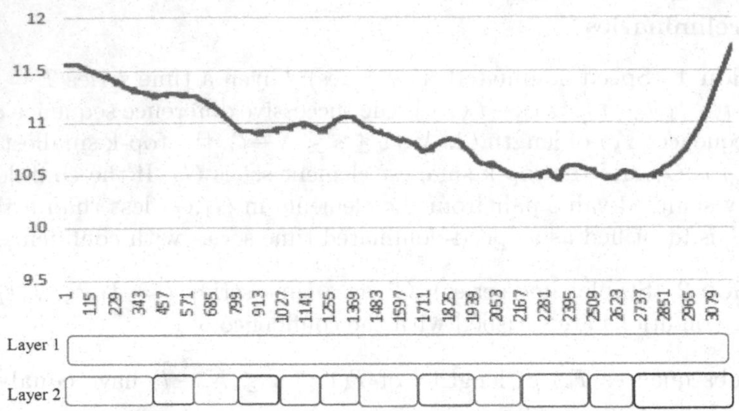

Fig. 1. An example of 2 layer multi-scale time series

alignment results with different color. The pattern-dominated alignment results are in three categories: green indicates a monotonically decreasing pattern, purple indicates stable pattern, red indicates a pattern that increases before decreasing, and the last segment is shown in black as this is classified as a speed-dominated sequence because it is not classified in any pattern.

2.2 Approach Overview

Problem 1 (Time series anomaly detection and data quality enhancing problem): For a time series $T = t_1, t_2, ..., t_N$ and a data quality evaluation function

$$g(t_k) = \begin{cases} 1, Normal\ value \\ 0, Abnormal\ value \end{cases} \tag{3}$$

our goal is to detect all timestamps $\{k\}$ that satisfy $g(t_k) = 0$ and find a repaired sequence $T_R = t_{r_1}, t_{r_2}, ..., t_{r_N}$ that satisfies $g(t_{r_i}) = 1$ for $1 \leq i \leq N$ while minimizing the repair cost $\sum_{i=1}^{N} \mid t_{r_i} - t_i \mid$.

In Fig. 2, we propose an anomaly detection and data quality enhancing framework for massive time series. We can obtain the results of the above problem by describing the decision process of MTSQE for an input time series in Algorithm 1.

Algorithm 1 demonstrates the anomaly detection and repair strategy for time series. The following section focus on the framework decision process, the construction of alignment, will give us a clearer understanding of the algorithm's process for time series.

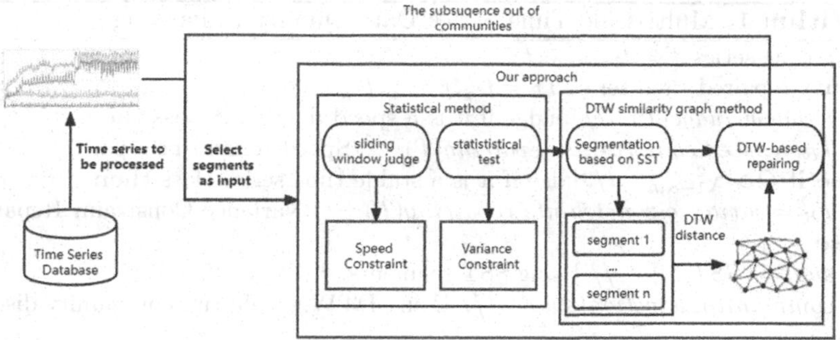

Fig. 2. Workflow of our approach

3 Methodology

3.1 Graph Construction and Pattern Alignment

For a time series T that does not pass the sliding window judge and the statistical test, we map the subsequence $T_{s,e}$ obtained from each segment of the SST split into a point V in the undirected graph.

The DTW distance between Ts,e into the edge weights E:

$$\{T_{s_i,e}\} \rightarrow V = \{v_i\} \qquad \{dtw(T_{s_i,e}, T_{s_j,e})\} \rightarrow E = \{e_{ij}\} \tag{4}$$

Problem 1 is converted into a community discovery problem on graph $G(V, E)$, labeling the patterns of time series segments. Note that G is a complete graph, so to avoid high complexity and unnecessary noise effects we prune the edges with too large DTW distances. We prune the edges with weight larger than the mean of all DTW distances, obtaining a new graph G'.

We use the DOCWN algorithm [8] for initial community selection and expansion on G'. The goal of DOCWN is to solve the set of points $C = \{c\}$ as communities in two steps, initialization and expansion of communities.

Definition 5 (node strength): Node strength is defined as the sum of the out-degrees of the point v_i:

$$k_v = node\ strength(v) = \sum weight(v, u) \quad u \in V, i \neq j \tag{5}$$

Definition 6 (Belonging degree): Belonging degree is defined as the sum of the weights of the edges of a point v_i and the points in its community c divided by its Node strength:

$$belonging\ degree(v_i) = \frac{\sum_j weight(v_i, v_j)}{node\ strength(v_i)} \qquad v_i \in c,\ v_j \in c,\ i \neq j \tag{6}$$

indicates the probability that a segment of subsequence belongs to the pattern represented by this community.

Algorithm 1. Multi-scale Time Series Data Quality Enhancing

Input: time series $T = t_1, t_2, ..., t_N$
Output: repaired time series $T_R = t_{r_1}, t_{r_2}, ..., t_{r_N}$

if *window_judge*(T) //Judge if it is a speed dominant class **then**
  T_R = *speed_constraints_repairing*(T) //Speed constraint repair;
else if $Q > \chi^2_{1-\alpha,h}$ //Judge if it is a stable time series class **then**
  T_R = *variance_constraints_repairing*(T) // Variance Constraint Repair;
else
  subT = $SST(T)$ // Using SST segments;
  community_detection(*subT*) // Using DTW graph, run community discovery;

  for *community* $\subseteq$ *subT* // Repair as a communities **do**
    T_R = *repair*(*community*);
  end for
  for *subt* $\in$ *subT* // Solve the subsequences outside the community recursively **do**
    if *subtnotincommunity* **then**
      $T_R = MTSQE(subt)$;
    end if
  end for
end if
return T_R;

Definition 7 (Modularity): Modularity is the data quality evaluation of the community, defined as

$$Q = \frac{1}{2m} \sum_{c \in C} \sum_{v \in V} \delta_{cu} \delta_{uv} (A_{uv} - \frac{k_u k_v}{2m}) \tag{7}$$

where A is the adjacency matrix of G', the Boolean value δ_{cu} indicates whether point u belongs to community C, k_v is the node strength, and m is the total number of edges in G'. This metric measures the degree of internal similarity of a class of subsequences, and ideally, each subsequence joining should contribute to the increase of Q. Generally, the size of the graphs is $O(logN)$, so the number of communities will not be large.

3.2 Data Quality Evaluation and Enhancing

Definition 8 (Data Quality index): For a calculated community $C \subseteq V$ on a complete graph $G(V, E)$, we define data quality index of node v_i in community C as:

$$Q(v_i) = \sum_{v_j \in C, i \neq j} weight(v_i, v_j) \tag{8}$$

Using the guiding segments found with high data quality, we can consider some nodes with high DTW matching costs of the sequence to be tested as abnormal nodes. When the DTW matching cost is over 2 times of the average cost, we consider this node to be an low-quality node.

Algorithm 2. weighted graph building and overlapping communities detecting

Input: k sequence segments $s_1, s_2, s_3, ..., s_k$ obtained by X

Output: for each sequence segment, the label $L = l_1, l_2, l_3, ..., l_k$

```
sum = 0
for i from 1 to k do
    for j from 1 to k do
        if i ≠ j then
            w_ij = dtw(s_i, s_j);
            sum = sum + w_ij;
        end if
    end for
end for
mean = sum/k    //Construction of basic graph;
for w in W do
    if w > mean then
        w = inf    //The edge weights greater than the mean are pruned;
    end if
end for  //
for num from 1 to max    do
    if len(L)/k < c    then
        L = initial(W);
        L = expand(W, L);
    end if
end for
```

After detecting abnormal nodes, we want to find a repair value of the abnormal node that minimizes the DTW distance. We remove the abnormal node from this sequence and observe the dtw matching path.

Case 1: If there is a mismatch in a node group of the guide sequence, the repair value of this node is directly set to the mean of the mismatched node group.
Case 2: If there is no mismatch, set the repair value to an unknown value that equals the mean of the two groups.

We can see a running example in Fig. 3, in the abnormal subsequence B and the guide subsequence C, if B is changed to $<0, 2, 10, -9, 0>$, where -9 is the abnormal node, we directly formulate the equation $(x + 0)/2 = 0$ and solve it, getting $x = 0$ as the repair value. This is the second case above.

Suppose $C' = <0, 0, 9, 10, 0>$, 9 is detected as an abnormal node. We replace 9 and can consider adding a node that matches $B[2] = 2$. In that case, we formulate the equation $(0 + x)/2 = 2$, and the result is $C'[3] = 4$, which means C' is seemingly $<0, 0, 4, 10, 0>$. But when we consider the case of matching $B[3] = 10$, we formulate the equation $(10 + x)/2 = 10$, and the result is $C'[3] = 10$ and $C' = <0, 0, 10, 10, 0>$. Based on the minimum repair cost principle, we keep the second result.

To sum up, we propose a community-wide anomaly detection and repair strategy based on double cost detection and minimum repair.

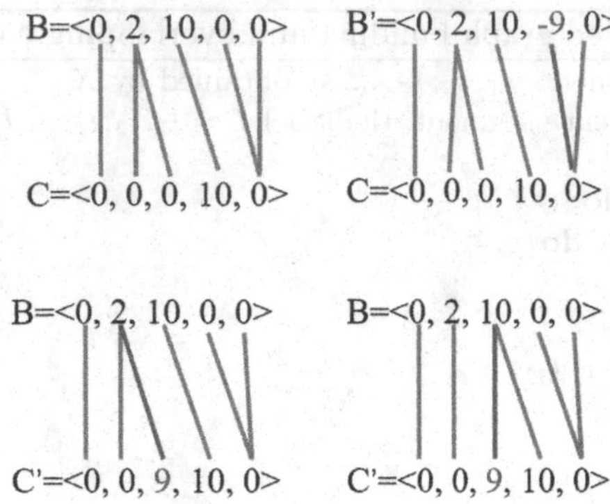

Fig. 3. An example of DTW matching

Algorithm 3. community repairing

Input: k sub sequence segments in community $S = s_1, s_2, s_3, ..., s_k$
Output: k sub sequence segments repaired $R = r_1, ..., r_k$

while repaired sequence segments exist **do**
  Select the sub sequence with the smallest $Q(v_i)$ in S as *Good_vertex*
  while *Good_vertex* has neighbors not repaired **do**
    Select the nearest neighbor of *Good_vertex* as *Bad_vertex* //greedy strategy
    $M = mean(dtwdistance(g_i))$ //calculate the mean path cost
    for $g \in G$ **do**
      if $dtwdistance(g_i) > 2{*}dtw(sp, sq)/\sqrt{l_1 * l_2}$ //double cost detection **then**
        $R = R \cup Repair(s_1[i])$
      end if
    end for
  end while
end while
return R

4 Evaluation

4.1 Experimental Setup

Datasets. We evaluate our approach with datasets as follows.

FPP Dataset: A temperature control system dataset from a large-scale Fossil-Fuel Power Plant. We process and analyze 351 batches of sequences on 1050K time points. Each batch consists of 2000 consecutive timestamps with more regularity and less fluctuations [6].

CHS Dataset: Consists of time series captured by pressure, power, vibration, ow, efficiency, and temperature sensors from a Complex Hydraulic System containing 43680 attributes and 2205 instances [7].

Our experiments were conducted on a computer with the Intel Xeonr Gold 5218 CPU @ 2.30 GHz (16 cores) and 512 GB DDR4 memory.

4.2 Case Study

The time series in Fig. 1 are divided into 10 segments.

Intuitively, the templates worth mining in this time series are several "hills". After dividing it into 10 segments with SST, since the graph is relatively small, after iterating a round of community, we discover:

At the beginning, we set *belonging degree* to 0.3, and the subsequences of No. 5–10 are divided into one category. Let's take a look at the shape of No. 5–10 (Fig. 4):

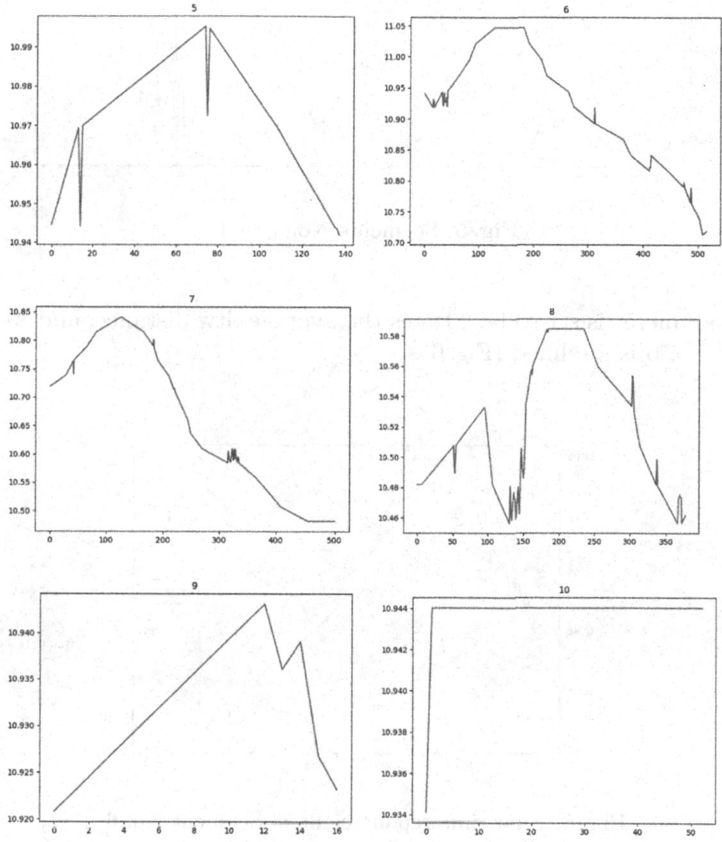

Fig. 4. Segments No. 5 to 10

This is exactly the "hill" we are looking for, and we find all the "hills" in Fig. 1. Other subsequences can basically be regarded as stable and speed-dominant, which are left to the next layer (Fig. 5):

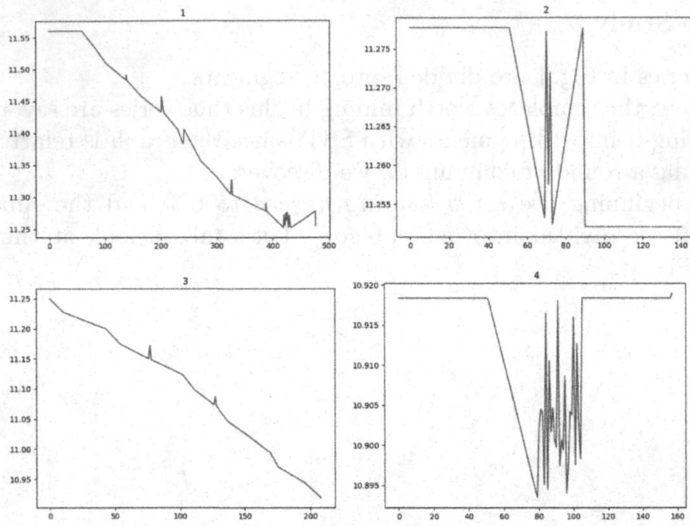

Fig. 5. Segments No. 1 to 4

The parameter is set to be 2 times the average dtw distance, and we observe that only No. 6 is modified (Fig. 6).

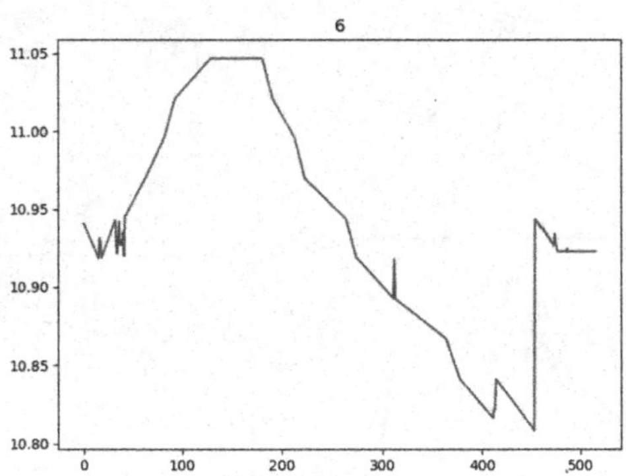

Fig. 6. First time repair result of Segment No. 6

After using the minimum repair principle, it can be repaired to be normal, but this implementation strategy of dtw still has problems such as excessive cost of the end node, resulting in poor effect of the algorithm (Fig. 7).

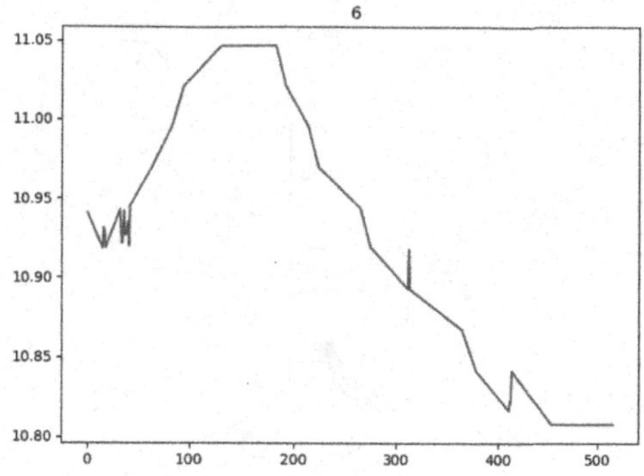

Fig. 7. Second time repair result of Segment No. 6

4.3 Adjustments and Experimental Results

For the problems above, we make the following adjustments to our method:

(1) Since we notice the conflict between dtw distance and DOCWN, we use the method of taking the reciprocal of dtw to re-experiment and try the effect. Round1: Using the original strategy, we only set the edge weights to the reciprocal of the dtw distance: Observe that the time series No. 4, 5, 9 and 10 are clustered into a group.

Round2: On the basis of round1, change the *initialize* to the node with the smallest *node strength.* Observe that only No.8 is a group alone.
It can be seen that using the reciprocal to construct the edge weight or changing the process of *initialize* can obtain reasonable results.

(2) For the problem that end cost is too large, we use double dtw mean detection to improve dtw.

In the process of unequal-length dtw matching, if the state transition equation is constructed by the distance without considering the timestamp, it will cause serious end mismatch problem, and the abnormal node can not be effectively detected. That is, in the process of matching the short sequence *ts* and the long sequence *TS*, there is always a problem that the matching cost is too high at the nodes in the end of the *TS*. As a result, when detecting and improving data quality, it always focus on the nodes in the end and not on the real abnormal nodes.

For subsequences with more anomalies, first repair the abnormal nodes with *screen.* Then we try to repair the result of matching the dtw of its subsequences with its own by 2 times, obtaining better results. The method is called inv-2DTW (Fig. 8).

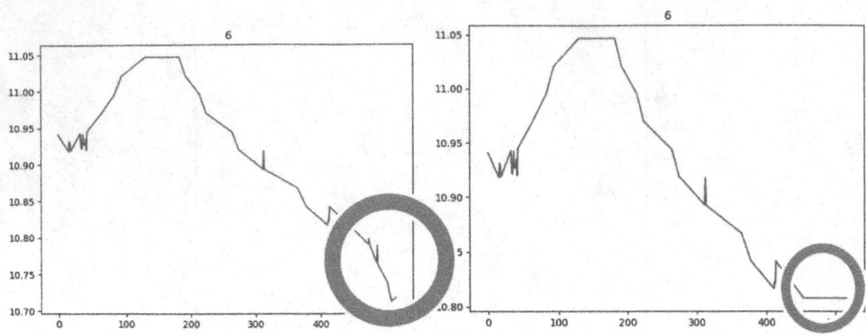

Fig. 8. Mismatching at the end

Based on the experiments above, for data fragments with high-frequency anomalies and obvious high-frequency subsequences, we use the original method and inv-2DTW method to detect abnormal patterns respectively. We compare the average time of detection and repair, accuracy and recall of each exception cell on abnormal data and decisive abnormal data, which are shown in Table 1.

Table 1. Results of experiment on high-frequency sequence group

Method	Anomaly detection accuracy	Anomaly detection recall	Data repair accuracy	Average detection and repair time
DTW method	81.3%	79.2%	84.2%	0.93s
Inv-2DTW method	72.4%	71.2%	75.1%	0.71s

Table 2 shows the results of experiment for data segments with low-frequency anomalies and obvious high-frequency subsequences.

For data segments with high-frequency anomalies and obvious high-frequency subsequences, DTW method has significantly higher accuracy and faster detection speed, but will lose a certain amount of running time. Inv-2DTW method is also significantly better than DTW method in the detection accuracy and recall on low-frequency abnormal subsequences.

Table 2. Results of experiment on low-frequency sequence group

Method	Anomaly detection accuracy	Anomaly detection recall	Data repair accuracy	Average detection and repair time
DTW method	76.5%	81.1%	86.1%	0.41s
Inv-2DTW method	79.4%	87.2%	79.0%	0.47s

5 Conclusion

In this paper, we propose a pattern alignment-based method, combining multiple constraints for data quality enhancing. We adapt DTW and SST for representing pattern of time series segments at different scale. To tackle the problem of pattern-guided time series data quality enhancing, we use community search to conduct pattern alignment, propose a guided matching strategy for repairing anomalies. It is shown in experimental results that our approach can work with satisfying accuracy and time cost in face of different kinds and scales of anomalies. In future work, we plan to consider user interests, experimental results and regression models to fit the parameter, such as ratio of different similarity metric in edge and vertices weight, probability density function, etc.

ACKNOWLEDGMENTS. The project is supported by State Grid Research Project "Study on Intelligent Analysis Technology of Abnormal Power Data Quality based on Rule Mining" (5700-202119176A-0-0-00)

References

1. Song, S., et al.: SCREEN: stream data cleaning under speed constraints. In: ACM SIGMOD International Conference on Management of Data ACM (2015)
2. Yin, W., Yue, T., Wang, H., Huang, Y., Li, Y.: Time series cleaning under variance constraints. In: Liu, C., Zou, L., Li, J. (eds.) DASFAA 2018. LNCS, vol. 10829, pp. 108–113. Springer, Cham (2018). https://doi.org/10.1007/978-3-319-91455-8_10
3. Sadik, S., Gruenwald, L., Leal, E.: Wadjet: finding outliers in multiple multi-dimensional heterogeneous data streams. In: 2018 IEEE 34th International Conference on Data Engineering (ICDE), Paris, pp. 1232–1235 (2018)
4. Liang, Z., Wang, H., Ding, X., Tianyu, M.: Industrial time series determinative anomaly detection based on constraint hypergraph. Knowl. Based Syst. **233**, 107548 (2021)
5. Wang, X., Wang, C.: Time Series Data Cleaning: A Survey (2020)
6. Ding, X., Wang, H., Su, J., et al.: Cleanits: a data cleaning system for industrial time series. Proc. VLDB Endow. **12**(12), 1786–1789 (2019)
7. Helwig, N., Pignanelli, E., Schutze, A.: Condition monitoring of a complex hydraulic system using multivariate statistics. Conf. Rec. IEEE Instrum. Measur. Technol. Conf. **2015**, 210–215 (2015)
8. Tong, C., Xie, Z., Mo, X., Niu, J., Zhang, Y.: Detecting overlapping communities of weighted networks by central figure algorithm. In: ComComAP, pp. 7–12 (2014)

Research on Feature Extraction Method of Data Quality Intelligent Detection

Weiwei Liu[1,2(✉)], Shuya Lei[1,2], Xiaokun Zheng[1,2], and Xiao Liang[1,2]

[1] Artificial Intelligence on Electric Power System State Grid Corporation Joint Laboratory (GEIRI), Beijing 102209, China
oweiwlo@163.com

[2] Global Energy Interconnection Research Institute Co. Ltd., Beijing 102209, China

Abstract. Data quality intelligent detection feature extraction method was studied in the paper. The text segmentation model, word clustering, similarity calculation and other methods were applied to the treatment of data asset list, Data quality detection feature key word library and data asset feature list were generated, and then data quality detection was performed. The data knowledge in the data asset list was firstly used to extract the data characteristics and precipitate the business knowledge. Besides, the method adaptability was firstly studied base on different data type. Moreover, general data quality detection was carried out intended for a large number of discrete data in this work. The results showed that, the efficiency was improved by automatically data feature extraction based on data asset list other than manual works. And the shortage of incomplete statistics and insufficient accuracy of feature extraction was covered. In addition, the generality of data quality detection was furtherly improved and, the blind scanning range of data quality detection was reduced, leading to significant improvement of the efficiency and the accuracy of data quality intelligent detection.

Keywords: Data asset · Word segmentation · Feature

1 Introduction

The global economic and social development is deeply affected by the digital process. The data is explosively growing and gathering. It have been continuously integrated into all aspects of the company which promoted industrial innovation and upgrading. In the era of digital economy the data has become key production factors and important resources for countries and enterprises. In this case, data governance has be-come a key problem restricting the development of digitization. At the meanwhile data quality anomaly detection is an important premise of data governance.

Many scholars at home and abroad have carried out a series of research on data anomaly detection. In the literature review of multidimensional time series anomaly detection algorithms [1], Hu min divides the research content of multidimensional time series anomaly detection into three aspects: dimension reduction,

U. K. Rage et al. (Eds.): DASFAA 2022 Workshops, LNCS 13248, pp. 376–386, 2022.
https://doi.org/10.1007/978-3-031-11217-1_27

time series pattern representation and anomaly pattern detection in logical order, and summarizes the mainstream algorithms to comprehensively show the current research status and characteristics of anomaly detection. The paper [3] proposes an improved abnormal data detection method, which is based on the correlation coefficients between process variables. A novel label propagation- based outlier detection algorithm is proposed in the paper of Outlier Detection Based on Label Propagation [2] by Zhao man. The graph model is adopted for implementing multiple label propagations. Thus, the difference in structure between normal data and outliers will be identical to the difference of label confidence between them.

The large amount of data, huge data types and complex abnormal data are the characteristics of power data. The amount of power data is huge which generated by more than 100 million equipment and 100 million subjects covered all industries and fields [11] in power grid; The specialty of power data is large, involving the whole links of "generation, transmission, transformation, distribution and utilization" such as development, construction, dispatching, operation and marketing; The quality requirements of power data are high, which is exactly accurate in all segments such as power production, metering and billing, power marketing and so on. Therefore, the power industry puts forward higher requirements for data anomaly detection [10].

In the paper of Research on Abnormal Electricity Detection Method Based on Multi-model by Stacking Ensemble Learning [4], taking the residential electricity data as the research object, and based on the analysis of different characteristics of users' habits, it studies the imbalance processing technology and classification prediction algorithm. Gou Xuan proposed a short-term electricity price prediction model based on empirical mode decomposition (EMD) and LSTM neural network. And the sequential electricity price prediction model of long-term and short-term memory neural network (LSTM) to realize the short-term electricity price prediction [8]. In the re-search on the data cleaning and anomaly identification method of dissolved gas in oil based on multi-layer architecture, Liu Yunpeng proposed a data cleaning and anomaly identification method of dissolved gas in oil based on multi-layer architecture to realize the trend anomaly detection in time series data [9].

At present, the mainstream detection processing of anomaly data includes data preprocessing, data exploration and data quality detection. Some detection rules are built in. At the same time, business rules are configured by professional. The main-stream detection methods of abnormal data are researched through algorithms. Academia mostly carries out research on abnormal data quality detection of specific scene data by algorithms, and the relevant research generally focuses on time series data algorithm [5–7], and there is little research on discrete data. In terms of learning paradigm, related algorithms can be divided into machine learning, deep learning and integrated learning; Whether to use label data can be divided into unsupervised learning [12–15], supervised learning [16–18], semi supervised learning, etc. The corresponding specific scenarios include power quality anomaly detection, user power anomaly detection, power metering anomaly detection, power equipment anomaly detection, network anomaly detection, etc. However, few scholars have studied algorithms suitable for different data types from the perspective of data types, so as to achieve higher universality of the algorithm and realize comprehensive anomaly detection of multi system and multi data.

In short, the current mainstream data anomaly detection processes and methods are often realized by exploring a large amount of data and artificially configuring rich detection rules or through targeted in-depth study of specific data quality detection algorithms for specific business scenarios. The advantage is that the anomaly detection of specific data is very accurate. Because the analysis of suspected problems is given through a large number of calculations and the results are judged by professional. But the disadvantage is large workload. A large number of business personnel from different business directions are needed to participate together. Meanwhile the cooperation of technology and business need a long time.

Based on the above reasons, there are three main problems in power data detection. First, through comprehensive data exploration of a large number of data, sampling or statistical analysis, the results not only requires a long execution time in the early stage, but also requires a large number of people to participate in the identification of suspected problems in the later stage. Second, rich detection rules have to be configured to achieve comprehensive detection. So a large number of business personnel and data developers are required. Third, the research of the algorithm is aimed at a certain kind of specific scene, which is not universal.

There are three significant differences between this paper and the previous research. One is the use of the data knowledge in the data asset list for the first time to replace the manual rule configuration which can automatically discover the business knowledge in different business directions from the data asset list. And business knowledge can be automatically and intelligently precipitated. Second, this paper studies algorithms suitable for different data types for the first time, realizes higher universality of the algorithm, and realizes comprehensive anomaly detection of multi system and multi data. Three, the research is to deal with a large number of discrete data and to apply to a large amount of data.

This paper makes full use of the data knowledge in the data asset list to identify all kinds of data. The method extracts data quality features and rules by using text word segmentation model, word clustering, similarity calculation. And then it can realize the general data quality detection of all types of discrete data by combining metadata and data content information.

2 Data Quality Intelligent Detection Method Based on Data Asset List

2.1 The Problem of Power Data

Power data can be divided into structured data and unstructured data in form. Unstructured data, including image, video, voice, text and other data, is not within the scope of this paper. This paper focuses on the general and comprehensive anomaly detection of structured data.

The type classification of structured data has different classifications from different angles. The classification of this paper is based on the perspective of quality detection, and the data types are divided into enumerable data, numerical data, time series data, text data, date data and coded data.

After analyzing the quality problems of various types of data in power data, the ab-normal data quality problems are analyzed from the commonness of data and different types of data, and the causes of the problems are analyzed (see Fig. 1).

data type	problem	example	reason
enumerable data	Illegal value	eg. "TYPE_CODE". "YXZT". "XS". "WJYXS". "SORT_CODE". "RATED_CURRENT". "PREPAY_FLAG". "CON_MODE". "DISP_MODE" etc. All of the above fields of enumeration type are constrained by the range. But some fileds are filled by illegal value.	It may be a system design problem. The system should have selected enumerable data in the form of pull-down options for business personnel to choose. But in the actual design, it may be necessary to manually input some data. As result some illegal values of enumerable data appear.
numerical data	Illegal value	eg. the value of "ZJHS" should be greater than 0, but actually some are 0. The value of "XSGLS" should be less than or equal to 1 million, but actually some are more then 1 million. The value of "QSLC" should be less then the value of "ZZLC", but actually this is not the only case.	The business input is not standardized and not validated the input. Most of them are filled manually. Then the problems appear.
time series data	sudden change	eg. In the tables of "E_MP_DAY_READ". "E_MP_DAY_DEMAND" etc. there are multiple days of data duplication or mutation.	The reasons are complex, which may be caused by acquisition device, acquisition channel, network and other reasons.
text data	Illegal filling	eg. the value of "LINENAME" should be "DYDJ"+"LINENAME"+"ID". But actually some are not filed according to this requirement. The value of "TDYY" is filled by "111". "123" which is meaningless.	The business input is not standardized and not validated the input. Most of them are filled manually. Then the problems appear.
date data	Illegal value or Logical error	eg. "JGRQ". "JHKGSJ". "ACCEPT_DATE". "APP_DATE". "CREATE_DATE". "FINISH_DATE" etc. All of the above fields should contain year. month and day, and should by less than current date.	The business input is not standardized and not validated the input. Most of them are filled manually. Then the problems appear.
coded data	Illegal value	eg. the length of "ASSET_NO" should be 12. but actually some are more then 12. The value of "WBSDH" should be number or char, but actually "123", "111111" are filled which should not appeare obviously.	The business input is not standardized and not validated the input. Most of them are filled manually. Sometimes data is truncated during transmission. Then the problems appear.
common problems	Null value	\	Some data is descard, but they are not labeled. The reasons may be system upgrade, migration and business logic change. The manual entry field is filled in blank. Non mandatory fields lead to a large number of null values.
	Inconsistent value	eg. The ID number is not identical on the date of birth. The return time is more than actual time. etc.	Some data is filled by default value because of system upgrade, migration and business logic change. The business is not validated. The business process and system process are different in order and time.

Fig. 1. The quality problems of various types of data.

2.2 Anomaly Detection Process

Aiming at the above sorted data quality anomaly problems, the anomaly detection process and method based on data assets are studied. The data quality anomaly detection process is shown in Fig. 2. Firstly, the key word library is generated, and then the key word library is segmented by using the text word segmentation model and the user-defined word library. After Chinese word segmentation statistics and clustering, the detection feature key word library is generated, and the detection feature key word library is mapped with the data asset list to generate data asset feature lists of tables and fields, After the data asset feature list is mapped with the rule, the data quality rule detection is finally performed.

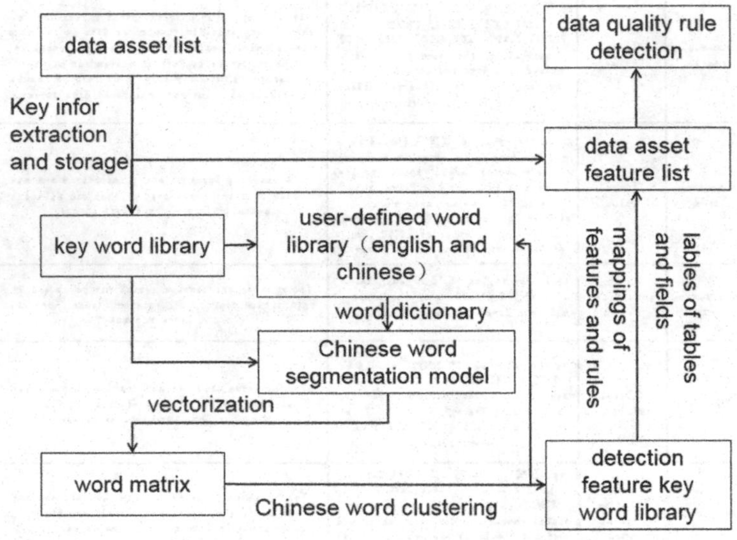

Fig. 2. Anomaly detection process.

2.3 Word Segmentation Based on Data Asset List

Key words library. The company's data asset list include the level, discipline/Department, system, Chinese name of table, English name of table, table description, English name of field, Chinese name of field, field description, field type, primary key, sensitive data. Key words library extract the key information of Chinese name of table, English name of table, Chinese name of field, English name of field and field description. Key word databases are generated respectively, including Chinese table name corpus, English table name corpus, Chinese field name corpus, English field name corpus and field description corpus (see Fig. 3). Preset data quality detection feature library. The data quality

table name	Table Chinese name	Table description	Table type	Field name	Field Chinese name	Field description	Field type
T_SB_ZWYC_GT	yun xing gan ta shu ju biao	gan ta jia kong shu xian lu zhong yong lai zhi cheng shu dian xian de zhi	TABLE	HCG	hu chen gao	hu chen gao	NUMBER
T_SB_ZWYC_GT	yun xing gan ta shu ju biao	gan ta jia kong shu xian lu zhong yong lai zhi cheng shu dian xian de zhi	TABLE	YXZT	yun xing zhuang tai	yun xing zhuang tai	NUMBER
T_SB_ZWYC_GT	yun xing gan ta shu ju biao	gan ta jia kong shu xian lu zhong yong lai zhi cheng shu dian xian de zhi	TABLE	SSQY	suo shu qu yu	suo shu qu yu	NUMBER
T_SB_ZWYC_GT	yun xing gan ta shu ju biao	gan ta jia kong shu xian lu zhong yong lai zhi cheng shu dian xian de zhi	TABLE	DYDJ	dian ya deng ji	dian ya deng ji	NUMBER
T_SB_ZWYC_GT	yun xing gan ta shu ju biao	gan ta jia kong shu xian lu zhong yong lai zhi cheng shu dian xian de zhi	TABLE	AXZJ	A xiang zhong ju	A xiang zhong ju	NUMBER
T_SB_ZWYC_GT	yun xing gan ta shu ju biao	gan ta jia kong shu xian lu zhong yong lai zhi cheng shu dian xian de zhi	TABLE	AXHDGD	A xiang heng dan gao du	A xiang heng dan gao du	NUMBER
T_SB_ZWYC_GT	yun xing gan ta shu ju biao	gan ta jia kong shu xian lu zhong yong lai zhi cheng shu dian xian de zhi	TABLE	SWID	shi wu ID	shi wu ID	NUMBER
T_SB_ZWYC_GT	yun xing gan ta shu ju biao	gan ta jia kong shu xian lu zhong yong lai zhi cheng shu dian xian de zhi	TABLE	ERPWBSBM	WBS bian ma	WBS bian ma	NUMBER
T_SB_ZWYC_GT	yun xing gan ta shu ju biao	gan ta jia kong shu xian lu zhong yong lai zhi cheng shu dian xian de zhi	TABLE	GTBH	gan ta bian hao	gan ta bian hao	NUMBER
T_SB_ZWYC_GT	yun xing gan ta shu ju biao	gan ta jia kong shu xian lu zhong yong lai zhi cheng shu dian xian de zhi	TABLE	DBDJ	dai biao dang ju	dai biao dang ju	NUMBER

Fig. 3. Key words library.

detection feature library is defined according to the data characteristics. The data is divided according to types and characteristics (see Fig. 4). Word segmentation and annotation. Firstly, the Chinese field name corpus needs accurately segment. Secondly data after word segmentation have to be marked according to the data quality characteristics, such as type, ownership, source, identification, method, account, primary key, serial number, name, examiner, collector, abbreviation, etc. Thirdly, the preset user-defined thesaurus is to be added into the user-defined thesaurus. Just like the above operation, Chinese table name corpus, English table name corpus, Chinese field name corpus, English field name corpus and field description corpus have to be respectively dealt according to perform full mode word segmentation.

2.4 Keyword Library of Detection Feature

The key steps of keyword library are word segmentation, word frequency sorting and keyword classification based on annotation. (see Fig. 5). The fields in the data asset list can be classified and mapped according to the key-word library. Then the data asset feature list can be generated, as shown in Fig. 6. At last we perform global data quality detection on the data based on the data asset feature list.

features of level 1	enumerable data	date data	coded data	text data
features of level 2	enumerable data	date data	coded data	manually fill
features of level 1	text data -name			
features of level 2	place name	person name	company name	equipment name
features of level 1	numerical data			
features of level 2	amount of money	ratio	positive parameter	integer
features of level 1	numerical data			
features of level 2	power specific values	index growth value	common sense value	
features of level 1	specific data			
features of level 2	Zip code	mailbox	telephone	address
features of level 1	specific data			
features of level 2	IP	Fax	port number	Longitude and latitude
features of level 1	specific data			
features of level 2	identity card	bank card No	social unified credit code	Organization code

Fig. 4. Preset data quality detection feature library.

2.5 Keyword Library of Detection Feature

(1) Feature extraction of numerical data

The algorithm is designed for the problem of data type classification. The algorithm is used to judge the specific type of numerical data according to the Chinese field name. The specific types of numerical data can be divided into "ratio", "common sense value", "power specific value", "amount type", "integer unit value", "positive parameter", "index growth value". And the else flag of numerical data is "non target".

In this paper, the quality feature extraction model of numerical data based on data asset list uses Bert_seq2seq framework and unilm scheme. And we use Roberta of Chinese pre-training model for text vectorization. Roberta is improved on the basis of Bert, which is better than Bert in parameter quantity and training data.

The word vector obtained by the pre-training model. Forecast classification is predicted through the linear classification layer. The model loss is calculated by using the cross entropy function. And the model is optimized by Adam. For the test of model performance, we select the accuracy of prediction as the evaluation indicator of the model. By testing on the real test data set, the accuracy of the model reaches 0.866 and achieves a good classification effect.

(2) Feature extraction of text data

Text data includes name and text. Name includes person name, company name, equipment name, etc. Text mainly refers to manually entered text, including remarks, reason, purpose, etc. In this paper, the manual input text is only auxiliary information without quality requirements because there is no constraint on filling in. Therefore, this paper focuses on the quality anomaly detection of per-son names in name text.

The name recognition algorithm uses Jieba automatic word segmentation, and rule discrimination to recognize the text. The main function of Jieba is to do Chinese word segmentation. It can perform simple word segmentation, parallel

features of level 1	features of level 2	key words
enumerable data	enumerable data	type, mode, grade, status, whether or not, identification, ownership, nature, model, classification, type, structure, phase number, sign, medium, format, category, form, source, gender and grade
date data	date data	date, time, date of manufacture, date of birth
text data-name	place	place, Place of Origin
	name	contact person, name, legal representative, construction personnel and responsible person
	company	manufacturer, manufacturer, unit, organization, name, abbreviation, unit name, construction unit and team
	equipment	equipment name
numerical data	amount of money	tax amount, amount including tax, amount excluding tax, contract amount, payment amount, XX (actually received, actually paid, collection, repayment, write off, invoice, cost, overdraft, transaction, deduction, payment, integral) amount, quality assurance deposit, deduction, paid, (current month, available) budget, subsidy price
	ratio	return to work rate, return to production rate, growth rate, line loss rate, tax rate, currency exchange rate, proportion of electricity charge, recovery rate and multiple rate of electricity charge
	positive parameter	(power line, feeder) length, (external speaker, meter to terminal) distance, (antenna) height, multiple, score, unit price, file size, equipment price, cycle (month, rotation), response time (duration), electricity price
	integer	number of pieces, sets, times, users, number of operation terminals, arrival quantity, installation quantity, removal quantity, verification quantity, number of qualified verification, number of equipment, number of turns (voltage and current transformer through turns), (re inspection, sorting and power purchase) times, ordering quantity, warehousing quantity, inventory quantity, outbound quantity, number of pieces, days of power
	power specific values	rated voltage (kV), rated current (a), rated capacity (MVA), short-circuit voltage, short-circuit loss, resistance, reactance, impedance, branch transformation ratio, susceptance, conductance, frequency (acquisition frequency, electrical frequency), power supply, power consumption, number of resumption, electricity charge, arrears, balance of electricity charge, operation capacity, contract capacity, total power connection capacity
	index growth value	month on month growth and cumulative growth
	common sense value	temperature 、 humidity
coded data	coded data	No., PK, Sn, Sn, code
specific data	Zip code	Zip code
	mailbox	E-mail, mailbox
	telephone	mobile phone, contact number, number, phone number, mobile number
	address	address, mailing address and contact address
	IP	IP address, network address, web address
	Fax	Fax number
	port number	port number
	Longitude and latitude	longitude, latitude
	identity card	identity card
	bank card No	bank card No
	social unified credit code	social unified credit code
	Organization code	Organization code
text data	manually fill	remarks, reason and purpose

Fig. 5. Keyword library of detection feature.

word segmentation and command line word segmentation. Its function is not limited to this. At present, it also supports keyword extraction, part of speech tagging, word location query, etc. In this algorithm design, Jieba's simple word segmentation and part of speech tagging functions are used. However, there are some errors in Jieba word segmentation. For example, during the experiment, it is found that Jieba has judgment errors for rare surnames and names including common words. Therefore, we add surname database discrimination on the basis of Jieba part of speech tagging. The name is irregular. Intuitively, the last name part of the text exists in the last name database, and the text length is between 2–4. We have reason to think that the text is a name, otherwise it is not a name. Through such processing, the running speed of the algorithm is improved while ensuring the accuracy of the algorithm.

The precision, recall and F1 values commonly used in classification algorithms. They are selected as the evaluation indexes of the model. Tested on the real test data set, the precision of the model is 0.999, recall is 0.899 and F1 is 0.94.

(3) Feature extraction of text data

For other type data, rule and statistical methods are used to extract data features and detect anomalies. For example, the enumeration data adopts the statistical method to judge whether it is an enumeration value, and the date data and specific data use the rule method to realize exception detection. There are more mature rules in specific data, such as zip code, mobile phone number

table name	Table Chinese name	Table description	Table type	Field name	Field Chinese name	Field description	Field type	feature
T_SB_ZWYC_GT	yun xing gan ta shu ju biao	gan ta jia kong shu xian lu zhong yong lai zhi cheng shu dian xian de zhi	TABLE	HCG	hu chen gao	hu chen gao	NUMBER	numerical data (positive parameter)
T_SB_ZWYC_GT	yun xing gan ta shu ju biao	gan ta jia kong shu xian lu zhong yong lai zhi cheng shu dian xian de zhi	TABLE	YXZT	yun xing zhuang tai	yun xing zhuang tai	NUMBER	enumerable data
T_SB_ZWYC_GT	yun xing gan ta shu ju biao	gan ta jia kong shu xian lu zhong yong lai zhi cheng shu dian xian de zhi	TABLE	SSQY	suo shu qu yu	suo shu qu yu	NUMBER	enumerable data
T_SB_ZWYC_GT	yun xing gan ta shu ju biao	gan ta jia kong shu xian lu zhong yong lai zhi cheng shu dian xian de zhi	TABLE	DYDJ	dian ya deng ji	dian ya deng ji	NUMBER	enumerable data
T_SB_ZWYC_GT	yun xing gan ta shu ju biao	gan ta jia kong shu xian lu zhong yong lai zhi cheng shu dian xian de zhi	TABLE	AXZJ	A xiang zhong ju	A xiang zhong ju	NUMBER	numerical data (positive parameter)
T_SB_ZWYC_GT	yun xing gan ta shu ju biao	gan ta jia kong shu xian lu zhong yong lai zhi cheng shu dian xian de zhi	TABLE	AXHDGD	A xiang heng dan gao du	A xiang heng dan gao du	NUMBER	numerical data (positive parameter)
T_SB_ZWYC_GT	yun xing gan ta shu ju biao	gan ta jia kong shu xian lu zhong yong lai zhi cheng shu dian xian de zhi	TABLE	SWID	shi wu ID	shi wu ID	NUMBER	coded data
T_SB_ZWYC_GT	yun xing gan ta shu ju biao	gan ta jia kong shu xian lu zhong yong lai zhi cheng shu dian xian de zhi	TABLE	ERPWBSBM	WBS bian ma	WBS bian ma	NUMBER	coded data
T_SB_ZWYC_GT	yun xing gan ta shu ju biao	gan ta jia kong shu xian lu zhong yong lai zhi cheng shu dian xian de zhi	TABLE	GTBH	gan ta bian hao	gan ta bian hao	NUMBER	coded data
T_SB_ZWYC_GT	yun xing gan ta shu ju biao	gan ta jia kong shu xian lu zhong yong lai zhi cheng shu dian xian de zhi	TABLE	DBDJ	dai biao dang ju	dai biao dang ju	NUMBER	numerical data (positive parameter)

Fig. 6. The data asset feature list.

and identity card number. The number data is based on regularization and other methods to judge and detect abnormal data.

3 Results and Analysis

This paper the data from three systems such as production management system (PMS), marketing business system and Enterprise Resource Planning system (ERP) have carried out word segmentation and feature extraction for more than 200 tables, more than 3000 fields and field descriptions. There are 7 types of primary features and 19 types of secondary features, with a total of more than 300 keywords. Through the detection of more than 1 million data, more than 1000 data problems are found. The detection time is 3.56 h, which is far less than the time of manually configuring the rules of each field in each table in PMS, marketing, ERP and other systems.

As shown in Fig. 7, the average accuracy of date and text data is high, in which the accuracy of date data is 1, because the data of date type have the limited classes of the rules to performs exception detection. The average accuracy of the text class is 0.93. The text class only contains the name, address, zip code, mailbox and other information in the specific data, and does not contain the manually entered text which is not detected in this paper. The average accuracy of numerical data is low. Numerical data has strong business meanings. In addition to relying on the detection based on the data itself, it needs to be accurately detected in combination with the business.

Based on the anomaly detection of data asset list, this paper realizes the general detection of a large number of discrete data. And the found data quality

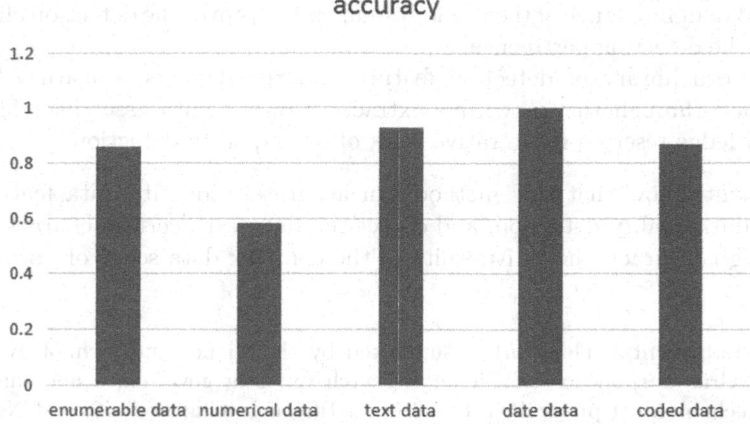

Fig. 7. Accuracy comparison.

problems are of great significance to the data quality optimization and data governance of actual business. For example, the filed "ppq" is in the table "G_TG_PQ" of the marketing business system. Its' characteristics are mapped into positive parameters and power specific values. The rules should be that values are greater than 0. But there are negative numbers in the data. Suspected problem data is problem data after confirmed by the business. The filed "mandt" is in the table "KONH" of the ERP system. Its' characteristics are mapped into the date type. But the rules do not conform to the date range. Data such as "40260303" appears. The filed "menge" is in the table "ESKL" of the ERP system. Its' characteristics are mapped into integer values. The detection rules are integer and positive, but there are negative numbers. The filed "cons_name" is in the table "c_cons" of the marketing business. Its' characteristics are mapped into name type. The detection rules are persons' name. But there is data such as "Qianye Guanling first-class corridor street lamp". They are all suspected to be problem data, and they are problem data after business confirmation.

4 Conclusions

For the first time, this paper based on data asset list applies text word segmentation model, word clustering, similarity calculation and other methods to generate keyword library of detection feature and to generate the data asset feature list. And then we performs data quality detection. The following conclusions are obtained:

(1) Data quality detection feature knowledges can be precipitated through related technology about word segmentation and feature extraction based on the information in the data asset list. They are valuable and accurate and provide an intelligent basis for data quality problem detection. This method reduces the

blind scanning range of the whole system and improves the detection efficiency and the detection pertinence.

(2) Keyword library of detection feature and the data asset feature list are formed through the knowledge extraction in the data asset list. They are knowledge reserves for iterative work of data quality detection.

Experiments show that this method can accurately identify data features for general data quality detection, and detect the data with certain characteristics. It has high accuracy and universality in the complex data scene of power multi system.

Acknowledgments. This work is supported by the science and technology project of State Grid Corporation of China:"Research on data governance and knowledge mining technology of power IOT based on Artificial Intelligence" (Grand No. 5700-202058184A-0-0-00).

References

1. Hu, M., Bai, X., Xu, W., Wu, B.: Literature review of anomaly detection algorithms for multidimensional time series. J. Comput. Appl. **40**(6), 1553 (2020)
2. Zhao, M., Zhao, Y., Zhu, Z.: Outlier detection based on label propagation. J. Data Acquis. Process. **34**(3), 331–340 (2019)
3. Pang, X., Huang, Y., Wang, Z., Yu, Y., Gao, S.: Multivariate process variables abnormal data segments detection based on correlation coefficient. Control Eng. China (1) (2020)
4. Kuang, M., Li, Y., Li, C., Cao, M.: Research on abnormal electricity detection method based on multi-model by stacking ensemble learning. Electric Power Sci. Eng. **37**(3), 23 (2021)
5. Ren, S., Zhang, J.: Overview of feature extraction algorithms for time series. J. Chin. Comput. Syst. (02) (2021)
6. Wen, Q., Gao, J., Song, X., et al.: RobustSTL: a robust seasonal-trend decomposition algorithm for long time series. In: Proceedings of the AAAI Conference on Artificial Intelligence, pp. 5409–5416 (2019)
7. Marteau, P.F.: Time warp edit distance with stiffness adjustment for time series matching. IEEE Trans. Pattern Anal. Mach. Intell. **31**(2), 306–318 (2009)
8. Gou, X., Xiao, X.: Short-term electricity price forecasting model based on empirical mode decomposition and LSTM neural network. J. Xi'an Univ. Technol. **36**, 129–134 (2020)
9. Liu, Y., Wang, Q., Xu, Z.: Research on data cleaning and abnormal recognition method of dissolved gas in oil based on multi-layer architecture. J. North China Electric Power Univ. (Nat. Sci. Ed.)

MAQTDS

Big Data Resources to Support Research Opportunities on Air Pollution Analysis in India

Sarath K. Guttikunda[1,2](✉)

[1] TRIP-C, Indian Institute of Technology, New Delhi, India
sguttikunda@urbanemissions.info
[2] Urban Emissions, New Delhi, India

Abstract. Most debates on air quality in India are (often) limited to big cities like Delhi, Mumbai, Kanpur, Pune, Hyderabad, and Kolkata, even though most of India's population lives in Tier-2, Tier-3, and smaller towns. There is little by way of local measurements for ground truthing or an assessment of sources contributing to air pollution problems in urban and rural areas or the growing health impacts associated with these pollution levels. The Air Pollution kNowledge Assessment (APnA) city program, launched in 2017, is an attempt to fill this lacuna of information, with an objective to create a baseline database for air pollution in Indian cities using open-access reanalysis data, satellite imagery, and satellite retrievals to inform policymakers as they evaluate the evolution of pollution and chart out strategies to improve air quality. This paper is based on the presentations delivered at two workshops - MAQTDS 2022, held as part of DASFAA 2022, and DCAAQ at BDA2021 - outlining an overview of air quality in India and opportunities for research to support air quality analysis using bigdata resources.

Keywords: India · Air quality · Bigdata · Satellite retrievals · NCAP · Air quality management · Emissions

1 Introduction – Air Quality in India

More than 50 Indian cities are ranked among the top 100 with the worst annual $PM_{2.5}$ averages, with Delhi taking the top spot among the capital cities worldwide in 2020 (https://www.iqair.com). Between 1998 and 2020 India's annual average $PM_{2.5}$ values have at least doubled [1]. On India's air quality index (AQI) scales, pollution levels over the Indo-Gangetic plain (IGP) moved from poor to very poor and severe conditions and the Central India region moved from moderate to poor conditions. At the administrative level, number of districts complying with India's annual ambient standard of 40 $\mu g/m^3$ dropped from 440 to 255 (out of 640 districts as per Census 2011) and number states dropped from 29 to 21 (out of 36, including union territories). Traditionally, these increases are observed over the cities. However, in the recent reanalysis databases which combine satellite retrievals, similar trends were observed over the rural areas. In these 23 years, total population complying with the annual ambient standard dropped from 60.5% to 28.4%, with most of this change coming from non-urban areas in IGP. In the

U. K. Rage et al. (Eds.): DASFAA 2022 Workshops, LNCS 13248, pp. 389–401, 2022.
https://doi.org/10.1007/978-3-031-11217-1_28

same period, the population exposed to poor, very poor, and severe AQI levels increased from 0.0% to 17.8%. In 2020, only a small portion of India's population lived in areas complying with World Health Organization (WHO)'s new guideline of 5 g/m^3.

According to the Global Burden of Disease (GBD) analysis, an estimated 1.2 million premature deaths in India can be traced back to exposure due to outdoor PM2.5 pollution levels [2]. According to GBD-Mapping Air Pollution Sources (MAPS) program, approximately 80% of the pollution originated from fossil fuel combustion and resuspended dust and the remainder coming from natural activities like sea salt, dust storms, and some agricultural activities [2, 3].

In 2019, India's Ministry of Environment Forests and Climate Change (MoEFCC) announced the National Clean Air Programme (NCAP) [4]. Under the programme, 132 non-attainment cities (i.e., cities that did not meet the annual ambient standard in 2017) were asked to prepare action plans to reduce their ambient PM2.5 pollution levels by 20–30% by 2024, compared to the pollution levels recorded in 2017. Individual cities have started to assimilate information on emissions and pollution loads to support the action plans.

In air quality management, general practice is to rely on monitoring data, which is basically snippets of information, both spatially and temporally. In India, there is an acute lack of ambient monitoring efforts in most cities to build a story just on that database. Using CPCB's own thumb rules, India requires 4000 stations across India and as of February 2022 there are only 340 in operation. Even the surveys and tests conducted to understand emissions are spread across temporally. For example, a pool of emission factors tests for vehicles was last conducted in 2010 as part of CPCB's 6-city study and there was one more round for a sample of vehicles in Pune in 2018. While these snippets of information are useful for ground truthing and expanding our understanding of the pollution loads and source strengths, the monitoring database needs to expand beyond the current capacity.

On the other side, we have the atmospheric modelling community, combining a larger pool of data from multiple resources including satellite retrievals and chemical transport models, helping us build patterns in emissions, pollution, and activity data, all in the hope of plugging the gaps in the monitoring data. Figure 1 presents a schematic of major components of air quality modelling. All of them are data intensive, computationally challenging, and require substantial personnel training to move forward from planning to execution. (a) Emissions modelling for both aggregate emissions and spatial/temporal allocations need a lot of data on source strengths, source locations, source emission control performance, and proxies for allocation of emissions at various scales (smaller the grid size, larger the need for proxies for finer distribution). (b) Meteorological modelling is streamlined with the existence of multiple global forecasting systems and agencies distributing the 3-dimensional fields to support scientific research and communications. For example, NASA's GFS and ESA's ECMWF. In India, one such system is maintained by the Indian Meteorological Department (IMD), which issues 10-day sub-regional forecasts and some feeds customized for fishery and farming communities. However, if the need is for meteorological data at a finer resolution, say 1-km over a city airshed, then downscaling models like WRF must be adapted, which require large computational capacity and personnel training. (c) Chemical transport modelling can vary in size and application depending on the requirements. Multiple models exist to accommodate these needs – models like CMAQ, CAMx, WRF-chem, GEOS-chem, and CHIMERE can help simulate multiple pollutants with full chemistry and evaluate the impacts of advection and chemistry at urban, regional, and global scales, including source apportionment; and models like inMAP and GAINS can help integrate chemical transport model results to evaluate scenarios and health impacts. (d) Validation is the central pillar of the whole modelling exercise, which is dependent on the monitoring data. There is no limit on data that can be used for validating and calibrating the modelling results, as long we have sample large enough to represent reality and represent the modelling domains spatially and temporally. (e) And finally, dissemination for public awareness and policy dialogue, which requires a completely different set of teams to take the message forward.

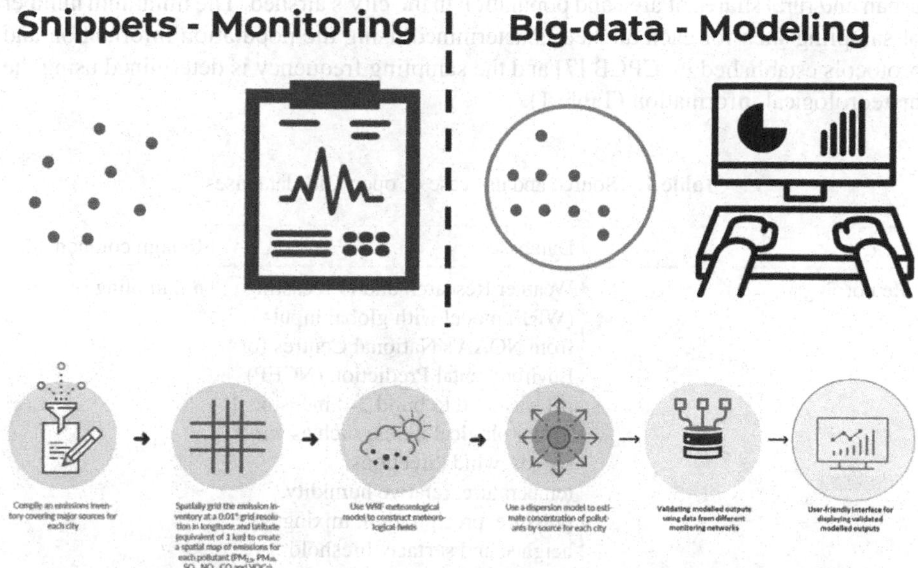

Fig. 1. Schematic of data management required for air quality analysis

While the kind of information gathered from monitoring and modelling exercises is different in shapes and sizes, both are integral pillars of an air quality management campaign, both needing snippets of information like surveys and pattern building from large (to very large) information databases. This paper presents a summary of research opportunities of using bigdata to study India's air quality.

2 Big Data Research

2.1 Support to Ambient Monitoring Efforts

As of February 2022, there are 340 continuous monitoring stations operating across India covering 174 cities with at least one station. Delhi (40), Mumbai (21), Bengaluru (10),

Ahmedabad (8) and Pune (8) are few cities with multiple stations. The total monitors count translates to 0.25 per million population and in most cases is not a representative sample for regulatory and research grade pollution analysis [5, 6]. This density factor is the lowest among the big countries - China (1.2), the USA (3.4), Japan (0.5), Brazil (1.8) and most European countries (2–3). In addition to the continuous stations, CPCB also operates 800 manual stations to collect 24-h average pollution levels for up to 104 days in a year.

Meteorology, population, and human settlements databases were accessed to support the monitoring network design under NCAP, starting with determining a city's representative airshed. A city's airshed is determined using urban-rural classifications, landuse information, and an understanding of the known emission sources in the immediate vicinity of the city's administrative boundary. Human settlements layer is used to estimate the urban and rural shares of area and population in the city's airshed. The minimum number of sampling sites for each airshed is determined using the population information and protocols established by CPCB [7] and the sampling frequency is determined using the meteorological information (Table 1).

Table 1. Source and use case of open GIS databases

Field	Database	Design component
Meteorology	Weather Research and Forecasting (WRF) model with global inputs from NOAA's National Centres for Environmental Prediction (NCEP) [8] was used to build 3-dimensional meteorological fields, such as wind speeds, wind directions, temperature, relative humidity, pressure, precipitation, mixing layer heights, and surface threshold velocities at 1-h temporal resolution for base year 2018	# Sampling seasons
Population	Census-India database at the district level [9] and Landscan of Oakridge National Laboratory [10] were used to create 0.01° resolution population database for the city airsheds. The raw databases are available at 30 s spatial resolution	# Sampling sites
Global Human settlement (GHS)	GHS layer of Landsat satellite imagery was used to designate the city airshed grids and the gridded population as urban and rural [11]	# Sampling sites

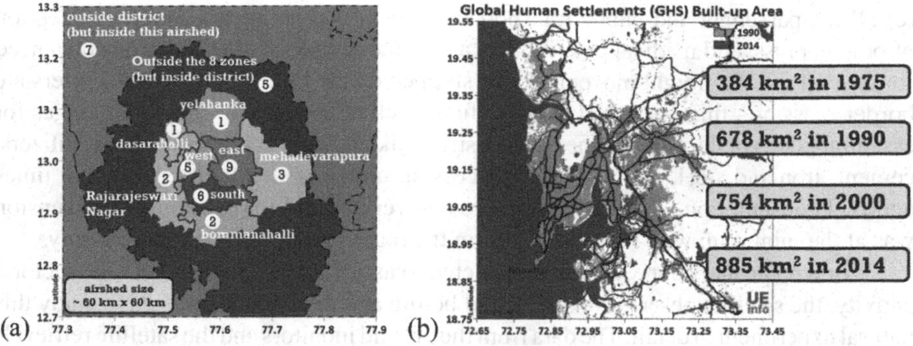

Fig. 2. (a) Allocation of monitoring sites by zones in Bengaluru (b) Expansion of built-up area in the Greater Mumbai airshed.

Overlapped with commercial activity density information in the form of number of hotels, hospitals, schools, parks, malls, markets, apartment complexes, industrial estates, worship sites, banks, eateries, fuel stations, and traffic spots like parking and stops, the recommended number of stations can be further assigned to a zone or a sub-district for better spatial representation – Fig. 2a presents an example for allocating 41 recommended stations in Bengaluru's 8 zones, peri-urban area surrounding the city administrative boundary, and the background rural areas in the airshed. The Human Settlement Built-Up maps can also be utilized to systematically shift the location of the monitoring sites as the city expands – Fig. 2b presents an example of expansion of the Greater Mumbai region, with the built-up area increasing from 384 km^2 in 1975 to 885 km^2 in 2014, which is a proxy for increasing demand for commercial and transport amenities, construction material, and together also increasing the demand for additional ambient monitoring for better representation of the activities.

2.2 Use of Satellite Retrievals

In 2020, COVID-19 lockdowns in March and April provided a glimpse into what is possible when the emissions are eliminated or reduced at all the known sources. In India, starting on March 24th, 2020, four lockdowns were announced (for 21, 21, 19, and 14 days), with the strictest regulations during the first lockdown period and slowly easing the restrictions by the end of the fourth. Thus resulting in better air quality across India with most improvements observed during the first period [12]. Lockdown periods featured the following regulations - (a) all the offices implemented work-from-home and all the schools, colleges, universities, training institutions, markets, malls, religious centers, and other public spaces were shut - this reduced most of the demand for passenger movement on the roads (b) all the shops and small-scale industries in the urban and rural areas were shut - with exceptions introduced after for essential food and medicine supply chains (c) all the construction activities were banned including brick manufacturing - this reduced the dust loading in the hotspots, debris movement, and construction freight movement (d) all the open waste burning activities were banned - this was possible since movement on the roads and inside/outside the residential communities was restricted

(e) all the passenger and public transport movement was stopped - with the exception of police, press, and medical practitioners and some with special permissions on a need for basis (f) all the freight movement was stopped on the highways and at the interstate border crossing - this was eased after the first week in response to supply shortages for essential goods in the cities (g) heavy industries (like power plants, refineries, fertilizers, cement, iron and steel, and other ore processing units) limited their operational times and fuel consumption loads, in response to a lower demand (h) road dust resuspension was at the minimum with reduced traffic on the roads and no construction activity.

While no primary surveys were conducted to ascertain these changes in the sectoral activity, the satellite observations provided before and after measurements to study this natural experiment in detail. The data from the ground monitors and the satellite retrievals also provided necessary data to study new research questions, which in the past would have been possible only in theory or lab experiments. Such as (a) impact of low emission densities on ozone photochemistry [13] (b) evaluation of NOx-VOC control regimes [14] (c) estimation of true background concentrations for cities, in the absence of all or most of the major emission sources [15].

The air quality during the lockdown periods is one example where the use of bigdata was demonstrated to explain the extreme lows and evaluation of daily trends. PM2.5 concentrations dropped across the country at the start of the lockdown periods and with every phase there is a marked increase in the average numbers was observed. On average, every lockdown period witnessed at least 25% drop across India, most (as high as 70%) coming from the cities. PM pollution is affected by all the known sources and all the regulations discussed above led to these drops [16, 17]. A climatological analysis of the satellite retrieval based AOD estimated a drop of 50% in the PM2.5 concentrations at the start of the lockdowns and slowing catching up to the decadal averages at the end of the 4th lockdown (see Fig. 3) [18]. The 4 lockdown periods ended on May 31st, 2020. Starting on June 1st, 2020, restrictions started to ease in phases, with individual states either continuing or easing them at their discretion. Similar databases are available for other pollutants – SO2, NO2, CO, HCHO, and Ozone, all of which can used to not only study the impact of chemistry and but also can be used to plug the gap in the monitoring efforts.

It is important to understand that the global models and satellite retrievals come with a lot of assumptions. While these broad insights are very helpful, limitations must be understood before applying these databases for public and policy use. Some of the limitations include (a) global models run at a coarse spatial resolution. For example, the reanalysis results presented in the introduction are from a model with 0.5-degree resolution, which cannot capture the core urban activities (b) the satellite retrievals used are just passes over India (a snapshot) and not geostationary with longer time stamps for a given day and (c) India's on-ground monitoring network is not wide enough to feed these models for representative calibration. Despite the limitations, satellite data retrievals during the COVID-19 pandemic and the associated lockdowns provided a new normal for Indian cities – a realization that "clean air" and "blue skies" is possible also during the times when it is not raining or windy. Some hard decisions are required to achieve such a reduction in the emissions at all the sources and a change in how the cities and regions manage air quality to sustain the benefits, and in this process bigdata can

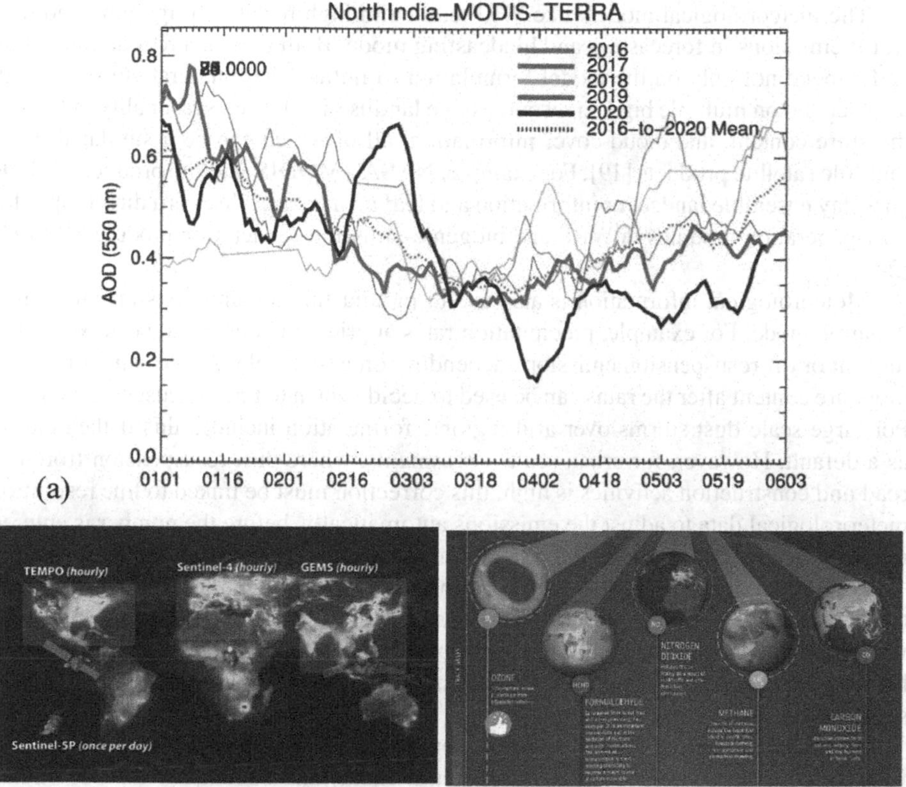

Fig. 3. (a) Satellite retrieval based AOD, averaged over North India for the lockdown period days between 2016 and 2020 [18] and [personal communication with co-author Dr Pawan Gupta, NASA] (b) Satellite clusters from NASA, ESA, and South Korea to support daily air quality analysis for multiple pollutants.

help nudge the change. Over Asia, GEMS system is expected to provide geostationary data for India in the coming years.

2.3 Use of Meteorological Data

Meteorology plays an integral part of pollution's ups and downs. It is responsible for the movement of emissions from source to receptor regions depending on the wind speeds and direction, for chemical evolution of the various pollutants within the gaseous phase and from gaseous to aerosol phase depending on the temperature, relative humidity, and pressure components which drive the chemical kinetics, and scavenging of pollution in the form of dry and wet deposition depending on wind speeds, landcover, and precipitation rates. While all these components are standardized in multiple chemical transport models, meteorology also plays a critical role in (a) emission modulation and (b) early pollution alert system.

The meteorological models like WRF can build high resolution lightning and dust storm emissions in forecasting and hindcasting mode. Both these sources are uncertain and depend not only on the model formulation to initiate these natural emissions, but also depend on multiple bigdata resources like landuse-landcover, seasonality in the soil moisture content, and cloud-cover information, all of which can be assimilated using multiple satellite products [19]. For example, NASA's MODIS satellite products include an 8-day ensemble landcover information and leaf area index, which is a direct input for dust-generation modules in WRF and biogenic-emissions generation module MEGAN [20].

Meteorological information is also useful in adjusting an emissions inventory in a dynamic mode. For example, precipitation rates at grid and hourly scale can be used to turn on or off resuspension emissions depending on a threshold. At the same time, soil moisture content after the rains can be used to decide when to turn on dust resuspension. For large scale dust storms over arid regions, formulation includes this if-then clause as a default. However, for urban scale assessments, where dust resuspension from on-road and construction activities is high, this correction must be linked to fine resolution meteorological data to adjust the emissions automatically, before the number is entered into the chemical transport model for further processing.

A similar correction can be employed in the space heating sector when the surface temperature and air temperature at 2 m drops below a threshold. In India, the during the winter months, surface temperature can drop to under 15 °C which triggers the need for heating. In most non-urban places, this demand is met by burning biomass, coal, and in some cases waste [21].

Summary of meteorological statistics for 2 cities is presented in Fig. 4 – Lucknow from North India and Hyderabad from South India (and summaries for 640 Indian districts is available at https://urbanemissions.info). For Lucknow, surface temperature is low for substantial number of hours during the winter nighttime, informing that the space heating emissions are an important part of Lucknow's inventory. For Hyderabad, while space heating is not a major component of its inventory, the wind directions alter significantly between the summer and the winter months, informing that the regional sources outside the city in the respective directions are important to track, since they have the right conditions to effect Hyderabad's air quality, as part of outside ("boundary") contributions at the chemical transport modelling stage [22, 23]. These deductions and dynamic adjustments are not possible from the use of just measurements at 1 or 10 locations in a city, but only possible when 3-dimensional high resolution meteorological modelling is conducted.

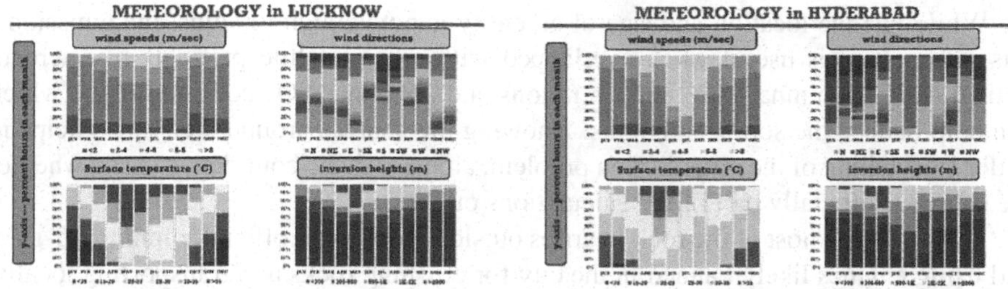

Fig. 4. Summary of wind speed, wind direction, temperature, and mixing heights as % hours in various bins in each month for 2 cities – Lucknow from North India and Hyderabad South India

2.4 Use of Google Earth Services

High resolution image processing is fast becoming an integral resource with easy access to multiple satellite feeds and algorithms to retrieve information for immediate use, not only among the air pollution modelling community, but also other areas like flood management and water resource management [24–26]. One free resource is Google Earth imagery, which has good spatial resolution to spot roads, landuse types, large industries, and large landfills. Below are examples of brick kilns spotted in Punjab and outside Mumbai, landfill in Mumbai, and rock quarries outside Pune (see Fig. 5).

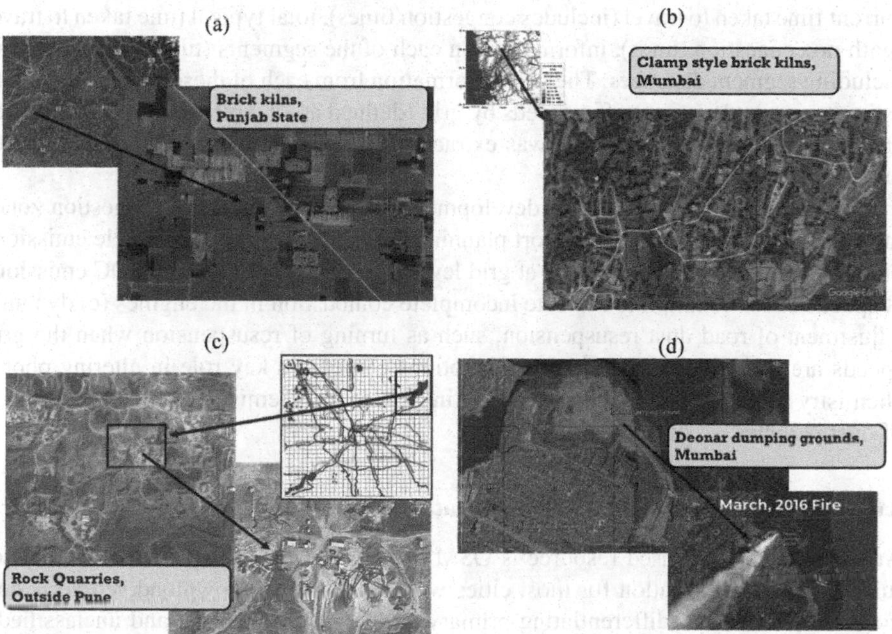

Fig. 5. Examples of activities spotted using Google Earth imagery (a) ~3000 fixed stack brick kilns in the state of Punjab (b) clamp style brick kilns outside Mumbai (c) ~11 km^2 of rock quarries outside Pune (d) ~2 km^2 landfill in Mumbai processing ~8000 tons per day waste from the city

While just the location and size of an entity is not enough to adjust the emissions, this information is useful when overlapped with other satellite products like aerosol optical depth, columnar NO_2 concentrations and columnar SO_2 concentrations, which can help deduce the source strengths. Knowing the sources around a city also helps in better storytelling of the air pollution problem, along with the confidence to say whether the source is officially part of the calculations or not.

For example, most of the rock quarries outside Pune are unofficial, running off-grid, and using engines likely banned in the city for crushing rocks and transporting locally. Most of the diesel used for this activity is not accounted in the official records of the city, but the emissions from the quarry activities will affect the air quality in the middle of the city.

For example, the clamp style brick kilns (showing outside Mumbai) are haphazardly placed around an area, with bricks piled along with biomass and coal mix to burn and bake. While the fixed stack kilns (showing in Punjab) are easy to spot, the clamp style kilns can only be mapped as an area and use it for back of the envelope calculations. This is the most inefficient way of manufacturing bricks and knowing where they are is the most useful information for air quality analysis and management.

2.5 Use of Google Maps Services

This is the only example which is a paid service. From the Google Maps distance API service, with each call, between 2 points, data can be extracted on total distance, total current time taken to travel (includes congestion times), total typical time taken to travel (with no congestion times), information on each of the segments (turns) along the way including segment distances. The base information from each of these calls is enough to estimate current average traffic speeds by grid (defined as ~1-km^2). For 30 Indian cities and non-Indian cities, such data was extracted and averaged at grid level for further processing (see Fig. 6).

Some applications include (a) development of speed profiles and congestion zones in the city to support urban transport planning (b) modulation of the vehicle emissions profile with average vehicle speed at grid level, such as higher CO and VOC emissions at speeds under 10 kmph to indicate incomplete combustion in the engines (c) dynamic adjustment of road dust resuspension, such as turning of resuspension when the grid speeds are under 10 kmph. The later 2 options can play a key role in altering photo-chemistry and ozone sensitivity to change in NO_x to VOC emission ratios and absence of dust particles for reactions.

2.6 Use of Open Street Maps (OSM) Database

An open and widely used resource is OSM database, which can provide several useful GIS layers information for most cities worldwide (https://download.geofabrik.de). For examples, roads (differentiating primary, secondary, motorable, and unclassified), railway lines, and commercial activity information in the form of number of hotels, hospitals, schools, parks, malls, markets, apartment complexes, industrial estates, worship sites, banks, eateries, fuel stations, and traffic spots like parking and stops. This is a crowd sourced database, so some level of ground truthing is advised before full use of the layers.

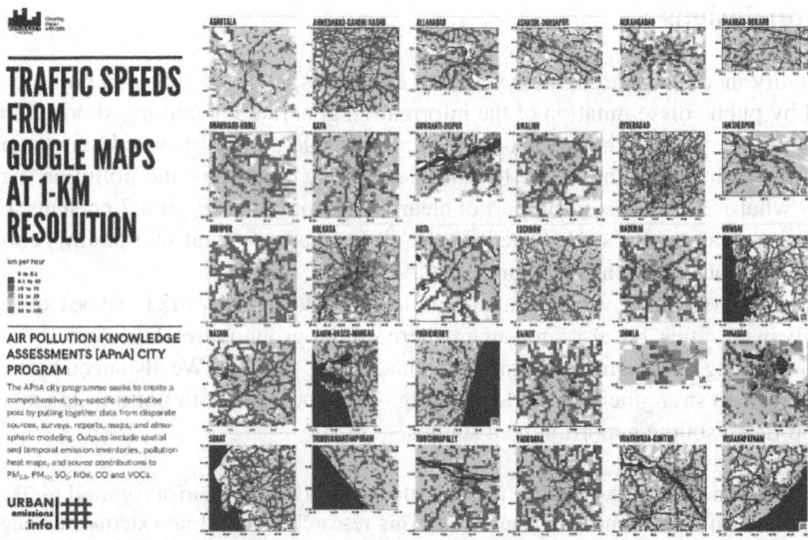

Fig. 6. Average traffic speeds at 1-km grid resolution for 30 Indian cities, extracted using the Google maps distance API service (Red colour indicates speeds under 15 kmph)

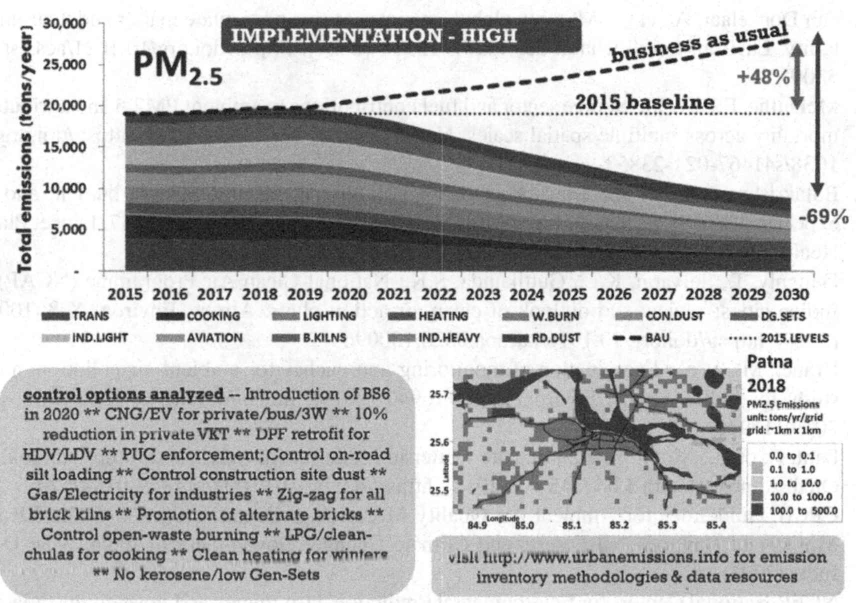

Fig. 7. Data integration to understand emission and pollution patterns and to support clean air action plans by evaluating their potential to reduce pollution (example: Patna under NCAP)

3 Conclusions

Air quality modelling through the stages of emissions, meteorology, and pollution, followed by public dissemination of the information generated is nothing short of art (see Fig. 1). At every stage, there is a lot of information (old and new) available in the public domain, which can be integrated to build defendable emissions and pollution maps to study "what-if" scenarios in support of clean air action plans. Figure 7 presents a summary of such an analysis conducted for the city of Patna – what will be impact of full implementation of actions listed under NCAP.

There is no second guessing that the ambient monitoring network must be expanded – not only in the cities, but also rural areas where similar growing trends are vividly visible. The monitoring data forms the basis for validating the bigdata. We also need more local level efforts to strengthen our understanding – this includes both bottom-up emissions and top-down source apportionment studies.

Acknowledgements. We would like to acknowledge and thank Dr Girish Agrawal for the invitation to present and submit this manuscript. This research received no external funding. The author declares no conflict of interest with the conference, conference organizers, and special issue editors.

References

1. van Donkelaar, A., et al.: Monthly global estimates of fine particulate matter and their uncertainty. Environ. Sci. Technol. **55**, 15287–15300 (2021). https://doi.org/10.1021/acs.est.1c05309
2. McDuffie, E.E., et al.: Source sector and fuel contributions to ambient PM2.5 and attributable mortality across multiple spatial scales. Nat. Commun. **12**, 3594 (2021). https://doi.org/10.1038/s41467-021-23853-y
3. Balakrishnan, K., et al.: The impact of air pollution on deaths, disease burden, and life expectancy across the states of India: the Global Burden of disease study 2017. Lancet Planet. Health **3**, e26–e39 (2019). https://doi.org/10.1016/s2542-5196(18)30261-4
4. Ganguly, T., Selvaraj, K.L., Guttikunda, S.K.: National Clean Air Programme (NCAP) for Indian cities: review and outlook of clean air action plans. Atmos. Environ. X **8**, 100096 (2020). https://doi.org/10.1016/j.aeaoa.2020.100096
5. Brauer, M., et al.: Examination of monitoring approaches for ambient air pollution: a case study for India. Atmos. Environ. **216**, 116940 (2019). https://doi.org/10.1016/j.atmosenv.2019.116940
6. Pant, P., et al.: Monitoring particulate matter in India: recent trends and future outlook. Air Qual. Atmos. Health **12**(1), 45–58 (2018). https://doi.org/10.1007/s11869-018-0629-6
7. CPCB. Guidelines for Ambient Air Quality Monitoring; Central Pollution Control Board, Ministry of Environment Forests and Climate Change, Government of India: New Delhi, India (2003)
8. NCEP. National Centers for Environmental Prediction. http://www.esrl.noaa.gov/psd/data/gridded/data.ncep.reanalysis.html. Accessed 15 Aug 2020
9. Census-India. Census of India 2011, The Governement of India, New Delhi, India (2011)
10. Rose, A.N.; McKee, J.J.; Urban, M.L.; Bright, E.A.; Sims, K.M.: LandScan 2018 (2019)

11. Pesaresi, M., et al.: GHS built-up grid, derived from Landsat, multitemporal (1975, 1990, 2000, 2014). European Commission, Joint Research Centre, JRC Data Catalogue (2015)
12. CPCB: Impact of lockdowns 25th March to 15th April on air quality (2020)
13. Kumar, A.H., Ratnam, M.V., Jain, C.D.: Influence of background dynamics on the vertical distribution of trace gases (CO/WV/O3) in the UTLS region during COVID-19 lockdown over India. Atmos. Res. **265**, 105876 (2022). https://doi.org/10.1016/j.atmosres.2021.105876
14. Rathod, A., Sahu, S.K., Singh, S., Beig, G.: Anomalous behaviour of ozone under COVID-19 and explicit diagnosis of O3-NOx-VOCs mechanism. Heliyon **7**, e06142 (2021). https://doi.org/10.1016/j.heliyon.2021.e06142
15. Beig, G., et al.: Towards baseline air pollution under COVID-19: implication for chronic health and policy research for Delhi, India. Current Sci. **119**, 00113891 (2020)
16. Gkatzelis, G.I., et al.: The global impacts of COVID-19 lockdowns on urban air pollution: a critical review and recommendations. Element. Sci. Anthrop. 9, 1–46 (2021). https://doi.org/10.1525/elementa.2021.00176
17. Ravindra, K., Singh, T., Biswal, A., Singh, V., Mor, S.: Impact of COVID-19 lockdown on ambient air quality in megacities of India and implication for air pollution control strategies. Environ. Sci. Pollut. Res. **28**(17), 21621–21632 (2021). https://doi.org/10.1007/s11356-020-11808-7
18. Sathe, Y., Gupta, P., Bawase, M., Lamsal, L., Patadia, F., Thipse, S.: Surface and satellite observations of air pollution in India during COVID-19 lockdown: implication to air quality. Sustain. Cities Soc. **66**, 102688 (2021). https://doi.org/10.1016/j.scs.2020.102688
19. Tinmaker, M.I.R., et al.: Relationships among lightning, rainfall, and meteorological parameters over oceanic and land regions of India. Meteorol. Atmos. Phys. **134**(1), 1–11 (2021). https://doi.org/10.1007/s00703-021-00841-x
20. Sindelarova, K., et al.: High-resolution biogenic global emission inventory for the time period 2000–2019 for air quality modelling. Earth Syst. Sci. Data **14**, 251–270 (2022). https://doi.org/10.5194/essd-14-251-2022
21. Chowdhury, S., Dey, S., Guttikunda, S., Pillarisetti, A., Smith, K.R., Di Girolamo, L.: Indian annual ambient air quality standard is achievable by completely mitigating emissions from household sources. Proc. Natl. Acad. Sci. USA **116**, 10711–10716 (2019). https://doi.org/10.1073/pnas.1900888116
22. Guttikunda, S.K., Nishadh, K.A., Jawahar, P.: Air pollution knowledge assessments (APnA) for 20 Indian cities. Urban Climate **27**, 124–141 (2019). https://doi.org/10.1016/j.uclim.2018.11.005
23. UEinfo: Air Pollution knowledge Assessments (APnA) city program covering 50 airsheds and 60 cities in India (2019). https://www.urbanemissions.info
24. Chithra, K., Binoy, B.V., Bimal, P.: Spatial mapping of the flood-affected regions of Northern Kerala: a case study of 2018 Kerala floods. J. Indian Soc. Rem. Sens. **50**, 677–688 (2021). https://doi.org/10.1007/s12524-021-01485-5
25. Goel, R., Miranda, J.J., Gouveia, N., Woodcock, J.: Using satellite imagery to estimate heavy vehicle volume for ecological injury analysis in India. Int. J. Inj. Contr. Saf. Promot. **28**, 68–77 (2021). https://doi.org/10.1080/17457300.2020.1837886
26. Lee, J., et al.: Scalable deep learning to identify brick kilns and aid regulatory capacity. Proc. Natl. Acad. Sci. **118**, e2018863118 (2021). https://doi.org/10.1073/pnas.2018863118

Air Quality Data Collection in Hyderabad Using Low-Cost Sensors: Initial Experiences

N. Chandra Shekar[1(✉)], A. Srinivas Reddy[1], P. Krishna Reddy[1], Anirban Mondal[2], and Girish Agrawal[3]

[1] IIIT Hyderabad, Hyderabad, India
chandra.shekar@research.iiit.ac.in
[2] Ashoka University, Delhi, India
[3] O.P. Jindal Global University, Sonipat, India

Abstract. Exposure to ambient Particulate Matter (PM) of air pollution is a leading risk factor for morbidity and mortality. The most common approach for air quality monitoring is to rely on environmental monitoring stations, which are expensive to acquire as well as to maintain. Moreover, such stations are typically sparsely deployed, thereby resulting in limited spatial resolution for measurements. Recently, low-cost air quality sensors have emerged as an alternative for improving the granularity of monitoring. We are making an effort to explore the framework for air quality data collection by employing low-cost sensors. In this paper, we have reported our initial experiences and observations concerning PM data collection for four months starting from October 2021 in the city of Hyderabad in India.

Keywords: Air quality · Particulate matter · Low-cost sensors

1 Introduction

Air pollution concerns the contamination of ambient air by chemical, physical or biological agents, which change the natural characteristics of the air that we breathe. Researchers have found that Particulate Matter (PM) is one of the serious air pollutants. Excessive presence of PM in the air and long-term exposure to PM can cause severe health problems such as breathing issues, mortality, ischemic heart disease, and premature death [12,18]. Moreover, air pollution also leads to acid rains, global warming, and depletion of the ozone layer [9]. The problem is very serious in developing countries like India due to the rapid growth of infrastructure, industrialization, mining activity, and so on [10,15].

PM can be classified in various ways depending on chemical composition, shape, or surface area. Two types of PM play an important role in understanding the extent of air quality: PM of size $\leq$2.5 µm (PM2.5) and PM of size $\leq$10 µm (PM10). The former is referred to as fine particles and the latter is referred to

U. K. Rage et al. (Eds.): DASFAA 2022 Workshops, LNCS 13248, pp. 402–416, 2022.
https://doi.org/10.1007/978-3-031-11217-1_29

as coarse particles. Primarily, PM2.5 contains the secondarily formed aerosols (gas-to-particle conversion), combustion particles, and re-condensed organic and metal vapors. The PM10 particles usually contain earth crust materials and fugitive dust from roads and industries [19].

In India, presently, some air quality data is publicly available at Central Pollution Control Board [1,4]. Official air quality monitoring stations are deployed sparsely across cities in India. The equipment deployed by government pollution control boards, private industries, big research organizations, etc. [6,13,16] is expensive. The coverage of these monitors is limited only to a few locations, and there is an urgent need to increase the coverage of locations to improve the spatial resolution of PM data. Moreover, a number of these stations do not record air quality data continuously. An additional issue with the existing system is that it cannot provide the air quality data at specific locations desired by stakeholders. Hence, there is an urgent need to develop a low-cost air quality data collection framework by employing and managing low-cost air quality data collection equipment to meet the diverse air quality information requirements of the various stakeholders.

In the literature, research efforts [5–7,11,13,14,16,17] have been made to improve both spatial and temporal resolution of air pollution data using low-cost air monitoring systems. The work in [5] explores the potential of routine mobile monitoring with fleet vehicles for measuring time-integrated concentrations at high spatial resolution. The work in [14] discusses the deployment of low-cost air quality sensors collocated with a reference instrument and develops an efficient and accurate calibration algorithms to calibrate the low-cost sensors. Another work in [7] proposes a low-cost cooperative monitoring tool that allows knowing, in real-time, the concentrations of polluting gases in various areas of the given city. The results show that the low-cost sensors are not as accurate as the official data; however, they provide useful indications of air quality in a specific location. It has been reported in [6] that low-cost portable monitors offer an opportunity to improve both spatial and temporal resolution of air pollution data and even validate, fine-tune or improve the existing ambient air quality models.

As a part of research effort to develop the framework for air quality data collection by employing low-cost sensors, we have collected air quality data by deploying static monitors in Hyderabad, India. Moreover, we have also collected air quality data using portable monitors by selecting four routes in Hyderabad, India. In this paper, we have reported the initial experiences and observations of PM data collection for four months, starting from October 2021 to February 2022, in Hyderabad city, India. In the remaining part of this paper, we provide the details of data collection equipment, and data collection methods, and report the trends along with our initial experiences and observations.

2 Materials and Methodology

In this section, we first discuss the details of the equipment. Next, we explain the details of the selected locations and routes and present the types of equipment used, number of equipment types, and the data collection methodology.

Table 1. Details of equipment

Type	Name	Purpose	Count
Static	Airveda PM2510CTH	To capture PM2.5, PM10	4
Portable	Airveda PM2510	To captures PM2.5, PM10	2
	Garmin GPS etrex10	To capture Location	2
	Mobile phones with GPS	To capture location (Latitude, Longitude), download data	2
	Motor cycle	To travel along the routes)	2

2.1 Details of the Equipment

Table 1 depicts the equipment type, purpose, and the number of the equipment type. For collecting air quality data from the given location, we have selected four monitors [3] (Airveda PM2510CTH) to sense PM2.5 and PM10 data. For collecting air quality data from the given route, two portable monitors (Airveda PM2510) were selected. Each Airveda PM2510CTH monitor and Airveda PM2510 monitor shown in Fig. 1(i) consists of a high-quality laser sensor, individually calibrated for Beta Attenuation Monitor, which is the most advanced system for measuring ambient air quality for Indian conditions. To capture the location (latitude, longitude) data, while traveling on the route, we have used both Garmin GPS and GPS enabled mobile phones. In addition, mobile phones were used to download the data from the monitors using the Airveda App [2]. The Airveda PM2510 monitors are equipped with a built-in rechargeable battery that can work for 3.5 h for one full charge. We have employed a motorcycle fixed with the air quality monitor, Garmin GPS, and mobile phone to capture air quality data (PM2.5 and PM10) and location data from the points on the route.

2.2 Details of the Locations and Routes

We have fixed monitors at four locations, viz., IIIT Hyderabad (IIITH), Masjid Banda (MB), Secunderabad (SCB), and Sagar Complex (SC) in the Hyderabad city. Figure 1(iii) depicts four locations (highlighted as star shape in map) and four routes in Hyderabad city (Fig. 1(ii)). We have placed one air quality monitor at each location with power supply and internet connectivity. Table 2 contains the description of the region and the position of the monitor. In IIIT Hyderabad, the static monitor is fixed to a pole located at the main gate entrance. It is a semi-highway and semi-residential region. In Masjid Banda, the monitor is placed in the 2nd floor of one of the flat's balconies. This location is a fast-growing commercial and residential area with a significant light and heavy vehicle moment. In Secunderabad, the monitor is placed on the ground floor of the house. This location is also a fast-growing commercial area with many ongoing infrastructural development works. Finally, in Sagar Complex, the monitor is placed on the first floor of the house. This location is a sparse residential area with a decent amount of green carpet.

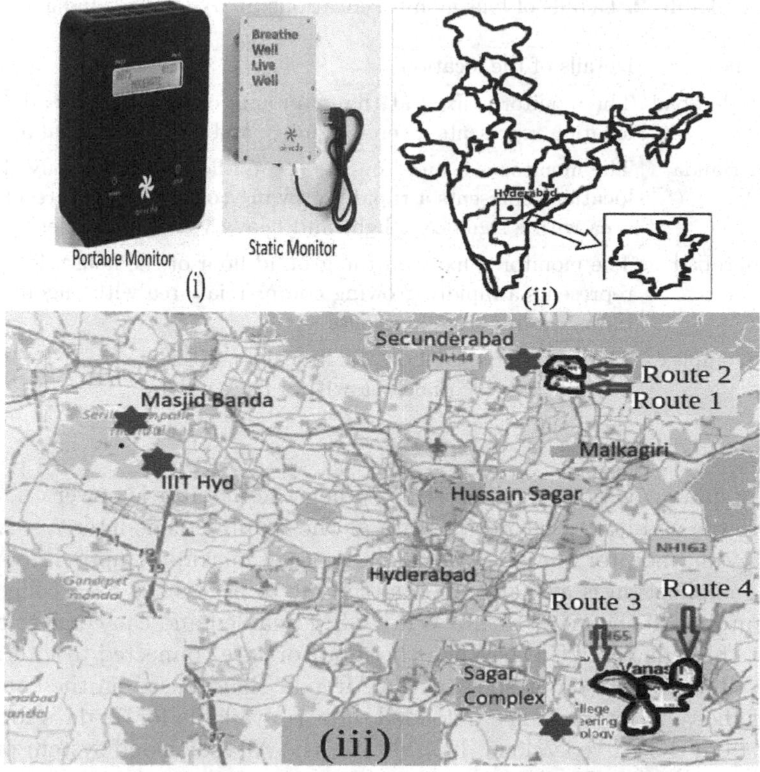

Fig. 1. (i) Photograph of a Monitor (ii) Hyderabad geographical location (iii) Locations of four monitors and the details of Four routes in Hyderabad, India.

We have also collected air quality data by traveling on fixed closed paths twice a week. Table 3 provides the details of four selected routes and refer these routes as Route 1, Route 2, Route 3, and Route 4. Figure 1(iii) shows the location of the routes on Hyderabad map. Figure 2 shows the paths of four routes. Each of the four routes vary between 5–7 km, with a crisscross. Each route consists of Market area, Main road and Residential area.

2.3 Data Collection Methodology

We explain the methods employed to collect the air quality data from the selected locations and the routes.

Data Collection from the Fixed Locations. Airveda monitors sense PM2.5 and PM10 values in $\mu g/m^3$ and computes air quality index (AQI) in the range of 0–500. In addition, it also senses the temperature value (in °C) and the humidity value (relative humidity (%)). In these monitors, the sensors can be programmed

Table 2. Details of four monitor locations in Hyderabad, India

Locations	Details of the location
IIIT Hyderabad	The monitor is fixed at the main gate of IIIT Hyderabad. The location represents a semi-highway and semi-residential area
Masjid Banda	The monitor is located on the second floor flat's balcony. The location represents a rapidly growing commercial and residential area with a significant light and heavy vehicle movement
Secunderabad	The monitor is fixed on the ground floor of the house. It represents a rapidly growing commercial area with ongoing infrastructural developments
Sagar Complex	The monitor is fixed on the first floor of the house. It represents a sparsely residential area with a park nearby

to collect data at various time intervals. When the monitor is power on, we can configure the monitor to record the data at the required time intervals. In addition, Airveda monitor contains a built-in battery backup of upto 1 h. Airveda software application is available on smartphone to access the data collected by the monitors. All the monitors are WiFi/GSM enabled and can connect to local Internet through WiFi/GSM. When the monitors are connected to internet, all the sensed data will be uploaded to the Airveda cloud based on the configured value of the time interval. The uploaded data is available for 30 days for downloading from the Airveda cloud. In addition, the Airveda cloud system provides past 30 days data in 30 min averages and hourly averages.

Data Collection from the Selected Routes. Like the monitors employed to collect data from the fixed locations, the monitors employed to collect the data from the selected routes also record air quality data and upload it to the Airveda cloud. In addition, these monitors will record the location data when connected to the GPS-enabled mobile phone. Note that the location data and time are recorded in synchronous with the recorded air quality data (i.e., GPS locations, date, time, and the corresponding AQI, PM2.5, PM10 are recorded).

The methodology to capture the air quality data from each route is as follows. The data is collected by traversing the given route by a motorcycle. The monitor is fixed to the motorcycle in a position that it can capture the environment at chest level height without any abstractions. We connect the monitor to our mobile phone (GPS and internet-enabled) using the Airveda Application through WiFi connectivity. In addition, we carry a Garmin GPS device to provide location data. We travel through the routes at a speed, of not more than 20kmph. Sometimes, we stop for a while at the places where a lot of activity happens like fire accidents, public gatherings, tea stalls, and bus stops to capture the air quality data at those locations at a higher temporal granularity. After covering the route, we download the captured data (date, time, latitude, longitude, AQI, PM2.5, PM10) using Airveda Application through email. Note that the mobile records the GPS

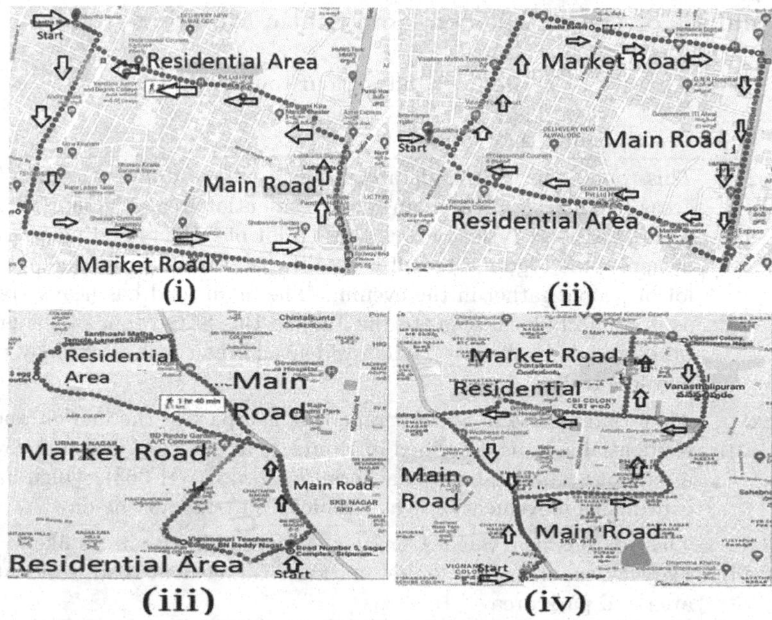

Fig. 2. Route details of four routes in Hyderabad (i) Route 1 near Secundrabad to Manchiryal highway (ii) Route 2 near Jubilee bus station (iii) Route 3 near Sagar complex (iv) Route 4 near Vanastalipuram

location, while the monitor records the AQI, PM2.5, and PM10 data because both the mobile and monitor are connected through WiFi.

Method to Map Mobile GPS Locations with the Garmin GPS Locations. Observe that the travel time for a route is different at different times we travel, due to variations in speed, stopping time, traffic, etc. The location points recorded by the Garmin GPS and mobile GPS were not the same because they are not connected (synchronization issue). We have mapped each GPS location along with PM2.5 and PM10 data captured by the mobile phone with the nearest Garmin GPS location point (reference point) in the route travel. To do this, we have mapped each location point with the nearest Garmin GPS location based on the distance function using the Haversine formula [8]. As a result, AQI, PM2.5 and PM10 values in each trip are available at the same location.

3 Results and Observations

We have reported analysis and observations for four months data collected after October 2021 for both four locations and four routes.

3.1 Analysis of Data Collected from Static Monitors

Table 3. Details of four routes

Route id	Description of the route
Route 1 (Fig. 2(i))	This route is near Secunderabad. It is a part of different types of areas like market area, main road, residential area. The market area road consists of rythu Bazar, where a lot of farmers and petty shop owners come together to sell vegetables and fruits, consequently, a lot of people gather in the evening. The main road has heavy traffic as it connects local areas to the Jubilee Bus Stand (one of the large buses stands in the city). The residential area consists of many parks and less traffic
Route 2 (Fig. 2(ii))	This route is also in Secunderabad. It is a part of the market and residential area as described in Route 1. In addition, Route 2 covers a part of Secunderabad - Macherial Highway (NH 363), which has a continuous movement of heavy vehicles throughout the day
Route 3 (Fig. 2(iii))	This route is near Sagar Complex. It covers diverse areas like highway, main road, colonies, market road, cement road, residential area and park area
Route 4 (Fig. 2(iv))	This route is in Vanastalipuram. This route is a part of areas like highways, main road to colonies, market road, junction, residential area, and vegetable/fruit market

Hourly Variation in a Day. We took the data from October 2021 to February 2022 for each location and computed averages for each corresponding hour of the day. For example, to plot data for 1:00 PM to 2:00 PM, we computed the average values of observations from 1:00 PM to 2:00 PM for each day, and then computed average of the daily average values. Figure 3(i) shows the variations in the average value of PM2.5 and Fig. 3(ii) shows the variations in the average values of PM10 during hourly time intervals of the day from October 2021 to February 2022. The graphs also show the safety level values specified by World Health Organization (WHO) and National Ambient Air Quality Standards (NAAQS).

It can be observed that PM2.5 and PM10 values are slightly high in the morning and night as compared to afternoon and evening. This is because PM2.5 and PM10 are trapped by the moisture in the air (humidity) and accumulate on the ground in the morning and night. The results also show that PM2.5 and PM10 values are high in Masjid Banda as compared to the values at other three places, and safety level values specified by both WHO and NAAQS. The likely reason is that there is a movement of a significant number of heavy vehicles and ongoing construction work. In the case of Sagar Complex and Secunderabad, these locations are semi-residential and semi-highway type areas, where a moderate amount of activity is present. These values are higher than WHO limit and close to NAAQS limit. In the case of IIIT Hyderabad, PM value is lower than that of other locations because this location is in the residential area. It can be observed that the values are higher than WHO limit and less than NAAQS limit.

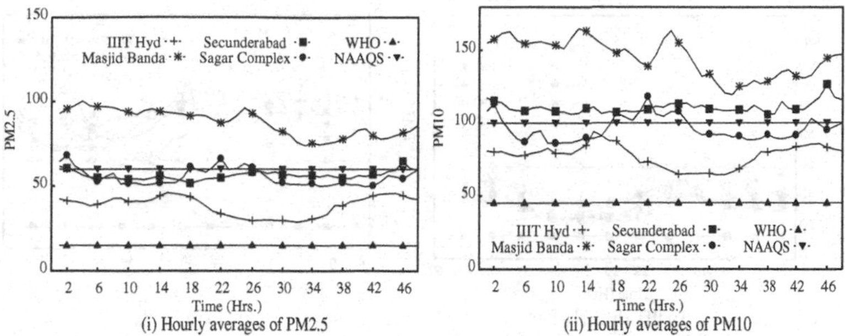

Fig. 3. Results of hourly variation in a day (In this plots, 0 indicates 12:00 AM and 23:59 indicates 11:59 PM).

Hourly Variation in the Weekdays. Figure 4 shows the variation in average value of PM2.5 and Fig. 5 shows the hourly variation of PM10 on all weeks days in four locations. For each day, we have plotted the daily average values by computing the hourly average values of PM2.5 and PM10. We have calculated average hourly values of corresponding week-day hourly values from October 2021 to February 2022. For instance, we have computed the average hourly values of all Mondays from October 2021 to February 2022.

The observation of PM2.5 trends in Fig. 4 are as follows. In the case of IIIT Hyderabad, the results show that there is not much variation in PM2.5 on working days and weekends. The likely reason is that the monitor is located in an area where a lot of software industries were located and during the Covid-19 pandemic, most of the offices had opted to work from home due to the lockdown. In the case of Masjid Banda, the results depict that the PM2.5 value is high during all working days due to ongoing infrastructural development work. Particularly on Thursday, we observe that the value of PM2.5 is high most likely because there is a market every Thursday and a lot of traffic in the market area. We also observe that the Secunderabad area has PM2.5 in the range of 70–80, which is higher than both IIIT Hyderabad and Sagar Complex. The likely reason is that Secunderabad is also a growing commercial area with infrastructural development works. However, in the case of Sagar complex monitor data, there is not much variation in PM2.5 values, because it is a sparsely occupied residential area and there is not much activity. The trend of PM10 values in Fig. 5 is similar to that of PM2.5.

Monthly Variation in the Weekdays. Figure 6 shows the monthly variation in PM2.5 values and Fig. 7 shows the monthly variation in PM10 from October 2021 to February 2022 for four monitors. For each day, we have calculated the daily average values by computing the hourly average values of PM2.5 and PM10. We have calculated the average daily values of the corresponding day in each month from October 2021 to February 2022. For instance, we have computed the average daily values of all Mondays for the given month.

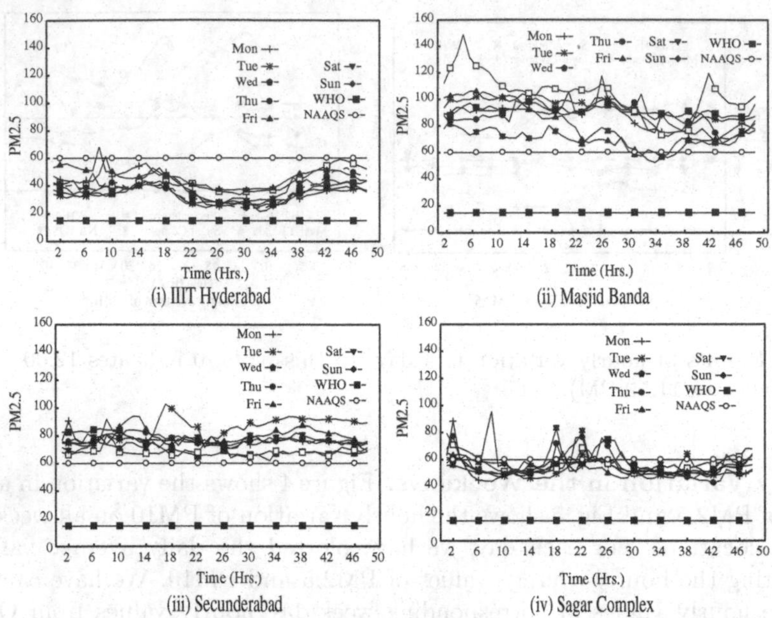

Fig. 4. Results of hourly variation of PM2.5 in week days.

The observation of PM2.5 trends in Fig. 6 is as follows. Similar to the preceding experiments, the value of PM in Masjid Banda is very high as compared to other locations. The PM2.5 values in Masjid Banda in December and January months are lower due to Covid-19 lockdown restrictions. In addition, we can also observe that there is not much difference in PM2.5 value in weekdays at all the four locations. From Fig. 6 we observe that the value of PM2.5 is high in December as compared to other months. It shows that the averages values of PM2.5 are high in November and December as compared to other months. One of the possible reason is due to the trapping of PM by the moisture in the air (snow) and the accumulation of PM on the ground due to heavyweight. The trend of PM10 values in Fig. 7 are similar to as that of PM2.5.

Variations in the Weekdays. Figure 8(i) shows the variation in average PM2.5 and Fig. 8(ii) shows the variation in average PM10 values in week days in all four locations. For this experiment, for each day, we have calculated the daily average values by computing the hourly average values of PM2.5 and PM10. We have calculated the average daily values of the corresponding day for all months from October 2021 to February 2022. For instance, we have computed the average daily values of all Mondays from October 2021 to February 2022.

The observation of PM2.5 trends in Fig. 8(i) is as follows. From the results, we observe that there is not much variation in the average value of PM2.5 on weekdays. We also observe that the value of PM2.5 in Masjid Banda is much

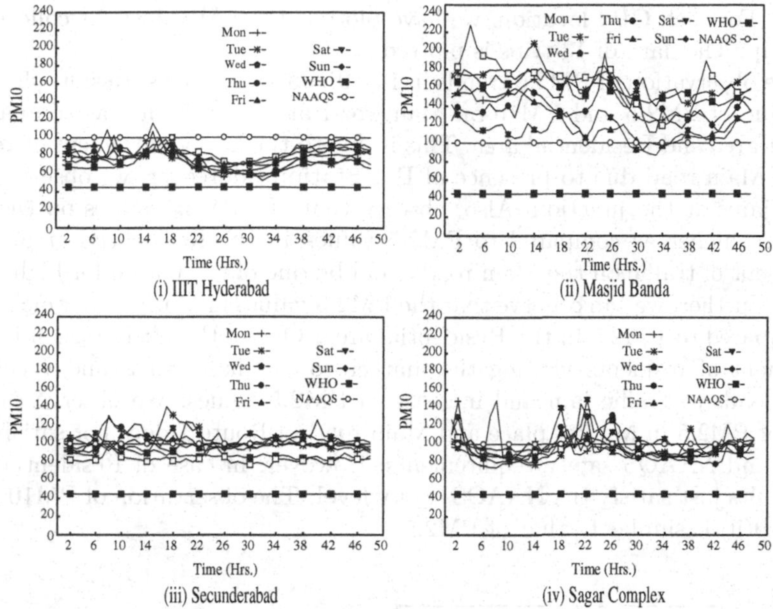

Fig. 5. Results of hourly variation of PM10 in week days.

higher compared to PM2.5 values in other locations. Moreover, we observe that the PM10 values in Masjid Banda and Secunderabad are comparable because both are rapidly developing areas with a lot of construction works. Observe that all the values of PM2.5 at four locations in Hyderabad are much above WHO safety level. However, it can be observed that PM2.5 values at both IIIT Hyderabad and Sagar Complex locations are below the NAAQS safety level. The observation of PM10 trends in Fig. 8(ii) is similar to that of PM2.5.

3.2 Analysis of Data Collected from the Selected Routes

We present the analysis of air quality data collected from selected four routes (refer Fig. 1 and Table 3). We have travelled once in a week in all four routes and captured location, PM2.5 and PM10 data from September 2021 to February 2022. Based on the location data, we have classified each route into three categories, namely, Market road, Main road, and Residential area. We now present the analysis of PM2.5 and PM10 values for Route 1 and Route 2. In order to plot the PM2.5 and PM10 values at each location, we have mapped each location (latitude, longitude) to a unique number starting from 1.

Analysis of Air Quality Data in Route 1. Figure 9(i) depicts the variation in PM2.5 and Fig. 9(ii) depicts the variation in PM10 in all the location points in Route 1. For this experiment, we mapped GPS locations to integers starting from 1. For each GPS location, we have plotted the PM values for collected for each trip. The data of 13 trips is plotted.

The observation of PM2.5 trends in Fig. 9(i) are as follows. Results show that the value of PM2.5 and PM10 on Main road are much higher as compared to Market area and Residential area. This is due to the heavy movement of vehicles on the Main road due to presence of Bus Station. Moreover, we observe heavy traffic jams at the junction. Also, observe that the PM2.5 values on the Main road are higher as compared to PM2.5 values in the Residential area. Heavy movement of traffic on the Main road could be one of the reason for high PM2.5 values. Further, we can observe that the PM2.5 values in Market are much higher as compared to PM2.5 in the Residential area. One of the likely reason is heavy movement of residents visiting the market to purchase fruits and vegetables, which leads to traffic jam and increase in PM2.5 values. We observe that the value of PM2.5 in Market place and Main road of Route 1 does not satisfy both WHO and NAAQS safety requirements. However, in case of Residential area, the results just satisfy the NAAQS safety level. The observation of PM10 trends in Fig. 9(ii) is similar to that of PM2.5.

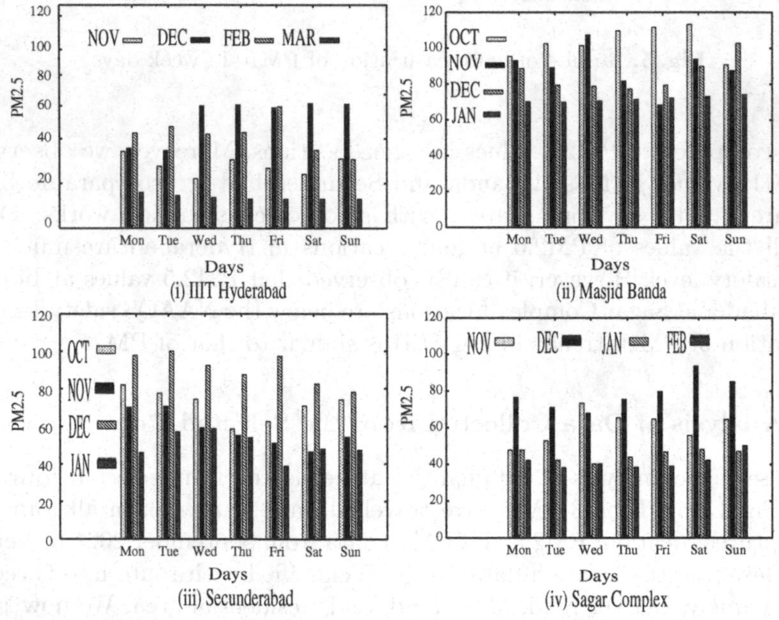

Fig. 6. Results of monthly variation of PM2.5 on all week days.

Analysis of Air Quality Data in Route 2. We have categorised Route 2 into Market area, Main road and Residential area. Figure 10(i) shows the variation in PM2.5 values in Route 2. The trend is similar to PM2.5 trend of Route 1. However, PM2.5 values in the Main Road of Route 1 are much higher as

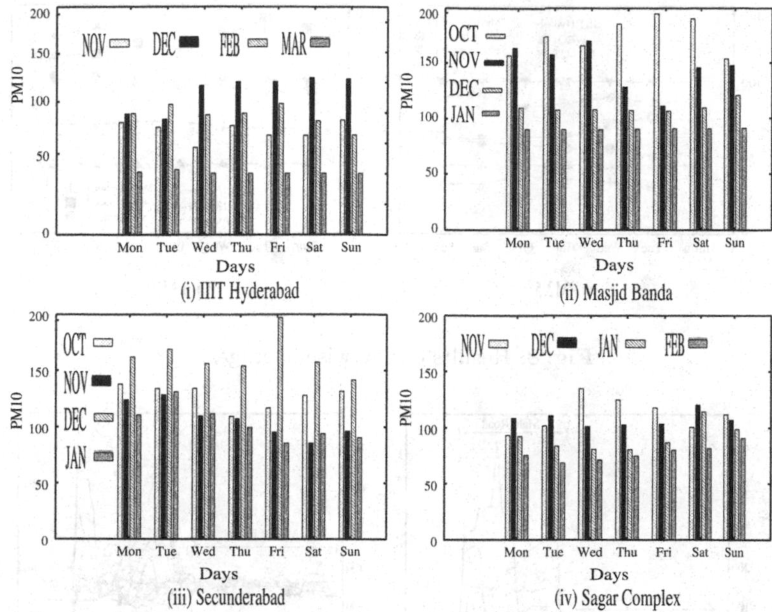

Fig. 7. Results of monthly variation of PM10 on all week days.

compared to the main road in Route 2. The likely reason being that the Main road in Route 1 is the part of National Highway. It can be observed that PM2.5 values in the Market place and the Main road of Route 2 does not satisfy both WHO and NAAQS safety limit. However, in case of Residential area, that values just satisfy the NAAQS limit. The observation of PM10 trends in Fig. 10(ii) is similar to that of PM2.5.

Analysis of Air Quality Data in Route 3 and 4. From the data analysis on Route 3 and Route 4, we have observed that the monitor which is used to collect the location, PM2.5 and PM10 was not sensing the correct values. So, we have not presented the results.

4 Implementation Issues

We discuss some implementation issues faced during the experiment.

- **Selection of locations to place monitors:** Some aspects that needed to be considered for each identified location were continuous power supply, Internet connectivity, and physical security of the monitor. The location should also need to be exposed to the ambient air. Suitable mounting arrangements and power plug-in sockets have to be made for fixing the sensors. In this project, we have placed the sensors on balconies in residential areas. When the number of sensors increases, restricting the sensor placement in the residential houses

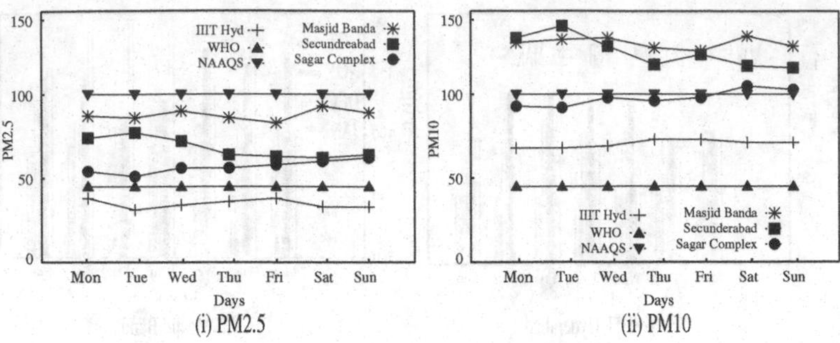

Fig. 8. Results of day-wise averages.

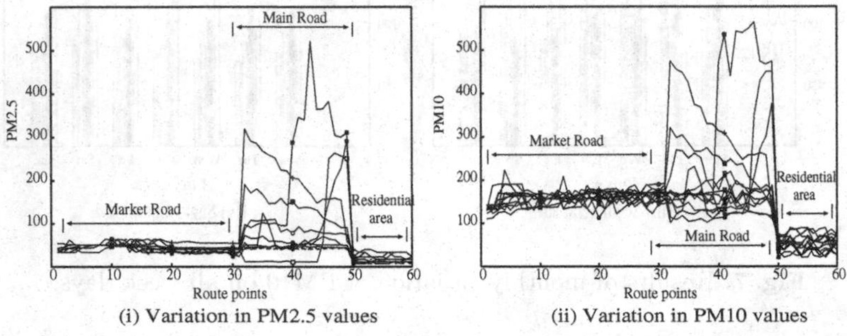

Fig. 9. Results of PM values of 13 trips for Route 1

may not be enough, and identifying appropriate locations in the target area is an implementation issue.

- **Capturing air quality with portable sensors:** After selecting the area and the route, the issue is to capture the air quality data using the portable monitor. We have to initialize the monitor before starting the trip. Also, longitude and latitude values have to be collected using Garmin GPS. The important requirement is the continuous internet connectivity and so the mobile phone with the hot-spot feature is essential. It has been observed that whenever there is an interruption, the GPS location could not be mapped and mapping air quality data becomes cumbersome. In this experiment, we have selected routes where an uninterrupted Internet connection is available through a mobile phone. On routes where Internet connection though the mobile phone is not available, using a GPS device is necessary.
- **Maintenance of Monitors:** Even though low-cost sensors are checked and calibrated at the manufacturer's laboratory, the user needs to check the calibration regularly, because the sensors are not particularly robust. We have carried out calibration checks by placing the monitor's side by side and checking whether all monitors reported the same value. The checking was done approximately every three weeks. If any of the monitors reported inconsistent values,

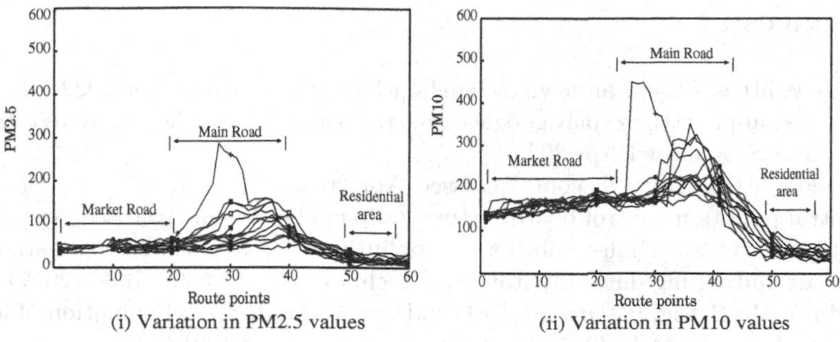

(i) Variation in PM2.5 values (ii) Variation in PM10 values

Fig. 10. Results of PM values of 13 trips for Route 2

it was sent back to the manufacturer for repair and re-calibration. Backup monitors should be available to avoid gaps in the data.

5 Conclusion

Long-term exposure to high PM can cause severe health problems. The problem is very serious in developing countries such as India. Cost-effective monitoring of air quality is an important research issue. Low-cost air quality sensors provide a cost-effective tool for monitoring air quality in areas beyond metro cities. As a part of the ongoing research effort to develop a framework for low-cost air quality data collection, we have collected air quality data by deploying monitors at four fixed locations and mobile monitors along four routes to collect data, which is representative of the range of urban forms in Hyderabad, India.

We have reported the initial experiences and observations of PM data collection for four months, from mid-October 2021 to mid-February 2022. From the analysis, we have observed that the values of PM2.5 and PM10 are generally high in neighborhoods where there is a lot of construction activity and heavy movement of vehicles. The values of PM2.5 and PM10 throughout the day at all four locations in Hyderabad are high. These values are significantly higher than the threshold value prescribed by WHO and in some cases exceed higher limits allowed by NAAQS. The initial results are interesting. These experiences and observations provide the impetus to continue our efforts to develop a cost-effective and robust air quality data collection framework.

Acknowledgements. The funding for this work is provided by the Mphasis laboratory for Machine Learning and Computational Thinking (ML2CT), Ashoka University, Sonipat, Haryana, India. Support is also provided by IIIT Hyderabad, India, and O.P.Jindal Global University, Sonipat, Haryana, India.

References

1. Airnow. https://www.airnow.gov/aqi/aqi-basics/. Accessed Apr 2022
2. Airveda app. https://play.google.com/store/apps/details?id=in.airveda&hl=en_IN&gl=US. Accessed Apr 2022
3. Airveda. https://airveda.com. Accessed Apr 2022
4. Central pollution control board. https://app.cpcbccr.com. Accessed Apr 2022
5. Apte, J.S., et al.: High-resolution air pollution mapping with google street view cars: exploiting big data. Environ. Sci. Technol. **51**(12), 6999–7008 (2017)
6. Badura, M., Batog, P., Drzeniecka-Osiadacz, A., Modzel, P.: Evaluation of low-cost sensors for ambient PM2. 5 monitoring. J. Sens. **2018** (2018)
7. Brienza, S., Galli, A., Anastasi, G., Bruschi, P.: A low-cost sensing system for cooperative air quality monitoring in urban areas. Sensors **15**(6), 12242–12259 (2015)
8. Chopde, N.R., Nichat, M.: Landmark based shortest path detection by using A* and haversine formula. Int. J. Innov. Res. Comput. Commun. Eng. **1**(2), 298–302 (2013)
9. Jacobson, M.Z.: Review of solutions to global warming, air pollution, and energy security. Energy Environ. Sci. **2**(2), 148–173 (2009)
10. Kumar, P., et al.: The nexus between air pollution, green infrastructure and human health. Environ. Int. **133**, 105181 (2019)
11. Lagerspetz, E., et al.: Megasense: feasibility of low-cost sensors for pollution hot-spot detection. In: IEEE 17th International Conference on Industrial Informatics, pp. 1083–1090 (2019)
12. Landrigan, P.J., et al.: The lancet commission on pollution and health. The Lancet **391**(10119), 462–512 (2018)
13. Lewis, A., Peltier, W.R., von Schneidemesser, E.: Low-cost sensors for the measurement of atmospheric composition: overview of topic and future applications. World Meteorological Organization (WMO No 1215) (2018)
14. Mahajan, S., Kumar, P.: Evaluation of low-cost sensors for quantitative personal exposure monitoring. Sustain. Urban Areas **57**, 102076 (2020)
15. Qing, W.: Urbanization and global health: the role of air pollution. Iran. J. Public Health **47**(11), 1644 (2018)
16. Rai, A.C., et al.: End-user perspective of low-cost sensors for outdoor air pollution monitoring. Sci. Total Environ. **607**, 691–705 (2017)
17. Reddy, C.R., et al.: Improving spatio-temporal understanding of particulate matter using low-cost IoT sensors. In: IEEE Annual International Symposium on Personal, Indoor and Mobile Radio Communications, pp. 1–7 (2020)
18. Saini, P., Sharma, M.: Cause and age-specific premature mortality attributable to PM2.5 exposure: an analysis for million-plus Indian cities. Sci. Total Environ. **710**, 135230 (2020)
19. WHO: Health aspects of air pollution with particulate matter, ozone and nitrogen dioxide: report on a WHO working group, Bonn, Germany. Technical report, WHO Regional Office for Europe, Copenhagen (2003)

Visualizing Spatio-temporal Variation of Ambient Air Pollution in Four Small Towns in India

Girish Agrawal[1(✉)], Hifzur Rahman[1], Anirban Mondal[2], and P. Krishna Reddy[3]

[1] O.P. Jindal Global University, Sonipat-Narela Road, Sonipat, Haryana 131001, India
gagrawal@jgu.edu.in

[2] Ashoka University, Rajiv Gandhi Education City, P.O. Rai Sonipat, Sonipat, Harayana 131029, India

[3] IIIT Hyderabad, Gachibowli, Hyderabad, Telangana 500032, India

Abstract. Air pollution is a major threat to human health in India. More than three-quarters of the people in India are exposed to pollution levels higher than the limits recommended by the National Ambient Air Quality Standards in India and significantly higher than those recommended by the World Health Organization. Despite the poor air quality, the monitoring of air pollution levels is limited even in large urban areas in India and virtually absent in small towns and rural areas. The lack of data results in a minimal understanding of spatial and temporal patterns of air pollutants at local and regional levels. This paper is the second in a planned series of papers presenting particulate air pollution trends monitored in small cities and towns in India. The findings presented here are important for framing state and regional level policies for addressing air pollution problems in urban areas, and achieve the sustainable development goals (SDGs) linked to public health, reduction in the adverse environmental impact of cities, and adaptation to climate change, as indicated by SDGs 3.9, 11.6 and 11.b.

Keywords: Ambient air pollution · Air pollution in small towns · Diurnal and weekly cycle of PM pollution · PM isopleths · Spatio-temporal averaging of data from mobile sensors · Sustainable development goals

1 Introduction

Exposure to ambient air pollution is a major threat to human health. Air pollution is caused by many factors such as increasing urbanization, industrial pollution, traffic emissions, agriculture, and energy usage [1]. Lim et al. [2] reported the significant effect of air pollution on global mortality. The 2017 data from the Global Burden of Disease study [3] provide new evidence regarding the significant effects of air pollution globally, placing it among the top ten risks confronted by human beings. Most cities worldwide cannot comply with the pollutant standards and have reported measurements that far exceed them, resulting in millions of premature deaths [1]. At the forefront of pollutants which exceed concentration limits are coarse and fine particulate matter (PM), defined

U. K. Rage et al. (Eds.): DASFAA 2022 Workshops, LNCS 13248, pp. 417–436, 2022.
https://doi.org/10.1007/978-3-031-11217-1_30

as particles with a nominal average diameter less than 10 μm (PM10) and 2.5 μm (PM2.5), respectively. The World Health Organization (WHO) report regarding ambient air pollution suggests that the annual mean concentration of PM2.5 or PM10 increased by more than 10% between 2010 and 2016 in at least 280 cities worldwide [4].

India has one of the highest annual average ambient particulate matter exposure levels in the world. Almost the entire country's population resides in areas that exceed the WHO Air Quality Guidelines, and the majority of the population resides in areas where even the less stringent limits set by the Indian National Ambient Air Quality Standard (NAAQS) [5] for PM are exceeded [6]. Air quality modeling by WHO indicates that the median exposure to PM2.5 in India is 66 $\mu g/m^3$ with lower and upper bounds of 45 $\mu g/m^3$ and 97 $\mu g/m^3$ [7].

Despite the poor air quality, the monitoring of air pollution levels is limited even in large urban areas in India and virtually absent in small towns and rural areas. The Central Pollution Control Board of the Government of India and its companion state-level boards currently maintain just under 350 ambient air quality monitoring stations (AAQMS). These numbers are insignificant for a country with a land area of 3.3 million km^2 and a population of over 1.3 billion, 34% of whom live in urban areas. Even on this sparse network, the availability of PM2.5 data is extremely limited as the National Air Quality Monitoring Program monitored only sulfur dioxide, nitrogen dioxide, and PM10 data before 2015. Real-time high-resolution pollutant concentration maps do not exist currently because they require a large amount of data, computing facilities, and high costs. This lack of data leads to a minimal understanding of spatial patterns of air pollutants at the local as well as regional levels, and hampers the ability of planners and administrators to assess the impact of interventions on air quality designed to meet SDG 11 requirements to "reduce the adverse per capita environmental impact of cities, including by paying special attention to air quality …."

Most of the air pollution data in India are from cities with populations greater than one million. According to the 2011 census of India [8], about 230 million of India's urban population lives in towns and cities with populations less than one million. The four cities selected for the present study, Darbhanga in Bihar, located in the Eastern Plains physiographic division of India, and Bhilai, Rajnandgaon, and Kanker in Chhattisgarh, located in the Eastern Plateau physiographic division of India. Darbhanga, Bhilai, Rajnandgaon, and Kanker have populations of about 400,000, 600,000, <200,000, and <50,000, respectively, per the 2011 census of India [8]. Ambient air pollution data for these cities are extremely sparse and limited. In recent years, relatively low-cost monitors have become available for measuring ambient particulate concentrations. For a duration of approximately 6 months, i.e., from early September 2021 through late February 2022 (the months with the worst pollution in north and central India), we used these low-cost monitors to measure particulate levels in the above mentioned four cities.

Lack of air quality studies for urban areas other than large metropolitan areas is a global problem, particularly in low to middle-income countries (LMICs). In LMICs, almost all air quality studies deal with large or metropolitan cities [9, 10]. There are hardly any studies from Asia and Africa dealing with small cities [11, 12]. Most air quality studies come from high-income countries (HICs), and recently from China [13]. Even in HICs, the most of the studies deal in detail with large cities [14–16].

The work reported here is an early step of a larger project with two linked goals: to provide the general public hyper-local air quality information, and to provide local administrators and planners the capability to evaluate the impact of interventions designed to improve air quality, and attain sustainable development goals in line with India's Smart City program goals. The cost for individual cities and towns to install and maintain a conventional air quality monitoring and assessment program runs into tens of millions of rupees, and requires having staff with specialized training. One of the goals of this project is to demonstrate that air quality assessment in small cities can be done with low-cost sensors, and so serve as the first step in developing and providing relatively inexpensive solutions for cities in LMICs to achieve SDG goals to substantially reduce the number of deaths and illnesses from air pollution (SDG 3.9), reduce the adverse per capita environmental impact of cities (SDG 11.6), and improve their ability to mitigate and adapt to climate change (SDG 11.b).

We report here on an initial step in generating pollutant concentrations maps – limited here to particulate matter concentrations – based on sparse, spatio-temporally varying data generated using a combination of mobile and static air quality monitors.

2 Methodology

2.1 Use of Low-Cost Air Quality Monitors

Our solution to overcome the lack of high-precision measurements of air quality is to adopt low-cost methods for robust pollution monitoring. Although these methods tend to yield lower quality data, they can be used in a significant number of locations simultaneously, thereby enabling the high-resolution assessment mapping of city pollution. Recent work by Genikomsakis and colleagues [17] has demonstrated that low-cost sensors work very well for collecting fine-grained spatio-temporal [particulate matter concentration] profiles in urban areas.

Compared with analytical instruments for measuring air pollutants, the sensors used in this study are less expensive and easier to deploy, operate, and manage. Retrieving data from the sensors is straightforward, and their automatic operation enables a widespread deployment. Data collected from the sensors can be managed, processed, and analyzed centrally, as well as shared with all the stakeholders.

The particular monitors used for collecting the data reported herein are commercial units that use such low-cost sensors. The monitors range in price from Rs. 15,000 to Rs. 40,000 ($\approx$USD 200 to 500), and have an operational and maintenance cost of about Rs. 5,000 for roughly 200 days of operation per year. The portable units can be charged using any domestic power source and used for 4–6 h before they require recharge. The static units can be mounted outdoors provided that a power source is available for the continuous measurement of ambient PM2.5 and PM10 concentrations. The technical specifications of these monitors are as follows:

- Measurement parameters: PM2.5, PM10 in μg/m^3
- Range of PM2.5: 0–999 μg/m^3
- Range of PM10: 0–1999 μg/m^3
- Minimum resolution of <0.3 μm

- Relative error Maximum of ±10% and ±10 μg/m^3
- Power voltage: 220 V
- 3000 mAh rechargeable battery. 6–8 h of battery backup
- Tested and Calibrated using a Beta attenuation monitor (BAM)

The operating principle of the sensors in these monitors is based on laser light scattering. A small fan draws in a stream of ambient air passing through a detection chamber. The PM in the air scatters the coherent light from the laser. A photodetector converts the scattered light into an electrical signal, which is then amplified and processed. The processing entails using existing correlations to calculate particulate concentrations in real time. The monitors were calibrated against a Beta Attenuation Monitor (BAM) by the vendor.

2.2 Locations of Air Quality Monitors With in Each Study Area

We installed six particulate concentration monitors at fixed locations in selected areas in the four study cities. Two monitors each were installed in Kanker and Darbhanga, and one each in Rajnandgaon and Bhilai. Portable monitors were used in each city to collect particulate matter concentration data along routes selected to pass through a range of urban morphologies: high-traffic, commercial zones; industrial zones; factory worker townships; peripheral, low-traffic, residential zones; and market areas with a relatively low volume of motorized traffic. Figure 1 shows the locations of the four cities on a map of India. Figures 2, 3, 4 and 5 show the location of each static monitor and the routes covered by portable monitors on the maps of the four cities.

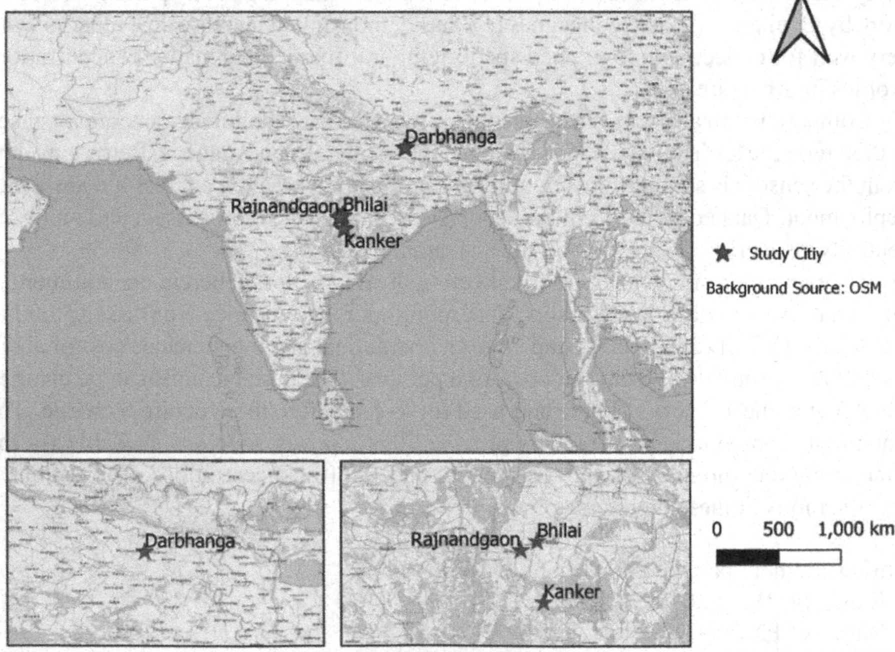

Fig. 1. Locations of Darbhanga, Rajnandgaon, Bhilai, and Kanker

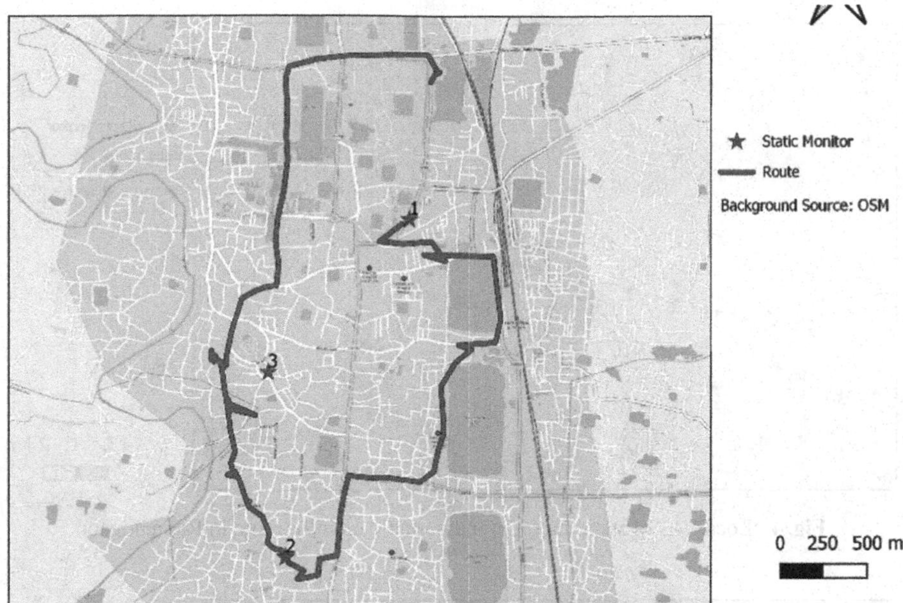

Fig. 2. Locations of static monitors and portable monitor route in Darbhanga

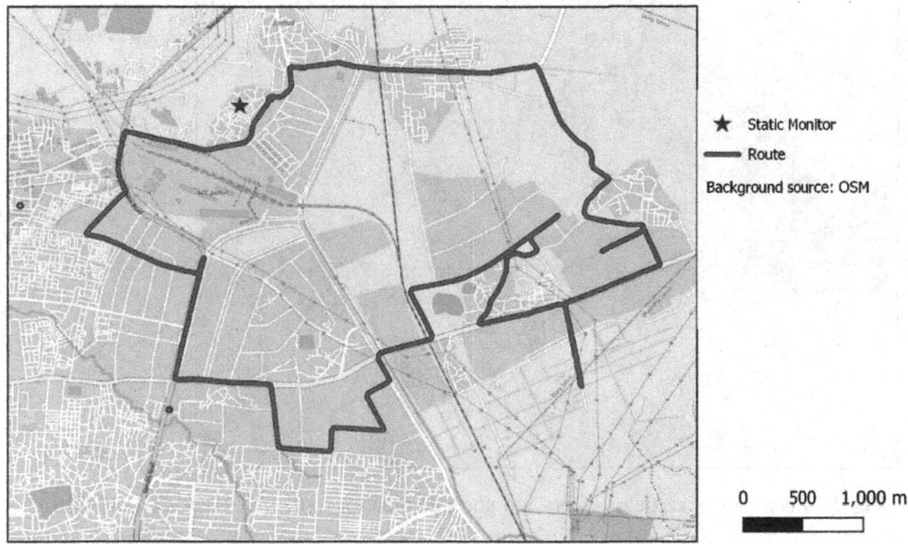

Fig. 3. Locations of static monitor and portable monitor route in Jamul area of Bhilai

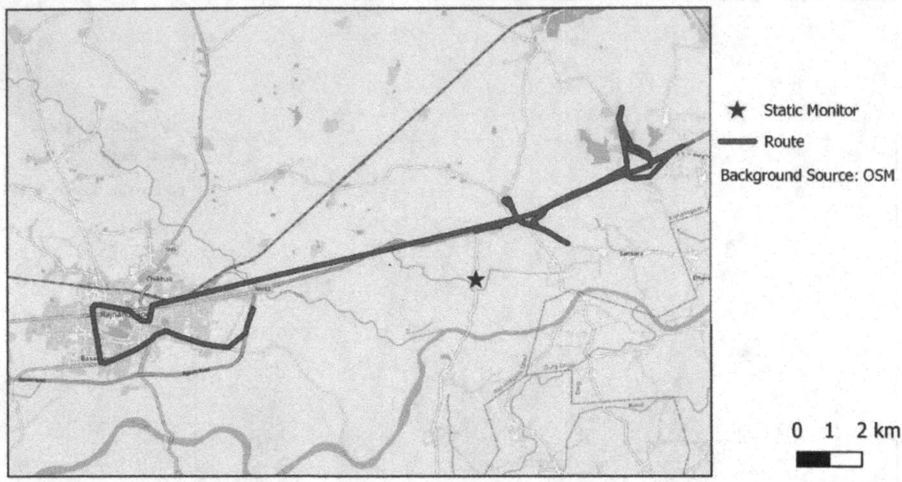

Fig. 4. Locations of static monitor and portable monitor route in Rajnandgaon

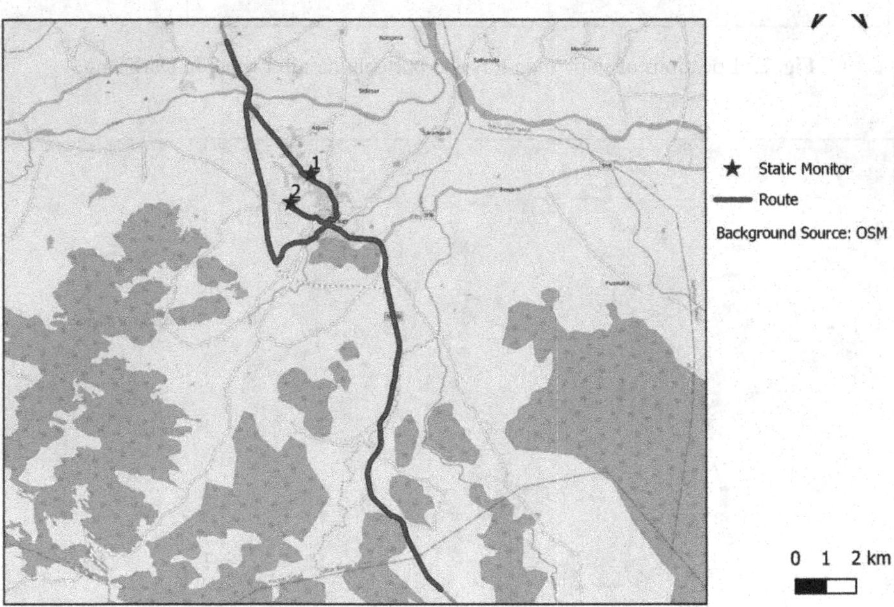

Fig. 5. Locations of static monitors and portable monitor route in Kanker city and district

3 Results and Discussion

3.1 Overview of Pollution Data

Our analysis and visualization was based on data collected from September 7, 2021 to February 28, 2022. The monitors were set to record a reading every minute and then upload the data to the server every 10 min. The data were averaged for varying durations, depending on visualization needs.

3.2 Daily Mean PM Variation in the Four Cities

To visually understand the daily changes in PM concentrations in ambient air, the daily mean data for the relevant areas in Darbhanga, Bhilai, Rajnandgaon, and Kanker is presented in Figs. 6, 7, 8 and 9, respectively. All four study areas exhibited similar rise and fall patterns for the PM2.5 and PM10 concentrations. This indicates that, for cities located in the Eastern Plains and Eastern Plateau physiographic divisions of India, a regional pattern of particulate pollution exists that dominates the local environment, at least for the winter and early spring seasons during which the data reported herein were collected. This pattern is consistent with the data collected from the monitors located in multiple locations in the National Capital Region of Delhi [18]. For almost all of the 180 ± 5 days of available data for Darbhanga and Bhilai, the PM pollution levels exceeded the NAAQS safe limits, whereas for Rajnandgaon and Kanker, the particulate concentration levels were above the safe limits for a period of about 100 days from early October 2021 to mid-January 2022. This is likely attributable to the fact that the first two cities are in heavily industrialized regions, whereas, Rajnandgaon, and more so Kanker, are in areas with light industry and a high forest cover.

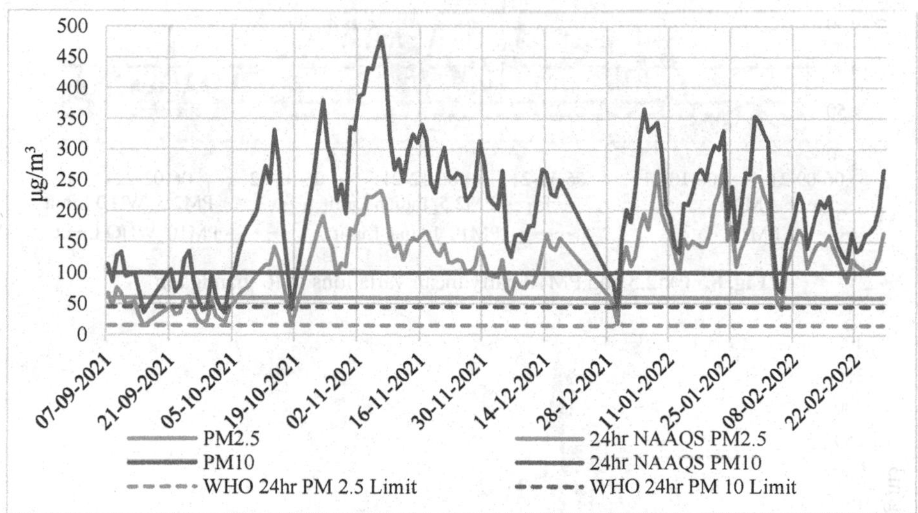

Fig. 6. PM2.5 and PM10 daily mean variations in Darbhanga

3.3 Monthly Mean PM Variation in the Four Cities

As another step in assessing the temporal variation of particulate matter concentrations in ambient air in small cities, the air quality data for the study period was averaged for each month and plotted. Monthly averages damp out the daily variations and give a better

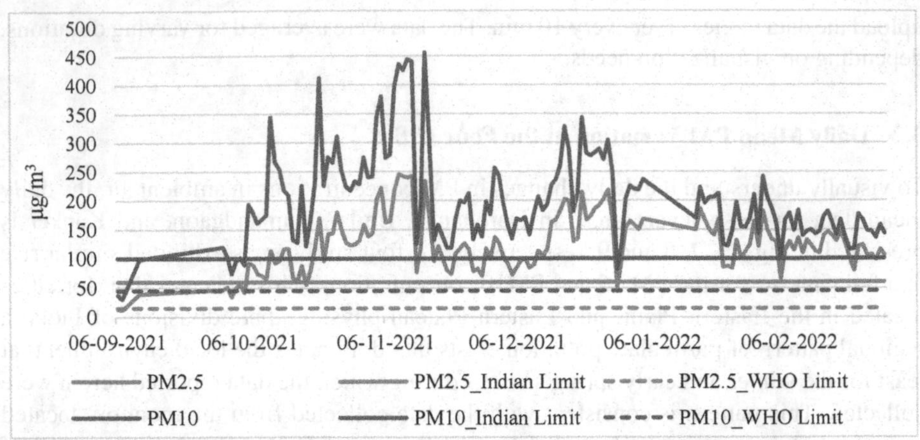

Fig. 7. PM2.5 and PM10 daily mean variations in Jamul area of Bhilai

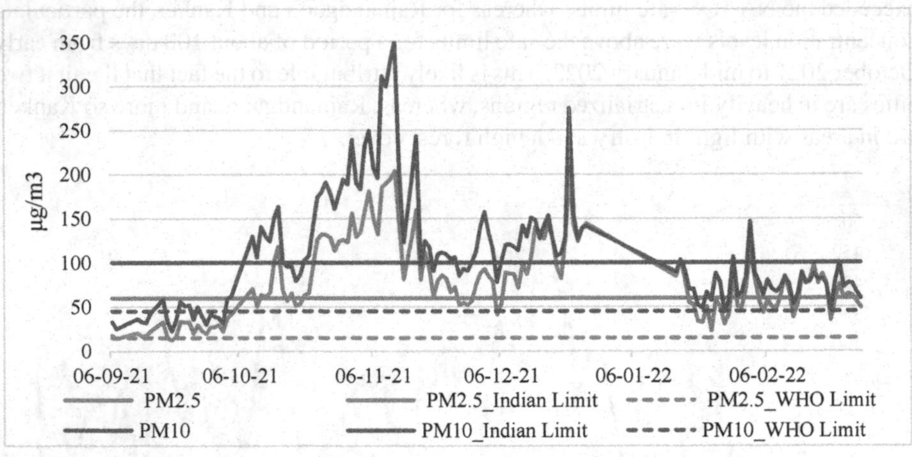

Fig. 8. PM2.5 and PM10 daily mean variations in Rajnandgaon

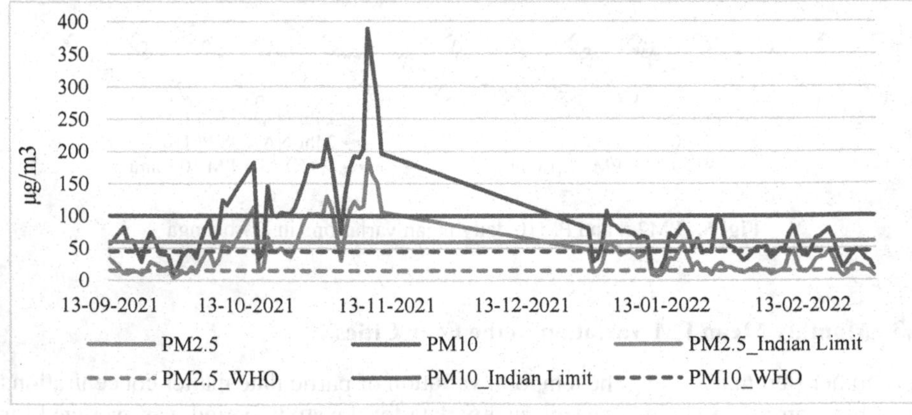

Fig. 9. PM2.5 and PM10 daily mean variations in Kanker

picture of the medium-term exposure to pollution. Figures 10, 11, 12 and 13 show the monthly means for PM2.5 and PM10, respectively.

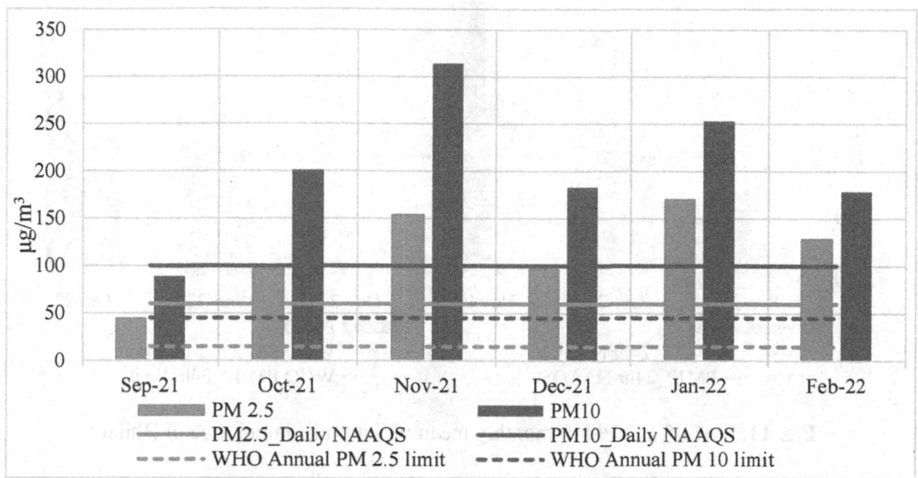

Fig. 10. PM2.5 and PM10 monthly mean variation in Darbhanga

3.4 Diurnal PM Variation in the Four Cities

Plots of diurnal variation allow us to determine the daily pattern of PM concentration variation in ambient air. To plot the diurnal variation, the recorded data for each half-hour for each day were averaged over the entire observation period, and are graphically depicted in Figs. 14, 15, 16 and 17 in terms of PM2.5 and PM10, respectively. The diurnal variation of particulate concentrations exhibit a similar cycle in three of the four study areas, with the exception of Kanker. The PM concentration levels reach a maximum in the evenings, from approximately 6 pm to 9 pm, and the minimum in the afternoons, between 1 pm and 4 pm. Although a similar pattern is also seen in Kanker, the variation is not so pronounced. The diurnal data plots show that the PM levels in the ambient air were consistently higher than the WHO daily average exposure safe limits, and even higher than the somewhat lax Indian NAAQS limits in three of the cties, again with the exception of Kanker, where the concentrations, although higher than the WHO limits, are within the NAAQS limits for most of the day.

That the PM2.5 and PM10 levels were high in the first half of the day and lower in the second half suggests that the particulate pollutants descended at night when the temperature was lower and rose as the temperature increased during the day, reaching a peak at mid-afternoon. This may be because a large proportion of the PM2.5 particles are volatile substances that tend to break down and rise as the ambient air temperatures increases. However, our current information regarding the constituents of the PM is insufficient to confirm this conjecture.

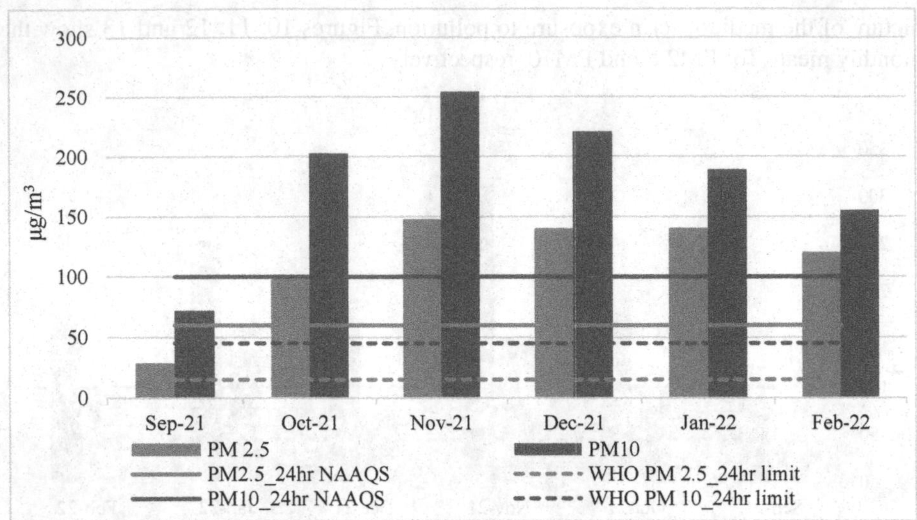

Fig. 11. PM2.5 and PM10 monthly mean variation in Jamul area of Bhilai

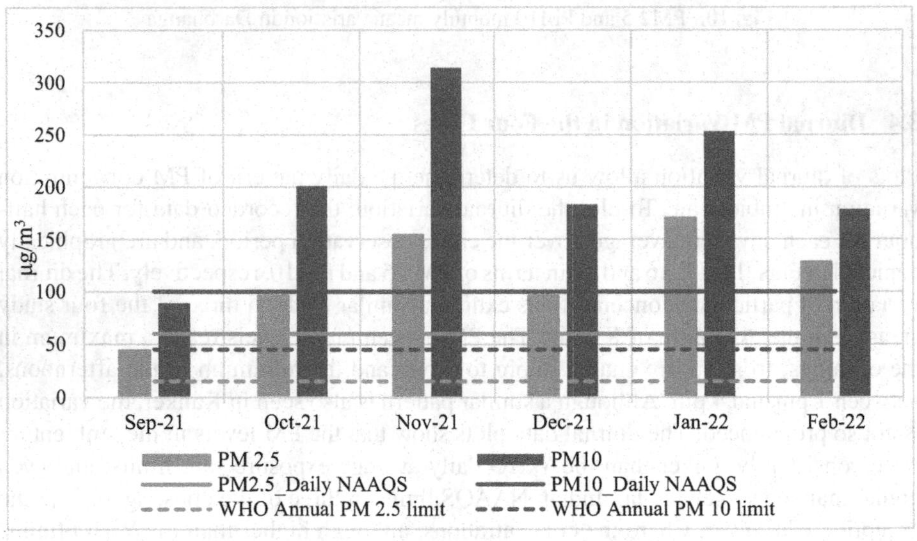

Fig. 12. PM2.5 and PM10 monthly mean variation in Rajnandgaon

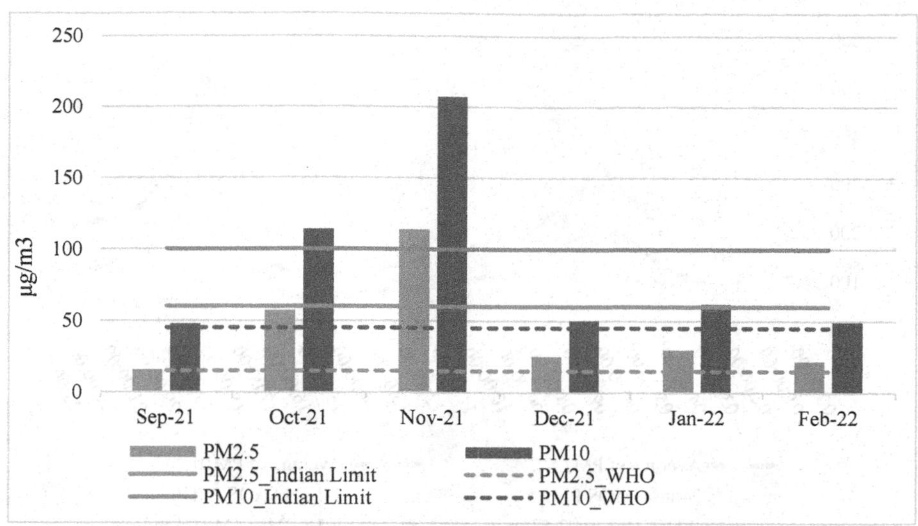

Fig. 13. PM2.5 and PM10 monthly mean variation in Kanker

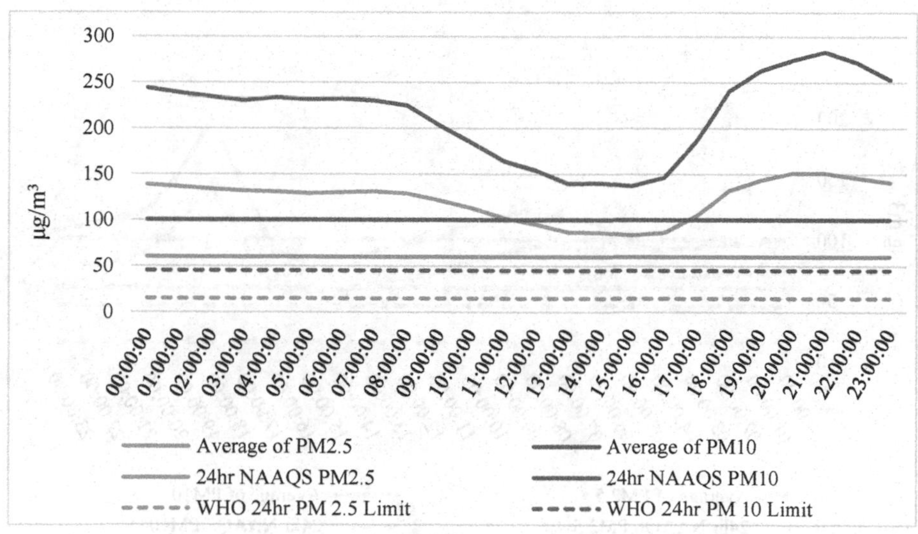

Fig. 14. Diurnal PM2.5 and PM10 variation in Darbhanga

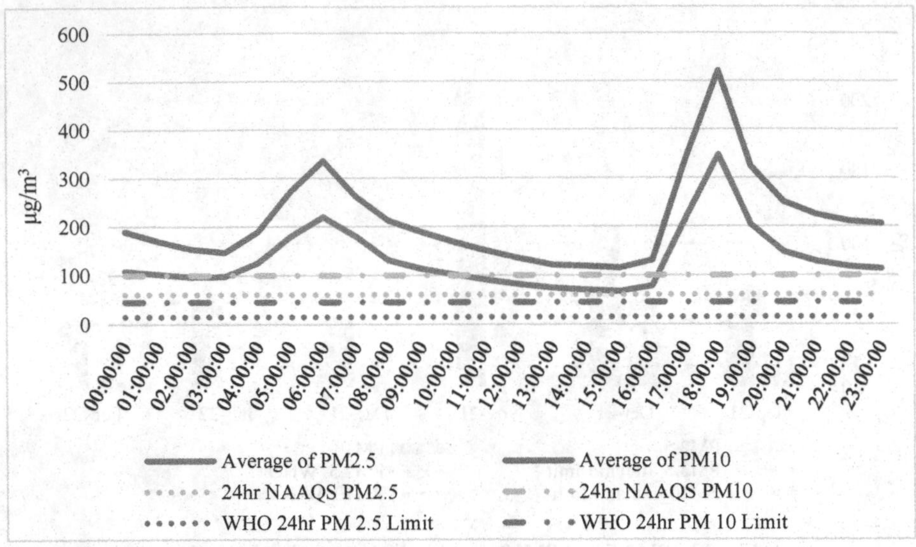

Fig. 15. Diurnal PM2.5 and PM10 variation in Jamul area of Bhilai

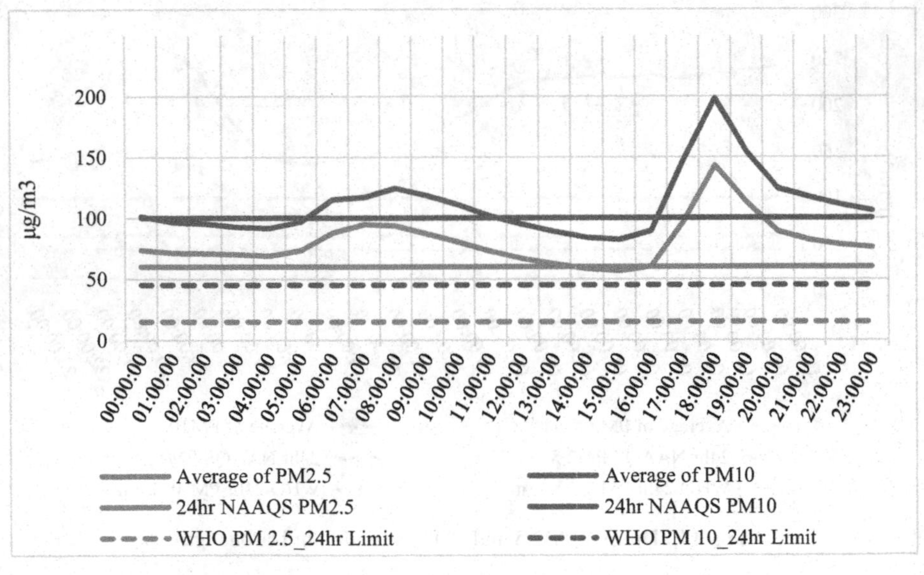

Fig. 16. Diurnal PM2.5 and PM10 variation in Rajnandgaon

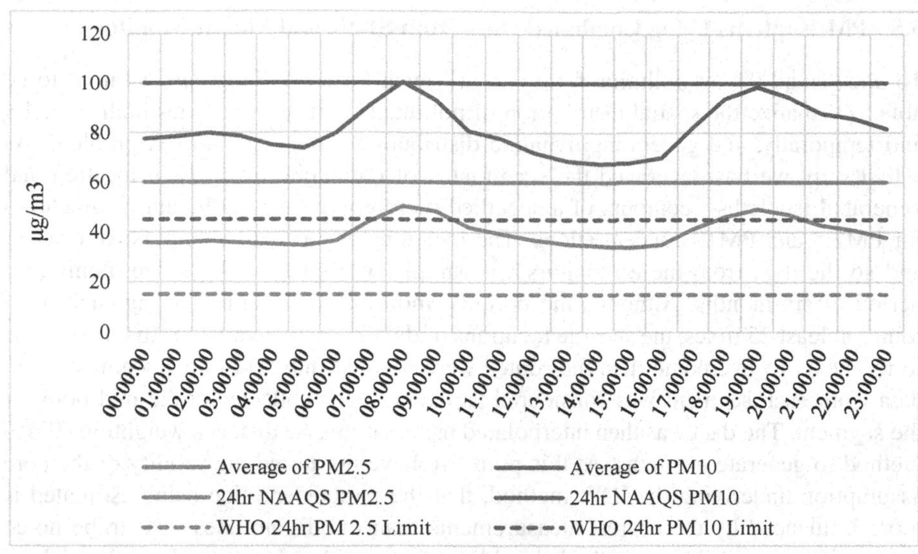

Fig. 17. Diurnal PM2.5 and PM10 variation in Kanker

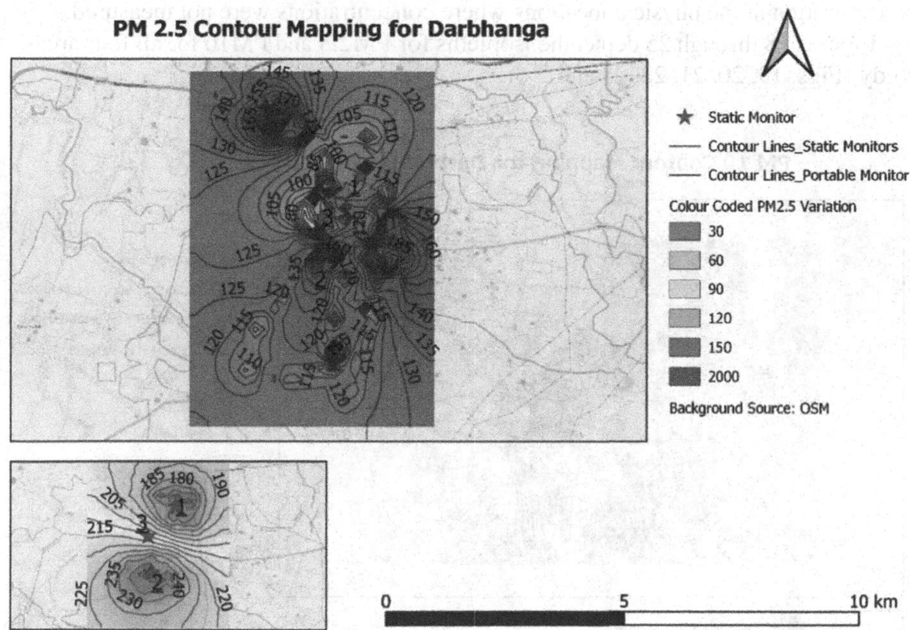

Fig. 18. Isopleths for PM2.5 data from Darbhanga

3.5 PM Isopleths Using Combined Data from Static and Mobile Monitors

To understand the air pollution for a region—even for a small region—we need to be able to visualize the spatial distribution of pollutants. Air quality varies both spatially and temporally, and generating dynamic distribution plots is a complex problem. As a first step, we have averaged the six months of data collected at each location and generated isopleths—contours of a specified meteorological or pollutant parameter—for PM2.5 and PM10 concentrations. The locations of the static monitors were fixed, and so the data from these monitors was simply averaged for the entire monitoring period of six months. Although the portable monitors were carried along each fixed routes at least 25 times, the specific locations of data collection varied from one traverse to the next. To overcome this, the routes were divided into 50-m long segments, and data from each segment was considered as having been collected at the mid-point of the segment. The data was then interpolated using an inverse distance weighting (IDW) method to generate isopleths. At this point we have not tested the validity of the core assumption underlying the IDW method, that the parameter value being estimated is more influenced by the nearest measurements than the distant ones. It is to be noted that an IDW interpolation method should not be considered as providing a model for pollutant distribution because it does not describe the data, and there is no underlying statistic to estimate the uncertainty associated with the prediction of particulate matter concentration at the physical locations where concentrations were not measured.

Figures 18 through 25 depict the isopleths for PM2.5 and PM10 for all four areas of study (Figs. 19, 20, 21, 22, 23 and 24).

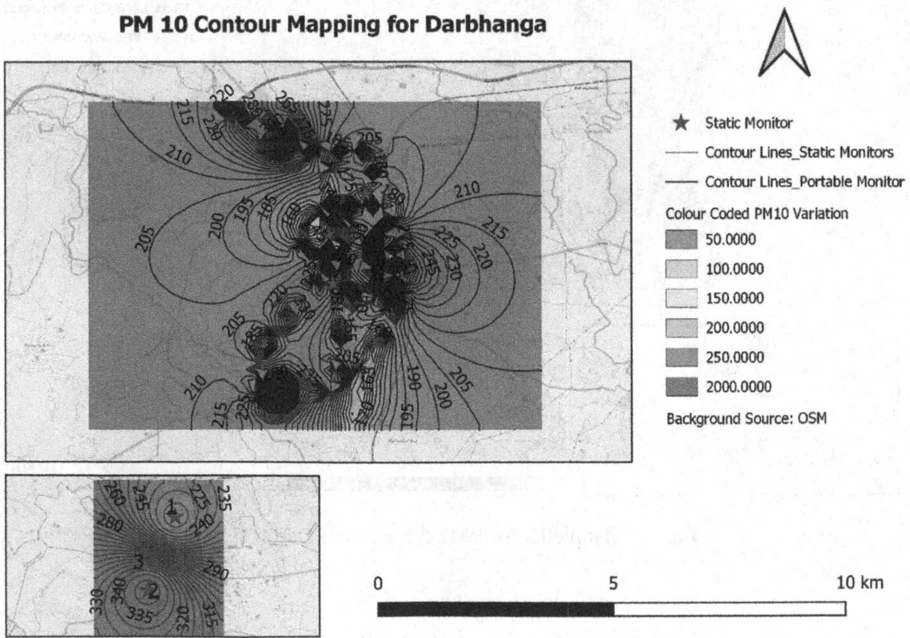

Fig. 19. Isopleths for PM10 data from Darbhanga

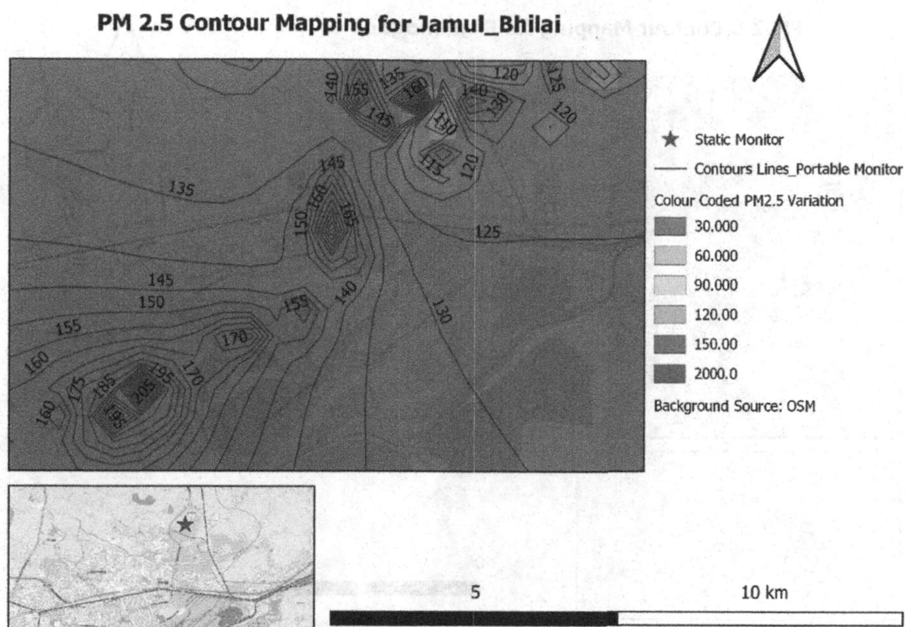

Fig. 20. Isopleths for PM2.5 data from Jamul area of Bhilai

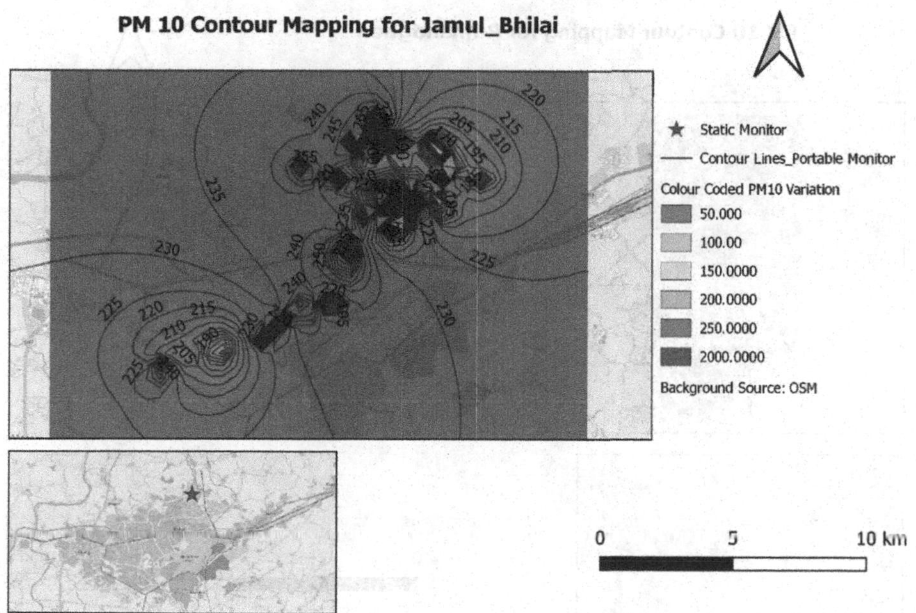

Fig. 21. Isopleths for PM10 data from Jamul area of Bhilai

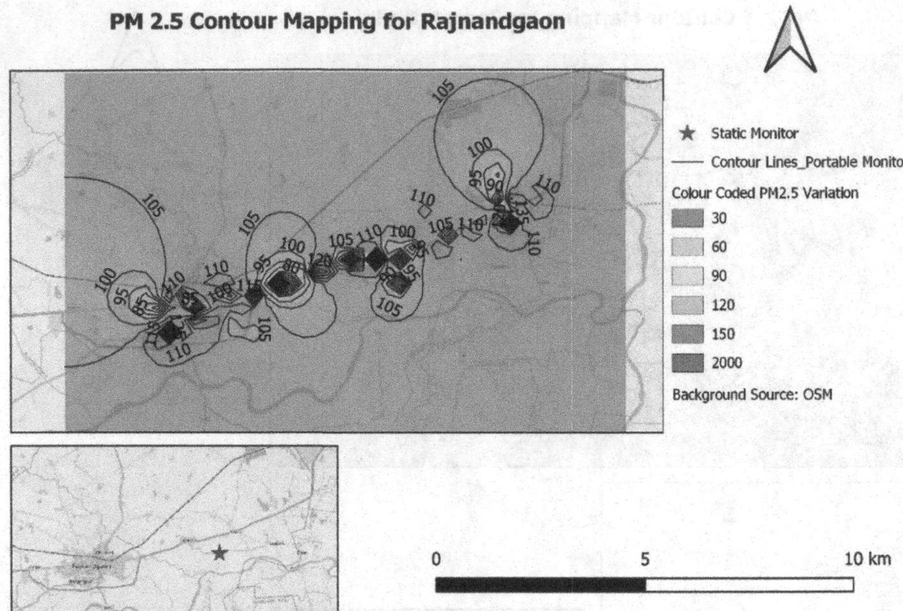

Fig. 22. Isopleths for PM2.5 data from Rajnandgaon

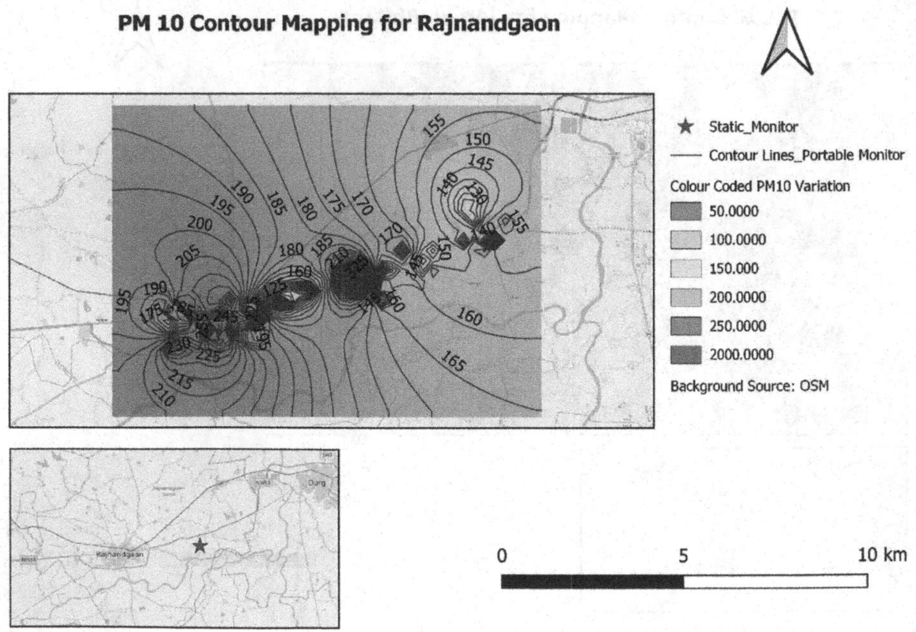

Fig. 23. Isopleths for PM10 data from Rajnandgaon

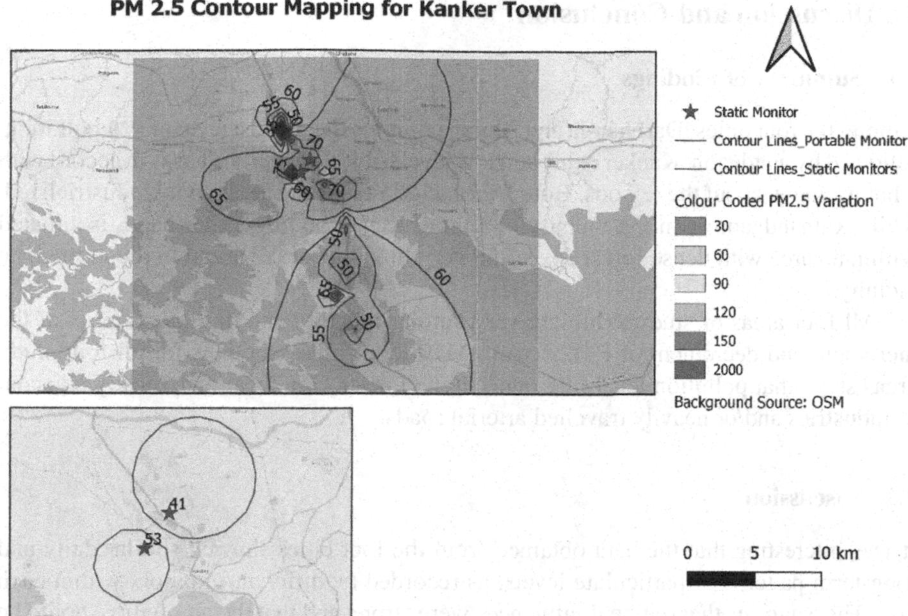

Fig. 24. Isopleths for PM2.5 data from Kanker

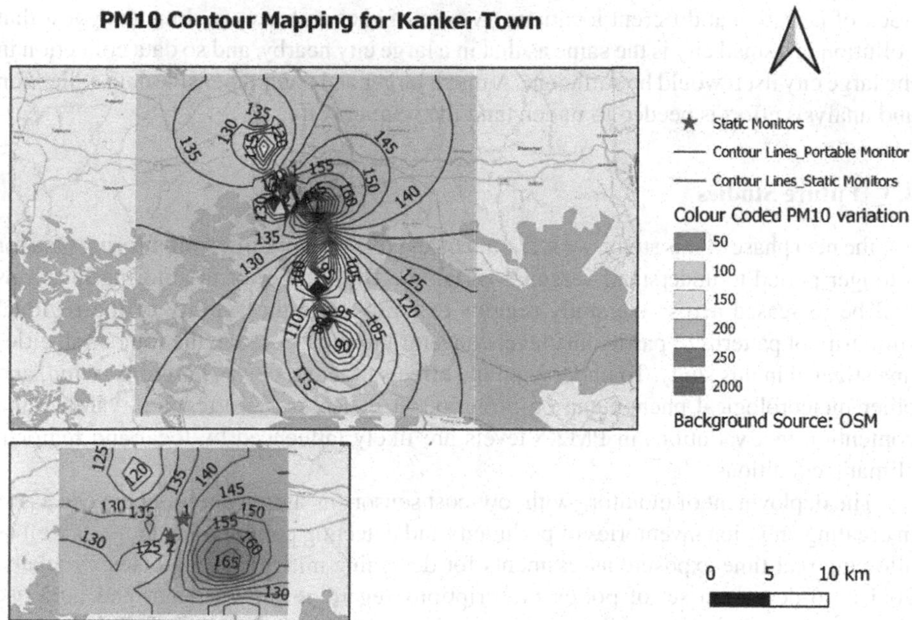

Fig. 25. Isopleths for PM10 data from Kanker

4 Discussion and Conclusion

4.1 Summary of Findings

Among the four cities, Darbhanga and Bhilai exhibited the highest level of PM pollution, followed by nandgaon. Kanker exhibited low levels of PM pollution. Ths is in accord with what we know about the regions. Both Darbhanga and Bholai are heavily industrialized, while Rajnandgaon is in the vicinity of Bhilai. Kanker, on the other hand, is situated within an area with dense forest cover and very little industrialization in the immediate vicinity.

All four areas of study exhibited very similar monthly and diurnal patterns of the increment and decrement of PM2.5 and PM10 levels. The isopleths for all four study areas show that pollution levels are concentrated in specific locations, mostly adjacent to industries and/or heavily travelled arterial roads.

4.2 Discussion

It was interesting that the data obtained from the four cities showed similar daily and long-term patterns of particulate levels, as recorded by different monitors within each city. This suggests that regional influences were strong and that the air quality should be monitored at the regional level as the pollution levels in very different size cities may be similar in spite of very different local activity levels because the influence of the wider region may be very large at present. It is possible that because of developments in the future, pollution levels may change in different cities and it would be useful to keep track of pollution at different locations. A note of caution here: we do not suggest that pollution in a small city is the same as that in a large city nearby, and so data collection in the large city itself would be sufficient. A much larger and comprehensive data collection and analysis effort is needed to understand the regional effect.

4.3 Future Studies

For the next phase of this study, we will collect and integrate pollution monitoring data for a longer period to understand seasonal variation. The number of monitoring locations will be increased across the study regions to corroborate the validity of the regional similarity of patterns of particulate level concentrations observed in the four small cities investigated in this study. To understand the effects of temperature, humidity, wind, and other meteorological phenomena on pollution, it is necessary to test and validate the contention that variations in PM2.5 levels are likely influenced by local and regional climatic conditions.

The deployment of monitors with low-cost sensors in significant numbers can assist in creating emission inventories of pollutants and detecting pollution hotspots, as well as allowing real-time exposure assessments for designing mitigation strategies. A related goal is to develop a set of policy prescriptions regarding urban layout and land use changes to maximize the attainment of sustainable development goals linked to public health, reduction in the adverse environmental impact of cities, and adaptation to climate change, as indicated by SDGs 3.9, 11.6 and 11.b, under India's smart cities initiative.

As suggested by one of our reviewers, as part of the next phase of this study, we will also investigate the utility of machine learning and data mining techniques to analyze air quality data from locations across various geographic regions and determine airshed boundaries, which would then allow expansion of the geographic range of air quality monitoring, even with a sparse network of monitors.

Acknowledgements. The funding for this work is provided by the Mphasis laboratory for Machine Learning and Computational Thinking (ML2CT), Ashoka University, Sonipat, Haryana, India. Support is also provided by IIIT Hyderabad, India, and O.P.Jindal Global University, Sonipat, Haryana, India. The authors also acknowledge the hard work and dedication of the field researchers who did the primary field data collection: Mr. Nohrit Mandavi in Kanker and surrounding areas, Ms. Namrata Banjare in Rajnandgaon, Mr. Durgesh Kumar in the Jamul area of Bhilai, and Ms. Gulrukh Fatima in Darbhanga.

References

1. Kumar, P., et al.: Ultrafine particles in cities. Environ. Int. **66**, 1–10 (2014). https://doi.org/10.1016/j.envint.2014.01.013
2. Lim, S.S., et al.: A comparative risk assessment of burden of disease and injury attributable to 67 risk factors and risk factor clusters in 21 regions, 1990–2010: A systematic analysis for the Global Burden of Disease Study 2010. Lancet (2012). https://doi.org/10.1016/S0140-6736(12)61766-8
3. Institute for Health Metrics and Evaluation (IHME): Findings from the Global Burden of Disease Study 2017, Lancet (2017)
4. WHO: WHO Global Ambient Air Quality Database, World Heal. Organ (2018)
5. CPCB: Guidelines for the Measurement of Ambient Air Pollutants (NAAQS), Cent. Pollut. Control Board, Gov. India (2009)
6. Greenstone, M., Nilekani, J., Pande, R., Ryan, N., Sudarshan, A., Sugathan, A.: Lower pollution, longer lives: life expectancy gains if India reduced particulate matter pollution. Econ. Polit. Wkly. **50**, 40–46 (2015)
7. WHO: Ambient air pollution: a global assessment of exposure and burden of disease (2016). https://apps.who.int/iris/handle/10665/250141
8. R.G. Census India, Census of India 2011: provisional population totals-India data sheet, Off. Regist. Gen. Census Comm. India. Indian Census Bur (2011)
9. Venkataraman, C., et al.: Indian network project on carbonaceous aerosol emissions, source apportionment and climate impacts (COALESCE). Bull. Am. Meteorol. Soc. **101**, E1052–E1068 (2020). https://doi.org/10.1175/BAMS-D-19-0030.1
10. Vafa-Arani, H., Jahani, S., Dashti, H., Heydari, J., Moazen, S.: A system dynamics modeling for urban air pollution: a case study of Tehran, Iran. Transp. Res. Part D Transp. Environ. **31**, 21–36 (2014). https://doi.org/10.1016/j.trd.2014.05.016
11. Karagulian, F., et al.: Contributions to cities' ambient particulate matter (PM): a systematic review of local source contributions at global level. Atmos. Environ. **120**, 475–483 (2015). https://doi.org/10.1016/j.atmosenv.2015.08.087
12. Mraihi, R., Harizi, R., Mraihi, T. and Bouzidi, M.T.: Urban air pollution and urban daily mobility in large Tunisia's cities. Renew. Sustain. Energy Rev. **43**, 315–320 (2015). https://doi.org/10.1016/j.rser.2014.11.022
13. Chen, R., et al.: Fine particulate air pollution and daily mortality: a nationwide analysis in 272 Chinese cities. Am. J. Respir. Crit. Care Med. **196**, 73–81 (2017). https://doi.org/10.1164/rccm.201609-1862OC

14. Baklanov, A., Molina, L.T., Gauss, M.: Megacities, air quality and climate. Atmos. Environ. **126**, 235–249 (2016). https://doi.org/10.1016/j.atmosenv.2015.11.059
15. Ilyas, S.Z., Khattak, A.I., Nasir, S.M., Qurashi, T., Durrani, T.: Air pollution assessment in urban areas and its impact on human health in the city of Quetta, Pakistan. Clean Technol. Environ. Policy **12**, 291–299 (2010). https://doi.org/10.1007/s10098-009-0209-4
16. Holman, C., Harrison, R., Querol, X.: Review of the efficacy of low emission zones to improve urban air quality in European cities. Atmos. Environ. **111**, 161–169 (2015). https://doi.org/10.1016/j.atmosenv.2015.04.009
17. Genikomsakis, K.N., Galatoulas, N.F., Dallas, P.I., Ibarra, L.M.C., Margaritis, D., Ioakimidis, C.S.: Development and on-field testing of low-cost portable system for monitoring PM2.5 concentrations. Sensors (Switzerland) **18** (2018). https://doi.org/10.3390/s18041056
18. Agrawal, G.: Diurnal patterns in particulate matter concentrations in the Delhi NCR, TBD (2020)

Author Index

Agrawal, Girish 402, 417

Bhise, Minal 347

Chandra Shekar, N. 402
Chen, Bofeng 229
Chen, Hao 363
Chen, Hui 229
Chen, Xu 214, 257
Cheng, Siyao 330

Diao, Yupeng 214
Draheim, Dirk 50
Du, Sheng 319
Duong, Hai 34

Fan, Tianlan 363
Feng, Jun 363
Feng, ZhiYong 175
Fournier-Viger, Philippe 34
Fraternali, Piero 104
Frigerio, Matteo 104

Gan, Wensheng 34
Gonzalez, Sergio Luis Herrera 104
Guttikunda, Sarath K. 389

Harris, Christopher G. 120
He, Yulin 21
Hong, Yinhao 319
Huang, Jianping 363
Huang, Joshua Zhexue 21
Huang, Qihang 21

Jia, Jiangkai 149, 160, 189
Jian, Li 229
Jiang, Yingshuo 244
Jin, Bo 301

Kaushik, Ashu 64
Kaushik, Minakshi 50
Khokhariya, Uday 135
Krishna Reddy, P. 402
Kumar, Shambhavi 135

Lai, Ruyi 257
Lei, Shuya 376
Leng, Jianquan 319
Li, Feng 330
Li, Huichao 149, 189
Li, Shuyi 189
Li, Wenhao 301
Li, Xiaoming 160
Li, Yanjun 229
Li, Yinlong 330
Lian, Defu 272
Liang, Xiao 376
Liang, Zheng 363
Lin, Jerry Chun-Wei 79
Liu, Dan 160
Liu, Jie 330
Liu, Shuncheng 214, 257
Liu, Weiwei 376
Liu, Ying 244
Lv, Jingwei 291

Ma, Jinchao 272
Malhotra, Ayushi 347
Mondal, Anirban 402, 417
Mu, Tiantong 203

Nouioua, Mourad 34

Pancholi, Nidhay 135
Parmar, Keyur 135
Patel, Darshi 91
Patel, Dhiren 91
Patel, Sakshee 347
Peious, Sijo Arakkal 50
Peng, Liangying 363

Qi, Ruoyan 244

Rahman, Hifzur 417
Reddy, P. Krishna 417
Righetti, Mattia 104

Shah, Kaushal 135
Shahin, Mahtab 50

Sharma, Rahul 50
Song, Wei 34
Song, Ying 203
Srinivas Reddy, A. 402
Su, Han 214, 257
Su, Yiteng 214
Sun, Ximin 149, 160, 189
Susan, Seba 64

Tiwari, Prayag 50
Truong, Tin 34

Vidyarthi, Ankit 50

Wang, Bo 203
Wang, Guoren 244
Wang, Hongkai 363
Wang, Hongzhi 363
Wang, Ke 79
Wang, Le 3
Wang, Mingda 189
Wang, Shuai 149, 160, 189
Wang, Xiangfeng 301
Wu, Hanling 330
Wu, Jimmy Ming-Tai 79
Wu, Linjuan 175
Wu, Tingyu 301
Wu, Youxi 34

Xia, Yuyang 257
Xu, Shengsheng 21
Xu, Zhi 257

Yadav, Nitish 347
Yang, Mingjie 203
Yang, Xiaochun 244
Yao, Shuzhen 291
Yu, Tianren 363

Zeng, Ximu 214, 257
Zhang, Bin 160
Zhang, Wei 301
Zhang, Xiaowang 175
Zhang, Zhongshuai 244
Zhao, Jiacheng 203
Zheng, Huanran 229
Zheng, Xiaokun 376
Zhou, Jing 149, 189
Zhou, Nan 291
Zhou, Tingliang 229
Zhu, Jiazheng 149, 175
Zhuang, Zhiqiang 175

Printed in the United States
by Baker & Taylor Publisher Services